THIRD SEMESTER CALCULUS
STUDENT SUPPLEMENT
4TH Edition

Mimi Rasky
Southwestern College

Revised
&
Expanded!

 *Anaphase Publishing*
4669 Cherokee Avenue, Suite E
San Diego, CA 92116-3654

THIRD SEMESTER CALCULUS
STUDENT SUPPLEMENT
4TH Edition

Mimi Rasky
Southwestern College

Copyright © 2007–2015 Mimi Rasky

ISBN 13: 978-0-945962-49-6

Book Design, Illustrations, Typesetting & Layout:
Lily Splane

For permission to use this work, contact Mimi Rasky at Southwestern College:
E-mail: mrasky@swccd.edu
Office phone: (619) 421-6700, ext. 5871

Printed in the United States of America

Anaphase Publishing
A DIVISION OF CYBERLEPSY MEDIA
4669 Cherokee Avenue, Suite E
San Diego, CA 92116-3654
WWW.CYBERLEPSY.COM

JULY 2015

TABLE OF CONTENTS

PREFACE

This *Calculus III Student Supplement, 4TH Edition* is meant to be an aid for the student. It is not intended to be a replacement for a main text in the course. Originally, this book was written to parallel Roland E. Larson's *Calculus, Seventh Edition*. However, this book can be used to supplement any third semester (multi-variable and vector analysis) calculus course.

This book has features and benefits to the student that are missing from other books and calculus supplements such as *Schaum's Outlines* and the like. Here are some advantages that this book has that others do not:

1) More examples that are completely worked out in detailed step-by-step fashion. I do not assume the student has excellent algebra skills (as most books do), so even the basic steps that include algebra concepts are worked out.

2) I use a conversational and colloquial style. I avoid the traditional technical and dry presentation approach found in most books on mathematics.

3) I have many more vector analysis examples needed for third semester calculus than the other books on the market. I also spend a lot of time and space discussing cylindrical and spherical coordinates, and include numerous examples on the basics of these as well as the vector analysis use of these coordinate systems.

4) The book includes a huge test bank with completely worked out examples as the last half of the book. Students have told me they really appreciate this.

Part I of this workbook contains study tips, hits and "tricks" for students that are usually not covered explicitly in calculus texts. I have come up with these based on my teaching experience. I have found that students need solutions worked out step-by-step; many texts assume (wrongly) that students can figure out the (missing) steps on their own. I also include examples of what *not* to do. Many instructors find that this is the "missing link" between the textbook and the student.

I have also found that students appreciate (and need) quite a bit of algebra review in their calculus classes. Although many instructors refuse to "backtrack" to cover algebra in a calculus course, I have found that students get the satisfaction of realizing their algebra skills actually have a place in higher levels of math. So, I take time (and space) to do this in many of my examples.

Parts II and **III** of the workbook contains tests with "worked-out" solutions. These tests have actually been used in my classes. Some of them are meant for in-class exams, others (especially the longer ones) are meant to be take-home exams. My students have told me they really appreciate the fact that my solutions contain lots of intermediate steps not found in other calculus student-solutions manuals.

Note to Instructors

I have found that many texts covering multi-variable calculus and vector analysis do not have challenging enough problems when it comes to limits, differentiation and integration. In other words, I like to see problems in this course that actually use concepts covered in second semester calculus: e.g. advanced integration techniques such as Partial Fraction Decomposition, Trigonometric Substitution, integrals involving powers of trig functions, Integration by Parts, as well as limits requiring L'Hôpital's Rule. Thus, my examples may seem more challenging than many found in basic texts.

I have made every effort to see that the answers are correct. However, I would appreciate very much hearing about any errors or other suggestions for improvement. Please e-mail me at mrasky@swccd.edu with any suggestions or comments you may have.

January 2007
Mimi Rasky
Southwestern College
900 Otay Lakes Road
Chula Vista, CA 91910

MIMI RASKY

PREFACE TO THE
4TH EDITION

A new section (3.9) was added entitled, "Lagrange Multipliers." The technique of Lagrange Multipliers is a powerful tool to tackle optimization problems where there is a constraint. It is also the only tool available to determine the extrema for a function of three variables, since the Second Partials Test makes sense only for functions of two variables. Additional exercises were added to correspond to this new material.

In addition, the Practice Final was revised and expanded. It now highlights all of the key concepts of this course (including Lagrange Multipliers) and gets the student to think about the material in its entirety and how it all ties together. This is obviously important when preparing for a comprehensive final exam!

Finally, I created a summary "flow-chart" to navigate through the (sometimes dangerous) waters of Line Integrals. It is contained at the end of the Green's Theorem section (section 5.4), and is a useful aid to understanding the different strategies one can use to solve these kinds of problems.

Again, I want to thank all of my Calculus III students for doing the "free editing" and alerting me to the mistakes and typos contained in the last edition. While I cannot promise that this edition is free of all "bugs," I can say that we are getting closer and closer to perfection with every new edition.

To students and instructors: Thank you for buying my book. I hope it helps you!

December, 2012
Mimi Rasky
Southwestern College
Chula Vista, CA

ACKNOWLEDGEMENTS

I would first like to acknowledge my husband, Steven Rasky, for his support and his love, especially when I got a little crazy trying to finish this book.

Secondly, I would like to thank Lily Splane, my editor, proofreader, and publisher for her amazing efforts in cleaning up and typesetting this book, as well as her artistic and creative ability with the cartoons!

Lastly, I would like to acknowledge all of my former and future Calculus III students. They are truly the source of inspiration for this book. I appreciate the editing and feedback from my former students, and hope that my future students find this book to be a valuable aid in their understanding of calculus. Thank you all!

—*Mimi Rasky*
December, 2007

PART I
STUDY TIPS & TRICKS

Chapter 1

Vectors and the Geometry of Space

Chapter Comments

This chapter contains absolutely no calculus! This means that there are no limits, no derivatives, and no integrals covered in this chapter. This does not mean that this chapter is unimportant. In fact, this chapter is the foundation for the entire semester. Your success this semester depends a lot on how well you grasp both the concepts and the terminology in this chapter. You will be introduced to *vectors* and vector operations to begin the chapter. The concept of vectors will be used in an introductory physics course, so if you haven't taken physics yet you are ahead of the game by being here in this class! Some of the exercises in our text will use vectors in real-life applications. The second concept covered in this chapter will be about lines and planes in space, and how to represent these symbolically. The third concept will be about *surfaces* and being able to recognize equations for surfaces. In your first semester of calculus, you worked in two dimensions. You looked at graphs in the *xy*-plane (represented by equations containing two variables), and learned to find slopes of those graphs (differentiation) and areas under those graphs (integration). This class will have you working in three dimensions, also known as space. The graphs in space are represented by equations involving three variables. In this class you will be finding slopes of surfaces (which are graphs in space), and you will

find volumes of solids bounded by these surfaces. In this course you will learn how to differentiate in 3-D to get those slopes, and you will learn how to integrate in 3-D to get those volumes. I know it is not easy to graph in 3-D, especially since you can only use your graphing calculator to graph equations of two variables that are functions of *x*. Some of you may have trouble at first trying to visualize what is going on in space (i.e., 3-D). Please do not worry, as I will not require that you sketch each and every surface that we work on in this course. I will, however, require that you know how to recognize equations for and sketch some "easy" surfaces in space, such as planes, spheres, and cylindrical surfaces.

The last concepts covered in this chapter are cylindrical and spherical coordinates, and the cylindrical and spherical coordinate systems. I hope that you remember what polar coordinates are from your last semester of calculus. If you do, the cylindrical coordinate system will be easy to learn, since it uses part of the polar coordinate system. The cylindrical coordinate system may be thought of as the polar coordinate system in three dimensions instead of two. Spherical coordinates are completely "brand-new."

1.1   Vectors in the Plane

A vector is a directed line segment—not just a "little pointed arrow." One vector is actually a family, or collection, of directed line segments, since vectors are not unique. Each and every vector has two things associated with it: 1) magnitude (length) and 2) direction (angle). There are two types of vectors we will talk about here: 1) geometric vectors and 2) algebraic vectors. A geometric vector does not actually have a "starting location." It can be thought of as a vector just sitting out in space somewhere. To give a geometric vector a starting point, we can move it (while maintaining its direction and magnitude) so that its initial point sits on the origin. When we translate a geometric vector in this way, we create an algebraic vector. An algebraic vector always has its starting point (aka initial point) at the origin. This position is known as *standard position*. The easy way to turn a geometric vector into an algebraic vector is shown in the example given after the next paragraph.

Since there are many types of notation used to denote vectors, it is important that you familiarize yourself with the way your text does it. Vector names are often denoted in **boldface** in textbooks, so you will be able to distinguish vectors from other constructs. When *you* denote vectors on your assignments, you will not use boldface. Instead, you will either put a small arrow over the name of the vector (e.g. $\vec{v}$) *or* you can put a "hat" over the name of the vector (e.g., $\hat{v}$). I will not give you full credit on problems where you fail to use proper notation such as this. Therefore, all handwritten vectors will require either "arrows" or "hats" above the name to denote a vector. Algebraic vectors can be denoted two ways: 1) component form or 2) as a linear combination of the standard unit vectors. In class, I will usually use component form instead of the linear combination of standard vectors. I have found that by doing this I avoid confusion as to what dimension I am in. It is not clear whether the vector $\mathbf{v} = 3\mathbf{j}$ (this is the linear combination of the standard unit vectors form) is a 2-dimensional or a 3-dimensional vector. However, if I write the vector instead as $\mathbf{v} = \langle 0,3,0 \rangle$, then it is clear that I am representing a 3-dimensional vector. You will still be responsible for knowing both forms of notation. The next example will show both forms.

EXAMPLE. The two points $C(-4, 2)$ and $D(-1, 6)$ can be connected to form a vector. Create the (algebraic) vector $\overrightarrow{DC}$ (not to be confused with $\overrightarrow{CD}$, which would be a vector going in the *opposite* direction.) Give your final answer using the two types of notation: 1) component form and 2) as a linear combination of the standard unit vectors, $\mathbf{i}$ and $\mathbf{j}$. *Note:* hereafter, this second form of notation will be referred to as "standard form" for short.

SOLUTION. To form an algebraic vector $\vec{v} = \langle x, y \rangle$ from a geometric vector, use the formulas $x = x_2 - x_1$ *and* $y = y_2 - y_1$, where the terminal point (that's the point with the little arrow on it) is represented by (x_2, y_2) and the initial point is represented by (x_1, y_1). So, here we have that the point C is the terminal point and the point D is the initial point because the notation $\overrightarrow{DC}$ was used:

$$x = x_2 - x_1 = -4 - (-1) = -3$$

$$y = y_2 - y_1 = 2 - 6 = -4$$

So, to represent the vector $\overrightarrow{DC}$ in component form, we have that $\overrightarrow{DC} = \langle -3, -4 \rangle$. Using the standard form, we have that $\overrightarrow{DC} = -3\mathbf{i} - 4\mathbf{j}$. *Note:* The vector $\overrightarrow{CD} = \langle 3, 4 \rangle$, or $\overrightarrow{CD} = 3\mathbf{i} + 4\mathbf{j}$.

Once you have an algebraic vector (in either component form or standard form), you can graph it simply by graphing the component form as a point (this will give you the terminal point of the vector), and connecting it to the origin. The arrow must go on top of the terminal point.

The two properties of vectors mentioned above (magnitude and direction) are very important. Suppose we want to find both the magnitude and direction of the vector in the example above. We can do this by using the following formulas. Please also note the notation with "double bars" is used for denoting magnitude.

Magnitude (length) for a 2-dimensional vector $\vec{v} = \langle x, y \rangle$: $\|\mathbf{v}\| = \sqrt{x^2 + y^2}$

Magnitude (length) for a 3-dimensional vector $\vec{v} = \langle x, y, z \rangle$: $\|\mathbf{v}\| = \sqrt{x^2 + y^2 + z^2}$

(You can extend this magnitude formula for an *n*-dimensional vector as well.)

Direction (angle) for a 2-dimensional vector $\vec{v} = \langle x, y \rangle$: $\theta = \tan^{-1}\left(\dfrac{y}{x}\right)$

(We will learn about direction in space in a later section.)

The quantities of magnitude and direction are called *scalars*. Scalars are pure numbers, as opposed to vectors, which are *not* represented by pure numbers. Remember that a vector has both magnitude and direction associated with it. Other examples of scalar quantities would be area, volume, mass, temperature, etc. Examples of vector quantities would be velocity, force, acceleration, tension, etc.

EXAMPLE. Give both the magnitude and direction for the vector $\vec{DC}$ found in the example above.

SOLUTION. Since $\vec{DC} = \langle -3, -4 \rangle$, we have that the magnitude is:

$$\left\| \vec{DC} \right\| = \sqrt{x^2 + y^2} = \sqrt{(-3)^2 + (-4)^2} = \sqrt{9 + 16} = \sqrt{25} = 5$$

Magnitude will always be a positive value! Why?

The direction (or angle) is $\theta = \tan^{-1}\left(\dfrac{y}{x}\right) = \tan^{-1}\left(\dfrac{-4}{-3}\right) \approx 0.93 \; radians \; (53.1°)$ This is *WRONG!* Why?

The angle given here is incorrect because the terminal point of the vector falls in the third quadrant. This result has the vector lying in the first quadrant, which is incorrect! The reason why the graphing calculator gives the incorrect result is because the range θ of the arctangent function lies between $-\dfrac{\pi}{2} \le \theta \le \dfrac{\pi}{2}$.

So, we need to think a little bit here. We will need to add π radians to the result so that we get the correct angle. So the *correct* final answer for the direction of our vector is:

$$\theta = \tan^{-1}\left(\dfrac{y}{x}\right) = \tan^{-1}\left(\dfrac{-4}{-3}\right) \approx 0.93 + \pi \; radians \approx 4.07 \; radians = 233.1°$$

Please know the properties of basic operations with vectors, most likely covered in a theorem in your text. You can add and subtract vectors, and multiply vectors with scalars very easily. The zero vector can be denoted as:

$$\vec{0} = \langle 0, 0 \rangle \text{ The zero vector has zero magnitude and any direction.}$$

Another important concept you need to know for vectors is the unit vector. Unit vectors are vectors that have a magnitude of 1. After studying section 1.2, you will be familiar with three unit vectors, which are the vectors **i**, **j**, and **k**. The component form for these standard unit vectors are:

$$\mathbf{i} = \langle 1,0,0 \rangle \quad \mathbf{j} = \langle 0,1,0 \rangle \quad \mathbf{k} = \langle 0,0,1 \rangle \text{ (for three dimensions)}$$

You can always create a unit vector from any given vector by simply using the conversion formula:

$$\mathbf{u} = \frac{\mathbf{v}}{\|\mathbf{v}\|}$$

You are simply dividing the vector **v** with a scalar (its magnitude).

EXAMPLE. Form the unit vector **u** from the vector $\overrightarrow{DC} = \langle -3,-4 \rangle$. Express in component form.

SOLUTION. We already found that the magnitude of this vector is 5. So, to create the unit vector, we have that:

$$\mathbf{u} = \frac{\mathbf{v}}{\|\mathbf{v}\|} = \frac{\langle -3,-4 \rangle}{5} = \left\langle -\frac{3}{5}, -\frac{4}{5} \right\rangle$$

Suppose you are given a vector's magnitude, $\|\mathbf{v}\|$, and direction, θ, and you need to write the vector in component form? Then, you need to use the conversion formulas:

$$\mathbf{v} = \langle \|\mathbf{v}\|\cos\theta, \ \|\mathbf{v}\|\sin\theta \rangle . \text{ This is also referred to as "resolving a vector into its components."}$$

EXAMPLE. Write the vector that has a length of 3 and makes an angle of $\dfrac{\pi}{6}$ with the positive x-axis in component form.

SOLUTION. Use the conversion formula given above to obtain:

$$\mathbf{v} = \langle \|\mathbf{v}\|\cos\theta, \ \|\mathbf{v}\|\sin\theta \rangle = \left\langle 3\cos\frac{\pi}{6}, 3\sin\frac{\pi}{6} \right\rangle = \left\langle (3)\left(\frac{\sqrt{3}}{2}\right), (3)\left(\frac{1}{2}\right) \right\rangle = \left\langle \frac{3\sqrt{3}}{2}, \frac{3}{2} \right\rangle$$

EXAMPLE. A small cart is being pulled along a smooth horizontal floor with a 20-lb. force **F** making an angle of 45° to the floor. What is the effective force moving the cart forward?

SOLUTION. The effective force moving the cart *forward* will be only the horizontal component of the force **F**. That is, we will have:

$$\text{Effective force} = a = \|\mathbf{F}\|\cos\theta = (20)\cos 45° = (20)\left(\frac{\sqrt{2}}{2}\right) = 10\sqrt{2} \text{ lbs. (approximately 14.14 lbs.)}$$

Now, let's turn our attention to some more examples using force vectors. I find that many students who haven't had physics get stressed when they see these types of problems. If you are one of them, please note that it really is not that difficult if you realize that forces are simply vectors. Vector addition is just as valid with vectors as it is with problems using forces represented as vectors.

Vector Addition

Adding two vectors together is easy if they are already in component form. When you add two vectors together you obtain what is termed "the resultant vector." Please memorize this terminology! So, if you must add the vector $\mathbf{v} = \langle v_1, \ v_2 \rangle$ with the vector $\mathbf{u} = \langle u_1, \ u_2 \rangle$, then you will get a new vector (the resultant vector), which is denoted **u** + **v**. We can find the vector **u** + **v** (in component form) by adding in the following manner:

$$\mathbf{u} + \mathbf{v} = \langle u_1, \ u_2 \rangle + \langle v_1, \ v_2 \rangle = \langle u_1 + v_1, u_2 + v_2 \rangle$$

Geometrically, the resultant vector can be obtained in one of two ways:

1) The "tail-to-tip" method
2) The "parallelogram" method

I will demonstrate the two ways. Consider the two vectors **u** and **v** in the following diagram:

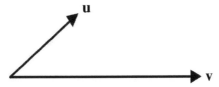

The "tail-to-tip" method will need one of the vectors tips to be moved (translated) so that it coincides with the other vector's tip. I will translate the vector **u** (while retaining both its direction and magnitude) so that its tip is "attached" to the tail of the vector **v** (see the diagram below):

The resultant vector can be found by starting at the tail of vector **v** and ending at the tip of vector **u** in the following way:

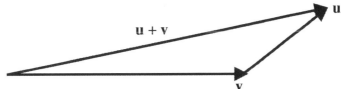

In using the "parallelogram" method to obtain the resultant vector, we simply align the vectors **u** and **v** so that they create two adjacent sides of a parallelogram (which they already are in my original diagram above) and the resultant vector (the sum) will be the diagonal of the parallelogram. Both ways are valid methods one can use to visualize a resultant vector.

Adding Vectors to Find the Resultant Vector (Finding Velocity)

EXAMPLE. An airplane is flying due east at 500 mph in still air, and encounters a 70 mph tailwind in the direction of 60° north of east. The airplane holds its compass heading due east, but, because of the wind, acquires a new ground speed and direction. What are they?

SOLUTION. They have given us the magnitudes of the two vectors here. Let **v** represent the velocity of the airplane, and let **w** represent the velocity of the wind. Keep in mind that speed and velocity are *not* the same. Velocity is a vector, possessing both a magnitude and a direction. Speed is always a scalar and is represented as the magnitude of a velocity vector. If $\|\mathbf{v}\|$ = the speed of the airplane in still air, and

$\|\mathbf{w}\|$ = the speed of the wind, then we have that $\|\mathbf{v}\| = 500$, and $\|\mathbf{w}\| = 70$. We need to find the magnitude and angle of the *resultant* vector, which is $\mathbf{v} + \mathbf{w}$; we are being asked to find the quantities $\|\mathbf{v} + \mathbf{w}\|$ and θ. To add the two vectors $\mathbf{v}$ and $\mathbf{w}$ together, we first need to represent each vector in component form, then we can add the components.

This diagram depicts the velocity vectors and resultant vector for this example:

$$\mathbf{v} = \langle \|\mathbf{v}\|\cos\theta, \|\mathbf{v}\|\sin\theta \rangle = \langle 500\cos 0, 500\sin 0 \rangle = \langle 500, 0 \rangle$$

$$\mathbf{w} = \langle \|\mathbf{w}\|\cos\theta, \|\mathbf{w}\|\sin\theta \rangle = \langle 70\cos 60°, 70\sin 60° \rangle = \langle 35, 35\sqrt{3} \rangle$$

We can find the resultant vector by using vector addition:

$$\mathbf{v} + \mathbf{w} = \langle 500 + 35, 0 + 35\sqrt{3} \rangle = \langle 535, 35\sqrt{3} \rangle$$

Many students make the mistake of claiming this as their final answer. We still need to find the magnitude and direction for this resultant vector!

$$\textbf{\textit{Magnitude is }} \|\mathbf{v} + \mathbf{w}\| = \sqrt{(535)^2 + \left(35\sqrt{3}\right)^2} = \sqrt{289{,}900} \approx 538.4$$

$$\textbf{\textit{Direction is }} \theta = \tan^{-1}\left(\frac{y}{x}\right) = \tan^{-1}\left(\frac{35\sqrt{3}}{535}\right) \approx 6.5°$$

Therefore, the new ground speed of the airplane is about 538.4 mph, and its new direction is about 6.5° north of east.

Adding Forces to Obtain a Resultant Vector

EXAMPLE. Three forces with magnitudes of 2 newtons, 3 newtons, and 4 newtons act on an object at angles of $-10°$, $140°$, and $200°$ with the positive x-axis. Find the direction and magnitude of the resultant force.

SOLUTION. The first step is to resolve all three vectors into their respective component forms. Then we will simply add all three vectors together. Finally, we will compute the magnitude and direction.

$$\mathbf{F_1} = \langle \|\mathbf{F_1}\|\cos\theta, \|\mathbf{F_1}\|\sin\theta \rangle = \langle 2\cos(-10), 2\sin(-10) \rangle \approx \langle 1.97, -0.35 \rangle$$

$$\mathbf{F_2} = \langle \|\mathbf{F_2}\|\cos\theta, \|\mathbf{F_2}\|\sin\theta \rangle = \langle 3\cos(140), 3\sin(140) \rangle \approx \langle -2.30, 1.93 \rangle$$

$$\mathbf{F_3} = \langle \|\mathbf{F_3}\|\cos\theta, \|\mathbf{F_3}\|\sin\theta \rangle = \langle 4\cos(200), 4\sin(200) \rangle \approx \langle -3.76, -1.37 \rangle$$

Resultant Force =

$$\mathbf{F_1} + \mathbf{F_2} + \mathbf{F_3} = \langle 1.97, -0.35 \rangle + \langle -2.30, 1.93 \rangle + \langle -3.76, -1.37 \rangle = \langle 1.97 + -2.30 + -3.76, \; -0.35 + 1.93 + -1.37 \rangle$$
$$= \langle -4.09, 0.21 \rangle$$

We can find the magnitude and direction:

$$\textbf{\textit{Magnitude}} = \|\mathbf{F_1} + \mathbf{F_2} + \mathbf{F_3}\| = \sqrt{(-4.09)^2 + (0.21)^2} \approx 4.10$$

$$\textbf{\textit{Direction}} = \theta = \tan^{-1}\left(\frac{y}{x}\right) = \tan^{-1}\left(\frac{0.21}{-4.09}\right) \approx -2.93° + 180° = 177.06°$$

Therefore, the resultant force has a magnitude of 4.1 newtons and a direction of 177.06°. Note that we needed to add 180° to the direction to have our vector end up in the second quadrant (since the arctangent function only gives results in either the first or fourth quadrants, remember?).

Static Equilibrium

EXAMPLE. Static equilibrium means that the sum of all forces will add to zero. To carry a 100-lb. weight, two workers lift on the ends of short ropes tied to an eyelet on the top center of the cylinder. One rope makes a 20° angle away from the vertical and the other a 30° angle.

(a) Find the rope's tension if the resultant force is vertical.

(b) Find the vertical component of each worker's force.

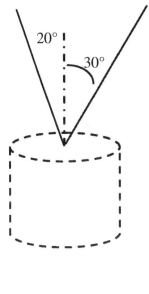

SOLUTION. First we will draw a force diagram.

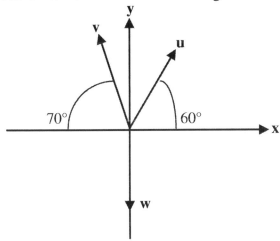

We have three forces **u**, **v**, and **w** whose resultant force adds to zero (because of static equilibrium). We resolve each into component form, keeping in mind that we want to find the magnitudes for **u** and **v**. (These are the unknown tensions.)

$$\mathbf{w} = \langle 0, -100 \rangle \quad \mathbf{u} = \langle \|\mathbf{u}\|\cos 60°, \|\mathbf{u}\|\sin 60° \rangle \quad \mathbf{v} = \langle -\|\mathbf{v}\|\cos 70°, \|\mathbf{v}\|\sin 70° \rangle$$

The zero vector has the form $\mathbf{0} = \langle 0,0 \rangle$

Summing the vectors and adding to the zero vector, we have:

$$\mathbf{u} + \mathbf{v} + \mathbf{w} = \mathbf{0}$$

$$\langle 0, -100 \rangle + \langle \|\mathbf{u}\|\cos 60°, \|\mathbf{u}\|\sin 60° \rangle + \langle -\|\mathbf{v}\|\cos 70°, \|\mathbf{v}\|\sin 70° \rangle = \langle 0, 0 \rangle$$

$$\langle 0 + \|\mathbf{u}\|\cos 60° + -\|\mathbf{v}\|\cos 70, \ -100 + \|\mathbf{u}\|\sin 60° + \|\mathbf{v}\|\sin 70 \rangle = \langle 0, 0 \rangle$$

We equate the components and obtain a linear system of two equations and two unknowns. In case it is unclear to you, the unknowns in this system are $\|\mathbf{u}\|$ and $\|\mathbf{v}\|$.

$$\|\mathbf{u}\|\cos 60° - \|\mathbf{v}\|\cos 70° = 0$$

$$-100 + \|\mathbf{u}\|\sin 60° + \|\mathbf{v}\|\sin 70° = 0$$

Replace the cosines and sines with the actual values, and we will have:

$$0.5\|\mathbf{u}\| - 0.34\|\mathbf{v}\| = 0$$

$$0.866\|\mathbf{u}\| + 0.94\|\mathbf{v}\| = 100$$

Solve this system of equations using either the substitution method or the elimination method you learned from a basic algebra course. I will use the substitution method, and first solve for $\|\mathbf{u}\|$ using the first equation to get:

$$\|\mathbf{u}\| = \frac{0.34}{0.5}\|\mathbf{v}\|$$

Substitute this into the second equation to get a single equation with a single unknown (namely, the magnitude of the vector $\mathbf{v}$):

$$(0.866)\left(\frac{0.34}{0.5}\|\mathbf{v}\|\right) + 0.94\|\mathbf{v}\| = 100$$

$$0.59\|\mathbf{v}\| + 0.94\|\mathbf{v}\| = 100$$

$$1.53\|\mathbf{v}\| = 100$$

$$\|\mathbf{v}\| = \frac{100}{1.53} = 65.36$$

This means that the tension in the left-hand rope is approximately 212.4 lbs. Solving for the magnitude of $\mathbf{u}$, we have the tension on the other side:

$$\|\mathbf{u}\| = \frac{0.34}{0.5}\|\mathbf{v}\| = \frac{0.34}{0.5}(65.36) \approx 44.4 \text{ lbs.}$$

So, the answer for part **(a)** is 65.36 lbs. and 44.4 lbs. of tension in each rope.

For part **(b)**, we get the vertical component by using the formula for a vertical component:

$$\|\mathbf{v}\|\sin\theta = 65.36\sin 70° \approx 61.4$$

$$\|\mathbf{u}\|\sin\theta = 44.4\sin 60° \approx 38.45$$

The last example I want to show you for this section involves some calculus. I know I promised you that this section has no calculus, but this one has no *new* calculus, OK? You will learn some more physics-type of applications of vectors, which will help you when we get to Chapter 2, which covers vector-valued functions. So, I can guarantee you it will be worth a look!

Fact: When an object is moving along a path in the plane (or space), its velocity is tangent to the path. Moreover, if the object is speeding up or slowing down, forces are acting in both the tangent direction and the perpendicular (also called *normal*) to it. We will investigate motion along a path in the plane when we get to Chapter 2.

A vector is tangent or normal to a curve at a point P if it is parallel or normal, respectively, to the line that is tangent to the curve at P. In this next example, we will show how to find vectors for a differentiable curve $y = f(x)$ in the plane.

EXAMPLE. An object is moving along the curve $y = \dfrac{x^3}{2} + \dfrac{1}{2}$. Find *unit* vectors tangent and normal to the curve at the point $(1, 1)$.

SOLUTION. Keep in mind that the final answers must be *unit* vectors (i.e., they must have a magnitude of 1). First, let's focus on getting the vector tangent to the curve at the given point. We know (from first semester calculus) that the slope of the tangent line at the point can be found by finding the derivative of the function at that point. So, we have that:

$$y' = \frac{3x^2}{2} \quad \text{and} \quad \text{slope} = \text{m} = y'(1) = \frac{3(1)^2}{2} = \frac{3}{2}$$

Now, we have a point on the line and the slope. We can find an equation for this tangent line using the point slope formula that says, $y - y_1 = m(x - x_1)$ to get that:

$$y - 1 = \frac{3}{2}(x - 1)$$

$$y = \frac{3}{2}x - \frac{3}{2} + 1$$

$$y = \frac{3}{2}x - \frac{1}{2}$$

To get a vector that "lives" on this tangent line requires that we know two points that "live" on this line. We already have one of them, which is the point $(1, 1)$. To get another one, choose an arbitrary value for x, say $x = 2$, and plug this into our equation for the tangent line to get the corresponding y-coordinate. Then we have:

$$y = \frac{3}{2}(2) - \frac{1}{2} = 3 - \frac{1}{2} = \frac{5}{2}$$

Therefore, the point $\left(2, \dfrac{5}{2}\right)$ is also on our tangent line. We can convert this to an algebraic vector using our conversion formulas $x = x_2 - x_1$ *and* $y = y_2 - y_1$ to obtain our tangent vector.

It will not matter which point we choose as our initial point here, since the direction of the motion was not specified. Later, when we get to Chapter 2 on vector-valued functions, we will have to worry about which is which!

So, the tangent vector, which we shall call $\mathbf{v}$ is given by:

$$\mathbf{v} = \left\langle x, y \right\rangle = \left\langle x_2 - x_1, \ y_2 - y_1 \right\rangle = \left\langle 2 - 1, \ \frac{5}{2} - 1 \right\rangle = \left\langle 1, \ \frac{3}{2} \right\rangle$$

Now, we need to "normalize" this vector—to find the corresponding unit vector. We will use the formula $\mathbf{u} = \dfrac{\mathbf{v}}{\|\mathbf{v}\|}$, where the magnitude of **v** is found by:

$$\|\mathbf{v}\| = \sqrt{(1)^2 + \left(\frac{3}{2}\right)^2} = \sqrt{1 + \frac{9}{4}} = \sqrt{\frac{13}{4}} = \frac{\sqrt{13}}{2}$$

Now we can get the unit vector:

$$\mathbf{u} = \frac{\mathbf{v}}{\|\mathbf{v}\|} = \frac{\left\langle 1, \frac{3}{2} \right\rangle}{\frac{\sqrt{13}}{2}} \left(\frac{2}{\sqrt{13}}\right)\left\langle 1, \frac{3}{2} \right\rangle = \left\langle \frac{2}{\sqrt{13}}, \frac{3}{\sqrt{13}} \right\rangle$$

This vector **u** is tangent to the curve at (1, 1) because it has the same direction as **v**.

Of course, $-\mathbf{u} = -\left\langle \dfrac{2}{\sqrt{13}}, \dfrac{3}{\sqrt{13}} \right\rangle = \left\langle -\dfrac{2}{\sqrt{13}}, -\dfrac{3}{\sqrt{13}} \right\rangle$ which points in the opposite direction, is also tangent to the curve at (1, 1). As mentioned above, without specifying the direction of the motion, there is no reason to prefer one of these vectors to another.

How do we obtain a vector normal (perpendicular) to the curve at the point (1, 1)? First, find the equation of a line that is perpendicular to the tangent line. This is actually an easy algebra problem. We found the slope of the tangent line to be $\dfrac{3}{2}$. A line that will be perpendicular to this tangent line needs to have a slope that is the negative reciprocal of this, which is $-\dfrac{2}{3}$. So, again using the point-slope form for an equation of a line, we have that:

$$y - 1 = -\frac{2}{3}(x - 1)$$

$$y = -\frac{2}{3}x + \frac{2}{3} + 1$$

$$y = -\frac{2}{3}x + \frac{5}{3}$$

This is the equation for a line perpendicular to our tangent line. We need to find a vector that "lives" on the perpendicular, or normal, line. We need two points to form a vector. We already have that the point (1, 1) is on this normal line. To get another point, choose an arbitrary value for x, say $x = 2$, and plug this into our equation for the normal line to get the corresponding y-coordinate. Then we have:

$$y = -\frac{2}{3}(2) + \frac{5}{3} = -\frac{4}{3} + \frac{5}{3} = \frac{1}{3}$$

So, the point $\left(2, \dfrac{1}{3}\right)$ is another point on the normal line. We can convert this to an algebraic vector using our conversion formulas $x = x_2 - x_1$ *and* $y = y_2 - y_1$ to obtain our normal vector. It will not matter which point we choose as our initial point here, since the direction of the motion was not specified. Later, when we get to Chapter 2 on vector-valued functions, we will require that our normal vector be pointing perpendicular to the direction of the motion. The normal vector, which we shall call **n** is given by:

$$\mathbf{n} = \langle x, y \rangle = \langle x_2 - x_1, \ y_2 - y_1 \rangle = \left\langle 2-1, \ \frac{1}{3}-1 \right\rangle = \left\langle 1, \ -\frac{2}{3} \right\rangle$$

Finally, we create the unit vector in the same direction as this normal vector. Let's find the magnitude of the vector **n** first:

$$\|\mathbf{n}\| = \sqrt{(1)^2 + \left(-\frac{2}{3}\right)^2} = \sqrt{1+\frac{4}{9}} = \sqrt{\frac{13}{9}} = \frac{\sqrt{13}}{3}$$

Now we can get the unit vector:

$$\mathbf{u} = \frac{\mathbf{n}}{\|\mathbf{n}\|} = \frac{\left\langle 1, -\frac{2}{3} \right\rangle}{\frac{\sqrt{13}}{3}} = \left(\frac{3}{\sqrt{13}}\right)\left\langle 1, -\frac{2}{3} \right\rangle = \left\langle \frac{3}{\sqrt{13}}, -\frac{2}{\sqrt{13}} \right\rangle \text{ (answer)}$$

This concludes the first section!

EXERCISES: #1–6 all from **Quiz 2**, #2a, 4 from **Practice Midterm 1**, #1 from **Quiz 4.**

1.2 SPACE COORDINATES AND VECTORS IN SPACE

You will be formally introduced to the three-dimensional rectangular coordinate system in this section, as well as vectors in space. Fortunately, all of the properties and operations that apply to two-dimensional vectors apply to three-dimensional vectors also. So, you can still form unit vectors using the same formula. You have now three standard unit vectors **i**, **j**, and **k** (instead of just the two **i** and **j**). You can find magnitudes (lengths) by simply extending the magnitude formula to include the z-coordinate. You can multiply and divide vectors by scalars. You can still add or subtract vectors in the usual manner. The zero vector in space looks like $\mathbf{0} = \langle 0, 0, 0 \rangle$. That is, it has three components instead of only two. The distance and midpoint formulas you learned in your basic algebra course can now be extended to 3-space.

Recall the equation for a circle in the plane, centered at the point (h, k) that you learned in algebra class which is $(x-h)^2 + (y-k)^2 = r^2$, where r is the radius. The analogous construct in space is the sphere, whose standard equation is $(x-x_0)^2 + (y-y_0)^2 + (z-z_0)^2 = r^2$ whose center is (x_0, y_0, z_0), and where r is the radius.

The next example will require the standard equation of a sphere, as well as the distance and midpoint formulas in space.

EXAMPLE. Find the standard equation for the sphere where the points (2, 2, 1) and (2, –4, 4) form the endpoints of a diameter.

SOLUTION. First, I will find the center point, which happens to be the midpoint between the two given points. I will use the Midpoint Formula:

$$M = \left(\frac{x_1 + x_2}{2}, \frac{y_1 + y_2}{2}, \frac{z_1 + z_2}{2} \right) = \left(\frac{2+2}{2}, \frac{2+-4}{2}, \frac{1+4}{2} \right) = \left(2, -1, \frac{5}{2} \right)$$

To find the radius, I will first find the length of the diameter, and then divide that by two. To find the length of the diameter, I will use the distance formula for space:

$$d = \sqrt{(x_1 - x_2)^2 + (y_1 - y_2)^2 + (z_1 - z_2)^2} = \sqrt{(2-2)^2 + (2-(-4))^2 + (1-4)^2} = \sqrt{0 + 6^2 + (-3)^2} = \sqrt{36+9}$$
$$= \sqrt{45} = 3\sqrt{5}$$

The radius is one-half of this, which means that $r = \dfrac{3\sqrt{5}}{2}$.

Finally, I will substitute the values for the center and radius into the standard equation of the sphere to get:

$$(x-2)^2 + (y+1)^2 + \left(z - \frac{5}{2} \right)^2 = \left(\frac{3\sqrt{5}}{2} \right)^2$$

$$(x-2)^2 + (y+1)^2 + \left(z - \frac{5}{2} \right)^2 = \frac{45}{4}$$

The last "new" concept covered in this section is that of parallel vectors. We can say that two non-zero vectors are parallel if there exists some scalar c such that one vector equals the product of the scalar and the other vector. We can say that the vectors **u** and **v** are parallel if **u** = *c***v** *or* **v** = *c***u** (either one).

EXAMPLE. Determine which of the vectors are parallel to the vector $\mathbf{z} = \dfrac{1}{2}\mathbf{i} - \dfrac{2}{3}\mathbf{j} + \dfrac{3}{4}\mathbf{k}$

$$\textbf{(a) } \mathbf{v} = \left\langle -1, \frac{4}{3}, \frac{3}{2} \right\rangle \qquad \textbf{(b) } \mathbf{w} = \left\langle \frac{3}{4}, -1, \frac{9}{8} \right\rangle$$

SOLUTION. For the vectors **z** and **v** to be parallel, we need to find a scalar c such that either **z** = *c***v** OR **v** = *c***z** (it does not matter which one). It looks as if the scalar $c = -2$ might work for **v** = *c***z**, except that it fails for the last component. We know that we can rewrite the vector **z** in component form to make it easier to compare to the vector **v**:

$$\mathbf{z} = \frac{1}{2}\mathbf{i} - \frac{2}{3}\mathbf{j} + \frac{3}{4}\mathbf{k} = \left\langle \frac{1}{2}, -\frac{2}{3}, \frac{3}{4} \right\rangle$$

$$\text{Does } \cdot \left\langle -1, \frac{4}{3}, \frac{3}{2} \right\rangle = (-2) \left\langle \frac{1}{2}, -\frac{2}{3}, \frac{3}{4} \right\rangle ?$$

Check on a component-by-component basis:

$$-1 = (-2)\left(\frac{1}{2}\right)$$

$$\frac{4}{3} = (-2)\left(-\frac{2}{3}\right)$$

$$\frac{3}{2} = (-2)\left(\frac{3}{4}\right) \quad \textit{This is not correct!}$$

We need the scalar to work for all three components. Since it fails for one of them, we can say that the two vectors are *not* parallel (answer to part **(a)**).

Let us take a look at comparing the vectors **z** and **w**. These will be parallel, since **w** = *c***z**, where the scalar $c = \frac{3}{2}$. *(answer)* Check this out for yourself!

EXERCISES: #1, 2 all from **Quiz 1**, # 1–10 from **Quiz 2**, #1a from **Practice Midterm 1**

1.3 THE DOT PRODUCT OF TWO VECTORS

The Dot Product of two vectors is often referred to as the Scalar Product of two vectors, since the result is a scalar. It is also sometimes referred to as the "Inner Product" as well. We shall see in the next section that there is another type of product between vectors whose result is a vector, not a scalar. You will need to be able to distinguish between the two types of products for later. You also need to memorize the process of finding each type of product. The dot product possesses both commutative and distributive properties, so it is easy to work with. The property that is not so obvious that you also need to memorize (especially for later) is the fact that $\mathbf{v} \cdot \mathbf{v} = \|\mathbf{v}\|^2$. Read your text for the proof of this; it makes sense when you see it. Once you know how to find the dot product, you can find the angle between any two vectors **u** and **v** in space by using the formula:

$$\cos\theta = \frac{\mathbf{u} \cdot \mathbf{v}}{\|\mathbf{u}\|\|\mathbf{v}\|}$$

If the angle is given, then you can rearrange this formula and use it instead to find the dot product of the two vectors $\mathbf{u} \cdot \mathbf{v} = \cos\theta\|\mathbf{u}\|\|\mathbf{v}\|$.

Please be familiar with *both* versions of this formula.

The next important topic is about vectors being *orthogonal* with one another. The text claims that the words "perpendicular," "orthogonal," and "normal" all mean essentially the same thing. This is not altogether true, as I will explain in a moment. You do want to use the terms in the following manner:

We say that two vectors are *orthogonal*, two lines or planes are *perpendicular*, and a vector is *normal* to a given plane or line. Here is where I differ with the text: two vectors can be orthogonal, but not really be perpendicular if they are vectors in *n* dimensions, where *n* is greater than 3. For example, vectors in 4 dimensions are impossible to visualize, and this is exactly why we cannot even talk about vectors being perpendicular in 4 dimensions! We will not consider vectors in *n* dimensions in this course, but those of you going on to take higher levels of math and physics may want to take note.

We say two vectors are orthogonal if the dot product is equal to zero. If you again look at the formula $\cos\theta = \dfrac{\mathbf{u}\cdot\mathbf{v}}{\|\mathbf{u}\|\|\mathbf{v}\|}$, then this means the numerator would be zero, so that the cosine would equal zero also. Angles whose cosines are equal to zero will be odd multiples of $\dfrac{\pi}{2}$, or 90°. This is how we can come to the conclusion of vectors being perpendicular.

Direction Cosines

For a vector in the plane, we measure direction by the angle the vector makes with the positive *x*-axis. In space, we measure direction in terms of the three angles the vector makes with each of the standard unit vectors, **i**, **j**, and **k**. We will denote these three angles α, β, γ (read: "alpha," "beta," and "gamma"), and these are called "direction angles." The cosines of these angles are called the "direction cosines." The formulas for these are in the text! An example using the formulas is shown below.

Projections of Vector Components

In the last section, we looked at adding two vectors together to produce a resultant vector. Now, let us do the reverse: We will decompose a vector into the sum of two vectors. Specifically, we will decompose a vector into 2 *orthogonal* components. It will go like this:

Given a vector **u**, which is the sum of two vectors, $\mathbf{w}_1$ and $\mathbf{w}_2$, where $\mathbf{w}_1$ and $\mathbf{w}_2$ are orthogonal to each other; we have that $\mathbf{u} = \mathbf{w}_1 + \mathbf{w}_2$. Let **v** be another vector that is parallel to $\mathbf{w}_1$. Then $\mathbf{w}_1$ is called the projection of **u** onto **v** and can be found using the formula:

$$\mathbf{w}_1 = \mathbf{proj}_v\mathbf{u} = \left(\frac{\mathbf{u}\cdot\mathbf{v}}{\|\mathbf{v}\|^2}\right)\mathbf{v}$$

Note the new notation we are using, which is $\mathbf{proj}_v\mathbf{u}$ to denote this vector. The formula to find $\mathbf{w}_2$ is then easily found. We will use the fact that we already know we need $\mathbf{u} = \mathbf{w}_1 + \mathbf{w}_2$.

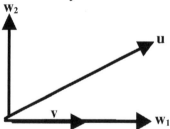

Since we just found $\mathbf{w}_1$ using the new projection formula, we will simply solve this "equation" for $\mathbf{w}_2$ by subtracting $\mathbf{w}_1$ from both sides and get that $\mathbf{w}_2 = \mathbf{u} - \mathbf{w}_1$.

Another way to write the sum $\mathbf{u} = \mathbf{w}_1 + \mathbf{w}_2$ would then be $\mathbf{u} = \mathbf{proj}_v\mathbf{u} + (\mathbf{u} - \mathbf{proj}_v\mathbf{u})$.

I want you to reconsider the formula for the projection of **u** onto **v**:

$$\mathbf{proj}_v\mathbf{u} = \left(\frac{\mathbf{u}\cdot\mathbf{v}}{\|\mathbf{v}\|^2}\right)\mathbf{v}$$

Note: The denominator for the scalar, which is $\|\mathbf{v}\|^2$, is the same as $\mathbf{v} \cdot \mathbf{v}$. You may use either.

Do you notice that the vector $\mathbf{v}$ appears more often than the vector $\mathbf{u}$ in this formula? I always like to remind my students that when memorizing this projection formula that the vector you are projecting *onto* will be "used" more times in the formula than the vector that is being projected!

EXAMPLE. A force $\mathbf{F} = 2\mathbf{i} + \mathbf{j} - 3\mathbf{k}$ N (N means newtons and is a measurement of force in physics) is applied to a spacecraft with velocity vector $\mathbf{v} = 3\mathbf{i} - \mathbf{j}$.

(a) Express the force $\mathbf{F}$ as a *sum* of a vector parallel to $\mathbf{v}$ and a vector orthogonal to $\mathbf{v}$. Use component forms in your answer.

(b) Find the angle between the force vector and the velocity vector. Round angle to the nearest hundredth radians.

(c) Find the direction cosines for the force vector $\mathbf{F}$. Then find the direction angles and round answers to the nearest hundredth radians.

(d) Show that the velocity vector and the orthogonal component of $\mathbf{F}$ are indeed perpendicular to each other.

SOLUTION.

(a) We need to write the force vector as a sum: $\mathbf{F} = \mathbf{proj}_{\mathbf{v}}\mathbf{F} + (\mathbf{F} - \mathbf{proj}_{\mathbf{v}}\mathbf{F})$.

We first find the projection of the force vector $\mathbf{F}$ onto the velocity vector $\mathbf{v}$ by using the formula:

$$\mathbf{proj}_{\mathbf{v}}\mathbf{F} = \left(\frac{\mathbf{F} \cdot \mathbf{v}}{\mathbf{v} \cdot \mathbf{v}}\right)\mathbf{v} = \left(\frac{\langle 2,1,-3\rangle \cdot \langle 3,-1,0\rangle}{\langle 3,-1,0\rangle \cdot \langle 3,-1,0\rangle}\right)\langle 3,-1,0\rangle = \left(\frac{(2)(3)+(1)(-1)+(-3)(0)}{(3)(3)+(-1)(-1)+(0)(0)}\right)\langle 3,-1,0\rangle = \left(\frac{5}{10}\right)\langle 3,-1,0\rangle$$

$$= \left\langle \frac{3}{2}, -\frac{1}{2}, 0\right\rangle$$

Now we find the orthogonal component, which is a vector orthogonal to $\mathbf{v}$. Presently, we will find the vector $(\mathbf{F} - \mathbf{proj}_{\mathbf{v}}\mathbf{F})$.

$$(\mathbf{F} - \mathbf{proj}_{\mathbf{v}}\mathbf{F}) = \langle 2,1,-3\rangle - \left\langle \frac{3}{2}, -\frac{1}{2}, 0\right\rangle = \left\langle 2 - \frac{3}{2}, 1 - \left(-\frac{1}{2}\right), -3 - 0\right\rangle = \left\langle \frac{1}{2}, \frac{3}{2}, -3\right\rangle$$

So, our final answer for part (a) will be $\mathbf{F} = \mathbf{proj}_{\mathbf{v}}\mathbf{F} + (\mathbf{F} - \mathbf{proj}_{\mathbf{v}}\mathbf{F}) = \left\langle \frac{3}{2}, -\frac{1}{2}, 0\right\rangle + \left\langle \frac{1}{2}, \frac{3}{2}, -3\right\rangle$.

The force $\mathbf{proj}_{\mathbf{v}}\mathbf{F} = \left\langle \frac{3}{2}, -\frac{1}{2}, 0\right\rangle$ is the effective force parallel to the velocity vector $\mathbf{v}$.

(b) The angle between the force and velocity vectors will be found using the formula:

$\theta = \cos^{-1}\left(\dfrac{\mathbf{F} \cdot \mathbf{v}}{\|\mathbf{F}\| \cdot \|\mathbf{v}\|}\right)$. We already found the dot product between $\mathbf{F}$ and $\mathbf{v}$, which is $\mathbf{F} \cdot \mathbf{v} = 5$.

Let us find each magnitude $\|\mathbf{F}\| = \sqrt{(2)^2 + (1)^2 + (-3)^2} = \sqrt{4+1+9} = \sqrt{14}$, and

$$\|\mathbf{v}\| = \sqrt{(3)^2 + (-1)^2 + (0)^2} = \sqrt{9+1} = \sqrt{10}$$

Now, we have the angle between the vectors $\theta = \cos^{-1}\left(\dfrac{\mathbf{F} \cdot \mathbf{v}}{\|\mathbf{F}\| \cdot \|\mathbf{v}\|}\right) = \cos^{-1}\left(\dfrac{5}{\sqrt{14}\sqrt{10}}\right) \approx 1.13$ radians.

(c) Direction cosines and angles. Using the formulas:

$\cos\alpha = \dfrac{a}{\|F\|} = \dfrac{2}{\sqrt{14}}$, where α is the angle between $\mathbf{F}$ and the x-axis, and α is the first component of $\mathbf{F}$.

Then $\alpha = \cos^{-1}\dfrac{2}{\sqrt{14}} \approx 1.01$ radians.

Continuing on: $\cos\beta = \dfrac{b}{\|F\|} = \dfrac{1}{\sqrt{14}}$, $\beta = \cos^{-1}\dfrac{1}{\sqrt{14}} \approx 1.30$ radians, and

$\cos\gamma = \dfrac{c}{\|F\|} = \dfrac{-3}{\sqrt{14}}$, $\gamma = \cos^{-1}\dfrac{-3}{\sqrt{14}} \approx 2.50$ radians, where β is the angle between $\mathbf{F}$ and the y-axis,

and γ is the angle between $\mathbf{F}$ and the z-axis.

(d) We need to show that the velocity vector $\mathbf{v}$ and the orthogonal component of $\mathbf{F}$, $\left(\mathbf{F} - \mathbf{proj}_v\mathbf{F}\right)$, are perpendicular. The quickest way to do this is to show that the angle between them is 90°, which we can do by showing that the cosine of the angle between them is equal to 0, and the way we will do *this* is to show that the dot product is equal to 0.

$$\mathbf{v} \cdot \left(\mathbf{F} - \mathbf{proj}_v\mathbf{F}\right) = \langle 3, -1, 0\rangle \cdot \left\langle \frac{1}{2}, \frac{3}{2}, -3\right\rangle = (3)\left(\frac{1}{2}\right) + (-1)\left(\frac{3}{2}\right) + (0)(-3) = \frac{3}{2} + \left(-\frac{3}{2}\right) = 0$$

Therefore, these two vectors are indeed orthogonal.

EXERCISES: #11–22, 26–28 all from **Quiz 2**, #1b, d, e, f, k, and 3 from **Practice Midterm 1**, #2 from **Quiz 4**

1.4 THE CROSS PRODUCT OF TWO VECTORS IN SPACE

In many physics and engineering problems one may be asked to find a vector that is orthogonal to two given vectors in space (simultaneously). You can do this by simply taking the cross product of these two vectors in space. The cross product is much different from the dot product. The dot product result is a scalar, while the cross product result is another vector. This new vector will be perpendicular to *both* of the vectors from which it is obtained. There are two key facts about the cross product I would like you to memorize:

1) The cross product only makes sense in space (three dimensions). So, there is no such thing as the cross product of two vectors in the plane (2 dimensions), or any other dimensional space.

2) The cross product between two vectors $\mathbf{u}$ and $\mathbf{v}$ is denoted $\mathbf{u} \times \mathbf{v}$, and can be found using the same method used to find the determinant of a 3×3 matrix. We will demonstrate this method in an example on the next page.

Some other handy geometric properties of the cross product are:

(a) $\|\mathbf{u} \times \mathbf{v}\| = \|\mathbf{u}\|\|\mathbf{v}\|\sin\theta$ = area of parallelogram having both vectors $\mathbf{u}$ and $\mathbf{v}$ as adjacent sides.

(b) $\mathbf{u} \times \mathbf{v} = \mathbf{0}$, the zero vector, if $\mathbf{u}$ and $\mathbf{v}$ are scalar multiples of each other. This means that $\mathbf{u} \times \mathbf{u} = \mathbf{0}$ also.

(c) $\mathbf{u} \times \mathbf{v} = -(\mathbf{v} \times \mathbf{u})$

EXAMPLE. Find a vector perpendicular to the plane containing the points $P(1, -1, 0)$, $Q(2, 1, -1)$, and $R(-1, 1, 2)$. Write your answer in component form. Then, obtain the area of the parallelogram having these points as three vertices.

SOLUTION. The vector obtained by taking the cross product of the vectors $\overrightarrow{PQ}$ and $\overrightarrow{PR}$ will be perpendicular to the plane. We first need to form the vectors $\overrightarrow{PQ}$ and $\overrightarrow{PR}$. Recall that you need to do "terminal point minus the initial point" to get a vector from two given points.

$$\overrightarrow{PQ} = "Q - P" = (2,1,-1) - (1,-1,0) = \langle (2-1),\ (1-(-1)),\ (-1-0) \rangle = \langle 1, 2, -1 \rangle$$

$$\overrightarrow{PR} = "R - P" = (-1,1,2) - (1,-1,0) = \langle (-1-1),\ (1-(-1)),\ (2-0) \rangle = \langle -2, 2, 2 \rangle$$

Now, we can find the cross product. We will use the method which is similar to finding the determinant of a 3×3 matrix.

$$\overrightarrow{PQ} \times \overrightarrow{PR} = \begin{vmatrix} \mathbf{i} & \mathbf{j} & \mathbf{k} \\ 1 & 2 & -1 \\ -2 & 2 & 2 \end{vmatrix} = \begin{vmatrix} 2 & -1 \\ 2 & 2 \end{vmatrix}\mathbf{i} - \begin{vmatrix} 1 & -1 \\ -2 & 2 \end{vmatrix}\mathbf{j} + \begin{vmatrix} 1 & 2 \\ -2 & 2 \end{vmatrix}\mathbf{k}$$

$$= \big((2)(2) - (2)(-1)\big)\mathbf{i} - \big((1)(2) - (-2)(-1)\big)\mathbf{j} + \big((1)(2) - (-2)(2)\big)\mathbf{k}$$

$$= 6\mathbf{i} + 0\mathbf{j} + 6\mathbf{k} = \langle 6, 0, 6 \rangle$$

PLEASE NOTICE THAT THERE IS A MINUS SIGN ("−") IN FRONT OF THE j VECTOR!

To find the area of the parallelogram, we will simply use the fact that the area of a parallelogram having two vectors as adjacent sides is equal to the magnitude of the cross product of these two vectors. That is, we will use the formula from above, which states:

$$\|\mathbf{u} \times \mathbf{v}\| = \|\mathbf{u}\|\|\mathbf{v}\|\sin\theta$$ = area of parallelogram having both vectors $\mathbf{u}$ and $\mathbf{v}$ as adjacent sides

Therefore, the area of the parallelogram having $\overrightarrow{PQ}$ and $\overrightarrow{PR}$ as adjacent sides will be the found by finding the magnitude of the cross product.

The magnitude of this vector, denoted $\left\|\overrightarrow{PQ} \times \overrightarrow{PR}\right\|$, will be:

$$\left\|\overrightarrow{PQ} \times \overrightarrow{PR}\right\| = \sqrt{6^2 + 0^2 + 6^2} = \sqrt{36 + 36} = \sqrt{72} = 6\sqrt{2} \text{ (answer)}$$

EXERCISES: #24, 25 from **Quiz 2**, #1c from **Practice Midterm 1**

1.5 LINES AND PLANES IN SPACE

After studying this section, you should meet four objectives:

    1) Write a set of parametric equations for a line in space.

    2) Write a linear equation for a plane in space.

    3) Sketch the plane given by a linear equation.

    4) Find the distance between points, planes, and lines in space.

You should not only be able to meet all four objectives, but do them using all previous knowledge of vectors.

While a line in two dimensions can be represented by a *single* equation of two variables (for example, $y = 2x - 3$), a line in three dimensions requires *three* separate equations and *four* variables total! This can be confusing at first. For those students who remember representing lines using parametric equations from their last semester of calculus, this won't be too hard. For those students who never learned parametric equations, this concept may require a bit more studying to grasp it. So, not only will we need the three variables x, y, and z, (for the three axes), we will need an additional variable—traditionally the letter "t" is used for this. Often times, this parameter may be used to represent time, the "fourth dimension." To represent a line in space with these three parametric equations, we require two pieces of information:

    1) A vector that is parallel to our line, say $\mathbf{v} = \langle a, b, c \rangle$ a.k.a. the "direction vector"

    2) A point that our line passes through, say $P(x_1, y_1, z_1)$

Using these two pieces of information, we obtain the three parametric equations that represent a *single* line in space: $x = x_1 + at$, $y = y_1 + bt$, and $z = z_1 + ct$

Note: The parametric equations for a line are *not* unique, because they depend on the point chosen, as well as the direction vector. The equations will be different if you use a different point that "lives" on the line. The equations may also be different if you use a *multiple* of the direction vector. Keep in mind that if the direction vector is given by $\mathbf{v} = \langle a, b, c \rangle$, you can also use any multiple of this direction and still be correct; you may use $k\mathbf{v} = k\langle a, b, c \rangle = \langle ka, kb, kc \rangle$ as your direction vector. This will result in a different set of parametric equations, but still be representing the same line! I will do examples involving parametric equations for lines in a little while. But first, I want to cover the (single!) equation for a plane in space. As mentioned above, we need a linear equation in two variables to represent a line in two dimensions. However, linear equations using three variables represent planes in three dimensions (space).

To represent a plane in space with a single linear equation (containing the three variables x, y, and z), we require two pieces of information:

    1) A vector that is *normal* (perpendicular) to our plane, say $\mathbf{n} = \langle a, b, c \rangle$

    2) A point that "lives" in our plane, say $P(x_1, y_1, z_1)$

So, we can say that any arbitrary point that also "lives" in our plane can be represented by $Q(x, y, z)$.

Then, the vector $\vec{PQ} = "Q - P" = \langle x, y, z \rangle - \langle x_1, y_1, z_1 \rangle = \langle x - x_1, y - y_1, z - z_1 \rangle$ will be a vector that "lives" on our plane. So, if the vector **n** is *normal* to our plane, then the dot product of the vector **n** with the vector $\vec{PQ}$ must be equal to zero—we require that $\mathbf{n} \cdot \vec{PQ} = 0$. This equation will actually give us our equation for our plane in the following way:

$$\mathbf{n} \cdot \vec{PQ} = 0$$
$$\langle a, b, c \rangle \cdot \langle x - x_1, y - y_1, z - z_1 \rangle = 0$$
$$a(x - x_1) + b(y - y_1) + c(z - z_1) = 0$$

This version of the equation is known as the standard equation of the plane. If we distribute and collect like terms, we obtain the equation $ax + by + cz = d$ (called the general form of the equation for a plane).

Please note that the coefficients of the general form of the plane are exactly the components of the vector that is normal to our plane. This "subtle" fact will be useful to you for problems on tests!

EXAMPLE #1. Find the equation of a plane in general form that passes through the points $(1, 1, -1)$ and $(2, 0, 2)$ and $(0, -2, 1)$.

SOLUTION. Recall the two pieces of information required to get the equation of a plane are 1) a point on the plane and 2) a vector normal to the plane. We already have plenty of points on the plane, but we are lacking a normal vector. To get this vector, we will first form two vectors that "live" on the plane. Then, we will take the cross product of these two vectors to obtain the normal vector to the plane. Let's name the points first: $P(1, 1, -1)$ and $Q(2, 0, 2)$ and $R(0, -2, 1)$. I will form the vectors $\vec{PQ}$ and $\vec{PR}$ first:

$$\vec{PQ} = "Q - P" = \langle 2, 0, 2 \rangle - \langle 1, 1, -1 \rangle = \langle 2 - 1, 0 - 1, 2 - (-1) \rangle = \langle 1, -1, 3 \rangle$$
$$\vec{PR} = "R - P" = \langle 0, -2, 1 \rangle - \langle 1, 1, -1 \rangle - = \langle 0 - 1, -2 - 1, 1 - (-1) \rangle = \langle -1, -3, 2 \rangle$$

Now, we will take the cross product:

$$\vec{PQ} \times \vec{PR} = \begin{vmatrix} \mathbf{i} & \mathbf{j} & \mathbf{k} \\ 1 & -1 & 3 \\ -1 & -3 & 2 \end{vmatrix} = \begin{vmatrix} -1 & 3 \\ -3 & 2 \end{vmatrix} \mathbf{i} - \begin{vmatrix} 1 & 3 \\ -1 & 2 \end{vmatrix} \mathbf{j} + \begin{vmatrix} 1 & -1 \\ -1 & -3 \end{vmatrix} \mathbf{k}$$

$$= ((-1)(2) - (-3)(3)) \mathbf{i} - ((1)(2) - (-1)(3)) \mathbf{j} + ((1)(-3) - (-1)(-1)) \mathbf{k}$$
$$= 7\mathbf{i} - 5\mathbf{j} - 4\mathbf{k}$$

AGAIN, PLEASE NOTICE THAT THERE IS A MINUS SIGN ("−") IN FRONT OF THE j VECTOR!

Therefore, a vector normal to our plane, which we shall denote **n**, will be $\mathbf{n} = \langle 7, -5, -4 \rangle$

We can obtain the equation for the plane using the standard form for the equation of a plane $a(x - x_1) + b(y - y_1) + c(z - z_1) = 0$. We have three different points to choose from on our plane, and it does not matter which one we choose. I will simply choose the point $P(1, 1, -1)$ to work with here. So, we have:

$$7(x-1) - 5(y-1) - 4(z - (-1)) = 0$$
$$7(x-1) - 5(y-1) - 4(z+1) = 0$$
$$7x - 7 - 5y + 5 - 4z - 4 = 0$$
$$7x - 5y - 4z = 6$$

And we are done!

Finding the Line of Intersection of Two Planes

EXAMPLE #2. *Fact:* Planes that are not parallel will intersect in a line. Given the two intersecting planes:

$$3x - 6y - 2z = 15 \quad and \quad 2x + y - 2z = 5$$

(a) Find a vector (in component form) parallel to the line of intersection of these two planes.

(b) Find parametric equations for the line of intersection for the two planes.

SOLUTION.

(a) The line of intersection of two planes is perpendicular to the planes' normal vectors $\mathbf{n}_1$ and $\mathbf{n}_2$. This means that we can find the vector that is perpendicular to both planes by simply finding the cross product of these two normal vectors; we just need to find $\mathbf{n}_1 \times \mathbf{n}_2$. First we need to find the component form for these two normal vectors. As mentioned above, the coefficients for the variables in each equation for the plane yield the components for the normal vector to each respective plane. That is:

$$\mathbf{n}_1 = \langle 3, -6, -2 \rangle \quad and \quad \mathbf{n}_2 = \langle 2, 1, -2 \rangle$$

Now we can find the cross product:

$$\mathbf{n}_1 \times \mathbf{n}_2 = \begin{vmatrix} \mathbf{i} & \mathbf{j} & \mathbf{k} \\ 3 & -6 & -2 \\ 2 & 1 & -2 \end{vmatrix} = \begin{vmatrix} -6 & -2 \\ 1 & -2 \end{vmatrix} \mathbf{i} - \begin{vmatrix} 3 & -2 \\ 2 & -2 \end{vmatrix} \mathbf{j} + \begin{vmatrix} 3 & -6 \\ 2 & 1 \end{vmatrix} \mathbf{k}$$

$$= ((-6)(-2) - (1)(-2))\mathbf{i} - ((3)(-2) - (2)(-2))\mathbf{j} + ((3)(1) - (2)(-6))$$

$$= 14\mathbf{i} + 2\mathbf{j} + 15\mathbf{k}$$

$$= \langle 14, 2, 15 \rangle$$

(b) To find the three parametric equations for this line of intersection, we need two pieces of information, which are:

(1) a vector in the direction of the line (a.k.a. "the direction vector")

(2) a point on the line.

We already have a vector pointing in the direction of this line. This is the result we got from part **a)** above, that is $\mathbf{n}_1 \times \mathbf{n}_2$. To find a point on the line, we find a point in common with both planes. Let's set $z = 0$ in the plane equations, and then solve for x and y simultaneously.

$$3x - 6y - 2(0) = 15$$
$$2x + y - 2(0) = 5$$
$$\begin{cases} 3x - 6y = 15 \\ 2x + y = 5 \end{cases}$$

We have a linear system with two equations and two unknowns here. I will multiply both sides of the second equation by 6, add the two equations together to eliminate the y-variable and solve for x:

$$\begin{cases} 3x - 6y = 15 \\ 6(2x + y) = (5)(6) \end{cases}$$

$$\begin{cases} 3x - 6y = 15 \\ 12x + 6y = 30 \end{cases}$$ Add these two equations together:

$$15x \qquad = 45$$

$$x = 3$$

Substitute this back into either of the two plane equations, and we have that $y = -1$. Therefore, a point in common with both planes is the point $(3, -1, 0)$.

We can use the three parametric equations for the line $x = x_1 + at$, $y = y_1 + bt$, and $z = z_1 + ct$, to get that $x = 3 + 14t$, $y = -1 + 2t$, and $z = 15t$ (answer).

EXAMPLE #3. Find the point of intersection between the line represented by:

$$x = \frac{8}{3} + 2t, \quad y = -2t, \quad \text{and} \quad z = 1 + t \text{ and the plane } 3x + 2y + 6z = 6$$

SOLUTION. Let the point of intersection be denoted P and have coordinates (x, y, z). This point P needs to "live" on the line so that the coordinates can also be represented as $\left(\frac{8}{3} + 2t, \ -2t, \ 1 + t \right)$.

Therefore, this point P will also lie in the plane if its coordinates satisfy the plane equation. So, let's substitute the coordinates into the plane equation in place of x, y, and z to get:

$$3\left(\frac{8}{3} + 2t \right) + 2(-2t) + 6(1 + t) = 6$$

After distributing and collecting like terms, we then have an easy linear equation in the variable t:

$$8 + 6t - 4t + 6 + t = 6$$
$$8t = -8$$
$$t = -1$$

This is not the answer, *OK?!* We want the *point* of intersection. The fact that $t = -1$ tells us at what *time* the plane and line intersect. Plug the value of $t = -1$ into the point P to get its coordinates:

$$\left(\frac{8}{3} + 2(-1), \ -2(-1), \ 1 + (-1) \right)$$
$$\left(\frac{2}{3}, 2, 0 \right)$$

The last topic in this section gives formulas to find distances between point and planes as well as between points and lines. Because these are pretty "formulaic," I will do only one example.

EXAMPLE. Find the distance between the planes $x + 2y + 6z = 1$ *and* $x + 2y + 6z = 10$.

SOLUTION. Recall the fact above that planes either intersect in a line or are parallel. I claim that these two planes are parallel. An easy way to tell is to notice that they both share the same normal vector, which can be "read" right from the coefficients of the variables $\mathbf{n} = \langle 1, 2, 6 \rangle$.

(*Note:* Two planes will be parallel if their normal vectors are parallel, and recall that two vectors are parallel if they are multiples of each other.)

So, if we can find two points (one in each plane) we can use the formula for the distance between a point and a plane and get our answer! If I set $y = z = 0$ in both plane equations, I will get two points that "live" in each of these planes. The point $P(1, 0, 0)$ is a point in the first plane and the point $Q(10, 0, 0)$ is a point in the second plane. Now, I will form the vector $\overrightarrow{PQ}$, and get that $\overrightarrow{PQ} = \langle 9, 0, 0 \rangle$. The distance between the two planes will be the magnitude of the projection of this vector onto the normal vector, $\mathbf{n}$. We will use the formula:

$$D = \left\| \text{proj}_{\mathbf{n}} \, \overrightarrow{PQ} \right\| = \frac{\left| \overrightarrow{PQ} \cdot \mathbf{n} \right|}{\| \mathbf{n} \|} = \frac{|(9)(1)|}{\sqrt{1^2 + 2^2 + 6^2}} = \frac{9}{\sqrt{41}} \quad \text{(answer)}$$

EXERCISES: #1, 2, 3 all from **Quiz 3**, #1g, h, i, j, 2b, 7 from **Practice Midterm 1**, #3 from **Quiz 4**

1.6 SURFACES IN SPACE

The two types of surfaces covered in this section are:

1) Quadric Surfaces

2) Cylindrical Surfaces

These are *not* the only types of surfaces in the world, however. (I skip surfaces of revolution.) There are an infinite number of types of surfaces out there, but we will restrict ourselves in this section to two special cases.

Quadric surfaces are the three-dimensional analog to conics in two dimensions. A quadric surface is the graph in space of a second degree equation in x, y, and z. The most general form is:

$$Ax^2 + By^2 + Cz^2 + Dxy + Eyz + Fxz + Gx + Hy + Jz + K = 0$$

…where A, B, C, and so on, are constants. The equation can be simplified by translation and rotation (i.e., eliminating the translation and rotation terms by getting $D = E = F = 0$) as in the two-dimensional cases. We will typically be interested in equations of the form:

$$Ax^2 + By^2 + Cz^2 + Gx + Hy + Jz + K = 0$$

The basic quadric surfaces are ellipsoids, paraboloids, elliptic cones, and hyperboloids. (We can think of spheres as special ellipsoids.) I will leave it to the reader to memorize each quadric type and its corresponding equation. I will not hold you responsible for graphing each of the surfaces. The two things to notice when distinguishing between quadrics are the signs *and* the coefficients of the terms x^2, y^2, z^2. Note that paraboloids have one second degree term missing; however, this variable will appear in its first degree.

Cylindrical surfaces are the other type of surface covered in this section, and they do not have a "general" equation. The way to recognize a cylindrical surface's equation is to notice that the equation will be missing one of the variables. The idea is to graph the equation (in two variables) as a curve in the plane of the two given variables, and then "extend" this curve throughout space in the direction of the missing variable. These straight line "extensions" are known as "rulings." The rulings will be parallel to the missing variable's axis. Therefore, the equation of a cylindrical surface in space will be equal to the equation of its generating curve. Do not be misled by the name—cylindrical surfaces are typically *not* cylinders (although they could be). Examples of cylindrical surface equations are:

$$z = y^2 \qquad z = \sin x \qquad y = x^2 \qquad x^2 + z^2 = 1$$

The first equation can be graphed by graphing the *curve* (that is a parabola) $z = y^2$ in the yz-plane, then "extending" it in 3-D where the rulings go in the direction parallel to the x-axis like so:

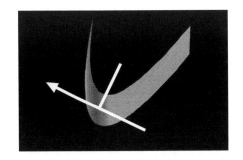

For complete examples, see the text and come to class!

EXERCISES: #4 from **Quiz 3**, #5 from **Practice Midterm 1**

1.7 CYLINDRICAL AND SPHERICAL COORDINATES

This section often occurs much later in many calculus texts. It is commonly not covered until the chapter on multiple integration. (Don't worry about what that is, yet!) Covering it long before we reach the chapter on multiple integration is nice because it "plants the seed" for students early in the semester and gets them thinking in terms of other coordinate systems. That way, we all won't have to stop what we are doing at that time and go over this stuff!

Recall the polar coordinate system that you learned in your last semester of calculus (I hope). You learned that the polar coordinate system is another way to represent two dimensions and that points are represented ordered pairs of the form (r, θ). You also learned conversion formulas between the polar coordinate system and the rectangular coordinate system. You also learned that polar representations of points are not unique since we can have that $(r, \theta) = (r, \theta + 2\pi k)$ $k \in \mathbb{Z}$. We can also have that $(r, \theta) = (-r, \theta + \pi)$.

Here in third semester calculus we will take a look at the polar coordinate analog of three dimensions, called the "cylindrical coordinate system." In the rectangular coordinate system, points are represented by ordered triples of the form (x, y, z). In the cylindrical coordinate system, points are represented by the ordered triple $P(r, \theta, z)$, where r and θ are exactly as they were defined for the polar coordinate system and the variable z is the directed distance from (r, θ) to out point P. (Here (r, θ) is the projection of our point P in the xy-plane.)

Conversion Formulas

FROM CYLINDRICAL TO RECTANGULAR	FROM RECTANGULAR TO CYLINDRICAL
$x = r\cos\theta$	$r^2 = x^2 + y^2$ or $r = \sqrt{x^2 + y^2}$
$y = r\sin\theta$	$\tan\theta = \dfrac{y}{x}$ or $\theta = \tan^{-1}\left(\dfrac{y}{x}\right)$
$z = z$	$z = z$

The nice thing for both conversions is that $z = z$! Keep in mind that converting points is different than converting *equations*. Some of my students have gotten confused on exams in previous years and gave me a point (i.e., ordered triple) for an answer when I asked them to convert an equation. We will do examples for both below. Keep these separate in your mind.

Cylindrical coordinates are excellent to use when representing surfaces that have the z-axis as the axis of symmetry. For example: cones, spheres, and cylinders.

The spherical coordinate system is yet another way to represent three dimensions. Here a point P will be represented by an ordered triple of the form (ρ, θ, ϕ). Let $\rho =$ the distance between the point and the origin, where $\rho \geq 0$. We know that θ is the same angle as before—that is, the angle θ used in cylindrical coordinates. Also, ϕ is the angle between the positive z-axis and the line segment $\overrightarrow{OP}$, where $0 \le \phi \le \pi$. (**Note:** the Greek letters ρ and ϕ are called "rho" and "phi," respectively.) I like to refer to ϕ as the "drop-down" angle in class.

Conversion Formulas

FROM SPHERICAL TO RECTANGULAR	FROM RECTANGULAR TO SPHERICAL
$x = \rho \sin\phi \cos\theta$	$\rho^2 = x^2 + y^2 + z^2$ or $\rho = \sqrt{x^2 + y^2 + z^2}$
$y = \rho \sin\phi \sin\theta$	$\tan\theta = \dfrac{y}{x}$ or $\theta = \tan^{-1}\left(\dfrac{y}{x}\right)$
$z = \rho \cos\phi$	$\phi = \arccos\left(\dfrac{z}{\sqrt{x^2 + y^2 + z^2}}\right)$

To change coordinates between cylindrical and spherical systems, use the following:

Conversion Formulas

FROM SPHERICAL TO CYLINDRICAL	FROM CYLINDRICAL TO SPHERICAL
$r^2 = \rho^2 \sin^2\phi$	$\rho = \sqrt{r^2 + z^2}$
$\theta = \theta$	$\theta = \theta$
$z = \rho \cos\phi$	$\phi = \arccos\left(\dfrac{z}{\sqrt{r^2 + z^2}}\right)$

The spherical coordinate system is useful primarily for surfaces in space that have a point of symmetry. Examples would be spheres, vertical half-planes, and half-cones.

OK—we are ready for some examples! First, I would like to show conversion for *points* between all three coordinate systems. Then, I'll do examples for converting *equations* between all three coordinate systems.

Converting Points

EXAMPLES. Complete the chart and convert the coordinates from one system to another among the rectangular cylindrical and spherical coordinate systems.

Rectangular	Cylindrical	Spherical
$\left(-2\sqrt{3},\, 2,\, 6\right)$		
	$\left(-3,\, \dfrac{\pi}{3},\, -2\right)$	
		$\left(10,\, \dfrac{5\pi}{4},\, \dfrac{2\pi}{3}\right)$

SOLUTION—ROW #1. Let's start by converting the point $\left(-2\sqrt{3},\, 2,\, 6\right)$ to cylindrical coordinates.

To find r, we know that $r^2 = x^2 + y^2$ or $r = \sqrt{x^2 + y^2}$.

I will use the second of these two formulas: $r = \sqrt{\left(-2\sqrt{3}\right)^2 + (2)^2} = \sqrt{(4\cdot 3)+4} = \sqrt{12+4} = \sqrt{16} = 4$.

We will find θ using the fact that $\theta = \tan^{-1}\left(\dfrac{y}{x}\right)$.

So, we have $\theta = \tan^{-1}\left(\dfrac{2}{-2\sqrt{3}}\right) = -30°\left(or\ -\dfrac{\pi}{6}\right)$.

I used my graphing calculator here in "degrees" mode. However, this answer for θ is *WRONG!* Why? Well, consider the original point again, which is $\left(-2\sqrt{3},\, 2,\, 6\right)$. Since the x-coordinate is negative and the y-coordinate is positive, this means that the projection of the point onto the xy-plane has an angle that is between 90° and 180°. The calculator gave us the wrong answer due to the fact that the range of the arctangent function gives angles that lie between −90° and +90° only. So, we need to make an adjustment. We need to add π radians (or 180°) to this answer to get the right Quadrant (Quadrant II).

Therefore, $\theta = -30° + 180° = 150°$, or $\dfrac{5\pi}{6}$ radians.

Now let's find the z-coordinate. But $z = z$ when converting from rectangular to cylindrical!

So, the final answer in cylindrical coordinates is $\left(4,\, \dfrac{5\pi}{6},\, 6\right)$.

We'll find the point using spherical coordinates. We start by using the formula $\rho = \sqrt{x^2 + y^2 + z^2}$.

Then we have $\rho = \sqrt{\left(-2\sqrt{3}\right)^2 + (2)^2 + (6)^2} = \sqrt{12+4+36} = \sqrt{52} = \sqrt{(4)(13)} = 2\sqrt{13}$.

The second coordinate will be the angle θ, which we found already. That is, $\theta = \dfrac{5\pi}{6}$.

We will find the angle ϕ using the formula:

$$\phi = \arccos\left(\frac{z}{\sqrt{x^2 + y^2 + z^2}}\right) = \arccos\left(\frac{6}{2\sqrt{13}}\right) = \arccos\left(\frac{3}{\sqrt{13}}\right) \approx 33.7° = 0.59 \text{ radians.}$$

Therefore, the final answer for spherical coordinates is $\left(2\sqrt{13}, \frac{5\pi}{6}, 0.59\right)$.

SOLUTION—ROW #2. Let's take a look at the cylindrical coordinate point $\left(-3, \frac{\pi}{3}, -2\right)$. Notice the radius, r, is a negative number. Recall that cylindrical coordinates are not unique. If we want r to be positive, we could add 180° (π radians) to the angle θ and still represent the same point. Our point could just as well be represented as $\left(+3, \frac{4\pi}{3}, -2\right)$. Let's work with this instead. Shall we?

We will convert the point to rectangular coordinates first. The x-coordinate will be:

$$x = r\cos\theta = 3\cos\left(\frac{4\pi}{3}\right) = 3\left(-\frac{1}{2}\right) = -\frac{3}{2}$$

The y-coordinate will be:

$$y = r\sin\theta = 3\sin\left(\frac{4\pi}{3}\right) = 3\left(-\frac{\sqrt{3}}{2}\right) = -\frac{3\sqrt{3}}{2}$$

The z-coordinate remains the same: $z = -2$.

Final answer for rectangular coordinates: $\left(-\frac{3}{2}, -\frac{3\sqrt{3}}{2}, -2\right)$.

To convert the cylindrical coordinates to spherical coordinates:

$$\rho = \sqrt{r^2 + z^2} = \sqrt{(3)^2 + (-2)^2} = \sqrt{9 + 4} = \sqrt{13}$$

The angle θ remains the same. The last coordinate ϕ will be found using the formula:

$$\phi = \arccos\left(\frac{z}{\sqrt{r^2 + z^2}}\right) = \arccos\left(\frac{-2}{\sqrt{13}}\right) \approx 2.16 \text{ radians}$$

Therefore, the final answer for spherical coordinates is:

$$\left(\sqrt{13}, \frac{\pi}{3}, 2.16\right).$$

SOLUTION—ROW #3 Let's take a look at the spherical coordinate point $\left(10, \dfrac{5\pi}{4}, \dfrac{2\pi}{3}\right)$.

We will convert to rectangular coordinates first. The x-coordinate will be:

$$x = \rho \sin \phi \cos \theta = 10 \sin\left(\frac{2\pi}{3}\right)\cos\left(\frac{5\pi}{4}\right) = 10\left(\frac{\sqrt{3}}{2}\right)\left(-\frac{\sqrt{2}}{2}\right) = -\frac{5\sqrt{6}}{2}$$

The y-coordinate will be $y = \rho \sin \phi \sin \theta = 10 \sin\left(\frac{2\pi}{3}\right)\sin\left(\frac{5\pi}{4}\right) = 10\left(\frac{\sqrt{3}}{2}\right)\left(-\frac{\sqrt{2}}{2}\right) = -\frac{5\sqrt{6}}{2}$.

The z-coordinate will be $z = \rho \cos \phi = 10 \cos\left(\frac{2\pi}{3}\right) = 10\left(-\frac{1}{2}\right) = -5$.

Therefore, the final answer for rectangular coordinates is $\left(-\dfrac{5\sqrt{6}}{2}, -\dfrac{5\sqrt{6}}{2}, -5\right)$.

Let's convert from spherical to cylindrical coordinates. We have that:

$$r = \rho \sin \phi = 10 \sin\left(\frac{2\pi}{3}\right) = 10\frac{\sqrt{3}}{2} = 5\sqrt{3}$$

The angle θ will remain the same. That is, $\theta = \dfrac{5\pi}{4}$.

The z-coordinate will be $z = \rho \cos \phi$, which we already found above: $z = -5$.

The final answer for cylindrical coordinates will be $\left(5\sqrt{3}, \dfrac{5\pi}{4}, -5\right)$.

Converting Equations Examples

EXAMPLE #1. Convert the equations from cylindrical to rectangular:

(a) $r = 5z$

(b) $\theta = \dfrac{\pi}{6}$

(c) $r^2 \sin 2\theta = 4z$

(d) $r = 2\cos\theta + 2\sin\theta$

(e) $r = (\csc\theta)e^{r^2 + z^2}$

SOLUTION.

(a) Using the conversion formula, we have that $r = 5z$ can be written as $\sqrt{x^2 + y^2} = 5z$. Although this answer is correct, it is traditional to attempt to rid the equation of radicals, if possible. Squaring both sides of the equation, we have:

$$\left(\sqrt{x^2 + y^2}\right)^2 = (5z)^2$$
$$x^2 + y^2 = 25z^2$$

The graph of this will be an elliptic cone.

(b) For the cylindrical equation $\theta = \dfrac{\pi}{6}$, we will take the tangent of both sides of the equation to get:

$$\tan\theta = \tan\left(\frac{\pi}{6}\right)$$

$$\frac{y}{x} = \frac{1}{\sqrt{3}}$$

We are done, except, we could just go ahead and solve for y to get a nice linear equation:

$$y = \frac{1}{\sqrt{3}}x$$

The graph of this will be a vertical half-plane hinged along the z-axis, making an angle of $30°$ with the positive x-axis.

(c) For the cylindrical equation $r^2 \sin 2\theta = 4z$, I will first use a trig ID (double angle formula) to rewrite $\sin 2\theta$ as $2\sin\theta\cos\theta$. Now our cylindrical equation looks like $r^2(2\sin\theta\cos\theta) = 4z$.

I am going to rearrange the factors on the left side of this equation: $2(r\sin\theta)(r\cos\theta) = 4z$.
But we know that according to the conversion formulas, $x = r\cos\theta$ and $y = r\sin\theta$. Replacing these in the cylindrical equation gives us the final answer as $2xy = 4z$. (The graph of this is a hyperbolic paraboloid.)

(d) The equation $r = 2\cos\theta + 2\sin\theta$ will require us to first "beef it up" by multiplying both sides of the equation by r. Now, ordinarily folks, it is *not* legal to multiply (or divide) an equation by a variable. The reason is because you are (potentially) multiplying or dividing by zero. We will make it legal here by assuming that the radius r is a positive value. We will have that:

$$(r)(r) = (2\cos\theta + 2\sin\theta)(r)$$
$$r^2 = 2r\cos\theta + 2r\sin\theta$$
$$x^2 + y^2 = 2x + 2y$$

This is the final answer.

How would you graph this surface in space? Well, recall that equations that are missing a variable are cylindrical surfaces. So, we would first need to graph the generating curve in the *xy*-plane. It will be a circle centered at the point (1, 1) with radius of $\sqrt{2}$, which we can see after we complete the square, like so:

$$x^2 + y^2 = 2x + 2y$$
$$x^2 - 2x + y^2 - 2y = 0$$
$$\left(x^2 - 2x + 1\right) + \left(y^2 - 2y + 1\right) = 2$$
$$(x-1)^2 + (y-1)^2 = 2$$

Then, the rulings would be parallel to the missing variable's axis, the *z*-axis. The result would be a cylinder.

(e) The cylindrical equation $r = (\csc\theta)e^{r^2 + z^2}$ will be easier if we first use a trig ID on the cosecant. We use the fact that $\csc\theta = \dfrac{1}{\sin\theta}$. Then, we will multiply both sides of the equation by $\sin\theta$ before applying the conversion formulas:

$$r = \left(\frac{1}{\sin\theta}\right)e^{r^2 + z^2}$$
$$r\sin\theta = e^{r^2 + z^2}$$
$$y = e^{x^2 + y^2 + z^2} \quad \text{(final answer)}$$

It would be hard to say what the graph of this surface looks like without a computer graphing system!

EXAMPLE #2. Convert the equations from spherical to rectangular:

    (a) $\rho = 10$

    (b) $\phi = \dfrac{\pi}{4}$

    (c) $\rho = 4\csc\phi\sec\theta$

    (d) $\rho = 2\sec\phi$

SOLUTION.

    (a) The equation $\rho = 10$ can be squared on both sides to get $\rho^2 = 100$. Then, we can apply conversion formulas: $x^2 + y^2 + z^2 = 100$ (answer). This is an equation for a sphere of radius 10.

    (b) We know conversion formulas for the cosine of the angle ϕ. So, let's take the cosine of both sides to begin, then we'll apply conversion formulas to the spherical equation $\phi = \dfrac{\pi}{4}$ and we'll have that:

$$\cos\phi = \cos\left(\frac{\pi}{4}\right)$$
$$\frac{z}{\sqrt{x^2 + y^2 + z^2}} = \frac{\sqrt{2}}{2}$$

Traditionally, the final answer should have radicals removed, if possible. Otherwise, we'd be done! So, let's square both sides of the equation, then simplify:

$$\left(\frac{z}{\sqrt{x^2+y^2+z^2}}\right)^2 = \left(\frac{\sqrt{2}}{2}\right)^2$$

$$\frac{z^2}{x^2+y^2+z^2} = \frac{2}{4}$$

$$z^2 = \left(\frac{1}{2}\right)\left(x^2+y^2+z^2\right)$$

$$2z^2 = x^2+y^2+z^2$$

$$0 = x^2+y^2-z^2$$

This equation's graph is a quadric; the surface is an elliptic cone.

(c) The spherical equation $\rho = 4\csc\phi\sec\theta$ needs to be rewritten using trig ID's first to get:

$$\rho = 4\left(\frac{1}{\sin\phi}\right)\left(\frac{1}{\cos\theta}\right)$$

Clear fractions by multiplying both sides of the equation by the product: $\sin\phi\cos\theta$ and we have $\rho\sin\phi\cos\theta = 4$.

Recall the conversion formula for the x-coordinate, which is $x = \rho\sin\phi\cos\theta$. But this is exactly what the left side of our equation is! So, we place the left side with "x" to get our final answer $x = 4$. This is the equation of a plane parallel to the yz-plane.

(d) The spherical equation $\rho = 2\sec\phi$ will be rewritten first (using a trig ID): $\rho = 2\left(\frac{1}{\cos\phi}\right)$

Next, clear fractions from this equation and then apply conversion formulas:

$$\rho\cos\phi = 2$$

$$\sqrt{x^2+y^2+z^2}\left(\frac{z}{\sqrt{x^2+y^2+z^2}}\right) = 2$$

$$z = 2$$

And we are done! This equation represents the graph of a plane that is parallel to the xy-plane.

We will convert rectangular equations to both cylindrical and spherical coordinates.

EXAMPLE #3. Convert the rectangular equations below to *both* cylindrical and spherical equations.

(a) $x^2 + y^2 + (z-1)^2 = 1$

(b) $x^2 + (y-3)^2 = 9$

(c) $x = 5$

SOLUTION.

(a) To convert the equation for the sphere $x^2 + y^2 + (z-1)^2 = 1$ to cylindrical coordinates is very easy. We use one formula and we are done: $r^2 + (z-1)^2 = 1$.

To convert to spherical coordinates takes a little more work. We apply the conversion formulas for the variables *x, y,* and *z,* and then expand and multiply:

$$(\rho \sin\phi \cos\theta)^2 + (\rho \sin\phi \sin\theta)^2 + (\rho \cos\phi - 1)^2 = 1$$
$$\rho^2 \sin^2\phi \cos^2\theta + \rho^2 \sin^2\phi \sin^2\theta + \rho^2 \cos^2\phi - 2\rho\cos\phi + 1 = 1$$

Next, notice that the first two terms share a factor of $\rho^2 \sin^2\phi$ in common. Factor that out:

$$\rho^2 \sin^2\phi (\cos^2\theta + \sin^2\theta) + \rho^2 \cos^2\phi - 2\rho\cos\phi = 0$$

Because we all know that $\cos^2\theta + \sin^2\theta = 1$ (don't we?).

We can write that $\rho^2 \sin^2\phi + \rho^2 \cos^2\phi - 2\rho\cos\phi = 0$.

We factor out the common factor of ρ^2 from the first two terms, and use the trig ID mentioned above again to get:

$$\rho^2 (\sin^2\phi + \cos^2\phi) - 2\rho\cos\phi = 0$$
$$\rho^2 \qquad\qquad\qquad = 2\rho\cos\phi$$

We can divide both sides by ρ by assuming that ρ is positive, and we won't lose any information because the final answer of $\rho = \cos\phi$ contains the case when ρ is equal to zero.

(b) The rectangular equation $x^2 + (y-3)^2 = 9$ can be rewritten first by expanding and multiplying:

$$x^2 + y^2 - 6y + 9 = 9$$
$$x^2 + y^2 - 6y = 0$$

Next, apply the cylindrical conversion formulas to get:

$$r^2 - 6r\sin\theta = 0$$
$$r^2 = 6r\sin\theta$$

Divide both sides by r to get:

$$r = 6\sin\theta \text{ (answer)}$$

To convert to spherical coordinates, we apply the conversion formulas and get:

$$(\rho\sin\phi\cos\theta)^2 + (\rho\sin\phi\sin\theta)^2 - 6\rho\sin\phi\sin\theta = 0$$
$$\rho^2\sin^2\phi\cos^2\theta + \rho^2\sin^2\phi\sin^2\theta - 6\rho\sin\phi\sin\theta = 0$$
$$\rho^2\sin^2\phi(\cos^2\theta + \sin^2\theta) - 6\rho\sin\phi\sin\theta = 0$$

Use the trig ID again!

So, we will have that:

$$\rho^2\sin^2\phi(1) - 6\rho\sin\phi\sin\theta = 0$$
$$\rho^2\sin^2\phi = 6\rho\sin\phi\sin\theta$$

Divide both sides by $\rho\sin\phi$ and then solve for ρ to get the final answer:

$$\rho\sin\phi = 6\sin\theta$$
$$\rho = \frac{6\sin\theta}{\sin\phi}$$
$$\rho = 6\sin\theta\csc\phi$$

(c) To convert the rectangular equation $x = 5$ to cylindrical coordinates is again straightforward and only requires one step:

$$r\cos\theta = 5 \text{ or } r = 5\sec\theta \text{ (answer)}$$

To convert to spherical, we write $\rho\sin\phi\cos\theta = 5$ or $\rho = \dfrac{5}{\sin\phi\cos\theta} = 5\sec\theta\csc\phi$ (answer).

This concludes the chapter!

EXERCISES: #6–10 from **Practice Midterm 1**, #1, 2, 3 from **Practice Midterm 2**

Chapter 2

VECTOR-VALUED FUNCTIONS

Chapter Comments

This chapter covers a brand–new-type family of functions called "vector-valued functions." They are not the same type of functions you all know and love from your previous years of algebra and calculus. Oh no, they are much different, as we shall soon see! We will also cover parameterization of curves in three dimensions. You (hopefully) recall studying parameterization of curves in two dimensions from your last semester of calculus. Remember sketching curves that had an orientation (direction)? Well, the graphs of these vector-valued functions result in curves in 3-D that also will have an orientation. After an introduction of vector-valued functions (which are a parameterization of sorts, as you will soon see), we will cover the following topics in the first two sections of this chapter involving vector-valued functions:

1) Finding Domains (and Continuity)
2) Limits
3) Differentiation
4) Integration

In the last three sections of the chapter, we will go on to discuss velocity and acceleration, tangent and normal vectors, arc length and curvature. Tangent and normal vectors will slow us down a bit, as finding them takes a bit of work. The exercises are a bit lengthy, and require both calculus and excellent algebra skills.

2.1 VECTOR-VALUED FUNCTIONS

Let's first talk about the type of functions you *are* familiar with—known as "real-valued functions." A real-valued function $y = f(x)$, has real number values x as its input, and real number values y as output. Here is a diagram representing the relationship between input and output. I call this diagram the "function machine" diagram:

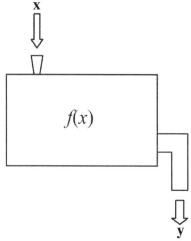

Vector-valued functions are different. A vector-valued function in two dimensions is represented using the notation $\mathbf{v}(t) = \langle f(t), g(t) \rangle$, where t is the parameter, and the component functions $f(t)$ and $g(t)$ are real-valued functions of the parameter t. Vector-valued functions have real number values as input, however, the output will be *vectors,* not real numbers! If you set an input of say, $t = -1$ into a vector-valued function, you will get a corresponding vector as output, say the vector:

$$\mathbf{v}(-1) = \left\langle 2, \frac{4}{5} \right\rangle$$

To sketch the graph of one of these vector-valued functions, you would need to "connect the tips" of the collection of output vectors. The result would be a curve. So, the "function machine" diagram for a vector-valued function would be:

t

$$\mathbf{v(t)} = \left\langle f(t), g(t) \right\rangle$$

$\langle f(t), g(t) \rangle$ (a vector!)

You can graph a two-dimensional vector-valued function using your graphing calculator in "parametric" mode. Just let $x = f(t)$ and $y = g(t)$. Be sure to show the orientation of the curve to get full credit. Also, if the range for the parameter t is not specified, it is your job to be sure and show the "full picture" of the curve and make sure the range of t has both negative and positive values in it. For example, I always try and make sure that the parameter t ranges from $[-10, 10]$.

Three-dimensional vector-valued functions use the notation $\mathbf{v}(t) = \langle f(t), g(t), h(t) \rangle$. The graphs of three-dimensional vector-valued functions will be curves in space, which are harder to sketch, of course!

Finding Domains of Vector-Valued Functions

I never assume students know how to find domains of functions, even though they were supposed to have learned it in their algebra and pre-calculus courses. Here is the "handy-dandy" procedure I use every time I am trying to find the domain of a vector-valued function.

"Handy-Dandy" Procedure to Find the Domain of Vector-Valued Function:

1) Find the domain of each component separately (see the procedure on next page)

2) Take the intersection of the domains for each component

"Handy-Dandy" Procedure to Find the Domain of a Real-Valued Function *(for a single component of a vector-valued function):*

1) First, assume all inputs are OK, that the parameter t can take on any value, so $t \in (-\infty, \infty)$.

2) If the component function contains an *even* radical (i.e., square roots, fourth roots, etc.) with an expression for its radicand, then eliminate all values for t that would make the radicand negative. (You do this because taking an even root of a negative yields a complex value, which you wish to avoid!)

3) If the component function contains a quotient with t in the denominator, eliminate all values of t that would make the denominator equal to zero.

4) If the component function contains a logarithm with an argument involving t, eliminate all values of t that would make the argument equal to zero or negative. (You do this because logarithm functions require positive-valued arguments!)

For combinations of **#1** through **#4**, take the intersection of the two sets. We'll do lots of examples!

Finding Domains of Vector-Valued Functions

EXAMPLES. Find the domains of the following vector-valued functions. Express your answers using interval notation to get full credit.

$$\textbf{(a)} \quad \mathbf{r}(t) = \sqrt[4]{3 - 5t}\,\mathbf{i} + \ln(t + 5)\,\mathbf{j} - \sin(t - 1)\,\mathbf{k}$$

$$\textbf{(b)} \quad \mathbf{r}(t) = \sqrt[3]{1 + t}\,\mathbf{i} + \frac{1}{\ln(3 - t)}\,\mathbf{j} + \frac{t - 1}{t + 6}\,\mathbf{k}$$

$$\textbf{(c)} \quad \mathbf{r}(t) = \frac{\sqrt{10 - t}}{t + 7}\,\mathbf{i} + |t + 2|\,\mathbf{j} + \frac{t}{6t^2 - t - 2}\,\mathbf{k}$$

$$\textbf{(d)} \quad \mathbf{r}(t) = \left\langle e^{t^2 - 4t + 3}, e^{\sqrt{2 + t}}, e^{1/t} \right\rangle$$

$$\textbf{(e)} \quad \mathbf{r}(t) = \left\langle \sqrt{t^2 - 4}, \frac{2}{1 + e^{t + 5}}, \frac{3}{\sin(\pi t)} \right\rangle$$

SOLUTION.

(a) The first component has an even radical, so we need to "worry." That is, we must eliminate all values of t that make the radicand negative. We need the inequality $3 - 5t \geq 0$ to be true. Solving for t, we have that $\frac{3}{5} \geq t$ or $t \leq \frac{3}{5}$. Consider the second component for the vector-valued function, which is $\ln(t + 5)$. We require the argument of the logarithm, which is $t + 5$, to be a positive value. So, we need the inequality $t + 5 > 0$ to be true. Solving this inequality for t, we get that $t > -5$. Finally, we look at the last component $-\sin(t - 1)$. This function has no restrictions on its domain. That is, t can be any real number $t \in (-\infty, \infty)$. Now, we take the intersection of the three sets $t \le \dfrac{3}{5}$, $t > -5$, and $t \in (-\infty, \infty)$.

The values of t in common with all three sets will be $t \in \left(-5, \dfrac{3}{5}\right]$ (final answer).

(b) The first component of this vector-valued function is an odd radical: $\sqrt[3]{1+t}$. But odd radicals have no restriction on their domains. Remember, it is perfectly OK to take odd roots of negative numbers, just not even roots of negative numbers. Therefore, the domain for the first component will be $t \in (-\infty, \infty)$.

The second component is what I call a "double whammy." It not only has a logarithm function with variables in its argument, but a denominator with variables at the same time!

First let's take a look at the logarithm by itself. We require the argument to be positive, so we need the inequality $3 - t > 0$ to be true. Solve for t to get that $3 > t$, which we can rewrite as $t < 3$.

Next consider the denominator of the quotient. We never want our denominators to be equal to zero. So, we ask ourselves the question: "Will the denominator of $\ln(3 - t) = 0$?"

We know that $\ln(1) = 0$. So, if $t = 2$, we will have $\ln(3 - 2) = \ln(1) = 0$. So, not only do we need t to be less than 3, we cannot have $t = 2$. We eliminate $t = 2$ from the domain.

At this point, we have the domain of the second component as the union of two separate intervals:
$$t \in (-\infty, 2) \cup (2, 3)$$

Now, we move on to consider the third component, or $\dfrac{t-1}{t+6}$.

We notice that the denominator will equal 0 if $t = -6$.

So, this value will be eliminated from the domain.

Therefore, the domain for the third component is $t \in (-\infty, -6) \cup (-6, \infty)$.

Now, we take the intersection of all three sets:
$$t \in (-\infty, \infty), \ t \in (-\infty, 2) \cup (2, 3), \text{ and } t \in (-\infty, -6) \cup (-6, \infty)$$

The final answer will be $t \in (-\infty, -6) \cup (-6, 2) \cup (2, 3)$.

(c) The first component contains two "worries." It has both an even radical as well as a denominator, both of which contain the variable t. We will find the domain of each separately, then take the intersection.

The radical $\sqrt{10-t}$ will need its radicand to be non-negative. That is, we need the inequality $10 - t \geq 0$.

Solving for t, we have that $10 \geq t$, which we can write as $t \leq 10$.

The denominator of $t + 7$ cannot be equal to 0. So, we eliminate the value $t = -7$ from the domain.

Take the intersection of the two sets $t \leq 10$ and $t \neq -7$; we have that $t \in (-\infty, -7) \cup (-7, 10]$.

The second component of the vector-valued function is an absolute value function. This function has no restrictions on its domain. That is, $t \in (-\infty, \infty)$ for the function $|t + 2|$.

The last component has a trinomial in its denominator. The trinomial is $6t^2 - t - 2$. We want to eliminate all values of t for which the denominator equals 0. So, let's find these values that need to be eliminated by solving the equation $6t^2 - t - 2 = 0$. Fortunately, this trinomial factors nicely over the set of integers.

Do you know how you factor trinomials *by hand?* Please review the "master product method." Conscientious calculus students show their work. The trinomial $6t^2 - t - 2$ factors as $(2t + 1)(3t - 2)$.

So, we have the equation $(2t + 1)(3t - 2) = 0$. We use the ZPP (Zero Product Property) to solve. Setting each individual factor equal to 0:

$$(2t + 1) = 0, \quad (3t - 2) = 0$$
$$2t = -1 \qquad \quad 3t = 2$$
$$t = -\frac{1}{2} \qquad \quad t = \frac{2}{3}$$

Remember, these are the values of t that we *don't* want! So, eliminate these from the domain, and the domain for the second component will be:

$$t \in \left(-\infty, -\frac{1}{2}\right) \cup \left(-\frac{1}{2}, \frac{2}{3}\right) \cup \left(\frac{2}{3}, \infty\right)$$

Finally, we take the intersection of the domains for all three components. The intersection of the three sets:

$t \in (-\infty, -7) \cup (-7, 10]$, $t \in (-\infty, \infty)$, and $t \in \left(-\infty, -\frac{1}{2}\right) \cup \left(-\frac{1}{2}, \frac{2}{3}\right) \cup \left(\frac{2}{3}, \infty\right)$

will be $t \in (-\infty, -7) \cup \left(-7, -\frac{1}{2}\right) \cup \left(-\frac{1}{2}, \frac{2}{3}\right) \cup \left(\frac{2}{3}, 10\right]$

(d) The first component, $e^{t^2 - 4t + 3}$, has no restrictions on its domain. That is, $t \in (-\infty, \infty)$ is the domain for this function.

The second component *will* have a restriction on its domain, since the argument of the exponential function has an even radical containing an expression. The radical $\sqrt{2 + t}$ will require its radicand, $2 + t$, to be greater than or equal to 0.

Solving this inequality $2 + t \geq 0$, we have that $t \geq -2$. The last component, $e^{1/t}$, will also have a restriction on its domain, since the argument of the exponential contains a quotient with an expression in its denominator. We don't want the denominator equal to 0, so we need $t \neq 0$.

We take the intersection of the three sets $t \in (-\infty, \infty)$, $t \geq -2$, and $t \neq 0$.

We get our final answer, which will be $t \in [-2, 0) \cup (0, \infty)$.

(e) The first component has an even radical, so we need to worry! We need the radicand $t^2 - 4$ to be greater than or equal to 0. We need to solve the polynomial inequality $t^2 - 4 \geq 0$. Remember that solving polynomial inequalities is a bit trickier than solving linear inequalities!

We first need to get the inequality into standard form: 0 on one side and everything else on the other. (Fortunately, this inequality already *is* in standard form!)

Then we factor the polynomial $(t + 2)(t - 2) \geq 0$. We "pretend" (temporarily) that this is an *equation* instead of an inequality to find the "critical numbers" for the polynomial inequality. That is, the solutions to the *equation* $(t + 2)(t - 2) = 0$, which are $t = -2$, and $t = 2$, are the critical numbers for the inequality. These critical numbers "slice up" the t-axis into three separate "test" intervals. We pick an arbitrary value out of each test interval and see which interval(s) satisfy the inequality.

So, the three "test intervals" are $(-\infty, -2)$, $(-2, 2)$, and $(2, \infty)$.

Pick $t = -3$, a value in the first test interval, and plug into the inequality $t^2 - 4 \geq 0$. Is the inequality: $(-3)^2 - 4 \geq 0$ true? *YES!* Therefore, the interval $(-\infty, -2)$ is in our domain.

Now, we pick an arbitrary value in the second test interval, say $t = 0$. Is the inequality: $(0)^2 - 4 \geq 0$ true? *NO!* Therefore, the second test interval is *not* part of our answer.

Finally, we pick an arbitrary value in the third interval, say $t = 3$. Is the inequality $(3)^2 - 4 \geq 0$ true? *YES!* So, the last test interval which is $(2, \infty)$ is in our answer.

Therefore, the domain for our first component function is $t \in (-\infty, -2) \cup (2, \infty)$.

The second component for the vector-valued function is $\dfrac{t}{1+e^{t+5}}$.

Although it has an expression in the denominator, we do not need to restrict the domain. The reason for this is because exponential functions are always positive; the exponential function e^{t+5} will *never* be equal to 0 or be negative.

Therefore, the entire denominator which is $1+e^{t+5}$, will always be a positive value. So, this second component has no restrictions on its domain, and $t \in (-\infty, \infty)$.

The last component for the vector-valued function is $\dfrac{3}{\sin(\pi t)}$.

We do not want the denominator equal to 0. That is, we want $\sin(\pi t) \neq 0$. Sine functions are equal to 0 when the arguments of the sine functions are multiples of p. So, we can write the domain of this last component of $\sin(\pi t)$ in the following way: $t \in (k, k+1)$, where $k \in \mathbf{Z}$. ($\mathbf{Z}$ denotes the set of all integers.) Another way you might see this domain written is the following:

$$t \in \cdots (-3, -2) \cup (-2, -1) \cup (-1, 0) \cup (0, 1) \cup (1, 2) \cup (2, 3) \cdots$$

The intersection of all three components' domains will be exactly the domain of the last component, except with two of the intervals lying between $t = -2$ and $t = 2$ deleted. So the answer is $t \in (k, k+1)$ where $k \in \mathbf{Z}, k \neq 0, \pm 1$.

The final answer is $t \in \cdots (-3, -2) \cup (-2, -1) \cup (1, 2) \cup (2, 3) \cdots$.

To find when a vector-valued function is continuous, you need simply only to find the intervals that make up the domain of the function.

Finding Limits of Vector-Valued Functions

The definition of the limit of a vector-valued function is: If $\mathbf{r}$ is a vector-valued function such that $\mathbf{r}(t) = f(t)\mathbf{i} + g(t)\mathbf{j} + h(t)\mathbf{k}$, then $\lim\limits_{t \to a} \mathbf{r}(t) = \left[\lim\limits_{t \to a} f(t)\right]\mathbf{i} + \left[\lim\limits_{t \to a} g(t)\right]\mathbf{j} + \left[\lim\limits_{t \to a} h(t)\right]\mathbf{k}$ provided f, g, and h have limits as $t \to a$.

The next examples are challenging since you need to be familiar with L'Hôpital's Rule.

EXAMPLE. Find the limits (if they exist) for the following vector-valued functions:

(a) $\quad \lim\limits_{t \to 0} \mathbf{r}(t) = \lim\limits_{t \to 0} \left(\dfrac{1 - \cos^2 t}{3t^2} \mathbf{i} + \left(e^t + t\right)^{1/t} \mathbf{j} - \sqrt{t + 15}\, \mathbf{k} \right)$

(b) $\quad \lim\limits_{t \to \infty} \mathbf{r}(t) = \lim\limits_{t \to \infty} \left(\dfrac{t^3 - 1}{4t^3 - t - 3} \mathbf{i} + \arctan t\, \mathbf{j} + t \tan \dfrac{1}{t} \mathbf{k} \right)$

(c) $\quad \lim\limits_{t \to \infty} \mathbf{r}(t) = \lim\limits_{t \to \infty} \left(\dfrac{t^2 - 1}{4t^3 - t - 3} \mathbf{i} + (\ln t)^{1/t} \mathbf{j} + \dfrac{\ln(t + 1)}{\log_2 t} \mathbf{k} \right)$

(d) $\quad \lim\limits_{t \to \infty} \mathbf{r}(t) = \lim\limits_{t \to \infty} \left(e^{-t^2} \mathbf{i} + \dfrac{\sin 3t}{t} \mathbf{j} + \sin t \mathbf{k} \right)$

(e) $\quad \lim\limits_{t \to 0^+} \mathbf{r}(t) = \lim\limits_{t \to 0^+} \left(e^t \mathbf{i} + 5\mathbf{j} + \ln t \mathbf{k} \right)$

SOLUTION.

(a) Always try to start all limit problems using direct substitution. So, after substituting $t = 0$ into the first component, we have:

$$\frac{1 - \cos^2(0)}{3(0)^2} = \frac{1 - 1}{0} = \frac{0}{0}. \; \textit{Do not write that } \frac{0}{0} = 0 \; ! \; \textit{That is totally false!}$$

Instead, know that anytime you run into the quotient $\dfrac{0}{0}$, you are probably going to use L'Hôpital's Rule.

This indeterminate form is one in which it is permissible to use L'Hôpital's Rule. After taking the derivative of the numerator and dividing it by the derivative of the denominator, we will again use direct substitution.

Note: Before differentiating $\cos^2 t$, rewrite it as $(\cos t)^2$, then apply the Chain Rule:

$$\lim_{t \to 0} \frac{1 - \cos^2 t}{3t^2} = \lim_{t \to 0} \frac{(-2)(\cos t)(-\sin t)}{6t} = \frac{(2)(\cos 0)(\sin 0)}{6(0)} = \frac{0}{0}$$

We need to apply L'Hôpital's Rule again! Notice that I will need to use the Product Rule to differentiate the numerator $2\cos t \sin t$:

$$\lim_{t \to 0} \frac{(2)(\cos t)(\sin t)}{6t} = \lim_{t \to 0} \frac{(-2\sin t)(\sin t) + (2)(\cos t)(\cos t)}{6} = \frac{(-2)\sin^2(0) + (2)\cos^2(0)}{6} = \frac{0 + (2)(1)}{6} = \frac{1}{3}$$

So, the first component's limit exists. Now, for the second component…

After direct substitution, we have $\lim_{t \to 0} \left(e^t + t\right)^{1/t} = \left(e^0 + 0\right)^{1/0} = (1 + 0)^\infty = 1^\infty$.

This result is an indeterminate form. (Please do *not* say this is equal to 1, because it isn't!) Unfortunately, this is not an indeterminate from that we can use L'Hôpital's Rule on. Instead, we need to apply what I call the "logarithmic technique." We start by assuming the limit exists, and call it "y." Then we take the natural logarithm of both sides of this new equation, like so:

$$y = \lim_{t \to 0} \left(e^t + t\right)^{1/t}$$

$$\ln y = \ln \left[\lim_{t \to 0} \left(e^t + t\right)^{1/t}\right]$$

We are able to "swap" the logarithm with the limit due to the linearity of both of these operators. After we do this, we then use a log property to simplify:

$$\ln y = \ln \left[\lim_{t \to 0} \left(e^t + t\right)^{1/t}\right]$$

$$\ln y = \lim_{t \to 0} \left[\ln \left(e^t + t\right)^{1/t}\right]$$

$$\ln y = \lim_{t \to 0} \left(\frac{1}{t}\right) \ln \left(e^t + t\right)$$

$$\ln y = \lim_{t \to 0} \frac{\ln \left(e^t + t\right)}{t}$$

Now, we will use direct substitution:

$$\ln y = \frac{\ln\left(e^0 + 0\right)}{0}$$

$$\ln y = \frac{\ln(1 + 0)}{0}$$

$$\ln y = \frac{0}{0}$$

Remember that $\frac{0}{0} \neq 0$! It simply means that we are allowed to apply L'Hôpital's Rule on the right side of the equation!

$$\ln y = \lim_{t \to 0} \frac{\left(\dfrac{1}{e^t + t}\right)\left(e^t + 1\right)}{1}$$

$$\ln y = \lim_{t \to 0} \frac{e^t + 1}{e^t + t}$$

$$\ln y = \frac{e^0 + 1}{e^0 + 0}$$

$$\ln y = \frac{2}{1}$$

$$e^2 = y$$

So, the limit exists, and it is equal to e^2.

Finally, consider the third component: $\lim_{t \to 0}\left(-\sqrt{t + 15}\right) = -\sqrt{0 + 15} = -\sqrt{15}$.

We needed only to use direct substitution. The final answer must be a vector, which is $\lim_{t \to 0} \mathbf{r}(t) = \left\langle \dfrac{1}{3}, e^2, -\sqrt{15} \right\rangle$.

(b) After applying direct substitution to the first component, we have the limit as:

$$\lim_{t \to \infty} \frac{t^3 - 1}{4t^3 - t - 3} = \frac{\infty^3 - 1}{4\infty^3 - \infty - 3} = \frac{\infty}{\infty}$$

This is an indeterminate form where it is permissible to apply L'Hôpital's Rule:

$$\lim_{t \to \infty} \frac{t^3 - 1}{4t^3 - t - 3} = \lim_{t \to \infty} \frac{3t^2}{12t^2 - 1} = \frac{3\infty^2}{12\infty^2 - 1} = \frac{\infty}{\infty}$$

Again, we need to apply L'Hôpital's Rule, $\lim_{t \to \infty} \dfrac{3t^2}{12t^2 - 1} = \dfrac{6t}{24t} = \dfrac{1}{4}$. So, the limit exists!

Using direct substitution on the second component, we have $\lim\limits_{t\to\infty}\arctan t = \arctan(\infty) = \dfrac{\pi}{2}$. I hope you remember why this is true from your last semester of calculus!

And finally, applying direct substitution on the third component, we have:

$$\lim_{t\to\infty} t\tan\frac{1}{t} = (\infty)\tan\left(\frac{1}{\infty}\right) = (\infty)\tan(0) = \infty \cdot 0$$

This last result is an indeterminate form, where it is *not* permissible to use L'Hôpital's Rule. Instead, we need to rewrite the expression as a quotient.

Recall from your last semester of calculus that multiplying an expression by t is the same as dividing by $\dfrac{1}{t}$. So, we rewrite the expression, then apply direct substitution.

$$\lim_{t\to\infty} t\tan\frac{1}{t} = \lim_{t\to\infty}\frac{\tan\dfrac{1}{t}}{\dfrac{1}{t}} = \frac{\tan\dfrac{1}{\infty}}{\dfrac{1}{\infty}} = \frac{\tan(0)}{0} = \frac{0}{0}$$

This is an indeterminate form where it is permissible to use L'Hôpital's Rule:

$$\lim_{t\to\infty}\frac{\tan\dfrac{1}{t}}{\dfrac{1}{t}} = \lim_{t\to\infty}\frac{\sec^2\left(\dfrac{1}{t}\right)\left(-t^{-2}\right)}{-t^{-2}} = \lim_{t\to\infty}\sec^2\left(\frac{1}{t}\right) = \sec^2(0) = \frac{1}{\cos^2(0)} = \frac{1}{1} = 1$$

The limit exists and is equal to 1. The final answer is a vector and has the form $\lim\limits_{t\to\infty}\mathbf{r}(t) = \left\langle \dfrac{1}{4}, \dfrac{\pi}{2}, 1 \right\rangle$.

(c) After direct substitution for the first component, we have:

$$\lim_{t\to\infty}\frac{t^2-1}{4t^3-t-3} = \frac{\infty^2-1}{4\infty^3-\infty-3} = \frac{\infty}{\infty}$$

We will apply L'Hôpital's Rule twice:

$$\lim_{t\to\infty}\frac{t^2-1}{4t^3-t-3} = \lim_{t\to\infty}\frac{2t}{12t^2-1} = \frac{2\infty}{12\infty^2-1} = \frac{\infty}{\infty}$$

$$\lim_{t\to\infty}\frac{2}{24t} = \frac{1}{12\infty} = \frac{1}{\infty} = 0$$

So, the limit exists and equals 0. For direct substitution on the second component:

$$\lim_{t\to\infty}(\ln t)^{1/t} = (\ln\infty)^{1/\infty} = (\infty)^0$$

But, the expression $(\infty)^0$ is an indeterminate form, where it is not permissible to use L'Hôpital's Rule. Instead, we need to use that logarithmic technique we used in a previous example. We start by assuming the limit exists (call it y), and then take the natural logarithm of both sides:

$$y = \lim_{t \to \infty}(\ln t)^{1/t}$$
$$\ln y = \ln\left[\lim_{t \to \infty}(\ln t)^{1/t}\right]$$
$$\ln y = \lim_{t \to \infty}\left[\ln(\ln t)^{1/t}\right]$$
$$\ln y = \lim_{t \to \infty}\frac{1}{t} \cdot \ln(\ln t)$$
$$\ln y = \lim_{t \to \infty}\frac{\ln(\ln t)}{t}$$

I used log properties after "swapping" the limit with the logarithm to simplify. We're ready for direct substitution again:

$$\ln y = \lim_{t \to \infty}\frac{\ln(\ln t)}{t} = \frac{\ln(\ln \infty)}{\infty} = \frac{\infty}{\infty}$$

This is an indeterminate form where it is permissible to use L'Hôpital's Rule. We do this:

$$\ln y = \lim_{t \to \infty}\frac{\ln(\ln t)}{t} = \lim_{t \to \infty}\frac{\left(\dfrac{1}{\ln t}\right) \cdot \left(\dfrac{1}{t}\right)}{1}$$
$$= \lim_{t \to \infty}\frac{1}{t \ln t} = \frac{1}{\infty \cdot \ln \infty} = \frac{1}{\infty} = 0$$

Remember, this is *not* the answer! We must do the last step here, which is solve for y:

$$e^0 = y$$
$$1 = y$$

The limit of the second component exists and is equal to 1. We consider the last component's limit:

$$\lim_{t \to \infty}\frac{\ln(t+1)}{\log_2 t} = \frac{\ln(\infty+1)}{\log_2 \infty} = \frac{\infty}{\infty}$$

This is an indeterminate form where it is permissible to use L'Hôpital's Rule. To do this, we must recall from our first semester calculus the rule for taking derivatives of logarithms with bases other than e:

$$\frac{d[\log_a x]}{dx} = \frac{1}{\ln a \cdot x}$$

$$\lim_{t \to \infty}\frac{\ln(t+1)}{\log_2 t} = \lim_{t \to \infty}\frac{\dfrac{1}{t+1}}{\dfrac{1}{(\ln 2)t}} = \lim_{t \to \infty}\frac{(\ln 2)t}{t+1} = \frac{(\ln 2)\infty}{\infty+1} = \frac{\infty}{\infty}$$

Again, this is an indeterminate form where it is permissible to use L'Hôpital's Rule:

$$\lim_{t \to \infty} \frac{(\ln 2)t}{t+1} = \lim_{t \to \infty} \frac{(\ln 2)(1)}{1} = \ln 2$$

The final answer is $\lim_{t \to \infty} \mathbf{r}(t) = \langle 0, 1, \ln 2 \rangle$.

(d) Using direct substitution on the first component: $\lim_{t \to \infty} e^{-t^2} = e^{-\infty} = 0$. This was easy!

Now, for the second component: $\lim_{t \to \infty} \frac{\sin 3t}{t} = \frac{\sin(3\infty)}{\infty} = \frac{\pm 1}{\infty} = 0$.

The expression $\sin(3t)$ oscillates between the values of 1 and –1 as $t \to \infty$. Ordinarily, this limit wouldn't exist, however the denominator "overpowers" this oscillation as it goes to infinity and takes the entire quotient to zero.

The third component does not have a denominator to overpower the oscillation of the sine function. Therefore, the limit of the third component does not exist. We can write $\lim_{t \to \infty} \sin t = d.n.e.$ (due to oscillation).

Fact: If even one component's limit does not exist when taking the limit of a vector-valued function, we can say the entire vector's limit does not exist. And that's the final answer here!

(e) The first component limit is easily obtained through direct substitution: $\lim_{t \to 0^+} e^t = e^0 = 1$.
The second component's limit is just as easy: $\lim_{t \to 0^+} 5 = 5$.

The last component gives us $\lim_{t \to 0^+} \ln t = -\infty$. Since this limit "blows up" (becomes infinite), we say that the limit of the entire vector does not exist (answer).

Plane Curves and Parameterization

So far, we have learned here in this course (and second-semester calculus) to represent graphs in 2-D and 3-D:

1) In the plane (2-D), we use:

   (a) a single equation involving two variables (traditionally, the x and y variables). The graph is a curve.

   (b) two equations using three variables x, y, and t, where t is the parameter. The graph is a curve having an orientation. For example:

   $$\begin{cases} x = f(t) \\ y = g(t) \end{cases}$$

   This notation is called a *set of parametric equations*.

2) In space (3-D), we have learned two things:

(a) a single equation involving three variables x, y, and z. The graph is a surface.

(b) a set of three parametric equations of the form $x = x_1 + at$, $y = y_1 + bt$, $z = z_1 + ct$. The graph is a line containing the point (x_1, y_1, z_1) in the direction of the direction vector $\mathbf{v} = \langle a, b, c \rangle$.

We can extend the concept of parameterization further in three dimensions. We are already familiar with parametric equations in 3-D for lines, but the line always has a special form for its parametric equations. Other curves in space (besides lines) can be represented as:

$$\begin{cases} x = f(t) \\ y = g(t) \\ z = h(t) \end{cases}$$

Does all of this have anything to do with vector-valued functions? You bet! Here's how: We take the set of parametric equations and simply form a vector-valued function from it.

In 2-D, for example, we take the two parametric equations $\begin{cases} x = f(t) \\ y = g(t) \end{cases}$ and write this as:

$$\mathbf{v}(t) = \langle f(t), g(t) \rangle$$

For three dimensions, we get a vector with three components: $\mathbf{v}(t) = \langle f(t), g(t), h(t) \rangle$.

I am going to start by giving a mini-review of second-semester calculus parameterization, but giving the final answers as vector-valued functions instead of just a set of two parametric equations. We will look at taking a single equation of the two variables x any y and writing it as a vector-valued function containing two components. Then, we'll do the reverse (which is called "eliminating the parameter"). First, I will discuss the basics of graphing in 2-D.

Once you are given a set of *two* parametric equations, you can either use your graphing calculator in "parametric mode" to graph the curve or graph it by hand. If you graph the curve by hand, then it helps to make a table with three columns: one each for x, y, and t. Whether you graph curves by hand or with your calculator, please be sure that the parameter (usually time t) varies enough so you can get the full picture. For example, if you were graphing the curve represented by the set of parametric equations:

$$\begin{cases} x = t^2 - 4 \\ y = t \end{cases}$$

...which can also be represented as a vector-valued function by $\mathbf{v}(t) = \langle t^2 - 4, t \rangle$, then be sure you have t vary so that you can see that the graph is a horizontal parabola. See diagram on the next page.

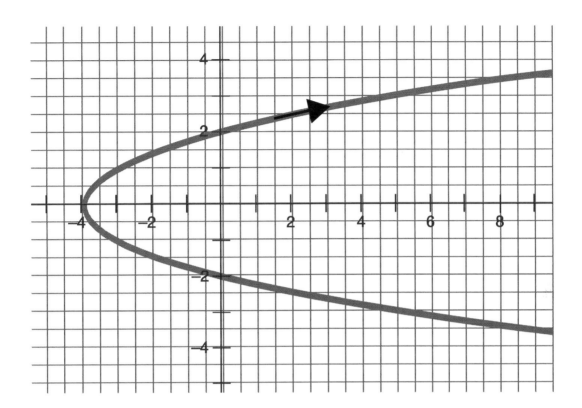

Let's say you have t start at $t = 0$ and go up to say, $t = 5$. Then, you would only see the upper-half of a parabola. This is not the whole picture, so I would not give you full credit. Instead, you would want to have t start at $t = -5$, and go up to $t = 5$. On your calculator, this means you need to go into the "WINDOW" or "RANGE" mode and change the parameter's minimum and maximum values.

Another student pitfall (besides problems with the starting and ending parameter values) is showing the orientation (or direction) for curves represented by parametric equations. Unlike regular graphs of plane curves (represented by a single equation), curves represented parametrically have a direction. You will show this by putting at least one arrow on your curve. See the couple of examples below to demonstrate this concept.

The curve represented by the parametric equations $\begin{cases} x = 3\cos\theta \\ y = 3\sin\theta \end{cases}$, which can also be represented as a vector-valued function by $\mathbf{v}(\theta) = \langle 3\cos\theta, 3\sin\theta \rangle$, is a circle of radius 3 oriented counter-clockwise, where the curve represented parametrically by $\begin{cases} x = 3\sin\theta \\ y = 3\cos\theta \end{cases}$, or $\mathbf{v}(\theta) = \langle 3\sin\theta, 3\cos\theta \rangle$, is a circle of radius 3 oriented clockwise. Please put at least one arrow on your graph so that I can tell what the orientation (direction) is.

The graph on the next page is the curve for the set of parametric equations $\begin{cases} x = 3\cos\theta \\ y = 3\sin\theta \end{cases}$.

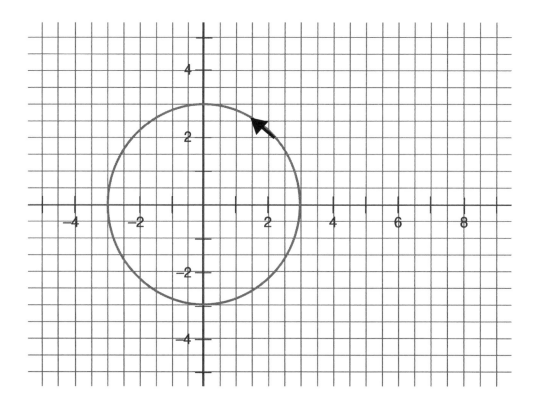

Notice that both sets of parametric equations in the last example give the same exact curve! Wow! The only difference is the orientation. This is the perfect time to go over the fact that parametric representations of curves are not unique. So, when you are looking up the answers for your homework, your answer may be different from what's in the back of the book, yet still be correct. A better example for two parametric representations of the same curve (where even the orientation is now the same) is the example:

$$\begin{cases} x = t^2 - 4 \\ y = t/2 \end{cases} \quad \text{and} \quad \begin{cases} x = 4t^2 - 4 \\ y = t \end{cases}$$

These can be represented by the vector-valued functions $\mathbf{v}(t) = \left\langle t^2 - 4, \dfrac{t}{2} \right\rangle$ and $\mathbf{v}(t) = \left\langle 4t^2 - 4, t \right\rangle$, respectively.

Both of these give curves that are exactly the same, including the orientation. Try it yourself! The difference between them is that the second curve is sketched out "faster" than the other one. So, the "moral of the story" is that parametric representations of curves are not unique!

The most challenging problem you may be asked to graph may have a curve traveling in *both* (opposite) directions! For example, the curve represented by the set of parametric equations:

$$\begin{cases} x = t^2 - 1 \\ y = 1 - t^2 \end{cases}, \text{ or } \mathbf{v}(t) = \left\langle t^2 - 1, 1 - t^2 \right\rangle \dots$$

…is a (diagonal) straight line that first goes to the left, stops at the point (–1, 1), and then travels in the reverse direction to the right. To see this on your calculator, first let the range for the parameter for t be on the interval [–5, 0] and observe the graph direction. Then, change the range for the parameter to the interval [0, 5], and again, observe the graph. Interesting, yes? So, your final answer will have the curve with two arrows on it, pointing in opposite directions!

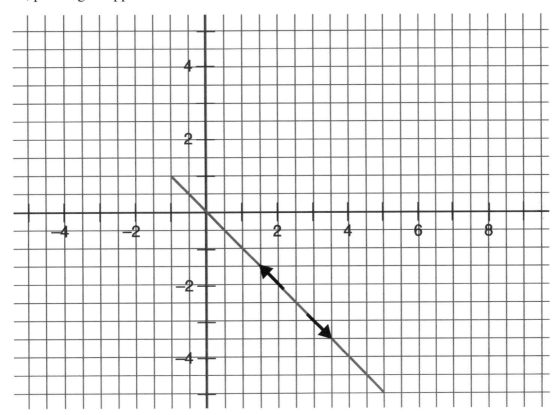

The last concept in this section is being able to parameterize a curve as well as eliminate the parameter. To parameterize, I simply let $x = t$, then plug $x = t$ into the equation for y, and I obtain a set of parametric equations! Of course, this technique only works for functions of x. For functions of y, let $y = t$, then substitute $y = t$ into the equation for x. For equations that are neither functions of x or y, this can be challenging. One strategy is to use trigonometric identities, but this may not always work. So, let's do a couple of examples of parameterizing:

EXAMPLE. Parameterize the equation $y = x^3 + 1$ (This is a function of x.) Write your answer as a vector-valued function.

SOLUTION. Let $x = t$, then we have that $y = t^3 + 1$. So, your final answer will be $\mathbf{v}(t) = \left\langle t, t^3 + 1 \right\rangle$.

EXAMPLE. Parameterize the equation $x^2 + y^2 = 25$. (This is a neither a function of x or y.) The graph is a circle centered at the origin of radius 5. Write your answer as a vector-valued function.

SOLUTION. First, rewrite the equation (divide both sides by 25): $\dfrac{x^2}{25} + \dfrac{y^2}{25} = 1$.

Use the trig ID $\cos^2 \theta + \sin^2 \theta = 1$ and consider allowing $\cos\theta = \dfrac{x}{5}$ and $\sin\theta = \dfrac{y}{5}$.

Solving for the variables *x* and *y*, we have $x = 5\cos\theta$ *and* $y = 5\sin\theta$. (So, the parameter here is θ, not the usual time *t*.) So, the final answer is $\mathbf{v}(\theta) = \langle 5\cos\theta, 5\sin\theta \rangle$.

Let's move on to doing the reverse of parameterizing—eliminating the parameter. That is, given a vector-valued function, obtain the corresponding rectangular equation. One strategy to do this involves solving one of the parametric equations for the variable (usually *t*), then "back-substituting" into the other equation. If that doesn't work, you can try using a trig ID. I will show a couple of examples.

EXAMPLE. Eliminate the parameter and find the corresponding rectangular equation for the parametric equations $\mathbf{v}(t) = \langle \sqrt{t}, 1-t \rangle$. Restrict the domain, if necessary. Then graph it.

SOLUTION. I'll graph the vector-valued function first:

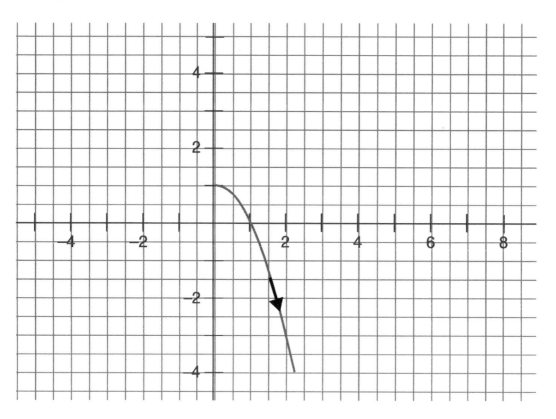

We can write the two parametric equation from this vector-valued function as $x = \sqrt{t}$ and $y = 1 - t$. Solve for *t* using the equation for *y* to get $t = 1 - y$.

Substitute this into the equation for *x* to get $x = \sqrt{1-y}$. We could leave it like this and restrict the domain by saying that we require that $y \le 1$, *or* we could write the equation as a function of *x* instead. That is, solve for *y* to get the equation:

$$(x)^2 = \left(\sqrt{1-y}\right)^2$$
$$x^2 = 1-y$$
$$y = 1-x^2$$

Restrict the domain so that we get the same graph as the one we got with the set of parametric equations (which is the right-half of a downward-facing vertical parabola). You must always make sure that your final rectangular equation's graph matches the graph for the vector-valued function! So, for this problem, we obtain a "match" of the graphs if we restrict the domain for the rectangular equation and require that $x \geq 0$. So, the final answer will be $y = 1 - x^2,\ x \geq 0$.

EXAMPLE. Eliminate the parameter and find the corresponding rectangular equation for the vector-valued function $\mathbf{v}(t) = \langle e^t, 2e^t + 1 \rangle$. Restrict the domain, if necessary.

SOLUTION. I'll sketch the graph for the vector-valued function first:

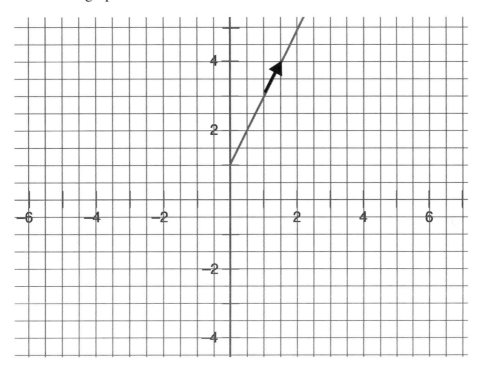

We can write equations for x and y as $\begin{cases} x = e^t \\ y = 2e^t + 1 \end{cases}$. Solve for e^t instead of t here.

Since we have that $x = e^t$, substitute this fact into the equation for y and then we can write that $y = 2x + 1$. This is the answer, except the graphs don't match at this point. This means that we also need to restrict the domain here, since the graph of the parametric equations is a line that needs $x > 0$.

$$\text{Final answer: } y = 2x + 1, x > 0.$$

EXAMPLE. Eliminate the parameter and find the corresponding rectangular equation for the vector-valued function $\mathbf{v}(\theta) = \langle 2 + 3\cos\theta, 5 + 4\sin\theta \rangle$. Restrict the domain, if necessary.

SOLUTION. Write the corresponding equations for x and y: $\begin{cases} x = 2 + 3\cos\theta \\ y = 5 + 4\sin\theta \end{cases}$

Solve for the trig functions cosine and sine instead of θ here. Then we have that:

$$\cos\theta = \frac{x-2}{3}, \quad \sin\theta = \frac{y-5}{4}$$

Substitute both of these into the trig ID that says $\cos^2\theta + \sin^2\theta = 1$, to get:

$$\left(\frac{x-2}{3}\right)^2 + \left(\frac{y-5}{4}\right)^2 = 1$$

$$\frac{(x-2)^2}{9} + \frac{(y-5)^2}{16} = 1$$

No need to restrict the domain, since this graph exactly matches the one for the parametric equations. This rectangular equation happens to be the standard equation of an ellipse, centered at the point (2, 5).

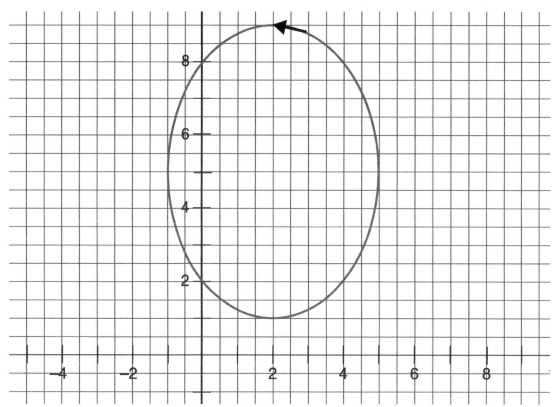

EXAMPLE. Represent the space curve found by intersecting the two given surfaces $x^2 + y^2 + z^2 = 16$ (a sphere) with the cylindrical surface $xy = 4$ as a vector-valued function. Use the parameterization $x = t$, first octant only. Restrict the domain, if necessary, for the parameter t.

SOLUTION. Substituting $x = t$ into the cylindrical surface's equation, we can solve for y:

$$ty = 4$$

$$y = \frac{4}{t}$$

Now, we substitute the facts that $y = \dfrac{4}{t}$ and $x = t$ into the sphere equation, and solve for z:

$$t^2 + \left(\frac{4}{t}\right)^2 + z^2 = 16$$

$$z^2 = 16 - t^2 - \frac{16}{t^2}$$

Since we know to give the first octant intersection only, when we take the square root of both sides here we need only take the positive version for z (we can eliminate the "±" symbol for the final answer):

$$\sqrt{z^2} = \pm\sqrt{16 - t^2 - \frac{16}{t^2}}$$

$$z = +\sqrt{16 - t^2 - \frac{16}{t^2}}$$

We can also simplify this if we like, by first writing the radicand as a single quotient, and using laws of radicals:

$$z = \sqrt{16\left(\frac{t^2}{t^2}\right) - t^2\left(\frac{t^2}{t^2}\right) - \frac{16}{t^2}}$$

$$z = \sqrt{\frac{16t^2 - t^4 - 16}{t^2}} = \frac{\sqrt{16t^2 - t^4 - 16}}{\sqrt{t^2}}$$

$$z = \frac{1}{t}\sqrt{16t^2 - t^4 - 16}$$

The vector-valued function will now be $\mathbf{v}(t) = \left\langle t, \dfrac{4}{t}, \dfrac{1}{t}\sqrt{16t^2 - t^4 - 16}\right\rangle$.

A graph of this vector-valued function is shown below:

Recall from our discussion of finding domains that we always need to "worry" for even radicals. We need the radicand to be non-negative. So, we need to find the solution for the inequality:

$$16t^2 - t^4 - 16 \geq 0$$

Using the "POLY SOLVER" on your calculator, we find the critical numbers for this polynomial inequality. And remembering we only care about the first octant (and therefore the first quadrant), we will need the parameter t to be approximately within the range of the interval $[1.03, 3.86]$. If you do not have "POLY SOLVER" on your calculator, you may use the following method:

Graph the function $f(x) = 16x^2 - x^4 - 16$ on your calculator. The solutions to the equation $f(x) = 16x^2 - x^4 - 16 = 0$ can be found by finding the x-intercepts of the graph of the function. (These will be the critical numbers for the inequality.) Since we need the function to be greater than or equal to zero, the interval(s) on which the graph is positive (above the x-axis) and equal to zero (the x-intercepts) will be the solution for the inequality.

EXERCISES: #4–7 from **Quiz 4**, #4–6 from **Practice Midterm 2**

2.2 DIFFERENTIATION AND INTEGRATION
OF VECTOR-VALUED FUNCTIONS

This section is fun, because you get to review all of your differentiation and integration rules you learned from your first two semesters of calculus, and just write your answers as vectors! The only "weird" part is when you do indefinite integrals. Instead of the Constant of Integration "C" that shows up in every answer from first and second semester calculus, you need the *vector* of integration, which has constants as components. So, we will always write $\vec{C}$ or $\langle C_1, C_2, C_3 \rangle$ in our final answer for indefinite integrals!

I will do some examples for this section. My examples are more challenging than many texts, and I will expect you to be able to handle this level of difficulty in the test bank for this book. The reason for this is because I want you to show me all of those wonderful differentiation and integration rules you learned in your first year of calculus!

EXAMPLE. Find $\mathbf{r}'(t)$ for the following vector-valued functions:

(a) $\mathbf{r}(t) = \left\langle \sin^5(t^2 + 5), \ln \dfrac{t^3 e^t}{1-t}, \arcsin(2t) \right\rangle$

(b) $\mathbf{r}(t) = \left\langle \dfrac{2t}{\sqrt[3]{t+2}}, \tan \sqrt{1-t}, t\cos(7t) \right\rangle$

SOLUTION.

(a) We will need the Chain Rule for the first component, log properties followed by the log rule and exponential rule for the second component, and the arcsine rule for the last component. I will rewrite the first component in order for you to see that the chain Rule is required:

$$\sin^5(t^2 + 5) = \left[\sin(t^2 + 5)\right]^5$$

Now, I will differentiate:

$$\frac{d\left[\sin(t^2 + 5)\right]^5}{dt} = 5\left[\sin(t^2 + 5)\right]^4 \cdot \cos(t^2 + 5) \cdot 2t = (2t)5\sin^4(t^2 + 5)\cos(t^2 + 5) = 10t\sin^4(t^2 + 5)\cos(t^2 + 5)$$

For the second component, I will simplify the expression first using log properties:

$$\ln\frac{t^3 e^t}{1-t} = \ln(t^3 e^t) - \ln(1-t) = \ln t^3 + \ln e^t - \ln(1-t) = 3\ln t + t\ln e - \ln(1-t) = 3\ln t + t - \ln(1-t)$$

Recall that $\ln(e) = 1$. I will differentiate:

$$\frac{d\left[3\ln t + t - \ln(1-t)\right]}{dt} = 3\left(\frac{1}{t}\right) + 1 - \left(\frac{1}{1-t}\right)(-1) = \frac{3}{t} + 1 + \frac{1}{1-t}$$

I could simplify this further by writing as a single quotient, but I will leave that to the reader!

Finally, I use the arcsine rule to differentiate the last component:

$$\frac{d\left[\arcsin(2t)\right]}{dt} = \frac{2}{\sqrt{1-(2t)^2}} = \frac{2}{\sqrt{1-4t^2}}$$

The final answer is a vector of the form:

$$\mathbf{r}'(t) = \left\langle 10t\sin^4(t^2 + 5)\cos(t^2 + 5), \frac{3}{t} + 1 + \frac{1}{1-t}, \frac{2}{\sqrt{1-4t^2}}\right\rangle$$

(b) To differentiate the first component of the vector-valued function, I will need both the Quotient Rule and the Chain Rule:

$$\frac{d\left[\dfrac{2t}{\sqrt[3]{t+2}}\right]}{dt} = \frac{d\left[\dfrac{2t}{(t+2)^{1/3}}\right]}{dt} = \frac{(2)(t+2)^{1/3} - \dfrac{1}{3}(t+2)^{-2/3}(2t)}{\left[(t+2)^{1/3}\right]^2} = \frac{2(t+2)^{1/3} - \dfrac{2t}{3(t+2)^{2/3}}}{(t+2)^{2/3}}$$

This is a complex fraction, and although it is correct, it could use simplification.

I will multiply this expression by a "special version of 1" which is the quotient $\dfrac{3(t+2)^{2/3}}{3(t+2)^{2/3}}$.

This will eliminate the fraction in the numerator of the overall fraction:

$$\frac{2(t+2)^{1/3} - \dfrac{2t}{3(t+2)^{2/3}}}{(t+2)^{2/3}}\left(\frac{3(t+2)^{2/3}}{3(t+2)^{2/3}}\right) = \frac{6(t+2) - 2t}{3(t+2)^{4/3}} = \frac{4(t+3)}{3(t+2)^{4/3}} \quad \textit{Whew!}$$

To differentiate the second term, we need to use the tangent derivative rule:

$$\frac{d\left[\tan\sqrt{1-t}\right]}{dt} = \frac{d\left[\tan(1-t)^{1/2}\right]}{dt} = \sec^2(1-t)^{1/2} \cdot \frac{1}{2}(1-t)^{-1/2}(-1) = -\frac{\sec^2(1-t)^{1/2}}{2(1-t)^{1/2}}$$

Finally, we need the Product Rule to differentiate the last component:

$$\frac{d\left[t\cos(7t)\right]}{dt} = (1)\cos(7t) - 7\sin(7t)(t) = \cos(7t) - 7t\sin(7t)$$

We write the final answer as a vector:

$$\mathbf{r}'(t) = \left\langle \frac{4(t+3)}{3(t+2)^{4/3}}, \; -\frac{\sec^2(1-t)^{1/2}}{2(1-t)^{1/2}}, \; \cos(7t) - 7t\sin(7t) \right\rangle$$

Let's move on to finding the open intervals on which a vector-valued function is smooth.

A vector-valued function is smooth if its graph has no sharp corners. The graph of a vector-valued function has sharp corners if the derivative of the vector-valued function equals the zero vector. So, you do not even have to graph the function to determine if it has sharp corners or not! Simply take the derivative, set it equal to zero vector, and see if there are any values for t that would make the two vectors equal. If not, then the graph will be smooth everywhere (on its domain, of course).

EXAMPLE. Find the open intervals on which the curve given by the vector-valued function is smooth. Write your final answer using interval notation.

(a) $\quad \mathbf{r}(t) = \left\langle 3t^3 - 4t, \; 4t^3 + \dfrac{5}{2}t^2 - 2t, \; \dfrac{3}{2}t^2 + 2t \right\rangle$

(b) $\quad \mathbf{r}(t) = \left\langle e^{-t}, \; \dfrac{1}{t}, \; 3t \right\rangle$

SOLUTION.

(a) The derivative is $\mathbf{r}'(t) = \left\langle 9t^2 - 4, \; 12t^2 + 5t - 2, \; 3t + 2 \right\rangle$.

We set the derivative equal to the zero *vector*: $\mathbf{r}'(t) = \mathbf{0}$

$$\left\langle 9t^2 - 4, \; 12t^2 + 5t - 2, \; 3t + 2 \right\rangle = \left\langle 0, 0, 0 \right\rangle$$

We will have to set each component equal to zero and solve the three separate equations simultaneously:

$$9t^2 - 4 = 0 \qquad\qquad 12t^2 + 5t - 2 = 0 \qquad\qquad 3t + 2 = 0$$

$$(3t + 2)(3t - 2) = 0 \qquad (3t + 2)(4t - 1) = 0 \qquad 3t = -2$$

$$3t + 2 = 0,\, 3t - 2 = 0 \qquad 3t + 2 = 0,\, 4t - 1 = 0 \qquad t = -\frac{2}{3}$$

$$t = -\frac{2}{3},\ t = \frac{2}{3} \qquad\qquad t = -\frac{2}{3},\ t = \frac{1}{4}$$

The only value for t that satisfies all three equations is the value for t when $t = -\frac{2}{3}$.

Therefore, the vector-valued function $\mathbf{r}(t)$ is *not* smooth when $t = -\frac{2}{3}$.

The final answer is: The vector-valued function is smooth on the intervals $t \in \left(-\infty, -\frac{2}{3}\right) \cup \left(-\frac{2}{3}, \infty\right)$

(b)   The derivative is: $\mathbf{r}'(t) = \left\langle -e^{-t}, -\frac{1}{t^2}, 3\right\rangle$. Since both exponential functions and constant functions can never be equal to zero, this derivative can never be equal to the zero vector. So, the vector-valued function is smooth for all values of t in its domain.

That is, when $t \in \left(-\infty, 0\right) \cup \left(0, \infty\right)$.

Integrating Vector-Valued Functions

Again, my exam problems will be more challenging than the homework exercises because I want to test how well you learned the integration techniques from your second semester of calculus. I will be looking for your skills in the following:

- Integration by Parts
- Partial Fraction Decomposition
- Trigonometric Substitution
- Trigonometric Integrals
- Change of Variables
- Polynomial Long Division

- Substitution
- Complete the Square
- Arcsine Rule
- Arctan Rule
- The "Split-Up Trick," etc.

EXAMPLE. Evaluate the indefinite integrals.

(a) $\int \left\langle \dfrac{t}{\sqrt{8t^2+1}}, \sin(2t)e^{\cos(2t)}, \cos^2(3t) \right\rangle dt$

(b) $\int \left\langle \dfrac{1}{(t^2+1)(2+\tan^{-1}t)}, \ln(t+1), (16+t^2)^{-3/2} \right\rangle dt$

(c) $\int \left\langle \dfrac{t^3+t^2}{t^2+t-2}, \dfrac{2t-1}{t^2+4}, \dfrac{\sin t}{1+\sin t} \right\rangle dt$

(d) $\int \left\langle \dfrac{1}{\sqrt{4t-t^2-3}}, t^2 e^{-3t}, t\sqrt{1-t} \right\rangle dt$

SOLUTION.

(a) The first component only needs general substitution. We let $u = 8t^2+1$, $du = 16t\,dt$. This means we need an additional factor of 16 in the integrand (provided we divide it out):

$$\int \frac{t}{\sqrt{8t^2+1}}\,dt = \frac{1}{16}\int \frac{16t}{\sqrt{8t^2+1}}\,dt = \frac{1}{16}\int \frac{du}{\sqrt{u}} = \frac{1}{16}\int u^{-1/2}\,du = \frac{1}{16}\left(\frac{u^{1/2}}{1/2}\right)+C_1 = \frac{1}{8}u^{1/2}+C_1 = \frac{1}{8}\left(8t^2+1\right)^{1/2}+C_1$$

The second component will use the general exponential rule. The substitution we will use is that $u = \cos(2t)$, $du = -2\sin(2t)dt$. We insert a factor of -2 (and divide it out):

$$\int \sin(2t)e^{\cos(2t)}\,dt = \frac{1}{-2}\int -2\sin(2t)e^{\cos(2t)}\,dt = -\frac{1}{2}\int e^u\,du = -\frac{1}{2}e^u+C_2 = -\frac{1}{2}e^{\cos(2t)}+C_2$$

The last component requires the use of a trig ID (power reduction formula):

$$\cos^2(3t) = \frac{1+\cos(2\cdot 3t)}{2} = \frac{1+\cos(6t)}{2}$$

$$\int \cos^2(3t)\,dt = \int \frac{1+\cos(6t)}{2}\,dt = \frac{1}{2}\int (1+\cos(6t))\,dt = \frac{1}{2}\left[t+\int\cos(6t)\,dt\right]$$

To integrate $\cos(6t)$, we use substitution, where $u=6t$, $du=6dt$. This means we need a multiple of 6 (and divide it out):

$$\frac{1}{2}\left[t+\int\cos(6t)\,dt\right] = \frac{1}{2}t+\left(\frac{1}{2}\right)\left(\frac{1}{6}\right)\int 6\cos(6t)\,dt = \frac{1}{2}t+\left(\frac{1}{12}\right)\sin(6t)+C_3$$

Our final answer is a vector:

$$\left\langle \frac{1}{8}(8t^2+1)^{1/2}+C_1,\ -\frac{1}{2}e^{\cos(2t)}+C_2,\ \frac{1}{2}t+\frac{1}{12}\sin(6t)+C_3 \right\rangle$$

You can also write this as:

$$\left\langle \frac{1}{8}(8t^2+1)^{1/2},\ -\frac{1}{2}e^{\cos(2t)},\ \frac{1}{2}t+\frac{1}{12}\sin(6t)\right\rangle + \left\langle C_1, C_2, C_3\right\rangle$$

$$=\left\langle \frac{1}{8}(8t^2+1)^{1/2},\ -\frac{1}{2}e^{\cos(2t)},\ \frac{1}{2}t+\frac{1}{12}\sin(6t)\right\rangle + \mathbf{C}$$

...where $\mathbf{C}$ is the constant of integration *vector*.

(b) The first component requires substitution, where we will let $u = 2 + \tan^{-1}t$, $du = \dfrac{1}{t^2+1}dt$. Then our integral will have the form:

$$\int \frac{1}{(t^2+1)(2+\tan^{-1}t)}dt = \int \frac{1}{u}du = \ln|u| + C_1 = \ln\left|2+\tan^{-1}t\right| + C_1$$

In the second component, we will use integration by parts, where our choices for *u, dv*, etc., are:

$$u = \ln(t+1), \quad dv = 1\,dt$$

$$du = \frac{1}{t+1}dt, \quad v = t$$

Applying the integration by parts formula $\int u\,dv = uv - \int v\,du$, we then have:

$$\int \ln(t+1)dt = t\ln(t+1) - \int t\left(\frac{1}{t+1}\right)dt$$

This last integral may be integrated in one of two ways:

1) Change of variables, or

2) Polynomial long division.

I will show both ways. Let's use Change of Variables first. We will let $u = t+1$, $du = dt$, $t = u-1$. We substitute all three of these "ingredients" into the integral to get:

$$\int \frac{t}{t+1}dt = \int \frac{u-1}{u}du = \int\left[\frac{u}{u}-\frac{1}{u}\right]du = \int\left[1-\frac{1}{u}\right]du = u - \ln|u| + C_2 = (t+1) - \ln|t+1| + C_2$$

Our final answer for the integral of the second component is:

$$\int \ln(t+1)dt = t\ln(t+1) - \left[(t+1) - \ln|t+1|\right] + C_2 = t\ln(t+1) - (t+1) + \ln|t+1| + C_2$$

We could also collect like terms and write this as:

$$\int \ln(t+1)dt = -(t+1) + (t+1)\ln|t+1| + C_2$$

Now, let us take a look at the same integral of $\int \frac{t}{t+1} dt$ and do it using the polynomial long division approach. Since we have an improper rational expression (the degree of the polynomial in the numerator is either equal to or greater than the degree of the polynomial in the denominator), we must perform polynomial long division:

$$
\begin{array}{r}
1 \phantom{{}+1} \\
t+1 \overline{)\, t \phantom{{}+11}} \\
\underline{t+1} \\
-1
\end{array}
$$

Now, rewrite the integrand as $Quotient + \dfrac{remainder}{divisor} = 1 + \dfrac{-1}{t+1}$.

So we integrate $\int \frac{t}{t+1} dt = \int \left[1 - \frac{1}{t+1}\right] dt = t - \ln|t+1| + C$.

Then our final answer for this component will be:

$$\int \ln(t+1)dt = t\ln(t+1) - \left[t - \ln|t+1|\right] + C_2 = t\ln(t+1) - t + \ln|t+1| + C_2 = (t+1)\ln|t+1| - t + C_2$$

And yet, this answer is a little different from our first answer for this component! What gives? Well, this happens sometimes, and it is OK. How to check your answer? You differentiate, of course, and make sure this result matches the original integrand. I claim both answers share the same derivative, so both answers are correct! It just means that the graphs of the functions for each of these answers have the same slope at each point.

The last component of the integrand can be rewritten as:

$$\int (16+t^2)^{-3/2} dt = \int \frac{1}{(16+t^2)^{3/2}} dt = \int \frac{1}{\left(\sqrt{16+t^2}\right)^3} dt$$

Why did we rewrite it like this? Well, we need to use the technique of trigonometric substitution that requires a radical with either a sum or difference of squares as its radicand. Now, we draw the corresponding right triangle and get the three "ingredients" to change the variables in the integrand from t to θ. Because we have a *sum* of squares in the integrand, $\sqrt{16+t^2}$ will serve as the hypotenuse of our right triangle:

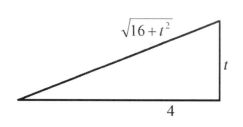

$$\tan\theta = \frac{opp}{adj} = \frac{t}{4} \qquad \sec\theta = \frac{hyp}{adj} = \frac{\sqrt{16+t^2}}{4}$$

Solving for both t and the radical, we get that:

$$t = 4\tan\theta \qquad \sqrt{16 + t^2} = 4\sec\theta$$

$$dt = 4\sec^2\theta d\theta$$

Now substitute these three ingredients into the integrand:

$$\int \frac{1}{\left(\sqrt{16 + t^2}\right)^3}\, dt = \int \frac{1}{\left(4\sec\theta\right)^3}\, 4\sec^2\theta d\theta = \frac{1}{4^2}\int \frac{1}{\sec\theta}\, d\theta = \frac{1}{16}\int \cos\theta d\theta = \frac{1}{16}\sin\theta + C_3$$

We are not done! We need the final answer to have the original variable of t in it, not the variable θ. Therefore, we need to "back-substitute" at this point. We look at our right triangle diagram and figure out what $\sin\theta$ is:

$$\sin\theta = \frac{opp}{hyp} = \frac{t}{\sqrt{16 + t^2}}$$

Now, we have that the integral of the third component is:

$$\left(\frac{1}{16}\right)\left(\frac{t}{\sqrt{16 + t^2}}\right) + C_3 = \frac{t}{16\sqrt{16 + t^2}} + C_3$$

The final answer for this problem is a vector:

$$\left\langle \ln\left|2 + \tan^{-1} t\right| + C_1,\ -(t+1) + (t+1)\ln|t+1| + C_2,\ \frac{t}{16\sqrt{16 + t^2}} + C_3 \right\rangle$$

$$= \left\langle \ln\left|2 + \tan^{-1} t\right|,\ -(t+1) + (t+1)\ln|t+1|,\ \frac{t}{16\sqrt{16 + t^2}} \right\rangle + \mathbf{C}$$

(c) The first component of this problem is a rational function that is improper. Recall that a rational function has the form:

$$f(x) = \frac{polynomial}{polynomial}$$

An improper rational function is one where the degree of the polynomial in the numerator is either equal to or greater than the degree of the polynomial in the denominator. We first need to perform polynomial long division, and then since the denominator is a polynomial that is factorable over the set of integers, we then perform partial fraction decomposition.

We will do the division: $\dfrac{t^3 + t^2}{t^2 + t - 2}$.

$$\begin{array}{r} t \\ t^2 + t - 2 \overline{\smash{)}\, t^3 + t^2 } \\ \underline{t^3 + t^2 - 2t} \\ 2t \end{array}$$

We write the result of the long division on the form:

$$Quotient + \frac{\mathrm{Re}\,mainder}{Divisor}$$

This means our integrand will be $t + \dfrac{2t}{t^2 + t - 2}$.

$$\int \frac{t^3 + t^2}{t^2 + t - 2}\,dt = \int \left[t + \frac{2t}{t^2 + t - 2}\right]dt = \frac{t^2}{2} + \int \frac{2t}{t^2 + t - 2}\,dt = \frac{t^2}{2} + \int \frac{2t}{(t+2)(t-1)}\,dt$$

Let us consider only the integral, where we will perform partial fraction decomposition:

$$\int \frac{2t}{(t+2)(t-1)}\,dt = \int \left[\frac{A}{t+2} + \frac{B}{t-1}\right]dt$$

To find the unknowns, A and B, we set the integrands equal:

$$\frac{2t}{(t+2)(t-1)} = \frac{A}{t+2} + \frac{B}{t-1}$$

Clear fractions from this equation to obtain the "Basic Equation":

$$2t = A(t-1) + B(t+2)$$

Let $t = 1$ to find B:

$$2(1) = A(1-1) + B(1+2)$$
$$2 = \qquad\quad B(3)$$
$$\frac{2}{3} = B$$

Now, we let $t = -2$ to find A:

$$2(-2) = A(-2-1) + B(-2+2)$$
$$-4 = A(-3)$$
$$\frac{4}{3} = A$$

So, we can integrate now:

$$\frac{t^2}{2} + \int \frac{2t}{(t+2)(t-1)}\,dt = \frac{t^2}{2} + \int \left[\frac{A}{t+2} + \frac{B}{t-1}\right]dt = \frac{t^2}{2} + \frac{4}{3}\int \frac{1}{t+2}\,dt + \frac{2}{3}\int \frac{1}{t-1}\,dt = \frac{t^2}{2} + \frac{4}{3}\ln|t+2| + \frac{2}{3}\ln|t-1| + C_1$$

The second component will have us do the "split-up trick":

$$\int \frac{2t-1}{t^2+4}\,dt = \int \left[\frac{2t}{t^2+4} - \frac{1}{t^2+4}\right]dt$$

We will use the Log Rule on the first term and the Arctan Rule on the second term:

$$\int \left[\frac{2t}{t^2+4} - \frac{1}{t^2+4} \right] dt = \ln\left(t^2+4\right) - \frac{1}{2}\arctan\left(\frac{t}{2}\right) + C_2$$

This last component is challenging. We first need to multiply the integrand by a "special version of 1," which involves the Pythagorean Conjugate of the denominator:

$$\int \frac{\sin t}{1+\sin t}\, dt = \int \left(\frac{\sin t}{1+\sin t}\right)\left(\frac{1-\sin t}{1-\sin t}\right) dt = \int \frac{\sin t - \sin^2 t}{1-\sin^2 t}\, dt = \int \frac{\sin t - \sin^2 t}{\cos^2 t}\, dt$$

Now, we will use the split-up trick, followed by u-substitution on the first term, and a trig ID on the second term:

$$\int \frac{\sin t - \sin^2 t}{\cos^2 t}\, dt = \int \frac{\sin t}{\cos^2 t}\, dt - \int \frac{\sin^2 t}{\cos^2 t}\, dt = \int \frac{\sin t}{(\cos t)^2}\, dt - \int \tan^2 t\, dt$$

For the first term where we use u-substitution, we let $u = \cos t$, $du = -\sin t\, dt$. (Note that we require a minus sign, so we will "put one in, provided we divide one out.")

$$\int \frac{\sin t}{\cos^2 t}\, dt = -\int \frac{-\sin t}{(\cos t)^2}\, dt = -\int \frac{du}{u^2} = -\int u^{-2} du = -\left(\frac{u^{-1}}{-1}\right) = +\frac{1}{u} = \frac{1}{\cos t} = \sec t$$

For the other integral, we use a trig ID again, where we write:

$$\int \tan^2 t\, dt = \int \left(\sec^2 t - 1\right) dt = \tan t - t$$

So, our final component has the integral:

$$\sec t - \left(\tan t - t\right) + C_3 = \sec t - \tan t + t + C_3$$

The final answer for this problem is:

$$\left\langle \frac{t^2}{2} + \frac{4}{3}\ln|t+2| + \frac{2}{3}\ln|t-1| + C_1,\ \ln(t^2+4) - \frac{1}{2}\arctan\left(\frac{t}{2}\right) + C_2,\ \sec t - \tan t + t + C_3 \right\rangle$$

(d) This last problem requires the Arcsine Rule on its first component, but not before we complete the square in the radicand:

$$\int \frac{1}{\sqrt{4t - t^2 - 3}}\, dt = \int \frac{1}{\sqrt{-3 + M - \left(t^2 - 4t + M\right)}}\, dt$$

The "M" that I both added and subtracted in the expression for the radicand is what I refer to as the "Magical Mystery Number." We find it by taking the coefficient of the "$4t$" term, dividing it by 2, and then squaring:

$$M = \left(\frac{4}{2}\right)^2 = 2^2 = 4$$

Then, we have our integral resembling the form for the Arcsine Rule:

$$\int \frac{1}{\sqrt{-3 + M - \left(t^2 - 4t + M\right)}}\,dt = \int \frac{1}{\sqrt{(-3+4)-(t^2-4t+4)}}\,dt = \int \frac{1}{\sqrt{1-(t-2)^2}}\,dt$$

Recall the Arcsine Rule for Integrals:

$$\int \frac{1}{\sqrt{a^2 - u^2}}\,du = \arcsin\left(\frac{u}{a}\right) + C$$

In our integral, we have that $a = 1$, and $u = t - 2$, $du = dt$.

So, the answer here is $\arcsin(t-2) + C_1$.

The second component needs us to perform repeated integration by parts on the integral:

$$\int t^2 e^{-3t}\,dt$$

We identify u, dv, etc first:

$$u = t^2 \qquad dv = e^{-3t}\,dt$$

$$du = 2t\,dt \qquad v = -\frac{1}{3}e^{-3t}$$

One of the biggest mistakes I see my students make doing an integral like this one, is when they obtain the "v" by accidentally differentiating instead of integrating. Please don't do this! Now, we use the integration by parts formula and find that:

$$\int t^2 e^{-3t}\,dt = \left(t^2\right)\left(-\frac{1}{3}e^{-3t}\right) - \int \left(-\frac{1}{3}e^{-3t}\right) 2t\,dt$$

I will rewrite and simplify a bit:

$$\int t^2 e^{-3t}\,dt = -\frac{1}{3}t^2 e^{-3t} + \frac{2}{3}\int t e^{-3t}\,dt$$

The last integral needs integration by parts again. The choices for u, dv, etc., this time are:

$$u = t \qquad dv = e^{-3t}\,dt$$

$$du = dt \qquad v = -\frac{1}{3}e^{-3t}$$

Applying the integration by parts formula again, we have:

$$\int t^2 e^{-3t} dt = -\frac{1}{3}t^2 e^{-3t} + \frac{2}{3}\left[(t)\left(-\frac{1}{3}e^{-3t}\right) - \int -\frac{1}{3}e^{-3t} dt\right]$$

$$= -\frac{1}{3}t^2 e^{-3t} - \frac{2}{9}te^{-3t} + \frac{2}{9}\int e^{-3t} dt$$

$$= -\frac{1}{3}t^2 e^{-3t} - \frac{2}{9}te^{-3t} + \frac{2}{9}\left(-\frac{1}{3}e^{-3t}\right) + C_2$$

$$= -\frac{1}{3}t^2 e^{-3t} - \frac{2}{9}te^{-3t} - \frac{2}{27}e^{-3t} + C_2$$

The last component, $\int t\sqrt{1-t}\, dt$, requires Change of Variables.
We let $u = 1-t,\ du = -dt,\ t = 1-u$ and rewrite the integral now:

$$-\int (1-u)\sqrt{u}\, du = -\int (1-u)u^{1/2}\, du = -\int \left(u^{1/2} - u^{3/2}\right) du = -\left(\frac{u^{3/2}}{3/2} - \frac{u^{5/2}}{5/2}\right) + C_3 = \frac{-2}{3}(1-t)^{3/2} + \frac{2}{5}(1-t)^{5/2} + C_3$$

The final answer is:

$$\left\langle \arcsin(t-2) + C_1,\ -\frac{1}{3}t^2 e^{-3t} - \frac{2}{9}te^{-3t} - \frac{2}{27}e^{-3t} + C_2,\ -\frac{2}{3}(1-t)^{3/2} + \frac{2}{5}(1-t)^{5/2} + C_3 \right\rangle$$

Wow!

My second-to-last example for this section will be evaluating a definite integral:

EXAMPLE. Evaluate the integral:

$$\int_1^2 \left[(6-6t)\mathbf{i} + 3\sqrt{t}\,\mathbf{j} + \left(\frac{4}{t^2}\right)\mathbf{k} \right] dt$$

SOLUTION. First, we find anti-derivatives:

$$\left[6t - 3t^2\right]_1^2 \mathbf{i} + \left[3\left(\frac{2}{3}\right)t^{3/2}\right]_1^2 \mathbf{j} - \left[\frac{4}{t}\right]_1^2 \mathbf{k}$$

Now, we apply the Fundamental Theorem of Calculus:

$$\left[\left(6(2) - 3(2)^2\right) - \left(6(1) - 3(1)^2\right)\right]\mathbf{i} + \left[2(2)^{3/2} - (2)(1)^{3/2}\right]\mathbf{j} - \left[\frac{4}{2} - \frac{4}{1}\right]\mathbf{k}$$

$$= \left[(0) - (3)\right]\mathbf{i} + \left[(2)^{5/2} - (2)\right]\mathbf{j} - [2-4]\mathbf{k}$$

$$= -3\mathbf{i} + (2^{5/2} - 2)\mathbf{j} + 2\mathbf{k} = \left\langle -3,\ 2^{5/2} - 2,\ 2 \right\rangle$$

— 68 —

My last example will be solving initial value problems for vector functions.

EXAMPLE. Find $\mathbf{r}(t)$ if you are given that $\mathbf{r}'(t) = \dfrac{3}{2}(t+1)^{1/2}\mathbf{i} + e^{-t}\mathbf{j} + \dfrac{1}{t+1}\mathbf{k}$ and $\mathbf{r}(0) = \mathbf{k}$

SOLUTION. We need to integrate first to get $\mathbf{r}(t)$:

$$\mathbf{r}(t) = \int \mathbf{r}'(t)dt = \int \left[\frac{3}{2}(t+1)^{1/2}\mathbf{i} + e^{-t}\mathbf{j} + \frac{1}{t+1}\mathbf{k}\right]dt = \left[(t+1)^{3/2} + C_1\right]\mathbf{i} + \left[-e^{-t} + C_2\right]\mathbf{j} + \left[\ln|t+1| + C_3\right]\mathbf{k}$$

We find the constants of integration C_1, C_2, and C_3 by using the fact that $\mathbf{r}(0) = \mathbf{k}$, which we can rewrite as:
$$\mathbf{r}(0) = \langle 0, 0, 1\rangle$$

Substituting the fact that $t = 0$ into our answer thus far for $\mathbf{r}(t)$:

$$\mathbf{r}(0) = \left[(0+1)^{3/2} + C_1\right]\mathbf{i} + \left[-e^{-0} + C_2\right]\mathbf{j} + \left[\ln|0+1| + C_3\right]\mathbf{k} = 0\mathbf{i} + 0\mathbf{j} + 1\mathbf{k}$$

We now equate each component to get the constants of integration:

$$1^{3/2} + C_1 = 0 \qquad -e^{-0} + C_2 = 0 \qquad \ln|1| + C_3 = 1$$
$$C_1 = -1 \qquad\qquad C_2 = 1 \qquad\qquad C_3 = 1$$

Our final answer is $\mathbf{r}(t) = \left[(t+1)^{3/2} - 1\right]\mathbf{i} + \left[-e^{-t} + 1\right]\mathbf{j} + \left[\ln|t+1| + 1\right]\mathbf{k}$.

If we write the answer in component form instead, then we have:
$$\mathbf{r}(t) = \left\langle (t+1)^{3/2} - 1, \ -e^{-t} + 1, \ \ln|t+1| + 1\right\rangle$$

EXERCISES: #7–12 from **Practice Midterm 2**, #10 from **Quiz 4**

2.3 VELOCITY AND ACCELERATION

The key points in this section are:

1) If a vector-valued function represents motion of a particle, then the velocity of the particle is also a vector and is the first derivative of the motion vector. (The velocity vector at any time t will always be tangent to the motion vector.) Speed is a scalar and can be found by taking the magnitude of the velocity vector.

2) The acceleration of the particle is also a vector and can be found by taking the second derivative of the motion vector (a.k.a. the first derivative of the velocity vector).

3) Projectile motion is a common application of vector functions used in modeling motion in the plane. It is a special case of motion and has specific formulas for the components.

EXAMPLE. The acceleration function is given by $\mathbf{a}(t) = (-\cos t)\mathbf{i} + (-\sin t)\mathbf{j}$, and $\mathbf{v}(0) = \mathbf{j} + \mathbf{k}$, $\mathbf{r}(0) = \mathbf{i} + 2\mathbf{j} + 3\mathbf{k}$. Find the velocity and position vectors, then find both the speed and the position when $t = 2$.

SOLUTION. We need to integrate acceleration to find velocity:

$$\mathbf{v}(t) = \int \mathbf{a}(t)dt = \int [(-\cos t)\mathbf{i} + (-\sin t)\mathbf{j}]dt = (-\sin t + C_1)\mathbf{i} + (\cos t + C_2)\mathbf{j} + C_3\mathbf{k}$$

To find the constants of integration, we use the initial condition of velocity that was given, which is $\mathbf{v}(0) = \mathbf{j} + \mathbf{k}$. Substitute $t = 0$ into the velocity function and equate to the given initial condition:

$$\mathbf{v}(0) = (-\sin 0 + C_1)\mathbf{i} + (\cos 0 + C_2)\mathbf{j} + C_3\mathbf{k} = 0\mathbf{i} + \mathbf{j} + \mathbf{k}$$

$$0 + C_1 = 0 \qquad 1 + C_2 = 1 \qquad C_3 = 1$$
$$C_1 = 0 \qquad\quad C_2 = 0$$

So, we have the answer for the velocity function:

$$\mathbf{v}(t) = (-\sin t)\mathbf{i} + (\cos t)\mathbf{j} + \mathbf{k}$$

We can write this using component form also:

$$\mathbf{v}(t) = \langle -\sin t, \ \cos t, \ 1 \rangle$$

Now, we can find speed, which is equal to the magnitude of velocity:

$$\text{Speed} = \|\mathbf{v}(t)\| = \sqrt{(-\sin t)^2 + (\cos t)^2 + (1)^2} = \sqrt{\sin^2 t + \cos^2 t + 1} = \sqrt{1+1} = \sqrt{2}$$

Here, speed is a constant, and is always equal to $\sqrt{2}$, regardless of the time t. (**Note:** Speed is not always a constant.)

Now, we find the position function by integrating the velocity function:

$$\mathbf{r}(t) = \int \mathbf{v}(t)dt = \int [(-\sin t)\mathbf{i} + (\cos t)\mathbf{j} + \mathbf{k}]dt = (\cos t + C_1)\mathbf{i} + (\sin t + C_2)\mathbf{j} + (t + C_3)\mathbf{k}$$

To find these new constants of integration, we use the initial condition for position that was given, which was $\mathbf{r}(0) = \mathbf{i} + 2\mathbf{j} + 3\mathbf{k}$. Substitute $t = 0$ into the position function and equate to the given initial condition:

$$\mathbf{r}(0) = (\cos 0 + C_1)\mathbf{i} + (\sin 0 + C_2)\mathbf{j} + (0 + C_3)\mathbf{k} = \mathbf{i} + 2\mathbf{j} + 3\mathbf{k}$$

$$1 + C_1 = 1 \qquad 0 + C_2 = 2 \qquad C_3 = 3$$
$$C_1 = 0 \qquad\qquad C_2 = 2$$

The position function is therefore given by:

$$\mathbf{r}(t) = (\cos t)\mathbf{i} + (\sin t + 2)\mathbf{j} + (t + 3)\mathbf{k}$$

We can write this in component form as well:

$$\mathbf{r}(t) = \langle \cos t,\ \sin t + 2,\ t + 3 \rangle$$

The position at time $t = 2$ can be found by simply substituting $t = 2$ into the position function:

$$\mathbf{r}(2) = \langle \cos 2,\ \sin 2 + 2,\ 2 + 3 \rangle = \langle \cos 2,\ \sin 2 + 2,\ 5 \rangle$$

Projectile Motion

EXAMPLE. A projectile is fired at a height of three feet from the ground with an initial velocity of 900 feet per second and at an angle of $45°$ above the horizontal.

    (a)   Find the vector-valued function describing the motion.

    (b)   Find the maximum height.

    (c)   How long was the projectile in the air?

    (d)   How far did it go? (That is, what was its range?)

SOLUTION.

    (a)   By the Theorem on Projectile Motion, the vector function has the form:

$$\mathbf{r}(t) = \left\langle v_0 \cos\theta\, t,\ h + v_0 \sin\theta\, t - \frac{1}{2}gt^2 \right\rangle$$

We use $g = -32$ feet per second per second. Using the values given, the motion function is:

$$\mathbf{r}(t) = \left\langle 900\cos(45°)t,\ 3 + 900\sin(45°)t - \frac{1}{2}(32)t^2 \right\rangle = \left\langle 900\left(\frac{\sqrt{2}}{2}\right)t,\ 3 + (900)\left(\frac{\sqrt{2}}{2}\right)t - 16t^2 \right\rangle$$

$$= \left\langle 450\sqrt{2}t,\ 3 + 450\sqrt{2}t - 16t^2 \right\rangle$$

(b)   To find the maximum height, we use the fact that maximum height will be achieved when the vertical component of velocity is equal to zero. Therefore, we need to find the velocity function first (by differentiating the motion function):

$$\mathbf{v}(t) = \frac{d[\mathbf{r}(t)]}{dt} = \frac{d\left\langle 450\sqrt{2}t,\ 3 + 450\sqrt{2}t - 16t^2 \right\rangle}{dt} = \left\langle 450\sqrt{2},\ 450\sqrt{2} - 32t \right\rangle$$

The vertical component of velocity is the second component. We set this equal to zero and solve for time t:

$$450\sqrt{2} - 32t = 0$$
$$450\sqrt{2} = 32t$$
$$\frac{450\sqrt{2}}{32} = t$$
$$\frac{225\sqrt{2}}{16} = t$$
$$t \approx 19.9 \text{ seconds}$$

Next, we substitute this value for time t into the vertical component of motion to get the maximum height:

$$3 + 450\sqrt{2}(19.9) - 16(19.9)^2 \approx 6331.12 \text{ feet}$$

Now, to find out how long it was in the air, we find out how long it took to hit the ground. The height of the projectile when it hits the ground is zero. So, we set the vertical component of the motion function equal to zero, and solve for t. We must then solve a quadratic equation using the quadratic formula:

$$3 + 450\sqrt{2}t - 16t^2 = 0$$
$$t = \frac{-b \pm \sqrt{b^2 - 4ac}}{2a}$$
$$t = \frac{-450\sqrt{2} \pm \sqrt{\left(450\sqrt{2}\right)^2 - 4(-16)(3)}}{(2)(-16)}$$
$$t \approx 39.8 \text{ seconds} \quad t \approx \text{-}0.005 \text{ seconds}$$

We discard the negative solution for the quadratic equation, since that is impossible. Therefore, the projectile is in the air for approximately 39.8 seconds, and the range is found by substituting this value of time into the horizontal component for the motion function:

$$\text{Range} = 450\sqrt{2}(39.8) \approx 25{,}328.5 \text{ feet}$$

EXAMPLE. A projectile is fired from the ground at an angle of 8° with the horizontal. Find the minimum velocity necessary if the projectile is to have a range of 50 meters.

SOLUTION. We use $g = -9.8$ meters per second per second for this problem. This problem involves solving a two-equation, two-unknowns non-linear system. We know that the position function (a.k.a. motion function) has the form:

$$\mathbf{r}(t) = \left\langle v_0 \cos(8°)t,\, 0 + v_0 \sin(8°)t - \frac{1}{2}(9.8)t^2 \right\rangle$$

We are given the fact that the range of the projectile needs to be 50 meters. We set the horizontal component of motion equal to 50, and solve for t:

$$v_0 \cos(8°)t = 50$$

$$t = \frac{50}{v_0 \cos 8°} \quad \text{(This is one equation with two unknowns in it.)}$$

For this value of time, the vertical component of motion will be equal to zero. When the projectile achieves this range of 50 meters, it will be hitting the ground. So, we set the vertical component of motion equal to 0:

$$v_0 \sin(8°)t - 4.9t^2 = 0$$

(This is the second equation of the non-linear system containing two unknowns.)

Thus, the non-linear system of equations looks like:

$$\begin{cases} t = \dfrac{50}{v_0 \cos 8°} \\ v_0 \sin(8°)t - 4.9t^2 = 0 \end{cases}$$

Plug the first equation into the second equation to eliminate the variable t and solve for the other unknown variable, which is v_0:

$$v_0 \sin(8°)\left(\frac{50}{v_0 \cos 8°}\right) - 4.9\left(\frac{50}{v_0 \cos 8°}\right)^2 = 0$$

$$\frac{50 \sin 8°}{\cos 8°} - \frac{4.9(2500)}{(v_0)^2 (\cos 8°)^2} = 0$$

$$\frac{50 \sin 8°}{\cos 8°} = \frac{4.9(2500)}{(v_0)^2 (\cos 8°)^2}$$

$$\frac{\cos 8°}{50 \sin 8°} = \frac{(v_0)^2 (\cos 8°)^2}{(4.9)(2500)}$$

$$\left(v_0\right)^2 = \frac{(4.9)(2500)}{(50)\sin 8° \cos 8°}$$

$$v_0 = \sqrt{\frac{(4.9)(2500)}{(50)\sin 8° \cos 8°}}$$

$$v_0 \approx 42.2 \ m/\sec$$

EXERCISES: #8, 9, 11, 12, 13, 14, from **Quiz 4**, #15 from **Practice Midterm 2**

2.4 TANGENT AND NORMAL VECTORS

The main idea for tangent vectors is to expound on a fact we already know (or should already know). That is, the derivative of a motion vector (which is the velocity vector) will yield vectors that are tangent to the line of motion. So, to make a *unit* tangent vector (recall that a unit vector is one whose magnitude is equal to 1), we simply use the formula to create a unit vector $\mathbf{u}(t)$, which is:

$$\mathbf{u}(t) = \frac{\mathbf{w}(t)}{\|\mathbf{w}(t)\|}$$

Therefore the *unit tangent vector*, denoted $\mathbf{T}(t)$, is found by finding the velocity vector and making it a unit vector. So, we will have:

$$\mathbf{T}(t) = \frac{\mathbf{r}'(t)}{\|\mathbf{r}'(t)\|}$$

Although there are infinitely many vectors that are orthogonal to $\mathbf{T}(t)$, we are interested in a special one. This special one is called the *unit normal vector*, denoted $\mathbf{N}(t)$, which can be found directly from $\mathbf{T}(t)$ itself. Further, this unit vector is not only orthogonal to $\mathbf{T}(t)$, but (when graphed) will be a vector that always points in the direction the object in motion is turning. That means $\mathbf{N}(t)$ will always point in the direction of the concave side of the curve of motion.

To find $\mathbf{N}(t)$, we use the formula $\mathbf{N}(t) = \dfrac{\mathbf{T}'(t)}{\|\mathbf{T}'(t)\|}$.

Please do *not* use a method that involves taking the second derivative of the motion function. This is a common student error! In other words, $\mathbf{N}(t) \neq \dfrac{\mathbf{r}''(t)}{\|\mathbf{r}''(t)\|}$. Please make a note of this, OK?

Now, the unit normal vector $\mathbf{N}(t)$ is usually challenging to obtain due to nasty algebra that results from having to apply the Quotient Rule (to take the derivative of the unit tangent vector). You will find this out by doing just a couple of homework exercises. I will do an example below as well.

Because the unit tangent vector and unit normal vector are always orthogonal, their dot product will be equal to 0. That is $\mathbf{N}(t) \cdot \mathbf{T}(t) = 0$. This is especially handy if you are given $\mathbf{T}(t)$ as a plane vector. You can find the unit normal vector without using the formula (and thereby avoid all of the challenging algebra that goes with it). So, suppose you are given (or you found that) the unit tangent vector was $\mathbf{T}(t) = \langle x(t), y(t) \rangle$. Then, in order for the dot product to be equal to 0, we could just write that the unit normal vector will be one of two vectors. The two choices you will have are $\mathbf{N}(t) = \langle y(t), -x(t) \rangle$ or $\mathbf{N}(t) = \langle -y(t), x(t) \rangle$. Check out that both will be orthogonal to $\mathbf{T}(t)$:

$$\mathbf{N}(t) \cdot \mathbf{T}(t) = 0??$$

$$\langle y(t), -x(t) \rangle \cdot \langle x(t), y(t) \rangle = y(t)[x(t)] + [-x(t)]y(t) = 0$$

$$\langle -y(t), x(t) \rangle \cdot \langle x(t), y(t) \rangle = -y(t)[x(t)] + [x(t)]y(t) = 0$$

So, yes, it turns out that both choices are indeed orthogonal to the unit tangent vector, $\mathbf{T}(t)$.

However, although both will work, only *one* of these choices for $\mathbf{N}(t)$ will be correct. Which one? It will depend on the problem. The quickest way to find out (without having to use the formula, of course) is to graph the corresponding motion function first. Then, graph both choices for the unit normal vector at an arbitrary time t. The vector that points in the direction the object is turning (that is, toward the concave side of the curve of motion) will be the one you will choose, i.e., $\mathbf{N}(t)$.

Note: You cannot use this neat trick for vectors in space (3-D). Instead, you will have to go through the challenge of finding the unit normal vector $\mathbf{N}(t)$ the hard way—by using the formula and doing lots of nasty algebra, etc., etc.

A concept not discussed in some texts, yet discussed in other calculus texts is that of the "Binormal Vector." This is yet another vector that is orthogonal to *both* the unit tangent and unit normal vectors. The binormal vector is denoted $\mathbf{B}(t)$ and can be found by taking the cross-product of the unit tangent and unit normal vectors—$\mathbf{B} = \mathbf{T} \times \mathbf{N}$. We can use the right-hand rule to determine what direction the binormal vector is pointing. Together, the three vectors $\mathbf{T}, \mathbf{N}$, and $\mathbf{B}$ make up what is referred to as the "**TNB** Frame."

The last major concept covered in this section concerns the acceleration vector, $\mathbf{a}(t) = \mathbf{r}''(t)$. It turns out that the acceleration vector can be related to both the unit tangent and unit normal vectors in the following way: The acceleration vector lies in the plane determined by the unit normal and unit tangent vectors! Another way to say this is that part of the acceleration acts in the line of motion (i.e. in the direction of $\mathbf{T}(t)$) and part of the acceleration acts perpendicular to the line of motion (i.e., in the direction of $\mathbf{N}(t)$). *Wow!* Amazing! What this means for us is that the acceleration vector $\mathbf{a}(t)$ can be written as a linear combination of the unit tangent and unit normal vectors *instead* of as a linear combination of the standard vectors $\mathbf{i}, \mathbf{j}$, and $\mathbf{k}$. So, we can write $\mathbf{a}(t) = a_T\mathbf{T}(t) + a_N\mathbf{N}(t)$, where the scalars a_T and a_N can be found using the formulas $a_T = \mathbf{a} \cdot \mathbf{T}$ and $a_N = \mathbf{a} \cdot \mathbf{N}$. The two scalars a_T and a_N are referred to as the tangential and normal components of acceleration, respectively.

An alternate formula you may use for finding the normal component of acceleration, a_N, is:

$$a_N = \sqrt{\|\mathbf{a}\|^2 - [a_T]^2}$$

Oftentimes, this alternative formula is easier to use than the other formula, because you do not need to find the (challenging) unit normal vector, $\mathbf{N}(t)$, first.

EXAMPLE. Given the space curve $\mathbf{r}(t) = \left\langle t, 3t^2, \dfrac{t^2}{2} \right\rangle$ at the time $t = 2$.

    (a) Find $\mathbf{T}(t)$ at the specified time.

    (b) Find $\mathbf{N}(t)$ at the specified time.

    (c) Find both a_T and a_N at the specified time.

    (d) Write the acceleration vector $\mathbf{a}(t)$ as a linear combination of the unit tangent and unit normal vectors at the specified time.

    (e) Find a set of parametric equations for the line tangent to the space curve at the point $P\left(-1, 3, \dfrac{1}{2}\right)$.

    (f) Find a vector that is orthogonal to both the unit tangent and the unit normal vectors at the specified time.

SOLUTION.

    (a) First, we take the derivative of the given position vector $\mathbf{r}(t)$ to get the velocity vector:

$$\mathbf{r}'(t) = \langle 1, 6t, t \rangle$$

    Next, we find the magnitude of this velocity vector (a.k.a. speed):

$$\|\mathbf{r}'(t)\| = \sqrt{(1)^2 + (6t)^2 + (t)^2} = \sqrt{1 + 36t^2 + t^2} = \sqrt{1 + 37t^2}$$

    Now, we can form the unit tangent vector using the formula:

$$\mathbf{T}(t) = \frac{\mathbf{r}'(t)}{\|\mathbf{r}'(t)\|} = \frac{\langle 1, 6t, t \rangle}{\sqrt{1 + 37t^2}} = \left\langle \frac{1}{\sqrt{1 + 37t^2}}, \frac{6t}{\sqrt{1 + 37t^2}}, \frac{t}{\sqrt{1 + 37t^2}} \right\rangle$$

    At the specified time, $t = 2$, we have that the unit tangent vector is:

$$\mathbf{T}(2) = \frac{\mathbf{r}'(2)}{\|\mathbf{r}'(2)\|} = \left\langle \frac{1}{\sqrt{1 + 37(2)^2}}, \frac{6(2)}{\sqrt{1 + 37(2)^2}}, \frac{(2)}{\sqrt{1 + 37(2)^2}} \right\rangle = \left\langle \frac{1}{\sqrt{149}}, \frac{12}{\sqrt{149}}, \frac{2}{\sqrt{149}} \right\rangle$$

    (b) To find the unit normal vector $\mathbf{N}(t)$, we must first take the derivative of the unit tangent vector, $\mathbf{T}(t)$ —we need to find $\mathbf{T}'(t)$. This is going to take us a while, since we need to use the Quotient Rule for Derivatives on the second and third components of this vector. The first component will not require the Quotient Rule, since the numerator is a constant.

    If we first rewrite the first component as $\left(1 + 37t^2\right)^{-1/2}$, we can then differentiate using the Chain Rule:

$$\frac{d\left[\left(1 + 37t^2\right)^{-1/2}\right]}{dt} = -\frac{1}{2}\left(1 + 37t^2\right)^{-3/2} \cdot \left(74t\right) = \frac{-(74t)}{2(1 + 37t^2)^{3/2}} = \frac{-37t}{(1 + 37t^2)^{3/2}}$$

Now, we will differentiate the second component using the Quotient Rule. Keep in mind that to differentiate the denominator during this process requires the Chain Rule. So, we need both the Quotient and Chain Rules to do this part of the problem. Thus, this differentiation of the second component is what I call a "Double Whammy!"

$$\frac{d\left[\dfrac{6t}{(1+37t^2)^{1/2}}\right]}{dt} = \frac{(6)(1+37t^2)^{1/2} - \dfrac{1}{2}(1+37t^2)^{-1/2}(74t)(6t)}{\left[\sqrt{1+37t^2}\right]^2} = \frac{6\sqrt{1+37t^2} - \dfrac{222t^2}{\sqrt{1+37t^2}}}{1+37t^2}$$

Now, although this is correct, it must be simplified so that it no longer is a complex fraction (i.e., a fraction containing fractions). I will multiply the quotient by a "special version of 1," which is:

$$\frac{\sqrt{1+37t^2}}{\sqrt{1+37t^2}} \text{ to clear the fraction:}$$

$$\left(\frac{6\sqrt{1+37t^2} - \dfrac{222t^2}{\sqrt{1+37t^2}}}{1+37t^2}\right)\frac{\sqrt{1+37t^2}}{\sqrt{1+37t^2}} = \frac{6(1+37t^2) - 222t^2}{(1+37t^2)\sqrt{1+37t^2}} = \frac{6+222t^2-222t^2}{(1+37t^2)^{3/2}} = \frac{6}{(1+37t^2)^{3/2}}$$

Now, we will differentiate the last component of the unit tangent vector:

$$\frac{d\left[\dfrac{t}{(1+37t^2)^{1/2}}\right]}{dt} = \frac{(1)(1+37t^2)^{1/2} - \dfrac{1}{2}(1+37t^2)^{-1/2}(74t)(t)}{\left[\sqrt{1+37t^2}\right]^2} = \frac{\sqrt{1+37t^2} - \dfrac{37t^2}{\sqrt{1+37t^2}}}{1+37t^2}$$

Again, we need to simplify this complex fraction:

$$\left(\frac{\sqrt{1+37t^2} - \dfrac{37t^2}{\sqrt{1+37t^2}}}{1+37t^2}\right)\frac{\sqrt{1+37t^2}}{\sqrt{1+37t^2}} = \frac{(1+37t^2) - 37t^2}{(1+37t^2)\sqrt{1+37t^2}} = \frac{1+37t^2 - 37t^2}{(1+37t^2)^{3/2}} = \frac{1}{(1+37t^2)^{3/2}}$$

What we have so far is that $\mathbf{T}'(t) = \left\langle \dfrac{-37t}{(1+37t^2)^{3/2}}, \dfrac{6}{(1+37t^2)^{3/2}}, \dfrac{1}{(1+37t^2)^{3/2}} \right\rangle$.

To find the unit normal vector, $\mathbf{N}(t)$, we still need to find the magnitude of $\mathbf{T}'(t)$, since the formula to find $\mathbf{N}(t)$ is:

$$\mathbf{N}(t) = \frac{\mathbf{T}'(t)}{\|\mathbf{T}'(t)\|}$$

We do that now:

$$\|\mathbf{T}'(t)\| = \sqrt{\left(\frac{-37t}{(1+37t^2)^{3/2}}\right)^2 + \left(\frac{6}{(1+37t^2)^{3/2}}\right)^2 + \left(\frac{1}{(1+37t^2)^{3/2}}\right)^2}$$

$$= \sqrt{\frac{(-37t)^2}{(1+37t^2)^3} + \frac{36}{(1+37t^2)^3} + \frac{1}{(1+37t^2)^3}}$$

$$= \sqrt{\frac{37^2 t^2 + 36 + 1}{(1+37t^2)^3}} \;=\; \sqrt{\frac{37(37t^2+1)}{(1+37t^2)^3}} \;=\; \sqrt{\frac{37}{(1+37t^2)^2}} \;=\; \frac{\sqrt{37}}{1+37t^2}$$

Therefore, the unit tangent vector $\mathbf{N}(t)$ is:

$$\mathbf{N}(t) = \frac{\mathbf{T}'(t)}{\|\mathbf{T}'(t)\|} = \frac{\left\langle \dfrac{-37t}{(1+37t^2)^{3/2}},\; \dfrac{6}{(1+37t^2)^{3/2}},\; \dfrac{1}{(1+37t^2)^{3/2}} \right\rangle}{\dfrac{\sqrt{37}}{1+37t^2}}$$

$$= \frac{1+37t^2}{\sqrt{37}} \left\langle \frac{-37t}{(1+37t^2)^{3/2}},\; \frac{6}{(1+37t^2)^{3/2}},\; \frac{1}{(1+37t^2)^{3/2}} \right\rangle$$

$$= \left\langle \frac{-\sqrt{37}t}{(1+37t^2)^{1/2}},\; \frac{6}{\sqrt{37}(1+37t^2)^{1/2}},\; \frac{1}{\sqrt{37}(1+37t^2)^{1/2}} \right\rangle$$

$$= \left\langle \frac{-\sqrt{37}t}{\sqrt{1+37t^2}},\; \frac{6}{\sqrt{37(1+37t^2)}},\; \frac{1}{\sqrt{37(1+37t^2)}} \right\rangle$$

To find the unit normal vector at the specified time $t = 2$, we have that:

$$\mathbf{N}(2) = \left\langle \frac{-\sqrt{37}(2)}{\sqrt{1+37(2)^2}},\; \frac{6}{\sqrt{37(1+37(2)^2)}},\; \frac{1}{\sqrt{37(1+37(2)^2)}} \right\rangle$$

$$= \left\langle \frac{-2\sqrt{37}}{\sqrt{149}},\; \frac{6}{\sqrt{5513}},\; \frac{1}{\sqrt{5513}} \right\rangle$$

(c) To find the tangential and normal components of acceleration, a_T and a_N, we use the formulas. Both formulas require the acceleration vector itself, $\mathbf{a}(t) = \mathbf{r}''(t)$. We found the velocity vector in part (a) above, which was $\mathbf{v}(t) = \mathbf{r}'(t) = \langle 1, 6t, t \rangle$. So we take the derivative once more to get the acceleration vector $\mathbf{a}(t) = \mathbf{r}''(t) = \langle 0, 6, 1 \rangle$. So, the acceleration at the specified time $t = 2$ will still be $\mathbf{a}(2) = \mathbf{r}''(2) = \langle 0, 6, 1 \rangle$. The tangential component of acceleration at the specified time $t = 2$ can now be found:

$$a_T = \mathbf{a} \cdot \mathbf{T} = \langle 0, 6, 1 \rangle \cdot \left\langle \frac{1}{\sqrt{149}}, \frac{12}{\sqrt{149}}, \frac{2}{\sqrt{149}} \right\rangle = (0)\left(\frac{1}{\sqrt{149}}\right) + (6)\left(\frac{12}{\sqrt{149}}\right) + (1)\left(\frac{2}{\sqrt{149}}\right)$$

$$= 0 + \frac{72}{\sqrt{149}} + \frac{2}{\sqrt{149}} = \frac{74}{\sqrt{149}}$$

To find the normal component of acceleration, I will use the formula $a_N = \mathbf{a} \cdot \mathbf{N}$, since I already found the unit normal vector, $\mathbf{N}(2)$. Now, we can find a_N:

$$a_N = \mathbf{a} \cdot \mathbf{N} = \langle 0, 6, 1 \rangle \cdot \left\langle \frac{-2\sqrt{37}}{\sqrt{149}}, \frac{6}{\sqrt{5513}}, \frac{1}{\sqrt{5513}} \right\rangle = 0 + \frac{36}{\sqrt{5513}} + \frac{1}{\sqrt{5513}} = \frac{37}{\sqrt{5513}} = \frac{37}{\sqrt{37}\sqrt{149}} = \frac{\sqrt{37}}{\sqrt{149}}$$

(d) We can write the acceleration vector as a linear combination of the unit tangent and unit normal vectors in the following way:

$$\mathbf{a} = a_T \mathbf{T} + a_N \mathbf{N} = \frac{74}{\sqrt{149}} \mathbf{T} + \frac{\sqrt{37}}{\sqrt{149}} \mathbf{N}$$

(e) To find a set of parametric equations for the line tangent to the space curve at the point $P\left(-1, 3, \frac{1}{2}\right)$, we first need to find out the time, t, at this point.

Since we are given the position function as $\mathbf{r}(t) = \left\langle t, 3t^2, \frac{t^2}{2} \right\rangle$, we can see that $t = -1$ at the given point P.

Now, I hope you recall that the two pieces of information needed to write the parametric equations for a line are: 1) a point on the line, and 2) a direction vector. We already have the first piece of info: the point on the line. A direction vector at this point may be found by taking the derivative of the position function; the velocity function is the direction vector, since the velocity vector is always a vector that is tangent to the line of motion. We found the velocity vector above, which was $\mathbf{v}(t) = \mathbf{r}'(t) = \langle 1, 6t, t \rangle$. At the time $t = -1$, we have the velocity vector (a.k.a. the direction vector for our line) as:

$$\mathbf{v}(-1) = \mathbf{r}'(-1) = \langle 1, 6(-1), -1 \rangle = \langle 1, -6, -1 \rangle$$

We use the formulas for parametric equations of a line in space, which are:

$$x = x_1 + at, \ y = y_1 + bt, \text{ and } z = z_1 + ct$$

...where the point on the line has coordinates (x_1, y_1, z_1)...
and the direction vector for the line is given by $\mathbf{v} = \langle a, b, c \rangle$.

Therefore, our parametric equations for our line are given by:

$$x = -1 + t, \ y = 3 - 6t, \text{ and } z = \frac{1}{2} - t \text{ (answer)}$$

(f) Since the binormal vector is a vector that is orthogonal to *both* the unit tangent and unit normal vectors, we need to find **B**. To do this, we take the cross-product $\mathbf{B} = \mathbf{N} \times \mathbf{T}$.

$$\mathbf{B} = \mathbf{T} \times \mathbf{N} = \left\langle \frac{1}{\sqrt{149}}, \frac{12}{\sqrt{149}}, \frac{2}{\sqrt{149}} \right\rangle \times \left\langle \frac{-2\sqrt{37}}{\sqrt{149}}, \frac{6}{\sqrt{5513}}, \frac{1}{\sqrt{5513}} \right\rangle$$

$$= \begin{vmatrix} \mathbf{i} & \mathbf{j} & \mathbf{k} \\ \dfrac{1}{\sqrt{149}} & \dfrac{12}{\sqrt{149}} & \dfrac{2}{\sqrt{149}} \\ \dfrac{-2\sqrt{37}}{\sqrt{149}} & \dfrac{6}{\sqrt{37}\sqrt{149}} & \dfrac{1}{\sqrt{37}\sqrt{149}} \end{vmatrix}$$

$$= \left[\frac{12}{149\sqrt{37}} - \frac{12}{149\sqrt{37}} \right]\mathbf{i} \;-\; \left[\frac{1}{149\sqrt{37}} - \frac{-4\sqrt{37}}{149} \right]\mathbf{j} \;+\; \left[\frac{6}{149\sqrt{37}} - \frac{-24\sqrt{37}}{149} \right]\mathbf{k}$$

$$= \quad 0\mathbf{i} \qquad\quad - \left[\frac{149}{149\sqrt{37}} \right]\mathbf{j} \qquad\quad + \left[\frac{6 + (24)(37)}{149\sqrt{37}} \right]\mathbf{k}$$

$$= \left\langle 0, -\frac{1}{\sqrt{37}}, \frac{(6)(149)}{149\sqrt{37}} \right\rangle = \left\langle 0, -\frac{1}{\sqrt{37}}, \frac{6}{\sqrt{37}} \right\rangle$$

EXERCISES: #15–17 from **Quiz 4**, #1–4 from **Quiz 5**, #13, 14a from **Practice Midterm 2**

2.5 ARC LENGTH AND CURVATURE

I hope you can recall that you did learn three formulas for arc length when you took second-semester calculus. In the chapter on parameterization, we found that a curve represented by the parametric equations $x = f(t)$ *and* $y = g(t)$ has an arc length on the interval $t \in [a, b]$ of $s = \int_{a}^{b} \sqrt{[f'(t)]^2 + [g'(t)]^2}\, dt$.

This is also true for a plane curve represented by a (two-dimensional) vector function, $\mathbf{r}(t) = \langle f(t), g(t) \rangle$. For space curves represented by vector functions of the form $\mathbf{r}(t) = \langle f(t), g(t), h(t) \rangle$, the arc length formula can be extended as $s = \int_{a}^{b} \sqrt{[f'(t)]^2 + [g'(t)]^2 + [h'(t)]^2}\, dt$.

Note that the integrand is merely the magnitude of the velocity function (speed). Restated, since the velocity function is $\mathbf{v}(t) = \mathbf{r}'(t) = \langle f'(t), g'(t), h'(t) \rangle$, the speed (which is the magnitude of velocity) is $\|\mathbf{v}(t)\| = \|\mathbf{r}'(t)\| = \sqrt{[f'(t)]^2 + [g'(t)]^2 + [h'(t)]^2}$.

But this is exactly the integrand in the arc length formula. So, we can also write the arc length formula as:

$$s = \int_a^b \|\mathbf{r}'(t)\| \, dt$$

The arc length, s, is often used to parameterize curves (instead of time, t). Why would we want to use s instead of t to parameterize a curve? Well, because it is useful when studying geometric properties of the curve. A lot of interesting relationships become apparent for the curve! When the arc length s is used instead, we define the arc length *function, $s(t)$,* which almost looks the same as the arc length formula. See if you can tell the difference between the arc length *formula* and the arc length *function:*

$$\text{Arc Length } Function = s(t) = \int_a^t \sqrt{[f'(u)]^2 + [g'(u)]^2 + [h'(u)]^2} \, du$$

Note that this arc length function is non-negative. The upper limit of the integral is a variable, t, and the lower limit, a, is oftentimes set equal to 0.

Fact: If the curve is represented by a vector-valued function which is parameterized using this arc length parameter—denoted $\mathbf{r}(s)$—then the magnitude of the derivative of this vector function will always be equal to 1! That is, it will always be true that:

$$\|\mathbf{v}(s)\| = \|\mathbf{r}'(s)\| = 1$$

Moreover, if it is true that $\|\mathbf{r}'(t)\| = 1$, then the parameter t is really the parameter s!

Curvature

We can measure the curvature of a curve by finding out how "curvy" it is, or how sharply a curve bends. There are three formulas for curvature (denoted K) that I want you to memorize. One involves the arc length parameter, s. The second formula for K involves the parameter t. The last curvature formula can be used for curves in the plane that can be represented as functions of x.

Here are the three formulas for curvature, K:

1) $K = \left\| \dfrac{d\mathbf{T}}{ds} \right\| = \|\mathbf{T}'(s)\|$. You need to be given (or find) the vector function representing the curve parameterized using the arc length parameter; that is, $\mathbf{r}(s)$.

2) $K = \dfrac{\|\mathbf{T}'(t)\|}{\|\mathbf{r}'(t)\|}$. You need to be given (or find) the vector function representing the curve parameterized using t; that is, $\mathbf{r}(t)$.

3) $K = \dfrac{|y''|}{\left[1 + (y')^2\right]^{3/2}}$. You need to be given (or find) the function of x representing the plane curve, $y = f(x)$.

EXAMPLE. Consider the helix represented by the vector-valued function:

$$\mathbf{r}(t) = \left\langle 4(\sin t - t \cos t),\ 4(\cos t + t \sin t),\ \frac{3}{2}t^2 \right\rangle$$

(a) Express the length of the arc s on the helix as a function of t.

(b) Solve for t in the relationship derived from part (a) and substitute the result into the original set of parametric equations. This yields a parameterization of the curve in terms of the arc length parameter, s.

(c) Find the coordinates of the point on the curve when the length of the arc is $s = 4$.

(d) Verify that $\|\mathbf{r}'(s)\| = 1$.

(e) Find the curvature K of the curve in terms of s, the arc length parameter.

SOLUTION.

(a) We will use the arc length function formula, which is:

$$s(t) = \int_a^t \sqrt{[f'(u)]^2 + [g'(u)]^2 + [h'(u)]^2}\, du$$

To do this, we first need to find the velocity function, which will be:

$$\mathbf{r}'(t) = \left\langle 4[\cos t + ((-1)\cos t - t(-\sin t))],\ 4[-\sin t + ((1)\sin t + t \cos t)],\ 3t \right\rangle$$
$$= \left\langle 4t \sin t,\ 4t \cos t,\ 3t \right\rangle$$

Notice that I needed the Product Rule to differentiate the first two components of the position function.

Now, I can plug this into the arc length function. I will assume the lower limit of integration, a, is equal to zero ($a = 0$).

$$s(t) = \int_0^t \sqrt{[4u \sin u]^2 + [4u \cos u]^2 + [3u]^2}\, du$$

$$= \int_0^t \sqrt{16u^2 \sin^2 u + 16u^2 \cos^2 u + 9u^2}\, du$$

So, the final answer is that the arc length function will look like:

$$= \int_0^t \sqrt{16u^2(\sin^2 u + \cos^2 u) + 9u^2}\, du$$

$$s = \frac{5}{2}t^2$$

$$= \int_0^t \sqrt{16u^2 + 9u^2}\, du = \int_0^t \sqrt{25u^2}\, du = \int_0^t 5u\, du = \frac{5}{2}u^2 \Big|_0^t = \frac{5}{2}t^2$$

(b) To solve for t in the equation $s = \dfrac{5}{2}t^2$, we just need to do a bit of algebra:

$$\left(\frac{2}{5}\right)s = \left(\frac{2}{5}\right)\frac{5}{2}t^2$$

$$\frac{2}{5}s = t^2$$

$$\sqrt{\frac{2}{5}s} = \sqrt{t^2}$$

$$t = \sqrt{\frac{2}{5}s}$$

Now, we only need to substitute this back into the vector-valued function in place of t:

$$\mathbf{r}(s) = \left\langle 4\left(\sin\sqrt{\frac{2s}{5}} - \sqrt{\frac{2s}{5}}\cos\sqrt{\frac{2s}{5}}\right), 4\left(\cos\sqrt{\frac{2s}{5}} + \sqrt{\frac{2s}{5}}\sin\sqrt{\frac{2s}{5}}\right), \frac{3}{2}\left(\sqrt{\frac{2s}{5}}\right)^2 \right\rangle$$

$$= \left\langle 4\left(\sin\sqrt{\frac{2s}{5}} - \sqrt{\frac{2s}{5}}\cos\sqrt{\frac{2s}{5}}\right), 4\left(\cos\sqrt{\frac{2s}{5}} + \sqrt{\frac{2s}{5}}\sin\sqrt{\frac{2s}{5}}\right), \frac{3}{5}s \right\rangle$$

This is the answer!

(c) To find the coordinates of the curve when the arc length is 4, simply substitute $s = 4$ into the vector function from part (b) above:

$$\mathbf{r}(4) = \left\langle 4\left(\sin\sqrt{\frac{2(4)}{5}} - \sqrt{\frac{2(4)}{5}}\cos\sqrt{\frac{2(4)}{5}}\right), 4\left(\cos\sqrt{\frac{2(4)}{5}} + \sqrt{\frac{2(4)}{5}}\sin\sqrt{\frac{2(4)}{5}}\right), \frac{3}{5}(4) \right\rangle$$

$$= \left\langle 4\left(\sin\sqrt{\frac{8}{5}} - \sqrt{\frac{8}{5}}\cos\sqrt{\frac{8}{5}}\right), 4\left(\cos\sqrt{\frac{8}{5}} + \sqrt{\frac{8}{5}}\sin\sqrt{\frac{8}{5}}\right), \frac{12}{5} \right\rangle$$

$$\approx \langle 2.291, 6.029, 2.4 \rangle$$

(d) To verify that $\|\mathbf{r}'(s)\| = 1$, we first need to find $\mathbf{r}'(s)$. We need to differentiate the vector-valued function found in part (b) above. Before we do, I want to point out that "radicals are evil," in that we need to rewrite the expression $\sqrt{\dfrac{2s}{5}} = \sqrt{\dfrac{2}{5}}s^{1/2}$. The derivative of *just* this expression is:

$$\frac{d\left[\sqrt{\dfrac{2}{5}}s^{1/2}\right]}{ds} = \sqrt{\frac{2}{5}}\left(\frac{1}{2}s^{-1/2}\right)$$

This means that when we use the Chain Rule and Product Rules below we will need to be sure to differentiate with respect to the variable s and we will have this last expression as a factor:

$$\frac{dx}{ds} = 4\left(\sqrt{\frac{2}{5}}\left(\frac{1}{2}s^{-1/2}\right)\cos\sqrt{\frac{2s}{5}} - \left(\sqrt{\frac{2}{5}}\left(\frac{1}{2}s^{-1/2}\right)\cos\sqrt{\frac{2s}{5}} + \sqrt{\frac{2s}{5}}\sqrt{\frac{2}{5}}\left(\frac{1}{2}s^{-1/2}\right)\left(-\sin\sqrt{\frac{2s}{5}}\right)\right)\right)$$

$$= +4\left(\frac{1}{2}s^{-1/2}\right)\frac{2}{5}\sqrt{s}\sin\sqrt{\frac{2s}{5}} \quad = \quad \frac{4}{5}\sin\sqrt{\frac{2s}{5}}$$

$$\frac{dy}{ds} = 4\left(-\sqrt{\frac{2}{5}}\left(\frac{1}{2}s^{-1/2}\right)\sin\sqrt{\frac{2s}{5}} + \sqrt{\frac{2}{5}}\left(\frac{1}{2}s^{-1/2}\right)\sin\sqrt{\frac{2s}{5}} + \sqrt{\frac{2s}{5}}\sqrt{\frac{2}{5}}\left(\frac{1}{2}s^{-1/2}\right)\cos\sqrt{\frac{2s}{5}}\right)$$

$$= 4\left(\frac{1}{2}s^{-1/2}\right)\frac{2}{5}\sqrt{s}\cos\sqrt{\frac{2s}{5}} \quad = \quad \frac{4}{5}\cos\sqrt{\frac{2s}{5}}$$

$$\frac{dz}{ds} = \frac{3}{5}$$

Thus, we have that $\mathbf{r}'(s) = \left\langle \frac{4}{5}\sin\sqrt{\frac{2s}{5}}, \ \frac{4}{5}\cos\sqrt{\frac{2s}{5}}, \ \frac{3}{5} \right\rangle$.

Now, we find the magnitude of this vector:

$$\|\mathbf{r}'(s)\| = \sqrt{\left(\frac{4}{5}\sin\sqrt{\frac{2s}{5}}\right)^2 + \left(\frac{4}{5}\cos\sqrt{\frac{2s}{5}}\right)^2 + \left(\frac{3}{5}\right)^2}$$

$$= \sqrt{\frac{16}{25}\sin^2\sqrt{\frac{2s}{5}} + \frac{16}{25}\cos^2\sqrt{\frac{2s}{5}} + \frac{9}{25}}$$

$$= \sqrt{\frac{16}{25}\left(\sin^2\sqrt{\frac{2s}{5}} + \cos^2\sqrt{\frac{2s}{5}}\right) + \frac{9}{25}}$$

We can use the Trig Identity that tells us $\sin^2\theta + \cos^2\theta = 1$ and continue simplifying the radicand:

$$\|\mathbf{r}'(s)\| = \sqrt{\frac{16}{25}(1) + \frac{9}{25}} = \sqrt{\frac{25}{25}} = 1 \text{ (as it should!)}$$

(e) To find curvature K in terms of s, we use the formula $K = \left\|\dfrac{d\mathbf{T}}{ds}\right\| = \|\mathbf{T}'(s)\|$.

To find $\mathbf{T}'(s)$, we use the formula to find the unit tangent vector $\mathbf{T}(s)$ first:

$$\mathbf{T}(s) = \frac{\mathbf{r}'(s)}{\|\mathbf{r}'(s)\|} = \frac{\left\langle \frac{4}{5}\sin\sqrt{\frac{2s}{5}},\ \frac{4}{5}\cos\sqrt{\frac{2s}{5}},\ \frac{3}{5} \right\rangle}{1} = \left\langle \frac{4}{5}\sin\sqrt{\frac{2s}{5}},\ \frac{4}{5}\cos\sqrt{\frac{2s}{5}},\ \frac{3}{5} \right\rangle$$

$$\mathbf{T}'(s) = \left\langle \frac{4}{5}\sqrt{\frac{2}{5}}\left(\frac{1}{2}s^{-1/2}\right)\cos\sqrt{\frac{2s}{5}},\ -\frac{4}{5}\sqrt{\frac{2}{5}}\left(\frac{1}{2}s^{-1/2}\right)\sin\sqrt{\frac{2s}{5}},\ 0 \right\rangle$$

$$= \left\langle \frac{2\sqrt{2}}{5\sqrt{5}\sqrt{s}}\cos\sqrt{\frac{2s}{5}},\ -\frac{2\sqrt{2}}{5\sqrt{5}\sqrt{s}}\sin\sqrt{\frac{2s}{5}},\ 0 \right\rangle$$

Now, we can evaluate curvature:

$$K = \|\mathbf{T}'(s)\| = \sqrt{\left(\frac{2\sqrt{2}}{5\sqrt{5}\sqrt{s}}\cos\sqrt{\frac{2s}{5}}\right)^2 + \left(-\frac{2\sqrt{2}}{5\sqrt{5}\sqrt{s}}\sin\sqrt{\frac{2s}{5}}\right)^2 + 0^2}$$

$$= \sqrt{\frac{(4)(2)}{(25)(5)(s)}\cos^2\sqrt{\frac{2s}{5}} + \frac{(4)(2)}{(25)(5)(s)}\sin^2\sqrt{\frac{2s}{5}}}$$

$$= \sqrt{\frac{(4)(2)}{(25)(5)(s)}\left(\cos^2\sqrt{\frac{2s}{5}} + \sin^2\sqrt{\frac{2s}{5}}\right)}$$

$$= \sqrt{\frac{(4)(2)}{(25)(5)(s)}(1)} = \frac{2\sqrt{2}}{5\sqrt{5s}}$$

This concludes the chapter!

Exercises: # 5–12 from **Quiz 5**, #14b, c, 16 from **Practice Midterm 2**

Chapter 3

Functions of Several Variables

Chapter Comments

This chapter starts off with fresh material—in it, we will introduce the concept of functions of several variables and do calculus on them. There will be no vectors or vector-valued functions here at all. (Well, almost none. We need some vectors when we obtain directional derivatives in section 3.6.) We will bring back vectors, vector operations, and vector-valued functions when we get to Chapter 5 (Vector Analysis). The good news is that the terminology in this chapter is analogous in many ways to functions of a single variable that you studied in first semester calculus. However, instead of doing calculus in two dimensions, we will be doing the calculus in three dimensions!

The first two sections are fairly easy—we will look at domains and graphs of functions of several variables, followed by limits. Graphing in three dimensions is challenging (as we have already seen), so we will discuss level curves and contour maps To facilitate visualizing the surfaces. The limits are a little trickier than for functions of a single variable. In your first semester of calculus, you had only to make sure that the left-hand and right-hand limits

agreed in order for the limit to exist (for functions of a single variable). For a limit to exist for functions of several variables, the limit must agree along *all* approaches (of which there are an infinite number). Fortunately, continuity for functions of several variables parallels that for functions of a single variable.

Sections 3 through 6 are all about differentiating functions of several variables. This process is known as partial differentiation and is a little different from regular differentiation. The notation will be different as well. We will also go over how to differentiate an implicit function of three variables, a process known as "implicit differentiation," which is a concept you learned in your first semester of calculus for differentiating implicit functions in two variables.

The last few sections involve some applications of the partial derivative. The major application in these sections is finding extrema in space. The ideas and terminology again parallel what was done in the plane in your first semester of calculus.

3.1 Introduction to Functions of Several Variables

You are already familiar with functions of a single variable, so let's start there. Functions of x can always be written in the form $y = f(x)$. More precisely, if you are given an equation with two variables and you can solve for y, then you have a function of x. When you do this, you have what is referred to as an "explicit" function, where x is the *independent* variable and y is the *dependent* variable. The graphs of *equations* in two variables give curves in the plane. The domain of a *function* includes all values of x where the function $y = f(x)$ is defined. if you have an equation of two variables and you *cannot* solve for y, then you have what is referred to as an "implicit" function, which is *not* a function of x, so you may *not* write it in the form $y = f(x)$.

Functions of *two* variables, x and y, parallel that of a single variable. Functions of two variables can always be written in the form $z = f(x, y)$. Here, you may be given an equation with *three* variables instead of two. If you can solve for z, then you have an *explicit* function of x and y. Then, x and y are the independent variables, and z is the dependent variable. If you cannot solve for z, then you have an *implicit* function, and you may *not* write it in the form $z = f(x, y)$. The graph of an *equation* in three variables is a surface in space. The graph of a function of two variables (x and y) is a surface—and is the set of all points (x, y, z) such that the projection of the surface onto the xy-plane gives the domain (OK inputs) to the function. The domain of a *function* of two variables (x and y) are all points (x, y) in the xy-plane where the function $z = f(x, y)$ is defined.

To find domains of functions of several variables, we will continue to use the procedure to find domains outlined in the previous chapter on vector-valued functions. Thus, our first step will be to automatically assume all inputs are OK—in this case, all ordered pairs in the xy-plane. Then, we will eliminate all values of the domain (in this case, ordered pairs) that yield denominators equal to zero, negative even radicands, negative (or zero) logarithm arguments, etc. There's one extra rule I should have included (but didn't) as we shall see in an example below.

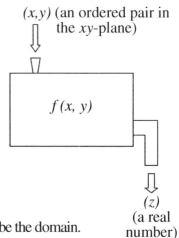

(x,y) (an ordered pair in the xy-plane)

$f(x, y)$

(z) (a real number)

EXAMPLE. Find the domain of the following functions. Use set notation to describe the domain.

(a) $\quad f(x, y) = \dfrac{\ln(x^2 - y^2 - 4)}{y}$

(b) $\quad |x + y|\sinh(x + y) - z = 0$

(c) $\quad z = \cos^{-1}(x + y)$

(d) $\quad f(x, y) = \dfrac{e^{2/x}}{x + y}$

(e) $\quad \sqrt{x^3 + y^3} = ze^{xy}$

SOLUTION.

(a) First, assume everything is OK. Then, notice we have variables in the argument of a logarithm, so we need to handle that. We want the argument of a logarithm to always be positive. So, we can write $x^2 - y^2 - 4 > 0$. We could leave it like this, but it is customary to go one step further (add 4 to both sides of the inequality) to obtain $x^2 - y^2 > 4$. Second, notice that the denominator contains a variable. We do not want the denominator to ever be equal to zero. So, we write $y \neq 0$. To put this all together in set notation to get the final answer, we write $\left\{(x, y) \mid x^2 - y^2 > 4, \ y \neq 0\right\}$.

(b) First, we need to rewrite this equation so we have an *explicit* function of x and y. Solving for z, we have that $z = f(x, y) = |x + y|\sinh(x + y)$. Let's find the domain for this function. After assuming that all inputs are OK, we notice that we do not have any "worries." That is, we have no even radicals, logarithms, or denominators with variables in them. Thus, the domain is the entire xy-plane! All ordered pairs are acceptable as inputs to this function. We write our final answer as $\{(x, y) \mid x \in \mathfrak{R}, \ y \in \mathfrak{R}\}$

(c) This is the problem where I neglected to write the extra rule into my procedure for finding domains in the last chapter. The inverse trigonometric functions of cosine and sine always require their arguments (domains) to be values between –1 and 1(inclusive). Why is this so? Well, consider the function of a *single* variable $y = \cos^{-1}(x)$. This is read: "The angle(s) y whose cosine value is x." So, what is the restriction on the domain here? Well, remember that cosine values oscillate between –1 and 1. (Think of the graph of the cosine function.) This means that we must have $-1 \le x \le 1$. So, for the problem at hand, which is finding the domain of the function of *two* variables $z = \cos^{-1}(x + y)$, we require the argument $x + y$ to be restricted to values between –1 and 1 (inclusive), and we write $-1 \le x + y \le 1$. Finally, we write the answer using set notation and we have $\{(x, y) \mid -1 \le x + y \le 1\}$

(d) This function is already written as a function of x and y, so no need to rewrite it. We have two "worries": two denominators with variables in them. The first denominator is contained in the power of the exponential function $\dfrac{2}{x}$. We do not want a denominator to be equal to zero, so we write $x \ne 0$. For the denominator of the entire quotient, we can write $x + y \ne 0$ or $x \ne -y$. Our final answer is then $\{(x, y) \mid x \ne 0, \ x \ne -y\}$

(e) First, we need to rewrite this equation as an explicit function of two variables, which means we need to solve for z.

$$\sqrt{x^3 + y^3} = ze^{xy}$$

$$z = \frac{\sqrt{x^3 + y^3}}{e^{xy}}$$

The domain of this function only requires a single restriction—concerning the even radical in the numerator. Why doesn't it require a restriction corresponding to the denominator equaling zero, since the denominator does have an expression? Well, the exponential function e^{xy} will always be a positive value, and has no restrictions for itself. Therefore, we only require that the radicand be greater than or equal to zero, and we write $x^3 + y^3 \ge 0$. In set notation, we have $\{(x, y) \mid x^3 + y^3 \ge 0\}$.

Level Curves

A graphical technique useful for describing the behavior of a function of two variables consists of sketching level curves. Level curves are often used in making topographic maps of rough terrain. To find the level curves of a function of two variables, we set the function equal to a constant, k, and examine the graphs resulting from different values for k. The view looking down from above the surface of these graphs (of the level curves) is referred to as a "contour map."

EXAMPLE. Describe the level curves for the function $f(x, y) = \sqrt{x^2 - y^2 + 5}$. A graph of this function is shown below:

(Multiple choice, pick one from below):

    (a)   The level curves are circles.

    (b)   The level curves are straight lines passing through the origin.

    (c)   The level curves are parabolas.

    (d)   The level curves are hyperbolas.

    (e)   None of the above.

SOLUTION. First, we set the function equal to a constant. Then, we simplify the resulting equation.

$$\sqrt{x^2 - y^2 + 5} = k$$
$$\left(\sqrt{x^2 - y^2 + 5}\right)^2 = (k)^2$$
$$x^2 - y^2 + 5 = k^2$$
$$x^2 - y^2 = k^2 - 5$$

Graphing this equation gives hyperbolas. A graph of a few of the level curves is shown below:

So, the answer is **(d)**. Had there been a "+" sign in between the x^2 and y^2 instead of the minus sign, then the level curves would have been circles. Had only one of the variables been squared, then the level curves would have been parabolas. If both variables had no power, then it would have been a linear equation, and the level curves would have been straight lines.

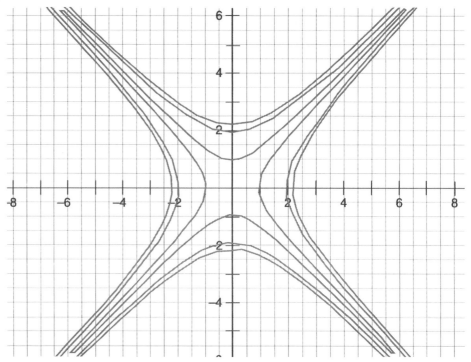

Level Surfaces

Level surfaces are the four dimensional analog to level curves for functions of three variables. I hope that you realize we cannot graph functions of three variables or more because we would need four dimensions or more (respectively). So, although we cannot visualize the graph for a function of three variables, $w = f(x, y, z)$, we can still graph its level *surfaces*.

EXAMPLE. Describe the level surfaces for the function $w = f(x, y, z) = 2y - 7z$.

(Multiple choice—pick the one that applies from below.)

    (a)   Planes parallel to the xy-plane

    (b)   Lines in the yz-plane

    (c)   Planes parallel to the z-axis

    (d)   Lines parallel to the x-axis

    (e)   None of the above

SOLUTION. We can eliminate the answers **(b)** and **(d)** immediately since our answer needs to be a surface, not a curve. We set the function equal to a constant and we get $2y - 7z = k$

The graphs of this equation for various values of the constant k would yield planes that are parallel to the x-axis. Therefore, the answer is **(e)** None of the above.

EXERCISES: #17, 18, 20, 21, from **Practice Midterm 2**, #13–19 from **Quiz 5**, #2 from **Practice Midterm 3**

3.2 LIMITS AND CONTINUITY

When taking limits for functions of two variables, you may *not* use L'Hôpital's Rule. However, you may use L'Hôpital's Rule on expressions in one variable. So, be careful!

EXAMPLE. Evaluate the limits (if they exist).

    (a) $\displaystyle \lim_{(x,y)\to(2,3)} \frac{3x - 2y}{6x^2 + 11xy - 10y^2}$

    (b) $\displaystyle \lim_{(x,y)\to(0,0)} \frac{10}{y^2 - x^2}$

    (c) $\displaystyle \lim_{(x,y)\to(\ln 5,0)} \frac{e^x \sin y}{y}$

SOLUTION.

(a)   We always try and do direct substitution first, as we have done for all limit problems in your first year of calculus classes. So, we do that here:

$$\lim_{(x,y)\to(2,3)}\frac{3x-2y}{6x^2+11xy-10y^2}=\frac{3(2)-2(3)}{6(2)^2+11(2)(3)-10(3)^2}=\frac{0}{24+66-90}=\frac{0}{0}$$

This result is an indeterminate form that we used to be able to apply L'Hôpital's Rule—but cannot as this is a function of two variables. Instead, let's try and simplify the expression.

It turns out the denominator is a factorable trinomial and we can reduce the quotient to lowest terms before re-evaluating! I will use the Master Product Method to factor the trinomial. I will require all students to be able to show me the steps to do this on their exams to receive full credit. The "Key Number" for this trinomial is the product of the lead coefficients of the two variables $(6)(-10)=-60$. Factors of -60 that add up to the middle coefficient (which is 11) are 15 and -4.

We rewrite the trinomial such that the middle term is split into two using these two factors:

$$\lim_{(x,y)\to(2,3)}\frac{3x-2y}{6x^2+11xy-10y^2}=\lim_{(x,y)\to(2,3)}\frac{3x-2y}{6x^2+15xy-4xy-10y^2}$$

Now, we factor by grouping, and then apply direct substitution to the reduced quotient:

$$=\lim_{(x,y)\to(2,3)}\frac{3x-2y}{\left(6x^2+15xy\right)+\left(-4xy-10y^2\right)}=\lim_{(x,y)\to(2,3)}\frac{3x-2y}{3x(2x+5y)-2y(2x+5y)}$$

$$=\lim_{(x,y)\to(2,3)}\frac{3x-2y}{(3x-2y)(2x+5y)}=\lim_{(x,y)\to(2,3)}\frac{1}{2x+5y}$$

$$=\lim_{(x,y)\to(2,3)}\frac{1}{2(2)+5(3)}=\frac{1}{4+15}=\frac{1}{19}$$

Therefore, the limit exists and is equal to $\dfrac{1}{19}$

(b)   We apply direct substitution for this limit to evaluate it and obtain:

$$\lim_{(x,y)\to(0,0)}\frac{10}{y^2-x^2}=\frac{10}{(0)^2-(0)^2}=\frac{10}{0}=d.n.e.$$

The final answer is: "The limit does not exist." You'll probably recall that back in second semester calculus class if we had a constant divided by zero we got infinity. This is still true, however, we will say the limit does not exist instead for functions of two variables.

(c) We first apply direct substitution and get an indeterminate form:

$$\lim_{(x,y)\to(\ln 5,0)} \frac{e^x \sin y}{y} = \frac{e^{\ln 5}\sin(0)}{0} = \frac{e^{\ln 5}(0)}{0} = \frac{0}{0}$$

Now, we are allowed to use L'Hôpital's Rule on the portion of the quotient in one variable only. But first we will use a property of limits, which tells us that the limit of a product is the same as the product of the limits:

$$\lim_{x\to c} A \cdot B = \lim_{x\to c} A \cdot \lim_{x\to c} B$$

We do this for the limit we are evaluating (and also apply L'Hôpital's Rule) to get:

$$\lim_{(x,y)\to(\ln 5,0)} e^x \cdot \lim_{(x,y)\to(\ln 5,0)} \frac{\sin y}{y} = \left(e^{\ln 5}\right)\frac{\sin(0)}{0} = (5)\frac{(0)}{0} = (5)\cdot \lim_{(x,y)\to(\ln 5,0)} \frac{\cos y}{1} = (5)\frac{\cos(0)}{1} = 5(1) = 5$$

Therefore, the limit exists and is equal to 5.

Showing that limits do not exist for other kinds of expressions can be challenging, as you will need to show that the limits do not agree on all paths approaching a point on the surface. For these kinds of problems, we will make it easy on you by providing a couple of paths of approach for the limit. Your job will be to point out that the limits for the two given approaches do not agree, and then that will be enough for you to be able to conclude: "The limit does not exist." More to the point: Getting multiple answers for a single limit means that the limit does not exist.

EXAMPLE. (Applying the 2 path test) Show that the limit $\displaystyle\lim_{(x,y)\to(0,0)} \frac{2x^2 y}{x^4 + y^2}$ does not exist by showing that the limit along two different paths gives different results. The two paths you will try are:

1) Along the parabola: $y = x^2,\ x \neq 0$.

2) Along the x-axis, which has the equation $y = 0$.

SOLUTION.

1) We end up using direct substitution, where $y = x^2$. So, instead of having $(x, y) \to (0, 0)$, we will have that $(x, x^2) \to (0, 0)$

$$\lim_{(x,x^2)\to(0,0)} \frac{2x^2 y}{x^4 + y^2} = \lim_{(x,x^2)\to(0,0)} \frac{2x^2(x^2)}{x^4 + (x^2)^2} = \lim_{(x,x^2)\to(0,0)} \frac{2x^4}{x^4 + x^4} = \lim_{(x,x^2)\to(0,0)} \frac{2x^4}{2x^4} = \lim_{(x,x^2)\to(0,0)} 1 = 1$$

Thus, along the path given, the limit is equal to 1.

2) Again, we use direct substitution, where $y = 0$. So, instead of having $(x, y) \to (0, 0)$, we will have that $(x, 0) \to (0, 0)$.

$$\lim_{(x,0)\to(0,0)} \frac{2x^2 y}{x^4 + y^2} = \lim_{(x,0)\to(0,0)} \frac{2x^2(0)}{x^4 + (0)^2} = \lim_{(x,0)\to(0,0)} \frac{0}{x^4} = 0$$

Thus, the approach along the *x*-axis gives a limit equal to zero.

Finally, we can conclude that the limit does not exist since we get two different answers along two different paths (answer).

Continuity

All you really need to notice is that a function of several variables is continuous on its domain (OK inputs). So, the answers for problems asking you to describe the points at which a function is continuous will be the same as the domain. I already did several examples above for finding domains of functions of two variables.

This concludes the section!

EXERCISES: #1–4 from **Quiz 6**, #19 from **Practice Midterm 2**, #1, 4, from **Practice Midterm 3**

3.3 PARTIAL DERIVATIVES

In this section you will learn how to differentiate functions of two variables and functions of three variables. This process is known as "partial differentiation." You will also learn how to do higher order partial derivatives.

When taking a partial derivative, it is important to treat all variables as constants, except for the one you are differentiating. So, knowing the two rules (from first semester calculus) dealing with differentiating constants is important. They are:

1) The Constant Rule: $\dfrac{d[c]}{dx} = 0$

2) The Constant Multiple Rule: $\dfrac{d[cf(x)]}{dx} = cf'(x)$

When you are taking a partial derivative, you are finding out how a function is affected by a change in one variable, while all other variables are held constant. A "real life" example might be a function of several variables used to model the economy. Suppose you are interested how the economy will be affected if taxes are raised (or cut), keeping all other variables such as unemployment rate, production rate, inflation, and so on constant, i.e., unchanged. The way you would do this is to take the partial derivative of the economy function with respect to the variable representing tax rates while treating all of the other variables as constants. The notation for partial differentiation is different from that of "regular differentiation." We use a stylized "∂" (similar to the lower-case Greek letter "δ" used in the limit definition). We use the regular lower-case "d" only when we do "regular" differentiation. I will take points away if you use incorrect notation on your exams! The definitions of partial derivatives come from limits, just as they did for your first semester of calculus.

How do you visualize a derivative for a function of two variables? What are we measuring when we take a partial derivative? We know how to visualize what a derivative gives us for a function of a single variable. You recall that a derivative gives us the slope of the tangent line to the graph of the function. The graph of a function of two variables is a surface in space. So, the two partial derivatives corresponding to this function (one for x, one for y) give the slopes of the surface at a point (x, y) in the direction of the x-axis and y-axis, respectively. A real life situation might be evaluating the steepness (i.e., slopes) of a mountain (a.k.a. surface) you are preparing to ski down in the northerly and easterly directions.

I will be testing not only your ability to do partial derivatives in this section, but your ability to recognize when to use the Product Rule, Quotient Rule, Chain Rule, Trigonometric and Inverse Trigonometric Rules, Exponential and Logarithmic Rules, etc. So, review all of your differentiation rules if you feel weak in this area!

EXAMPLE. Given the function: $z = \dfrac{x^3}{5y} + \dfrac{10y^4}{x} - 3x + 6xy\ln(x^2 + 2x + 8)$, find z_x and z_{xy}. Write final answers without negative exponents.

SOLUTION. I will first rewrite the function without quotients (using negative exponents) so that I may use the power rule when differentiating the first two terms. So, we have:

$$z = \frac{x^3}{5}y^{-1} + 10y^4x^{-1} - 3x + 6xy\ln(x^2 + 2x + 8).$$

Now, to find z_x, I need to treat the variable y as a constant. Note that the very last term will require the use of the Product Rule as well as the Logarithm Rule. Also note that we will not be using the Quotient Rule even though the first two terms were originally quotients. The reason for this is because the quotient's terms did not have *both* numerator and denominator as functions of x.

$$z_x = (3x^2)\left(\frac{y^{-1}}{5}\right) + 10(-1)x^{-2}y^4 - 3 + 6y\ln(x^2 + 2x + 8) + 6xy\left(\frac{2x+2}{x^2+2x+8}\right)$$

$$= \frac{3x^2}{5y} - \frac{10y^4}{x^2} - 3 + 6y\ln(x^2 + 2x + 8) + \frac{6xy(2x+2)}{x^2+2x+8}$$

To find the *mixed partial derivative*, z_{xy}, I need to take the derivative of z_x with respect to the variable y. z_{yx} is a second derivative. Functions of two variables have four second derivatives: Z_{xx}, Z_{xy}, Z_{yx}, and Z_{yy}. The fi*xy*t letter in the subscript tells you what variable you must differentiate with respect to the *first* derivative. Then, you use this result to take your second derivative.

Please note that the order of mixed partials is important. If they had asked for the mixed partial derivative z_{yx} instead, I would have first had to take the derivative with respect to y, then x. Of course, there is a Theorem, which states that: "A function of x and y will have that $z_{xy} = z_{yx}$ if the mixed partials themselves are continuous in an open disc."

Now, we will evaluate z_{xy}. I will use the version of z_x above that had negative exponents. Keep in mind that we will treat the variable x as a constant.

$$z_{xy} = \left(\frac{3x^2}{5}\right)(-1)y^{-2} - \left(\frac{10}{x^2}\right)(4y^3) + 6\ln(x^2 + 2x + 8) + \frac{6x(2x+2)}{x^2 + 2x + 8}$$

$$= -\frac{3x^2}{5y^2} - \frac{40y^3}{x^2} + 6\ln(x^2 + 2x + 8) + \frac{6x(2x+2)}{x^2 + 2x + 8}$$

Exercises: #5 from **Quiz 6**, #3, 6, 7 from **Practice Midterm 3**

3.4 Differentials

Recall from first semester calculus that for a function of a single variable, a differential is an approximation of a change in the function value. If you are given a function $y = f(x)$, then the actual change of the function from $x = a$ to $x = b$ (where a and b are assumed close to each other) is found by $\Delta y = f(b) - f(a)$, and the differential of the function is given by $dy = f'(x)dx$, where $dy \approx f'(a)\Delta x = f'(a)(b-a)$. The basic idea is that if the interval is small enough, the differential dy is a very good approximation for actual change Δy.

For functions of two variables $z = f(x, y)$, we have two first derivatives, $\dfrac{\partial z}{\partial x}$ and $\dfrac{\partial z}{\partial y}$. If we move from a point (x, y) to a point $(x + dx, y + dy)$ nearby, the resulting change in the function $z = f(x, y)$ is referred to as the "total differential" for the function, which has the form:

$$dz = \frac{\partial z}{\partial x}dx + \frac{\partial z}{\partial y}dy$$

This total differential will be a good approximation for the actual change in z, which can be evaluated as $\Delta z = f(x + dx, y + dy) - f(x, y)$

Two great real-life applications for differentials are measuring sensitivity to changes and error analysis, i.e., estimating propagated error involved in taking measurements.

Sensitivity to Change

EXAMPLE. The centripetal acceleration of a particle moving in a circle is $a = \dfrac{v^2}{r}$, where v is the velocity and r is the radius of the circle.

    (a)   Suppose the velocity is 2 m/s and the radius of the circle is 5 meters . How sensitive is the centripetal acceleration to small variations in the velocity and radius? That is, which variable, v or r, generates the most changes in the acceleration?

(b) Suppose that the velocity is now 2 m/s and the radius is equal to 5 meters, but that the velocity and radius are off by the amounts $dv = 0.1$ and $dr = -0.2$. Estimate the resulting absolute, relative, and percentage changes in the acceleration. Then, compare the actual change in acceleration to the absolute change in acceleration you just found. Are they close? (I hope so!)

(c) *Error Analysis*: Approximate the maximum percent errors in measuring the acceleration due to errors of 2% in v and 1% in r.

SOLUTION.

(a) As a function of velocity and radius, we have the acceleration function as $a(v,r) = \dfrac{v^2}{r}$.

The partial derivatives of the function are:

$$a_v = \frac{2v}{r} \text{ and } a_r = -\frac{v^2}{r^2}$$

(This one was found by first rewriting the acceleration function as $a = v^2(r^{-1})$ before taking the partial derivative.)

Now, we can write the total differential:

$$da = a_v dv + a_r dr = \frac{2v}{r} dv - \frac{v^2}{r^2} dr$$

We plug in $v = 2$ and $r = 5$ and get:

$$da = \frac{2(2)}{5} dv - \frac{(2)^2}{(5)^2} dr = \frac{4}{5} dv - \frac{4}{25} dr$$

Thus, a 1-unit change in v will change the acceleration, a, by about $\dfrac{4}{5}$ units and a 1-unit change in r will change a by about $-\dfrac{4}{25}$ units. So, the acceleration is almost five times more sensitive to a small change in v than it is to a small change in r. As a quality control engineer concerned with highway safety, you would want to be more concerned with driver's velocity than the radius of the road.

However, suppose we are working with different values for velocity and acceleration. Consider if we let velocity be 1m/s and the radius be 0.1 meter. The total differential da will look like:

$$da = \frac{2v}{r} dv - \frac{v^2}{r^2} dr = \frac{2(1)}{0.1} dv - \frac{(1)^2}{(0.1)^2} dr = 20dv - 100dr$$

Here, a 1-unit change in v will result in a 20-unit change in acceleration, and a 1-unit change in r will result in a -100-unit change in acceleration. The acceleration is five times more sensitive to a small change in r than it is to a small change in v.

The moral of the story here is that functions are most sensitive to changes in the variables that generate the largest partial derivatives at a given point.

(b) When we move from a given point (x_0, y_0) to a point nearby, we can describe the corresponding changes in the value of the function $z = f(x, y)$ in three different ways:

	ACTUAL	**ESTIMATE**
Absolute change	$\Delta z = \Delta f$	$dz = df$ (the total differential)
Relative change	$\dfrac{\Delta f}{f(x_0, y_0)}$	$\dfrac{df}{f(x_0, y_0)}$
Percentage change	$\dfrac{\Delta f}{f(x_0, y_0)} \times 100$	$\dfrac{df}{f(x_0, y_0)} \times 100$

To estimate absolute changes in the acceleration, we evaluate $da = \dfrac{2v}{r}\, dv - \dfrac{v^2}{r^2}\, dr$ at the given values and get:

$$da = \frac{2(2)}{5}(0.1) - \frac{(2)^2}{(5)^2}(-0.2) = \frac{4}{5}(0.1) + \frac{4}{25}(0.2) = 0.112 \, m/s^2$$

Now, we are going to find the relative changes in acceleration. We first need to evaluate the acceleration at the given point and we get that $a(2, 5) = \dfrac{2^2}{5} = \dfrac{4}{5}$. Now we form the quotient to find relative change:

$$\frac{da}{a(x_0, y_0)} = \frac{0.112}{\dfrac{4}{5}} = 0.14$$

Then we find percentage change by multiplying this value by 100 and get 14%.

The actual change in acceleration, Δa, can be found where the point (v, r) is $(2, 5)$ and the point nearby $(v + dv, r + dr) = (2 + 0.1, 5 - 0.2) = (2.1, 4.8)$. So, then we have actual change in acceleration as:

$$\Delta a = a(v + dv, r + dr) - a(v, r) = a(2.1, 4.8) - a(2, 5)$$
$$= \frac{(2.1)^2}{4.8} - \frac{(2)^2}{5}$$
$$\approx 0.119 \, m/s^2$$

Let us compare the actual change in acceleration with the total differential. How good of an approximation is da to Δa? Well the values 0.112 and 0.119 differ by approximately 0.007 (which is pretty good)!

(c) From the total differential $da = \dfrac{2v}{r}dv - \dfrac{v^2}{r^2}dr$, I hope you can see that in this case, maximum errors occurs for acceleration when the velocity error and radius error differ in sign.

Dividing the total differential (both sides of the equation) by the function a, which we know to be the same as $a = \dfrac{v^2}{r}$, we get the relative changes in a, which is $\dfrac{da}{a}$:

$$\frac{da}{a} = \frac{\dfrac{2v}{r}dv - \dfrac{v^2}{r^2}dr}{a}$$

$$= \frac{\dfrac{2v}{r}dv - \dfrac{v^2}{r^2}dr}{\dfrac{v^2}{r}}$$

$$= \left(\frac{2v}{r}dv - \frac{v^2}{r^2}dr\right)\left(\frac{r}{v^2}\right)$$

$$= \frac{2}{v}dv - \frac{1}{r}dr$$

So, if dv and dr differ in sign, the relative changes in a will be maximized.

Now, if the error in measuring velocity is 2%, this tell us that the quotient $\dfrac{dv}{v} = 0.02$. Similarly, if the error in measuring the radius is 1%, we have that $\dfrac{dr}{r} = 0.01$. For maximum error in acceleration, let us make this relative error in radius differ in sign from velocity, and we'll have that equal to –0.01 instead. We will substitute these values in to the relative change in acceleration found above:

$$\frac{da}{a} = \frac{2}{v}dv - \frac{1}{r}dr$$

$$= 2\frac{dv}{v} - \frac{dr}{r}$$

$$= 2(0.02) + 0.01 = 0.05$$

We create the maximum percentage error by multiplying this value by 100. We can conclude "The maximum error in acceleration from a 2% error measuring velocity and a 1% error measuring the radius will be 5%." (answer)

EXERCISES: #6, 7, from **Quiz 6**, #8, 9 from **Practice Midterm 3**

3.5 CHAIN RULES FOR FUNCTIONS OF SEVERAL VARIABLES

The two concepts in this section you will study are:

1) The Chain Rule for functions of several variables
2) Implicit differentiation for implicit functions involving three variables

The Chain Rule gives us a way to take the derivative of composite functions in higher dimensions. There are a couple of Theorems you need here. You will use the Theorem that is appropriate to the function given. These higher level versions of the Chain Rule are simply extensions of the Chain Rule you learned in your first semester of calculus, which basically stated is: "If $w = f(x)$ is a differentiable function of x and $x = g(t)$ is a differentiable function of t, then the derivative of w with respect to t, or $\dfrac{dw}{dt}$, can be calculated with the formula: $\dfrac{dw}{dt} = \dfrac{dw}{dx} \cdot \dfrac{dx}{dt}$."

The analogous formula for a function of two variables $w = f(x, y)$ is given by the following Theorem.

THEOREM:
Chain Rule for Functions of Two Independent Variables

If $w = f(x, y)$ is differentiable and x and y are differentiable functions of t, then w is a differentiable function of t and:

$$\frac{dw}{dt} = \frac{\partial w}{\partial x} \cdot \frac{dx}{dt} + \frac{\partial w}{\partial y} \cdot \frac{dy}{dt} \quad \text{where} \quad \frac{\partial w}{\partial x} \quad \text{and} \quad \frac{\partial w}{\partial y} \quad \text{are the partial derivatives for the function } w.$$

This theorem can be extended for functions of three variables.

THEOREM:
Chain Rule for Functions of Three Independent Variables

If $w = f(x, y, z)$ is differentiable and x, y, and z are differentiable functions of t, then w is a differentiable function of t and:

$$\frac{dw}{dt} = \frac{\partial w}{\partial x} \cdot \frac{dx}{dt} + \frac{\partial w}{\partial y} \cdot \frac{dy}{dt} + \frac{\partial w}{\partial z} \cdot \frac{dz}{dt}$$

You will need to find three partial derivatives when using this Theorem.

THEOREM:
Functions of Three Independent Variables

EXAMPLE. Find $\dfrac{dw}{dt}$ for the function of three variables $w = f(x, y, z) = xy + z$ given that $x = \cos t, \ y = \sin t, \ z = t$.

Note: You will *not* get credit for this problem if you substitute the values for x, y, and z into the function first and then differentiate with respect to t (even though this will indeed give you the correct answer). You *must* employ the Chain Rule to get full credit!

SOLUTION. We find the three partial derivatives for w first:

$$\frac{\partial w}{\partial x} = y \qquad \frac{\partial w}{\partial y} = x \qquad \frac{\partial w}{\partial z} = 1$$

Next, we find the derivatives for x, y, and z with respect to t:

$$\frac{dx}{dt} = -\sin t, \ \frac{dy}{dt} = \cos t, \ \frac{dz}{dt} = 1$$

Now we use the Chain Rule for functions of three independent variables:

$$\frac{dw}{dt} = \frac{\partial w}{\partial x} \cdot \frac{dx}{dt} + \frac{\partial w}{\partial y} \cdot \frac{dy}{dt} + \frac{\partial w}{\partial z} \cdot \frac{dz}{dt}$$
$$= (y)(-\sin t) + (x)(\cos t) + (1)(1)$$

DANGER! Do *not* let this be your final answer! Why not? Well, take a look at how many variables this result has. Since we are looking for the derivative of the function with respect to the variable t, we should only see that variable in the final answer! So, what to do next? Simple! We just "back-substitute" the given function for x, y, and z here:

$$\frac{dw}{dt} = (y)(-\sin t) + (x)(\cos t) + (1)(1)$$
$$= (\sin t)(-\sin t) + (\cos t)(\cos t) + 1$$
$$= -\sin^2 t + \cos^2 t + 1$$

We could also use a trig ID that tells us $1 - \sin^2 t = \cos^2 t$ and simplify this result further:

$$\frac{dw}{dt} = \cos^2 t + \cos^2 t = 2\cos^2 t \ \text{(answer)}$$

EXAMPLE. The length, width, and height of a rectangular box are changing at the rate of 2 meters per hour, 4.5 meters per hour, and 1.5 meters per hour respectively. Find the rate at which the surface area is changing at the instant the length is 11 meters, the width is 5 meters, and the height is 3 meters.

SOLUTION. We need to know what the surface area function is, since they did not tell us. The surface area for a rectangular box is $S.A. = 2LW + 2LH + 2WH$. This is a function of three variables. The independent variable for L, W, and H is time, t (in hours). We need to find the derivative of surface area with respect to time, t. This gives us the rate at which the surface area is changing with respect to time. We will use the Chain Rule, that tells us:

$$\frac{d[S.A.]}{dt} = \frac{\partial[S.A.]}{\partial L} \cdot \frac{dL}{dt} + \frac{\partial[S.A.]}{\partial W} \cdot \frac{dW}{dt} + \frac{\partial[S.A.]}{\partial H} \cdot \frac{dH}{dt}$$

We can find the partial derivatives, but not the regular derivatives, since we do not have separate functions for L, W, and H. We will simply plug in the values given above for those. We continue finding the rate of change of surface area by evaluating the partial derivatives:

$$\frac{d[S.A.]}{dt} = (2W + 2H) \cdot \frac{dL}{dt} + (2L + 2H) \cdot \frac{dW}{dt} + (2L + 2W) \cdot \frac{dH}{dt}$$

At this point, all we need to do are substitutions. We have that $L = 11$ m, $W = 5$ m, and $H = 3$ m.

We also have been told that:

$$\frac{dL}{dt} = 2\,m/hr, \quad \frac{dW}{dt} = 4.5\,m/hr, \quad \text{and} \quad \frac{dH}{dt} = 1.5\,m/hr$$

So, the surface area is changing at a rate of:

$$\frac{d[S.A.]}{dt} = (2(5) + 2(3)) \cdot (2) + (2(11) + 2(3)) \cdot (4.5) + (2(11) + 2(5) \cdot (1.5)$$

$$= (16)(2) + (28)(4.5) + (32)(1.5)$$

$$= 206\,m^2/hour$$

There are two more theorems for the Chain Rule for functions that are a little more complicated.

THEOREM:
Chain Rule for Functions for Two Independent Variables (r and s) and Two Intermediate Values (x and y)

If $w = f(x, y)$, $x = g(r,s)$, $y = h(r,s)$ and all three of these functions are differentiable, then w has partial derivatives with respect to r and s, given by the formulas:

$$\frac{\partial w}{\partial r} = \frac{\partial w}{\partial x} \cdot \frac{\partial x}{\partial r} + \frac{\partial w}{\partial y} \cdot \frac{\partial y}{\partial r}$$

$$\frac{\partial w}{\partial s} = \frac{\partial w}{\partial x} \cdot \frac{\partial x}{\partial s} + \frac{\partial w}{\partial y} \cdot \frac{\partial y}{\partial s}$$

You will need to find eight partial derivatives in all when using this Theorem. This theorem can be extended for functions of three intermediate values.

THEOREM:
Chain Rule for Functions for Two Independent Variables (r and s)
and Three Intermediate Values (x, y, and z)

If $w = f(x, y, z)$, $x = g(r,s)$, $y = h(r,s)$, and $z = k(r,s)$ and all four of these functions are differentiable, then w has partial derivatives with respect to r and s, given by the formulas:

$$\frac{\partial w}{\partial r} = \frac{\partial w}{\partial x} \cdot \frac{\partial x}{\partial r} + \frac{\partial w}{\partial y} \cdot \frac{\partial y}{\partial r} + \frac{\partial w}{\partial z} \cdot \frac{\partial z}{\partial r}$$

$$\frac{\partial w}{\partial s} = \frac{\partial w}{\partial x} \cdot \frac{\partial x}{\partial s} + \frac{\partial w}{\partial y} \cdot \frac{\partial y}{\partial s} + \frac{\partial w}{\partial z} \cdot \frac{\partial z}{\partial s}$$

You will need to find twelve partial derivatives in all when using this Theorem.

EXAMPLE. Given the function $w = \ln(x^2 + y^2 + z^2)$, $x = ue^v \sin u$, $y = ue^v \cos u$, $z = ue^v$

Find $\dfrac{\partial w}{\partial u}$ and $\dfrac{\partial w}{\partial v}$ as functions of u and v in two ways:

1) By using the Chain rule

2) By expressing w directly in terms of u and v before differentiating

Then, evaluate both $\dfrac{\partial w}{\partial u}$ and $\dfrac{\partial w}{\partial v}$ at the point $(u, v) = (-2, 0)$.

SOLUTION. We use the Theorem above where we need to take twelve partial derivatives in all To find the derivatives of w using the Chain Rule. In this problem, the independent variables are u and v instead of r and s as in the Theorem. So, the two partial derivatives for w have the form:

$$\frac{\partial w}{\partial u} = \frac{\partial w}{\partial x} \cdot \frac{\partial x}{\partial u} + \frac{\partial w}{\partial y} \cdot \frac{\partial y}{\partial u} + \frac{\partial w}{\partial z} \cdot \frac{\partial z}{\partial u}$$

$$\frac{\partial w}{\partial v} = \frac{\partial w}{\partial x} \cdot \frac{\partial x}{\partial v} + \frac{\partial w}{\partial y} \cdot \frac{\partial y}{\partial v} + \frac{\partial w}{\partial z} \cdot \frac{\partial z}{\partial v}$$

Now, we will actually evaluate each of these. Please observe that we need the Log Rule to find $\dfrac{\partial w}{\partial x}, \dfrac{\partial w}{\partial y}$ and $\dfrac{\partial w}{\partial z}$. We will also need the Product Rule on two of the partials $\dfrac{\partial x}{\partial u}$ and $\dfrac{\partial y}{\partial u}$.

$$\frac{\partial w}{\partial u} = \left(\frac{2x}{x^2 + y^2 + z^2} \right)\left(e^v \sin u + ue^v \cos u\right) + \left(\frac{2y}{x^2 + y^2 + z^2} \right)\left(e^v \cos u - ue^v \sin u\right) + \left(\frac{2z}{x^2 + y^2 + z^2} \right)\left(e^v\right)$$

$$\frac{\partial w}{\partial v} = \left(\frac{2x}{x^2+y^2+z^2}\right)\left(ue^v \sin u\right) + \left(\frac{2y}{x^2+y^2+z^2}\right)\left(ue^v \cos u\right) + \left(\frac{2z}{x^2+y^2+z^2}\right)\left(ue^v\right)$$

We are not done. Why not? Again, the final answers must contain one variable per derivative. So, we must "back-substitute" for the variables *x, y,* and *z*:

$$\frac{\partial w}{\partial u} = \frac{2ue^v \sin u(e^v \sin u + ue^v \cos u)}{\left(ue^v \sin u\right)^2 + \left(ue^v \cos u\right)^2 + \left(ue^v\right)^2} + \frac{2ue^v \cos u(e^v \cos u - ue^v \sin u)}{\left(ue^v \sin u\right)^2 + \left(ue^v \cos u\right)^2 + \left(ue^v\right)^2} + \frac{2ue^v(e^v)}{\left(ue^v \sin u\right)^2 + \left(ue^v \cos u\right)^2 + \left(ue^v\right)^2}$$

$$= \frac{2ue^{2v} \sin^2 u + 2u^2 e^{2v} \sin u \cos u + 2ue^{2v} \cos^2 u - 2u^2 e^{2v} \sin u \cos u + 2u\ e^{2v}}{u^2 e^{2v}(\sin^2 u + \cos^2 u + 1)}$$

$$= \frac{2ue^{2v}(\sin^2 u + \cos^2 u + 1)}{u^2 e^{2v}(1+1)} = \frac{2ue^{2v}(1+1)}{2u^2 e^{2v}} = \frac{2}{u}$$

Similarly, we have that:

$$\frac{\partial w}{\partial v} = \frac{2ue^v \sin u(ue^v \sin u)}{(ue^v \sin u)^2 + \left(ue^v \cos u\right)^2 + \left(ue^v\right)^2} + \frac{2ue^v \cos u(ue^v \cos u)}{(ue^v \sin u)^2 + \left(ue^v \cos u\right)^2 + \left(ue^v\right)^2} + \frac{2ue^v\left(ue^v\right)}{x^2+y^2+z^2}$$

$$= \frac{2u^2 e^{2v} \sin^2 u + 2u^2 e^{2v} \cos^2 u + 2u^2 e^{2v}}{u^2 e^{2v} \sin^2 u + u^2 e^{2v} \cos^2 u + u^2 e^{2v}}$$

$$= \frac{2u^2 e^{2v}(\sin^2 u + \cos^2 u + 1)}{u^2 e^{2v}(\sin^2 u + \cos^2 u + 1)} = 2$$

At the point $(u, v) = (-2, 0)$, we have $\dfrac{\partial w}{\partial u} = \dfrac{2}{-2} = -1$ and $\dfrac{\partial w}{\partial v} = 2$ (answer)

Now, we will turn our attention to implicit differentiation. Let's make sure that we all are able to distinguish between an explicit and implicit function. We need to be able to do this, since we only want to use implicit differentiation on implicit functions. Implicit functions (in two variables: *x* and *y*) are given by equations in which we cannot solve for *y*. Implicit functions (in three variables: *x, y,* and *z*) are given by equations in which we cannot solve for *z*.

Let's start with implicit functions in two variables. Consider an equation in *x* and *y*, where we cannot solve for *y*. Recall from first semester calculus that To find $\dfrac{dy}{dx}$, we took the derivative of both sides of the equation with respect to *x* (keeping in mind that we needed to use the Chain Rule on all terms involving the variable *y*), and then finally solve the equation for $\dfrac{dy}{dx}$.

Well, another strategy is the following:

1) Get the equation to be of the form $F(x, y) = 0$; get everything on one side and zero on the other.

2) Now, this equation defines y implicitly as a function of x, say $y = h(x)$.

3) Since $w = F(x, y) = 0$, it must be true that $\dfrac{dw}{dx} = 0$, right? So, let us use the Chain Rule we just studied to get that:

$$0 = \frac{dw}{dx} = F_x \cdot \frac{dx}{dx} + F_y \cdot \frac{dy}{dx}$$

$$= F_x \cdot (1) + F_y \cdot \frac{dy}{dx}$$

Now, assuming that $F_y = \dfrac{\partial w}{\partial y} \neq 0$, let us solve this equation for $\dfrac{dy}{dx}$:

$$0 = F_x \cdot (1) + F_y \cdot \frac{dy}{dx}$$

$$-F_x = F_y \cdot \frac{dy}{dx}$$

$$-\frac{F_x}{F_y} = \frac{dy}{dx}$$

And *that's* the formula for implicit differentiation!

EXAMPLE. Given the equation defining y as an implicit function of x:

$$x^3 + \sin(xy) = 2y^2 - xy + 10$$

Find $\dfrac{dy}{dx}$ in two different ways:

1) The technique you learned in first semester calculus

2) Using the formula above

SOLUTION.

1) Take the derivative of both sides of the equation:

$$\frac{d\left[x^3 + \sin(xy)\right]}{dx} = \frac{d\left[2y^2 - xy + 10\right]}{dx}$$

$$3x^2 + \cos(xy)\cdot\left(y + x\frac{dy}{dx}\right) = 4y\cdot\frac{dy}{dx} - \left(y + x\frac{dy}{dx}\right) + 0$$

$$3x^2 + y\cos(xy) + x\cos(xy)\cdot\frac{dy}{dx} = 4y\cdot\frac{dy}{dx} - y - x\frac{dy}{dx}$$

$$x\cos(xy)\cdot\frac{dy}{dx} - 4y\cdot\frac{dy}{dx} + x\frac{dy}{dx} = -3x^2 - y\cos(xy) - y$$

$$\frac{dy}{dx}\left(x\cos(xy) - 4y + x\right) = -3x^2 - y\cos(xy) - y$$

$$\frac{dy}{dx} = \frac{-3x^2 - y\cos(xy) - y}{x\cos(xy) - 4y + x}$$

2) First, get everything on one side and zero on the other to get the equation into the form:

$$F(x, y) = 0:$$

$$x^3 + \sin(xy) - 2y^2 + xy - 10 = 0$$

Now, we will use the formula above that says:

$$\frac{dy}{dx} = -\frac{F_x}{F_y}$$

To get the numerator, we take the partial derivative of $F(x, y)$ with respect to x:

$$F_x = 3x^2 + \cos(xy)\cdot y - 0 + y - 0 = 3x^2 + y\cos(xy) + y$$

The denominator is obtained by taking the derivative of $F(x, y)$ with respect to y:

$$F_y = 0 + \cos(xy)\cdot x - 4y + x - 0 = x\cos(xy) - 4y + x$$

Now, we form the quotient (remember the minus sign in front!):

$$\frac{dy}{dx} = -\frac{F_x}{F_y} = -\frac{3x^2 + y\cos(xy) + y}{x\cos(xy) - 4y + x}$$

This is the same answer as before (as it should be).

For implicit functions of three variables, we need *two* formulas instead of one. The theorem for this states that if we have an equation of the form $F(x, y, z) = 0$ that defines z as an implicit differentiable function of x and y, then the partial derivatives $\dfrac{\partial z}{\partial x}$ and $\dfrac{\partial z}{\partial y}$ are as follows:

$$\frac{\partial z}{\partial x} = -\frac{F_x(x, y, z)}{F_z(x, y, z)} \quad and \quad \frac{\partial z}{\partial y} = -\frac{F_y(x, y, z)}{F_z(x, y, z)}$$

Don't forget the minus signs!

EXAMPLE. Use the equation $z^3 + yz + y^3 = xy + 2$ to find the values of $\dfrac{\partial z}{\partial x}$ and $\dfrac{\partial z}{\partial y}$ at the point $(1, 1, 1)$.

SOLUTION. First, get everything on one side and zero on the other to get the equation into the form:

$$F(x, y, z) = 0 :$$
$$z^3 + yz + y^3 - xy - 2 = 0$$

Now, we find the partial derivatives F_x, F_y, and F_z:

$$F_x = 0 + 0 + 0 - y + 0 = -y$$
$$F_y = 0 + z + 3y^2 - x + 0 = z + 3y^2 - x$$
$$F_z = 3z^2 + y + 0 + 0 + 0 = 3z^2 + y$$

Next, we apply the formulas from above:

$$\frac{\partial z}{\partial x} = -\frac{F_x(x, y, z)}{F_z(x, y, z)} = -\frac{-y}{3z^2 + y} = \frac{y}{3z^2 + y}$$

$$\frac{\partial z}{\partial y} = -\frac{F_y(x, y, z)}{F_z(x, y, z)} = -\frac{z + 3y^2 - x}{3z^2 + y} = \frac{x - z - 3y^2}{3z^2 + y}$$

Finally, we evaluate both of these partial derivatives at the point $(1, 1, 1)$:

$$\left.\frac{\partial z}{\partial x}\right|_{(1,1,1)} = \frac{1}{3(1)^2 + 1} = \frac{1}{4}$$

$$\left.\frac{\partial z}{\partial y}\right|_{(1,1,1)} = \frac{1 - 1 - 3(1)^2}{3(1)^2 + 1} = \frac{-3}{4}$$

This concludes the section.

EXERCISES: #5, 10, 11 from **Practice Midterm 3**

3.6 DIRECTIONAL DERIVATIVES AND GRADIENTS

You already know that the partial derivatives of a function of two variables gives the slopes of the surface (a.k.a. the graph of the function of two variables) in the direction of the x and y axes. But what about slopes of the surface in other directions? Well, you can get these, and they are known as directional derivatives.

The directional derivative is best found by taking the dot product of the *gradient* of the function with the direction vector (normalized), where the *gradient* of a function of two variables is simply a vector whose components are the partial derivatives for the function, and the direction vector is a vector pointing in the direction you want the slope.

The notation for gradient of f is ∇f (read "del f"), whose form is $\nabla f(x,y) = \left\langle f_x , f_y \right\rangle$. Again, please notice that the gradient is a *vector*. Further, not only is the gradient a vector, but it is a vector in the plane for a function of two variables. So, although the graph of the function $f(x,y)$ is a *surface* in three dimensions, its gradient is a vector in two dimensions. So, how do they relate? Well, the gradient is a vector in the plane that is always orthogonal (perpendicular) to the level curves of the function $f(x,y)$. We can extend this line of thinking for functions of three variables $f(x,y,z)$. We cannot visualize the graph of a function of three variables since we would need four dimensions to do so. However, the gradient for a function of three variables is a vector of the form $\nabla f(x,y,z) = \left\langle f_x , f_y , f_z \right\rangle$ which is a vector in three dimensions. Here, the gradient will be a vector *normal* to the level surfaces for the function $f(x,y,z)$.

The directional derivative is a scalar. The notation for directional derivative in the direction of the unit vector $\mathbf{u}$ is $D_{\mathbf{u}} f = \vec{\nabla} f \cdot \mathbf{u} = \left\langle f_x , f_y \right\rangle \cdot \left\langle \cos\theta, \sin\theta \right\rangle = f_x \cos\theta + f_y \sin\theta$. A common student pitfall is that students forget to make sure that the direction vector is a unit vector. So, please remember to normalize the direction vector (if it isn't already) for every problem you do out of this section.

The last major point having to do with gradients is that functions increase most rapidly in the direction of the gradient (i.e., in the direction of ∇f) and decrease most rapidly in the direction *opposite* of the gradient, or $-\nabla f$. The actual rates of change in either of those directions will be equal to the magnitude of the gradient $|\nabla f|$, or the opposite of this $-|\nabla f|$, respectively.

EXAMPLE. Given the function $f(x,y,z) = 3xy^2 - 4yz + z^3$:

    (a)   Find the gradient of the function.

    (b)   Find the gradient of the function at the point $P(1, -2, 1)$

    (c)   Find the directional derivative for the function at the point $P(1, -2, 1)$ in the direction of $Q(-3, 2, 0)$.

    (d)   Find the directions in which the function will increase and decrease most rapidly at the point $P(1, -2, 1)$.

    (e)   Find the maximum value of the directional derivative at the point $P(1, -2, 1)$.

SOLUTION.

(a) The gradient is a vector: $\nabla f(x, y, z) = \left\langle f_x, f_y, f_z \right\rangle = \left\langle 3y^2, 6xy - 4z, -4y + 3z^2 \right\rangle$

(b) The gradient at the point $P(1, -2, 1)$ is:

$$\nabla f(1, -2, 1) = \left\langle 3(-2)^2, 6(1)(-2) - 4(1), -4(-2) + 3(1)^2 \right\rangle = \left\langle 12, -16, 11 \right\rangle$$

c) The direction vector needs to be found first. We form the vector $\overrightarrow{PQ}$ by using the technique from the very first chapter in this course. (Remember "terminal—initial?")

$$\overrightarrow{PQ} = \text{"}(-3, 2, 0) - (1, -2, 1)\text{"} = \left\langle -3 - 1, 2 - (-2), 0 - 1 \right\rangle = \left\langle -4, 4, -1 \right\rangle$$

Next, we need to make a unit vector out of this:

$$\mathbf{u} = \frac{\overrightarrow{PQ}}{\left\| \overrightarrow{PQ} \right\|} = \frac{\left\langle -4, 4, -1 \right\rangle}{\sqrt{(-4)^2 + (4)^2 + (-1)^2}} = \frac{\left\langle -4, 4, -1 \right\rangle}{\sqrt{33}} = \left\langle \frac{-4}{\sqrt{33}}, \frac{4}{\sqrt{33}}, \frac{-1}{\sqrt{33}} \right\rangle$$

The directional derivative at the point $P(1, -2, 1)$ in the direction of the unit vector $\mathbf{u}$ will simply be the dot product of the gradient at the point P and the unit vector $\mathbf{u}$.

$$\left(D_\mathbf{u} f \right)_{(1, -2, 1)} = \left\langle 12, -16, 11 \right\rangle \cdot \left\langle \frac{-4}{\sqrt{33}}, \frac{4}{\sqrt{33}}, \frac{-1}{\sqrt{33}} \right\rangle = (12)\left(\frac{-4}{\sqrt{33}} \right) + (-16)\left(\frac{4}{\sqrt{33}} \right) + (11)\left(\frac{-1}{\sqrt{33}} \right)$$

$$= \frac{-48}{\sqrt{33}} + \frac{-64}{\sqrt{33}} + \frac{-11}{\sqrt{33}} = -\frac{123}{\sqrt{33}}$$

(d) The direction of maximum increase for the function at the point P will be exactly in the direction of the gradient at the point P, which we found for the answer above in part **(b)**, which was:

$$\nabla f \big|_{(1, -2, 1)} = \left\langle 12, -16, 11 \right\rangle$$

The direction in which the function decreases most rapidly will be:

$$-\nabla f \big|_{(1, -2, 1)} = \left\langle -12, 16, -11 \right\rangle$$

(e) The maximum value of the directional derivative at the point will be equal to the magnitude of the gradient at that point:

$$\left\| \nabla f \right\|_{(1, -2, 1)} = \left\| \left\langle 12, -16, 11 \right\rangle \right\| = \sqrt{(12)^2 + (-16)^2 + (11)^2} = \sqrt{521}$$

EXERCISES: #12, 13, 14, 18 from **Practice Midterm 3**, #4 from **Quiz 8**

3.7   TANGENT PLANES AND NORMAL LINES

Mentioned in the previous section was the fact that the gradient for a function of three variables is a vector normal to the surface. We can use this fact to talk about a tangent plane to the surface. You remember talking about tangent *lines* to curves in two dimensions. Well, the analogous version of this for three dimensions is a tangent plane to a surface. Then, we can find the equation for this plane tangent to the surface at a given point on the surface. Recall from Chapter 1 that to find the equation of a plane you only need a point on the plane and a vector normal to the plane. The equation for a plane has the form:

$$a(x - x_0) + b(y - y_0) + c(z - z_0) = 0$$

…where the normal vector is the gradient evaluated at the point given:

$$\mathbf{v} = \langle a, b, c \rangle = \nabla f\big|_{(x_0, y_0, z_0)} = \langle f_x(x_0, y_0, z_0), f_y(x_0, y_0, z_0), f_z(x_0, y_0, z_0) \rangle$$

…and the point is given by:

$$P(x_0, y_0, z_0)$$

In addition, we can find the parametric equations for the normal line to the surface at the point $P(x_0, y_0, z_0)$ by using the formulas for parametric equations for a line in space:

$$x = x_0 + at, \ y = y_0 + bt, \ z = z_0 + ct,$$

…where $\mathbf{v} = \langle a, b, c \rangle = \nabla f\big|_{(x_0, y_0, z_0)} = \langle f_x(x_0, y_0, z_0), f_y(x_0, y_0, z_0), f_z(x_0, y_0, z_0) \rangle$ is the direction vector for the line (same as gradient).

The main thing to be on the lookout in this section is the format for the function and/or equation. If you are given an equation of three variables, then you need to rewrite the equation in the form $F(x, y, z) = 0$ before proceeding. However, if you are given a function of two variables $z = f(x, y)$, then subtract z from both sides of the equation:

$$0 = f(x, y) - z$$

Then you have that $F(x, y, z) = f(x, y) - z = 0$, and proceed normally. The reason behind doing all of this is because the gradient for a function of three variables is a vector that is normal to the level surfaces of the function $F(x, y, z)$. Remember, we cannot graph functions of three variables, but we can graph the level *surfaces* for a function of three variables. Recall also that to find the level surfaces for a function of three variables, we simply set the function equal to a constant $F(x, y, z) = c$. In this section, the constant c is equal to zero. Then after differentiating both sides of the equation $F(x, y, z) = 0$ and using the Chain Rule we learned from last section, we find that the derivative of $F(x, y, z)$ at a point can be written as the dot product of the gradient and the tangent vector to the curve on the surface containing the point like so:

$$\frac{d[F(x, y, z)]}{dt} = \frac{d[0]}{dt}$$

$$F_x \cdot \frac{dx}{dt} + F_y \cdot \frac{dy}{dt} + F_z \cdot \frac{dz}{dt} = 0$$

$$\langle F_x, F_y, F_z \rangle \cdot \left\langle \frac{dx}{dt}, \frac{dy}{dt}, \frac{dz}{dt} \right\rangle = 0$$

$$\nabla F \cdot \mathbf{r}'(t) = 0$$

Remember that if a dot product of two vectors is equal to zero, then they are orthogonal. In other words, we have that ∇F *and* $\mathbf{r}'(t)$ are orthogonal. Therefore, the gradient for the level surface is a vector normal to the plane tangent to the surface at the point, and the tangent vectors all lie in the tangent plane.

EXAMPLE. Find the equation of the tangent plane and the equation of the normal line to the surfaces at the given points.

    (a) $\cos(\pi x) + e^{xz} + yz = 4 + x^2 y$ at the point $P(0, 1, 2)$

    (b) $f(x, y) = \sqrt{9 - x^2 - y^2}$ at the point $P(1, 2, 2)$

SOLUTION.

    (a)   First, we get the equation in the form $F(x, y, z) = 0$:

$$\cos(\pi x) + e^{xz} + yz - 4 - x^2 y = 0$$

        Next, we find the gradient for $F(x, y, z)$:

$$\nabla F = \left\langle F_x, F_y, F_z \right\rangle = \left\langle -\pi \sin(\pi x) + ze^{xz} - 2xy, z - x^2, xe^{xz} + y \right\rangle$$

        We will evaluate this gradient at the point given and obtain the vector normal to the tangent plane:

$$\nabla F(0, 1, 2) = \left\langle -\pi \sin(\pi \cdot 0) + (2)e^{(0)(2)} - 2(0)(1), 2 - (0)^2, (0)e^{(0)(2)} + 1 \right\rangle = \left\langle 2, 2, 1 \right\rangle$$

        To get the equation of the tangent plane, we use the standard form for the equation of a plane:

$$a(x - x_0) + b(y - y_0) + c(z - z_0) = 0$$

$$2(x - 0) + 2(y - 1) + 1(z - 2) = 0$$
$$2x + 2y - 2 + z - 2 = 0$$
$$2x + 2y + z = 4$$

        Now, we can get the parametric equation for the line normal to the surface using the parametric equations:

$$x = x_0 + at, \quad y = y_0 + bt, \quad z = z_0 + ct$$

$$x = 0 + 2t, \quad y = 1 + 2t, \quad z = 2 + 1t$$
$$x = 2t, \quad y = 1 + 2t, \quad z = 2 + t$$

    (b)   The function given, which is $f(x, y) = \sqrt{9 - x^2 - y^2}$, needs to be rewritten. We first can replace the "$f(x, y)$" with "z." Then, we need to get everything on one side and zero on the other to get the equation in the form $F(x, y, z) = 0$:

$$z = \sqrt{9 - x^2 - y^2}$$

$$F(x, y, z) = \sqrt{9 - x^2 - y^2} - z = 0$$

Before I find the gradient for the function $F(x, y, z)$, I want to rewrite the expression with a radical using exponents to make it easier to differentiate (where I will use the Chain Rule).

$$F(x, y, z) = \left(9 - x^2 - y^2\right)^{1/2} - z = 0$$

Now, I will go ahead and find the gradient:

$$\nabla F = \left\langle F_x, F_y, F_z \right\rangle = \left\langle \frac{1}{2}\left(9 - x^2 - y^2\right)^{-1/2} \cdot \left(-2x\right), \ \frac{1}{2}\left(9 - x^2 - y^2\right)^{-1/2} \cdot \left(-2y\right), \ -1 \right\rangle$$

$$= \left\langle \frac{-x}{\sqrt{9 - x^2 - y^2}}, \frac{-y}{\sqrt{9 - x^2 - y^2}}, -1 \right\rangle$$

Evaluating this gradient at the point $P(1, 2, 2)$:

$$\nabla F(1, 2, 2) = \left\langle \frac{-1}{\sqrt{9 - 1^2 - 2^2}}, \frac{-2}{\sqrt{9 - 1^2 - 2^2}}, -1 \right\rangle$$

$$= \left\langle \frac{-1}{2}, -1, -1 \right\rangle$$

Now, we can find the equation for the tangent plane and the parametric equations for the normal line to the surface at the point $(1, 2, 2)$:

The equation for the tangent plane:

$$-\frac{1}{2}(x - 1) + (-1)(y - 2) + (-1)(z - 2) = 0$$

$$-\frac{1}{2}x + \frac{1}{2} - y + 2 - z + 2 = 0$$

I will multiply both sides of the equation by –2 to clear the fractions to make it easier on the eyes:

$$x - 1 + 2y - 4 + 2z - 4 = 0$$

$$x + 2y + 2z = 9$$

Finally, the parametric equations for the line normal to the surface at the point $(1, 2, 2)$:

$$x = 1 - \frac{1}{2}t, \ \ y = 2 - (1)t, \ \ z = 2 - (1)t$$

$$x = 1 - \frac{1}{2}t, \ \ \ y = 2 - t, \ \ \ z = 2 - t$$

This concludes the section.

EXERCISES: #1 from **Quiz 7**, #1, 2, 3 from **Quiz 8**, #15, 17 from **Practice Midterm 3**

3.8 EXTREMA OF FUNCTIONS OF TWO VARIABLES

This section is fun and challenging. You get to find extrema for functions of two variables. Recall from first semester calculus that finding extrema for a function of a single variable used the following procedure:

1) Take derivative of $y = f(x)$

2) Find the critical values, since extrema can only occur at critical values.

   (a) Set $f'(x) = 0$. Solve for x. Solution(s), if any, give critical values.

   (b) Set $f'(x) =$ undefined. Solve for x. Solution(s), if any, give critical values.

   If no critical values result, the function has no extrema. If function has at least one critical value, continue this procedure.

3) Perform the First Derivative Test or Second Derivative Test to determine if any of the critical values yield extrema (or points of inflection). Suppose $x = c$ is a critical value.

   (a) **For First Derivative Test:** Use the critical value(s) to divide the x-axis into test intervals. Make an interval Summary table and test an arbitrary value by evaluating the derivative at that arbitrary value for each test interval. If $f' > 0$, f is increasing on that test interval. If $f' < 0$, f is decreasing on that test interval. If the derivative changes sign, we have an extrema.

   (b) **For Second Derivative Test:** Evaluate the second derivative at each critical value. If $f''(c) > 0$, the function is concave upward and has a relative minimum at the critical value. If $f''(c) < 0$, the function is concave downward at the critical value and has a relative maximum at that value. If $f''(c) = 0$, the test is inconclusive.

4) Points of inflection can be found by solving the equation $f''(x) = 0$, and testing the intervals formed by using this solution as a divider of the x-axis.

Now, what about functions of two variables? There are some similarities in finding extrema. You will need to first take both partial derivatives for the function. Next, extrema may only occur at what is called "critical points." Critical points can be found in one of two ways:

1) Set both partial derivatives equal to zero simultaneously and solve the resulting system of equations— set $f_x = 0$ and $f_y = 0$, and find the values of x and y that make *both* of the partials equal to 0 at the same time.

2) Set one of the partials equal to undefined. If even one of the partials is undefined at a point, that yields a critical point (a.k.a. potential extrema).

   If no critical points result, the function has no extrema. If function has at least one critical point, continue this procedure. Suppose the point (a, b) is a critical point.

3) Perform the "D-Test" (a.k.a. the Second Partials Test). Evaluate the expression: $D = f_{xx} f_{yy} - [f_{xy}]^2$ at each critical point found (if any).

   (a) If $D > 0$, and $f_{xx}(a,b) > 0$, then the function has a relative minimum at the point (a, b).

   (b) If $D > 0$, and $f_{xx}(a,b) < 0$, then the function has a relative maximum at the point (a, b).

   (c) If $D < 0$, then the point (a, b) is a saddle point.

(d)   The test is inconclusive if $D = 0$.

The major obstacle students run into is solving the system of equations $f_x = 0$ and $f_y = 0$. One of three "systems" occur, and I test all three on my exams:

1)   The system of equations may not even be a genuine system in that one of the equations has the x-variable only in it, and the other equation has only the y-variable. So, whatever values you get form the critical point(s). And yes, it's possible to get more than one. For example, suppose you were trying to locate the critical points by solving the system:

$$f_x = x^4 - 1 = 0 \text{ and } f_y = y^3 + 8 = 0$$

To solve, you would find that:

$$x^4 - 1 = 0 \text{ and } y^3 + 8 = 0$$
$$x^4 = 1 \qquad\qquad y^3 = -8$$
$$\sqrt[4]{x^4} = \pm\sqrt[4]{1} \qquad \sqrt[3]{y^3} = \sqrt[3]{-8}$$
$$x = \pm 1 \qquad\qquad y = -2$$

Please note the Square Root Property applies to all even roots (only!).

So, what exactly are the critical points here? There are two of them: (1, –2) and (–1, –2).

2)   The second kind of system you may run into is a linear system of two equations (and two unknowns). These are easy to solve, especially since you solved a lot of these in your algebra days.

3)   The third kind of system is the most challenging to students. It is a non-linear system of equations. Most students may have had a glimpse of these in their pre-calculus class, but forgot the techniques involved. Most of the time the Substitution Method is employed to solve these kinds of systems. The test-bank exercises give many opportunities for you to try your skill at this.

EXAMPLES.   Find the extrema and saddle points, if any, for the following functions. Give all answers as ordered triples to get full credit.

(a) $f(x, y) = 5x^2 + 4xy - 2y^2 + 4x - 4y$

(b) $f(x, y) = 2x^3 + 3xy + 2y^3$

(c) $f(x, y) = x^4 + y^4 + 4xy$

(d) $f(x, y) = 3x^3 + 3x^2 + \dfrac{2}{5}y^5 - 8x - 32y + 7$

(e) $f(x, y) = x^2 y^2$

SOLUTIONS.

(a) First, we get the first partials $f_x(x, y) = 10x + 4y + 4$ *and* $f_y(x, y) = 4x - 4y - 4$.

Then, we simultaneously set them equal to zero and obtain a linear system of equations:

$$\begin{cases} 10x + 4y + 4 = 0 \\ 4x - 4y - 4 = 0 \end{cases} \qquad \Longrightarrow \qquad \begin{cases} 10x + 4y = -4 \\ 4x - 4y = 4 \end{cases}$$

We solve using the method of elimination. Adding the two equations together eliminates the y-variable and we get $14x = 0 \Rightarrow x = 0$. Substitute this back into either of the original equations to get the corresponding y-coordinate. I'll use the first equation: $10(0) + 4y + 4 = 0 \Rightarrow y = -1$.

Therefore, the solution to the system is the point $(0, -1)$. This is also the critical point. Extrema and saddle points may occur only at critical points. To determine if this critical point gives an extremum or saddle point, we need to perform the D-Test. We first form the expression for D:

$$D = (f_{xx})(f_{yy}) - (f_{xy})^2 = (10)(-4) - (4)^2 = -40 - 16 = -56 < 0.$$

When D is negative, the point must be a saddle point.

Final answer: We have a saddle point at $(0, -1, 2)$.

(b) First, we get the first partials: $f_x(x, y) = 6x^2 + 3y$ *and* $f_y(x, y) = 3x + 6y^2$.

Then, we simultaneously set them equal to zero and obtain a non-linear system of equations:

$$\begin{cases} 6x^2 + 3y = 0 \\ 3x + 6y^2 = 0 \end{cases}$$

To solve the system, we use the method of substitution. This method involves solving for one of the variables using one equation, and then substituting this into the other equation to eliminate a variable. I will solve for y using the first equation to get $3y = -6x^2 \Rightarrow y = -2x^2$. Substitute this into the second equation for y and get:

$$3x + 6(-2x^2)^2 = 0$$
$$3x + 6(4x^4) = 0$$
$$3x + 24x^4 = 0$$

This is a third degree polynomial equation. To solve, we will factor and apply the ZPP (Zero Product Property):

$$3x(1 + 8x^3) = 0$$

$$3x = 0 \quad or \quad 1 + 8x^3 = 0$$
$$x = 0 \qquad\qquad 8x^3 = -1$$
$$x^3 = -\frac{1}{8}$$
$$x = \sqrt[3]{-\frac{1}{8}}$$
$$x = -\frac{1}{2}$$

Since we have two values for x, we need to find the two corresponding y-values. I will separately substitute each of these into the first equation of our non-linear system to find these y-coordinates:

$$6(0)^2 + 3y = 0 \Rightarrow y = 0$$

So, the point $(0, 0)$ is a solution to the system.

$$6\left(-\frac{1}{2}\right)^2 + 3y = 0 \Rightarrow 6\left(\frac{1}{4}\right) + 3y = 0 \Rightarrow 3y = -\frac{3}{2} \Rightarrow y = -\frac{3}{2}\left(\frac{1}{3}\right) = -\frac{1}{2}$$

Therefore, the point $\left(-\frac{1}{2}, -\frac{1}{2}\right)$ is also a solution to the system.

The two critical points we need to perform the D-Test on are $(0, 0)$ and $\left(-\frac{1}{2}, -\frac{1}{2}\right)$
We form the expression for D:

$$D = \left(f_{xx}\right)\left(f_{yy}\right) - \left(f_{xy}\right)^2 = (12x)(12y) - (3)^2 = 144xy - 9$$

We test the point $(0, 0)$ first:

$$D(0, 0) = 144(0)(0) - 9 = -9 < 0$$

When D is negative, we have a saddle point. Since $f(0,0) = 0$, the point $(0, 0, 0)$ is a saddle point.

Next, we test the point $\left(-\frac{1}{2}, -\frac{1}{2}\right)$:

$$D\left(-\frac{1}{2}, -\frac{1}{2}\right) = 144\left(-\frac{1}{2}\right)\left(-\frac{1}{2}\right) - 9 = 36 - 9 = 27 > 0$$

When D is positive, we may have either a minimum or a maximum. We have to test $f_{xx}(x, y)$ at the critical point to determine what type of extremum we have:

$$f_{xx}\left(-\frac{1}{2}, -\frac{1}{2}\right) = 12\left(-\frac{1}{2}\right) = -6 < 0$$

Since D is positive and f_{xx} is negative at our critical point, we have a relative maximum. Since $f\left(-\dfrac{1}{2}, -\dfrac{1}{2}\right) = \dfrac{1}{4}$, the point $\left(-\dfrac{1}{2}, -\dfrac{1}{2}, \dfrac{1}{4}\right)$ is a relative maximum.

Final answer: Relative maximum at the point $\left(-\dfrac{1}{2}, -\dfrac{1}{2}, \dfrac{1}{4}\right)$ and saddle point at $(0, 0, 0)$.

(c) Given the function $f(x, y) = x^4 + y^4 + 4xy$, we find its first partials:

$$f_x(x, y) = 4x^3 + 4y \quad and \quad f_y(x, y) = 4x + 4y^3$$

Then, we simultaneously set them equal to zero and obtain a non-linear system of equations:

$$\begin{cases} 4x^3 + 4y = 0 \\ 4x + 4y^3 = 0 \end{cases}$$

To solve the system, we use the method of substitution. I will solve for y using the first equation to get:

$$4y = -4x^3 \Rightarrow y = -x^3$$

Substitute this into the second equation for y and get:

$$4x + 4\left(-x^3\right)^3 = 0 \Rightarrow 4x - 4x^9 = 0 \Rightarrow 4x(1 - x^8) = 0$$

Apply the ZPP to solve:

$$4x = 0 \quad or \quad 1 - x^8 = 0$$

This gives that $x = 0$ and/or:

$$1 = x^8 \Rightarrow \pm\sqrt[8]{1} = x \Rightarrow x = \pm 1$$

Remember the "Square Root Property" from Algebra class? If not, please remember that anytime you need to take an *even* root of both sides of an equation, you *must* put the "±" symbol on one side of the equation! I did that in that last step. Now, we have three answers for the x-coordinate 0, 1, and –1. To get the corresponding y-values, use the fact that we solved for y: $y = -x^3$. So, we have that when $x = 0$, $y = 0$ also, when $x = 1$, $y = -1$, and when $x = -1$, $y = 1$. So, the solutions to our non-linear system (as well as the critical points) are $(0, 0)$, $(1, -1)$, and $(-1, 1)$.

We form the expression for D:

$$D = \left(f_{xx}\right)\left(f_{yy}\right) - \left(f_{xy}\right)^2 = (12x^2)(12y^2) - (4)^2 = 144x^2y^2 - 16$$

We test the point $(0, 0)$ first:

$$D(0,0) = 144(0)^2(0)^2 - 16 = -16 < 0$$

Since D is negative, we have a saddle point. Since $f(0,0) = 0$, we have a saddle point at $(0, 0, 0)$.

We test the point $(1, -1)$:

$$D(1, -1) = 144(1)^2(-1)^2 - 16 = 128 > 0$$

Since D is positive, we need to evaluate $f_{xx}(x, y)$ at this critical point to determine the type of extremum:

$$f_{xx}(1,-1) = 12(1)^2 = 12 > 0$$

Since both D and $f_{xx}(x, y)$ are positive, we have a local minimum.

Since $f(1, -1) = -2$, we have a relative minimum at the point $(1, -1, -2)$.

Finally, we test the point $(-1, 1)$:

$$D(-1, 1) = 144(-1)^2(1)^2 - 16 = 128 > 0$$

Since D is positive, we need to evaluate $f_{xx}(x, y)$ at this critical point to determine the type of extremum:

$$f_{xx}(-1,1) = 12(-1)^2 = 12 > 0$$

Since both D and $f_{xx}(x, y)$ are positive, we have a local minimum.

Since $f(-1, 1) = -2$, we have another relative minimum at the point $(-1, 1, -2)$.

Final answer: Saddle point at $(0, 0, 0)$ and two relative minimums at the points $(-1, 1, -2)$ and $(1, -1, -2)$.

(d)   For the function $f(x, y) = 3x^3 + 3x^2 + \dfrac{2}{5}y^5 - 8x - 32y + 7$, we find the first partials:

$$f_x(x, y) = 9x^2 + 6x - 8 \quad and \quad f_y(x, y) = 2y^4 - 32y$$

Then, we simultaneously set them equal to zero and obtain a non-linear system of equations:

$$\begin{cases} 9x^2 + 6x - 8 = 0 \\ 2y^4 - 32 = 0 \end{cases}$$

BE CAREFUL! This is not a genuine system of equations, since the first equation has only the variable "x" and the second equation has only the variable "y." We will solve each equation separately, and then we will "mix and match" all of the answers obtained to form the critical points for the function. We solve the first equation:

$$9x^2 + 6x - 8 = 0 \implies 9x^2 + 12x - 6x - 8 = 0 \implies 3x(3x + 4) - 2(3x + 4) = 0 \implies (3x + 4)(3x - 2) = 0$$

I used the Master Product Method (also known as the Key Number Method) to factor the trinomial. The Key Number is found by multiplying the lead coefficient with the constant $(9)(-8) = -72$. The factors of this Key Number that sum up to the middle coefficient (that is 6 here in this trinomial) are 12 and –6. We use these factors to rewrite the middle term as a sum of two terms. Finally, we factor by grouping. Now, I apply the ZPP to the factored form of the equation to solve for x:

$$3x + 4 = 0 \quad or \quad 3x - 2 = 0$$

$$x = -\frac{4}{3} \quad or \quad x = \frac{2}{3}$$

Next, we find the solutions for the second equation:

$$2y^4 - 32 = 0$$

$$2(y^4 - 16) = 0$$

$$y^4 = 16$$

$$y = \pm\sqrt[4]{16} = \pm 2$$

So, we do the "mix and match" for all the possible combinations of x and y to form the critical points:

$$\left(-\frac{4}{3}, 2\right), \left(-\frac{4}{3}, -2\right), \left(\frac{2}{3}, 2\right), \left(\frac{2}{3}, -2\right)$$

We form the expression for D:

$$D = \left(f_{xx}\right)\left(f_{yy}\right) - \left(f_{xy}\right)^2 = (18x + 6)(8y^3 - 32) - (0)^2$$

We evaluate D at each of the critical points to test for extrema:

$$D\left(-\frac{4}{3}, 2\right) = \left[18\left(-\frac{4}{3}\right) + 6\right]\left[8(2)^3 - 32\right] = -576 < 0. \text{ We have a saddle point.}$$

Since $f\left(-\frac{4}{3}, 2\right) = \frac{-1589}{45}$, we have a saddle point at $\left(-\frac{4}{3}, 2, -\frac{1589}{45}\right)$.

We test the second critical point:

$$D\left(-\frac{4}{3}, -2\right) = \left[18\left(-\frac{4}{3}\right) + 6\right]\left[8(-2)^3 - 32\right] = 1728 > 0$$

Since D is positive, we must evaluate f_{xx} at this point to determine if we have a minimum or maximum:

$$f_{xx}\left(-\frac{4}{3}, -2\right) = 18\left(-\frac{4}{3}\right) + 6 = -18 < 0$$

Since f_{xx} is negative, we have a relative maximum at the point:

$$\left(-\frac{4}{3}, -2, \frac{3019}{45}\right)$$

We test the third critical point:

$$D\left(\frac{2}{3}, 2\right) = \left[18\left(\frac{2}{3}\right) + 6\right]\left[8(2)^3 - 32\right] = 576 > 0$$

Since D is positive, we must evaluate f_{xx} at this point to determine if we have a minimum or maximum:

$$f_{xx}\left(\frac{2}{3}, 2\right) = 18\left(\frac{2}{3}\right) + 6 = 18 > 0$$

Since *both* D and f_{xx} are positive, we have a relative minimum at the point:

$$\left(\frac{2}{3}, 2, -\frac{2129}{45}\right)$$

We test the final critical point:

$$D\left(\frac{2}{3}, -2\right) = \left[18\left(\frac{2}{3}\right) + 6\right]\left[8(-2)^3 - 32\right] = -1728 < 0$$

Since D is negative, we have *another* saddle point at:

$$\left(\frac{2}{3}, -2, \frac{2479}{45}\right)$$

Final answer: *two* saddle points located at:

$$\left(\frac{2}{3}, -2, \frac{2479}{45}\right) \text{ and } \left(-\frac{4}{3}, 2, -\frac{1589}{45}\right)$$

A relative minimum at:

$$\left(\frac{2}{3}, 2, -\frac{2129}{45}\right)$$

A relative maximum at:

$$\left(-\frac{4}{3}, -2, \frac{3019}{45}\right)$$

(e) The first partials for the function $f(x, y) = x^2 y^2$ are:

$$f_x(x, y) = 2xy^2 \quad and \quad f_y(x, y) = 2x^2 y$$

Setting both of these equal to zero simultaneously gives us the system:

$$\begin{cases} f_x(x, y) = 2xy^2 = 0 \\ f_y(x, y) = 2x^2 y = 0 \end{cases}$$

The only critical point we get is when $x = 0$ and $y = 0$. In other words, there are an infinite number of critical points that "live" on both of the x and y axes. Unfortunately, the D-Test is inconclusive here because:

$$D = (f_{xx})(f_{yy}) - (f_{xy})^2 = (2y^2)(2x^2) - (4xy)^2 = 4x^2 y^2 - 16x^2 y^2 = 0 \text{ when either } x = 0 \text{ OR } y = 0.$$

To figure out what is going on, we need to think about the function itself. Because the function $f(x, y) = x^2 y^2$ is never negative, we can determine that all of the critical points must be at a minimum. In fact, it is the absolute minimum for all regions! *(answer)*.

This concludes the concept of finding relative extrema and saddle points for functions of two variables!

Finding Absolute Extrema

You recall (from first semester calculus) when you needed to find absolute extrema for functions of a single variable: You needed to test the endpoints of the closed interval (which was given) as well as the critical values. In addition, the absolute extrema theorem guarantees an absolute maximum and an absolute minimum; the answer will never be "no extrema" for an absolute extrema problem (only for relative extrema problem is that answer possible). To find absolute extrema for a function of two variables, you not only have to find the critical points and test those out, but you need to evaluate the *border,* or boundary, of the region given. This border is analogous to the endpoints for the closed interval from first semester calculus.

EXAMPLE. Find the absolute extrema for the function $f(x, y) = x^2 + 2xy + y^2$ on the region given by:

$$R = \left\{ (x, y) \mid |x| \le 2, |y| \le 1 \right\}$$

SOLUTION. First, we had better get a handle on what this region looks like. It is a rectangle, where $-2 \le x \le 2$ and $-1 \le y \le 1$. (These are the solutions to the absolute value equations contained in the set for R.) We will need to test the boundary (border) values for this rectangle soon. These borders need corresponding equations, that are $x = 2$, $x = -2$, $y = 1$, and $y = -1$.

x	y	$f(x, y)$

I advise making a table for all absolute extreme problems that you encounter. Your table should start like this: ⇨

You will fill this table in as you go. The absolute maximum will be the largest value you'll find for $f(x, y)$, and the absolute minimum will be the smallest value you'll find for $f(x, y)$. Remember, although you may have a "tie" for both of these, you need at least one of each for your final answer!

Now, let's get the partials and find the critical point(s):

$$f_x = 2x + 2y = 0$$
$$f_y = 2x + 2y = 0$$

Both partials give the line $y = -x$; there are an infinite number of critical points and they all lie on the line $y = -x$. Keep in mind that we only need to consider the portion of this line that lies within our rectangle, R. Now, let's test these critical points. For the line $y = -x$, we have the function:

$$f(x, y) = f(x, -x) = x^2 + 2x(-x) + (-x)^2 = x^2 - 2x^2 + x^2 = 0$$

We use this newfound info to fill in our table:

x	y	$f(x, y)$
x	$-x$	0

Next, let's test one of the borders of our region, the rectangle. Along the line $y = 1$, we have that x stays within the range $-2 \le x \le 2$. Substituting $y = 1$ into our function, we have that:

$$f(x,1) = f(x, 1) = x^2 + 2x(1) + (1)^2 = x^2 + 2x + 1$$

We have a function of a single variable. Let's use our knowledge from first semester calculus, and take the derivative for this function and examine the resulting critical value(s), if any:

$$f'(x) = 2x + 2$$
$$f'(x) = 0??$$
$$2x + 2 = 0$$
$$x = -1$$

So, we need to take a look at the point $(-1, 1)$. (Remember, we are still on the border, that is, the line $y = 1$.) Plug this into our table:

x	y	$f(x, y)$
x	$-x$	0
-1	1	0

Before we leave this line $y = 1$, we need to examine the endpoints—on the range of $-2 \le x \le 2$. Evaluate the function at the points $(-2, 1)$ and $(2, 1)$, and put these into our table:

x	y	$f(x, y)$
x	$-x$	**0**
-1	1	**0**
-2	1	**1**
2	1	**9**

So, far, it appears that our absolute maximum will be the value of 9 at the point $(2, 1)$, and the absolute minimum will be the value of 0 at all points that lie on the line $y = -x$. But, we are not done yet! We need to evaluate the other three boundaries for the rectangle—one by one.

Let us take a look at the line $y = -1$. Again, the range for x will be $-2 \le x \le 2$. Substituting $y = -1$ into our function, we have that:

$$f(x, -1) = f(x, -1) = x^2 + 2x(-1) + (-1)^2 = x^2 - 2x + 1$$

We (again) have a function of a single variable. Let's use our knowledge from first semester calculus, and take the derivative for this function and examine the resulting critical value(s), if any.

$$f'(x) = 2x - 2$$
$$f'(x) = 0\,??$$
$$2x - 2 = 0$$
$$x = 1$$

-1	1	**0**
-2	1	**1**
2	1	**9**
1	-1	**0**
x	y	$f(x, y)$
x	$-x$	**0**

So, we need to take a look at the point $(1, -1)$. (Remember, we are still on the border—that is, the line $y = -1$.) Plug this into our table.

This point did not give us any new values for the function. That is, we already have the function equaling the value of 0; we still have that the absolute minimum occurs on the line $y = -x$. Let's look at the "endpoints" again for the line $y = -1$ that will be $(-2, -1)$ and $(2, -1)$.

Plug these into our table.

x	y	$f(x, y)$
x	$-x$	**0**
-1	1	**0**
-2	1	**1**
2	1	**9**
1	-1	**0**
-2	-1	**9**
2	-1	**1**

Wow! We have two points that are "tied" for absolute maximum Both the points $(-2, -1)$ and $(2, 1)$ give a value of 9 for the function—the largest value so far! We still have to check out the lines $x = 2$ and $x = -2$ on the border of the rectangle.

Let's start with the line $x = 2$. The range for y will be $-1 \le y \le 1$. Substituting $x = 2$ into our function, we have that:

$$f(x, y) = f(2, y) = 2^2 + 2(2)(y) + y^2 = 4 + 4y + y^2$$

We (again) have a function of a single variable. Let's use our knowledge from first semester calculus, and take the derivative for this function and examine the resulting critical value(s), if any.

$$f'(y) = 4 + 2y$$
$$f'(y) = 0??$$
$$4 + 2y = 0$$
$$y = -2$$

We will *not* check out the point $(2, -2)$. Why not? Well, this point lies outside the region altogether. (Graph it and see.) We only need to worry about points that lie within the rectangle given. And we already checked the points on the edges of this line, which are $(2, 1)$ and $(2, -1)$. They are already contained within our table, so we do not need to check them again.

Let's move on to the line $x = -2$. The range for y will still be $-1 \le y \le 1$. Substituting $x = -2$ into our function, we have that:

$$f(x, y) = f(-2, y) = (-2)^2 + 2(-2)(y) + y^2 = 4 - 4y + y^2$$

We (again) have a function of a single variable. Let's use our knowledge from first semester calculus, and take the derivative for this function and examine the resulting critical value(s), if any.

$$f'(y) = -4 + 2y$$
$$f'(y) = 0??$$
$$-4 + 2y = 0$$
$$y = 2$$

Same problem as before: The point $(-2, 2)$ lies outside of our rectangle.

So, the final answer for this problem is the following:

"The absolute maximum is 9 and occurs at the points $(-2, -1)$ and $(2, 1)$.

"The absolute minimum is 0 and occurs at the points that lie on the line $y = -x$ contained within the rectangle."

EXERCISES: #5–12 all from **Quiz 8**, #1–4 from **Quiz 9**, #19–25 all from **Practice Midterm 3**

SECTION 3.9   LAGRANGE MULTIPLIERS

EXAMPLE. Suppose you were asked to find the dimensions of a rectangle of maximum area that can be inscribed inside of an ellipse given by: $\dfrac{x^2}{4^2} + \dfrac{y^2}{3^2} = 1$. Let the point (x,y) be the vertex of the rectangle in the function $f(x,y) = (width)(length) = (2x)(2y) = 4xy$, as shown in the diagram.

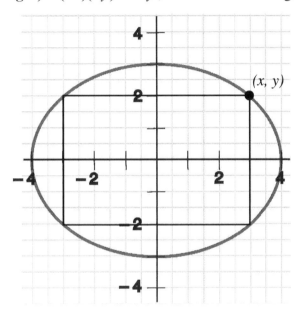

You want to find x and y such that $f(x,y)$ is a maximum, subject to the constraint that both x and y must lie on the boundary of the specified ellipse.

One way to solve this problem using Calculus I techniques would be to solve the constraint equation for y: $y = \dfrac{3}{4}\sqrt{16 - x^2}$, then substitute this value for y into the function $f(x,y)$ to get a function of a single variable $f(x) = 4x\left(\dfrac{3}{4}\sqrt{16 - x^2}\right)$.

Now, find the maximum of this function by setting the derivative equal to zero, etc., etc., the way you learned to do in your first year of calculus.

Because this problem is what's known as a "constrained optimization problem," we can solve it using another method that involves the use of variables called Lagrange Multipliers. (This method is named after Joseph-Louis Lagrange, a French mathematician who figured this all out at the tender age of 19.)

The Method of Lagrange Multipliers uses the following procedure:

1) If $f(x,y)$ has a maximum (or a minimum) subject to the constraint $g(x,y) = 0$, then this maximum or minimum will occur at one of the critical numbers of a *new* function defined by $L(x,y,\lambda) = f(x,y) - \lambda g(x,y)$. We will refer to the function $L(x,y,\lambda)$ as the Lagrange function in this book. The scalar multiple, λ (lambda), is known as the ***Lagrange Multiplier.***

Note: Be sure and rewrite your constraint equation so that it has the form $g(x, y) = 0$ before forming the Lagrange function.

2) Find all of the first partial derivatives for your Lagrange function.

3) Set all of the first partial derivatives equal to 0, and solve the system of equations. At this point, you will have a system resembling the following:

Call this system *:
$$\begin{cases} L_x = f_x(x, y) - \lambda g_x(x, y) = 0 \\ L_y = f_y(x, y) - \lambda g_y(x, y) = 0 \\ L_\lambda = 0 \quad - \quad g(x, y) = 0 \end{cases}$$

A few comments are in order at this moment. Notice that the very last first partial equation, $L_\lambda = 0$, can be rewritten so that it is simply a copy of the constraint equation given for the original problem. This will always be the case. Also, notice that if we rewrite the first and second equations by adding the term involving the Lagrange Multiplier to both sides our system of equations will now have the form:

Call this system **:
$$\begin{cases} f_x(x, y) = \lambda g_x(x, y) \\ f_y(x, y) = \lambda g_y(x, y) \\ g(x, y) = 0 \end{cases}$$

Another way to think about the first two equations is to recall our knowledge of gradients:

$$\nabla f(x, y) = \langle f_x(x, y), \ f_y(x, y) \rangle$$
$$\nabla g(x, y) = \langle g_x(x, y), \ g_y(x, y) \rangle$$

Now, what the first two equations of our system are saying is that the gradient of the function f (a vector) is simply equal to a scalar multiple of the gradient of the function g (another vector), where the scalar is the Lagrange Multpilier, λ. In other words: $\nabla f(x, y) = \lambda \nabla g(x, y)$.

Do you remember back in the first chapter of this book that when a vector is a scalar multiple of the other, it means the two vectors must be parallel? Well, they are! We desire the gradient of the function f (vector that is orthogonal to all of the level curves of f, remember?) to be parallel to the gradient of the function g (a vector orthogonal to the fixed curve given by the constraint equation $g(x, y)$). In other words: $\nabla f(x, y) \| \nabla g(x, y)$.

So, whether you solve the system above denoted *, or you solve the other system above denoted **, you are doing the same thing. The reason why I am mentioning *both* forms for the system is that different textbooks give one or the other, but rarely discuss both as being equivalent.

Geometrically speaking, we are maximizing $f(x, y)$ by finding the point at which one of the level curves for $f(x, y)$ is tangent to the curve given by the constraint function. To find the appropriate level curve that will be tangent, we use the fact that two curves are tangent at a point if and only if their velocity vectors are parallel at that point. But if a curve on the surface of $f(x, y)$ is given by a vector-valued function $\mathbf{r}(t) = \langle x(t), y(t) \rangle$, then for $f(x, y) = f(x(t), y(t))$ to have an extrema at a point on the curve, we would require that the derivative needs to be equal to zero at that point, i.e.:

$$\frac{d[f(x(t), y(t))]}{dt} = 0$$

Using the Chain Rule for Several Variables in a previous section we have that:

$$\frac{d[f(x(t),\, y(t))]}{dt} = \frac{\partial f}{\partial x} \cdot \frac{dx}{dt} + \frac{\partial f}{\partial y} \cdot \frac{dy}{dt} = 0$$

Notice that the sum of two products is really the dot product of two vectors, one of which is the gradient of $f(x, y)$ and the other is the velocity vector of $\mathbf{r}(t)$. That is:

$$\frac{d[f(x(t),\, y(t))]}{dt} = \frac{\partial f}{\partial x} \cdot \frac{dx}{dt} + \frac{\partial f}{\partial y} \cdot \frac{dy}{dt} = \left\langle \frac{\partial f}{\partial x}, \frac{\partial f}{\partial y} \right\rangle \cdot \left\langle \frac{dx}{dt}, \frac{dy}{dt} \right\rangle = \nabla f(x, y) \cdot \mathbf{r}'(t) = 0$$

But recall from Chapter 1 that if two vectors have a dot product equal to zero, then those two vectors are orthogonal! So, the gradient of $f(x, y)$ is orthogonal to the velocity vector. Similarly, the gradient of $g(x, y)$ will also be orthogonal to the velocity vector that is tangent to the curve given by the constraint equation. Since we are interested in the point at which the curves are tangent to each other (i.e., just touching) at which the velocity vectors for each curve are parallel to each other, then we can now conclude that the gradients of both $f(x, y)$ and $g(x, y)$ must be parallel to each other as well.

One last comment for this last step of our procedure before we actually complete our rectangle inscribed inside an ellipse problem: The system of equations you get when using the Method of Lagrange Multipliers are typically *not* linear systems. They can be quite challenging to solve. You have to be fairly clever to solve it at times.

First rewrite the constraint equation so it has the form: $g(x, y) = 0$. $\dfrac{x^2}{4^2} + \dfrac{y^2}{3^2} - 1 = 0$

Next, form the Lagrange function: $L(x, y, \lambda) = f(x, y) - \lambda g(x, y) \ldots$

$$L(x, y, \lambda) = 4xy - \lambda \left(\frac{x^2}{4^2} + \frac{y^2}{3^2} - 1 \right)$$

We find all three first partial derivatives for this Lagrange function and set them simultaneously equal to 0 to create a system of equations:

$$\begin{cases} L_x = 4y - \lambda \left(\dfrac{2x}{16} \right) = 0 \\[2mm] L_y = 4x - \lambda \left(\dfrac{2y}{9} \right) = 0 \\[2mm] L_\lambda = \ 0 \ \ - \left(\dfrac{x^2}{16} + \dfrac{y^2}{9} - 1 \right) = 0 \end{cases}$$

Solve the first equation for λ:

$$4y = \lambda\left(\frac{x}{8}\right)$$

$$4y\left(\frac{8}{x}\right) = \lambda$$

$$\frac{32y}{x} = \lambda$$

Substitute this into the second equation for λ:

$$4x - \left(\frac{32y}{x}\right)\left(\frac{2y}{9}\right) = 0$$

Multiply both sides of the equation by $9x$ to clear fractions, then solve it for x^2:

$$(9x)\left[4x - \left(\frac{32y}{x}\right)\left(\frac{2y}{9}\right)\right] = 0(9x)$$

$$36x^2 - 64y^2 = 0$$

$$x^2 = \frac{64}{36}y^2 = \frac{16}{9}y^2$$

Substitute this value for x^2 into the constraint equation: $\dfrac{x^2}{4^2} + \dfrac{y^2}{3^2} = 1$ to get that:

$$\frac{1}{4^2}\left(\frac{16}{9}y^2\right) + \frac{y^2}{3^2} = 1$$

$$\frac{2}{9}y^2 = 1$$

$$y^2 = \frac{9}{2}$$

$$y = \pm\sqrt{\frac{9}{2}} = \pm\frac{3}{\sqrt{2}}$$

Of course, we require that y be positive, so we have that $y = \dfrac{3}{\sqrt{2}}$.

To find the corresponding value for x, we use the fact that we found: $x^2 = \dfrac{16}{9}y^2 = \dfrac{16}{9}\left(\dfrac{9}{2}\right) = 8$.

So, take the square root of both sides, and choose the positive value for x to get: $x = \sqrt{8} = 2\sqrt{2}$.

So the dimensions of the rectangle are $2x \times 2y$ that are:

$$2x = 2\left(2\sqrt{2}\right) = 4\sqrt{2} \quad and \quad 2y = 2\left(\frac{3}{\sqrt{2}}\right) = \frac{6}{\sqrt{2}}$$

The actual maximum area of the inscribed rectangle is given by:

$$f(x,y) = (width)(length) = (2x)(2y) = 4xy = 4\left(2\sqrt{2}\right)\left(\frac{3}{\sqrt{2}}\right) = 24 \text{ square units}$$

Using the Method of Lagrange Multipliers and Three Variables.

You can extend the procedure we used in the previous problems to optimize functions of three variables. So, if you plan to optimize a function given by: $f(x,y,z)$ subject to a constraint: $g(x,y,z) = 0$, you form the Lagrange function: $L(x,y,z,\lambda) = f(x,y,z) - \lambda g(x,y,z)$.

Now, you will need to find all *four* first partial derivatives for the Lagrange function and set them all simultaneously equal to zero. Put differently, you will have a system of four equations and four unknowns of the form:

$$\begin{cases} L_x = f_x(x,y,z) - \lambda g_x(x,y,z) = 0 \\ L_y = f_y(x,y,z) - \lambda g_y(x,y,z) = 0 \\ L_z = f_z(x,y,z) - \lambda g_z(x,y,z) = 0 \\ L_\lambda = 0 \quad - \quad g(x,y,z) = 0 \end{cases}$$

The geometrical interpretation is similar as it was for optimizing a function of two variables. We are again stating that two vectors are parallel to each other: the gradient of *f*, which is a vector orthogonal to the level *surfaces* (not curves this time) of *f*, and the gradient of *g*, which is a vector orthogonal to the fixed *surface* given by the constraint equation. That is: $\nabla f(x,y,z) = \lambda \nabla g(x,y,z)$

EXAMPLE. A delivery company accepts only rectangular boxes whose length and girth (girth means the perimeter of a cross section) do not exceed 108 inches. Find the dimensions of an acceptable box of largest volume.

SOLUTION. Let *x, y*, and *z* represent the length, width, and height of the rectangular box, respectively. Then the girth is: $2y + 2z$. We want to maximize the volume of the box $f(x,y,z) = xyz$ subject to the constraint that length *plus* the girth does not exceed 108 inches: $x + 2y + 2z = 108$. We rewrite our constraint equation so it has the form: $g(x,y,z) = 0$ to get that: $x + 2y + 2z - 108 = 0$. We form the Lagrange function: $L(x,y,z,\lambda) = f(x,y,z) - \lambda g(x,y,z)$.

$$L(x,y,z,\lambda) = xyz - \lambda(x + 2y + 2z - 108)$$

Next, we find all four first partial derivatives for the Lagrange function and set them all simultaneously equal to zero:

$$\begin{cases} L_x = yz - \lambda(1) = 0 \\ L_y = xz - \lambda(2) = 0 \\ L_z = xy - \lambda(2) = 0 \\ L_\lambda = 0 \quad - \quad (x + 2y + 2z - 108) = 0 \end{cases}$$

Solve for λ using the first equation, and substitute it into the second equation:

$$\lambda = yz$$
$$xz - (yz)(2) = 0$$
$$xz = 2yz$$
$$x = 2y$$

Solve for λ using the second equation, and substitute it into the third equation:

$$2\lambda = xz$$
$$\lambda = \frac{xz}{2}$$
$$xy - \left(\frac{xz}{2}\right)(2) = 0$$
$$xy = xz$$
$$y = z$$

Next, since we know that $x = 2y$ and $y = z$, we must have that $x = 2z$. So substitute, $x = 2z$ and $y = z$ into the constraint equation $x + 2y + 2z = 108$, and then solve it for z:

$$(2z) + 2(z) + 2z = 108$$
$$6z = 108$$
$$z = \frac{108}{6} = 18 \; inches$$

Finally we have the answer to this problem. The dimensions of the box are:

$$x = 36 \text{ inches}, \; y = 18 \text{ inches}, \; z = 18 \text{ inches}$$

The actual maximum volume is $V = xyz = (36(18)(18) = 11,664$ cubic inches.

Could we have avoided using Lagrange Multipliers altogether to solve this problem? The answer is yes, but it is a bit more of a challenge. First, you need to reduce the function to be maximized: $f(x, y, z) = V = xyz$ so that it is a function of two variables *only*. To do this, you would solve the constraint equation for z to get:

$$x + 2y + 2z = 108$$
$$2z = 108 - x - 2y$$
$$z = \frac{108 - x - 2y}{2}$$

Then, the function to be maximized would be: $f(x, y) = xy\left(\dfrac{108 - x - 2y}{2}\right) = \dfrac{1}{2}\left(108xy - x^2y - 2xy^2\right).$

You would then use the techniques discussed in the previous section to find the maximum. In other words, you would first find the first partial derivatives and set them equal to zero simultaneously.

$$\begin{cases} f_x(x,y) = \dfrac{1}{2}\left(108y - 2xy - 2y^2\right) = \dfrac{y}{2}\left(108 - 2x - 2y\right) = 0 \\[2mm] f_y(x,y) = \dfrac{1}{2}\left(108x - x^2 - 4xy\right) = \dfrac{x}{2}\left(108 - x - 4y\right) = 0 \end{cases}$$

Then, you would solve the system to find the critical points. The system you would get for this problem would end up with two critical points. Finally you would perform the Second Partials test (a.k.a. The D-Test) to determine if the critical point yields a maximum.

Last word: The Method of Lagrange Multipliers is a powerful tool and can be used to determine the extrema for functions of two *or* three variables. No Second Partials Test is required. However, the function *must* be accompanied by a constraint equation. If there is no constraint equation supplied, then the Method of Lagrange will not be appropriate, and you must use the method(s) in either Calculus I or the previous section using the Second Partial Test instead, if possible. The only other downside is that there is no quick way to tell if the extrema found is a relative maximum or a relative minimum. The way to find out would be to start by plugging the solutions (critical points) for the system of equations back into the original function that you are trying to optimize. Then, choose "some" points nearby and plug those into your function as well. If the value of the function at your critical point is larger than all of the other values, you have a maximum; if smaller, then you have a minimum. Usually, the problems will specify whether you are to find a maximum or a minimum when you are asked to use the Method of Lagrange, so don't worry too much!

Exercises: problems #10–14 from **Quiz 8**, problems #3 and 4 from **Quiz 9**, problems #25–27 from **Practice Midterm 3**.

Chapter 4

MULTIPLE INTEGRATION

Chapter Comments

The most difficult part of multiple integration is setting up the limits. Sometimes, it will be easier to evaluate an integral by reversing the order of integration. To do this, you have to change the limits, which can be challenging. Although the order of integration does not affect the value of the integral, it does affect the difficulty of the problem. It will be important for you to sketch the region you are integrating over—to help determine the limits as well as determine the best choice for the order of integration.

The good news is that we covered polar coordinates back in second semester calculus; so double integrals using polar coordinates will be easy. We also covered both cylindrical and spherical coordinates earlier this semester, so triple integrals using these coordinate systems shouldn't be too much of a challenge. Sometimes, you will want to convert a triple integral from rectangular to one of these other coordinate systems to make it easier to evaluate.

I will be very interested in also testing your ability to use the advanced techniques of integration you learned in second semester calculus—integration by parts, integration involving inverse trigonometric functions, trigonometric substitution, partial fraction decomposition, improper integrals, etc. So, be ready!

4.1 ITERATED INTEGRALS AND AREA IN THE PLANE

The concept of double integrals used to find area in the plane is not too difficult to grasp (I hope). The integrand will only contain dx and dy for double integrals representing area in the plane. Double integrals that have more than just dx and dy in the integrand actually represent volume, as we shall see in the next section. An integral of the form:

$$\int_a^b \int_{f(x)}^{g(x)} dy\,dx = \text{area}$$

…and an integral of the form:

$$\int_a^b \int_{f(x)}^{g(x)} h(x, y)\,dy\,dx = \text{volume}$$

The hardest two things you need to learn how to do in this section are to 1) determine the limits and 2) to reverse the order of integration. It is important you understand that the outside limits must be constant with respect to *all* variables. (This will be true for triple integrals later as well.) In addition, the variable of integration cannot appear in the limits of integration.

To test you on these concepts, determine what is wrong (if anything) for the following integrals:

(*Note:* these integrals all have more than just dx and dy in the integrand, so they all are meant to represent volume, not area.)

(a) $\displaystyle\int_0^2\int_{y^2}^{2y}(4x+2y)\,dy\,dx$

(b) $\displaystyle\int_0^3\int_{\sqrt{x/3}}^{1}e^{y^3}\,dy\,dx$

(c) $\displaystyle\int_{-x}^{x}\int_{-3}^{0}2\,dx\,dy$

(d) $\displaystyle\int_0^{1/16}\int_{y^{1/4}}^{1/2}\cos(16\pi x^5)\,dx\,dy$

If you answered that *both* (a) and (c) are incorrectly set up, then you got it right! What is wrong with part (a)? Well, the variable of integration for the inside integral is the variable y. This means that the limits for that integral may not have the variable y in them, but this one does—which is wrong!

What is wrong for the integral in part (c)? The outside integral must *always* have constants for limits, and this one does not.

Let's evaluate the other two integrals in the example below.

EXAMPLE. Evaluate the double integrals:

(a) $\displaystyle\int_0^3\int_{\sqrt{x/3}}^{1}e^{y^3}\,dy\,dx$

(b) $\displaystyle\int_0^{1/16}\int_{y^{1/4}}^{1/2}\cos(16\pi x^5)\,dx\,dy$

(c) $\displaystyle\int_1^2\int_0^{y^2}e^{x/y^2}\,dx\,dy$

(d) $\displaystyle\int_0^\infty\int_0^\infty xye^{-(x^2+y^2)}\,dy\,dx$

SOLUTION.

(a)   The first two integrals of these examples will require the order of integration to be reversed. This is due to the fact that these integrals are not integrable using basic integration techniques. The first step for reversing order of integration is to sketch the region we are integrating over. Because the order of integration is "*dydx*," we know this means that a representative rectangle will be vertical. The outside integral has limits of 0 and 3 and its variable of integration is *x*. This means that we need to graph the lines $x = 0$ and $x = 3$. The inside integral has limits of $\sqrt{\dfrac{x}{3}}$ and 1 and its variable of integration is *y*. This means we need to graph the equations $y = \sqrt{\dfrac{x}{3}}$ and $y = 1$. The region in question lies within all four graphs. Here is a diagram containing the graphs of all four equations:

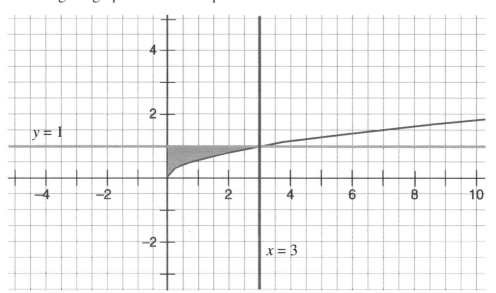

Can you determine what area to shade in that represents the region? This is the challenging part for many students. You need to shade in the part *above* the "curvy" graph of $y = \sqrt{\dfrac{x}{3}}$ (to the right of the *y*-axis), yet below the horizontal line $y = 1$. The reason for this is because the *lower* limit for *y* is $\sqrt{\dfrac{x}{3}}$ and the *upper* limit for *y* is 1.

Now, how to reverse the order of integration? Since we want the new order of integration to be "*dxdy*," the representative rectangle needs to be drawn horizontally. The limits for *y* need to be constants, since the outer integral must always have constants for limits. From the diagram, we see that the lower limit for *y* is 0 and the upper limit is 1. The lower limit for *x* is the *y*-axis (i.e., the equation $x = 0$), while the upper limit lies on the graph of $y = \sqrt{\dfrac{x}{3}}$. We need to rewrite this equation as a function of *y*—we need to solve this equation for *x*. Why do we need to do this? Remember that the limits for the integral that has *x* as its variable of integration must *not* contain the variable x itself. It can only have limits that are either constants or expressions containing the variable of an outside integral (which is y in this case).

So, let us solve for x: $\quad (y)^2 = \left(\sqrt{\dfrac{x}{3}}\right)^2$

$$y^2 = \frac{x}{3}$$

$$3y^2 = x$$

Therefore the upper limit for x is $3y^2$

Now, we can rewrite the entire double integral, so that the reversed order of integration is in place: $\qquad \displaystyle\int_0^1 \int_0^{3y^2} e^{y^3}\,dx\,dy$

This integral can be integrated. We integrate with respect to x first. Since the integrand contains only the expression e^{y^3} (which is a constant under the variable x), the anti-derivative will be xe^{y^3}. So, we have:

$$\int_0^1 e^{y^3}\left[x\right]_0^{3y^2} dy = \int_0^1 e^{y^3}\left(3y^2 - 0\right)dy = \int_0^1 e^{y^3}\,3y^2\,dy$$

We can easily integrate this using the general exponential rule, where $u = y^3$ and $du = 3y^2$.

Then, we will have: $\quad \displaystyle\int_0^1 e^{y^3}\,3y^2\,dy = e^{y^3}\Big|_0^1 = e^{1^3} - e^{0^3} = e - 1 \quad$ *(answer)*

(b) The integral $\displaystyle\int_0^{1/16} \int_{y^{1/4}}^{1/2} \cos(16\pi x^5)\,dx\,dy$ also needs the order of integration reversed. Again, the first

step is to sketch the region of integration. The four equations that need to be graphed are:

$$y = 0, \quad y = \frac{1}{16}, \quad x = y^{1/4}, \quad \text{and} \quad x = \frac{1}{2}$$

The equation $x = y^{1/4}$ can be graphed more easily if we solve for y first:

$$(x)^4 = \left(y^{1/4}\right)^4$$
$$x^4 = y$$

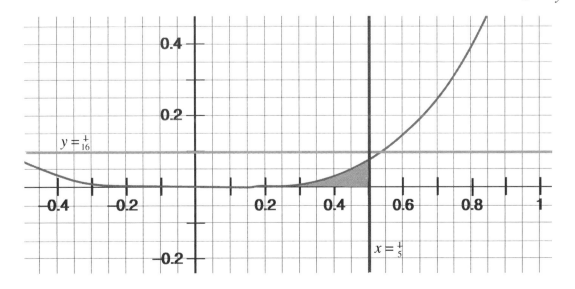

The region in question lies to the right of the parabola-shaped graph of $y = x^4$, but to the left of the vertical line $x = \dfrac{1}{2}$. To reverse the order of integration, we need the limits for x to be constant. We will have the lower limit as 0 and the upper limit as $\dfrac{1}{2}$. The limits for y range from $y = 0$ to $y = x^4$. Now, we can evaluate the double integral:

$$\int_0^{1/2} \int_0^{x^4} \cos(16\pi x^5)\,dy\,dx$$

We integrate with respect to y first, which is easy to do, since the cosine function has only the variable x in its argument.

$$\int_0^{1/2} \cos(16\pi x^5)[y]_0^{x^4}\,dx = \int_0^{1/2} \cos(16\pi x^5)(x^4 - 0)\,dx = \int_0^{1/2} x^4 \cos(16\pi x^5)\,dx$$

Now, we can use substitution, where $u = 16\pi x^5$ and $du = 80\pi x^4\,dx$. Since we need a factor of 80π. in the integrand, we will also need to divide it out, i.e., multiply the integral by $\dfrac{1}{80\pi}$. We evaluate the integral:

$$\frac{1}{80\pi} \int_0^{1/2} x^4 \cos(16\pi x^5)\,dx = \frac{1}{80\pi}\left[\sin(16\pi x^5)\right]_0^{1/2} = \frac{1}{80\pi}\left[\sin\left(16\pi\left(\frac{1}{2}\right)^5\right) - \sin(16\pi(0)^5)\right]$$

$$= \frac{1}{80\pi}\sin\left(\frac{\pi}{2}\right) = \frac{1}{80\pi}$$

(c) We start by using u-substitution with the exponential rule. We identify u as the exponent for the integrand's exponential function:

$$u = \frac{x}{y^2}$$

Next, we determine du, keeping in mind that the "du" will have us do partial differentiation with respect to the variable x (because we are currently integrating with respect to the x). So, we have that:

$$du = \frac{1}{y^2}\,dx$$

Because y is fixed when differentiating with respect to x, the expression $\dfrac{1}{y^2}$ is considered a constant. We are permitted to insert factors into an integrand so long as they represent constants and provided we multiply the front of the integral by its reciprocal. This may seem very strange because you were most likely told in both first- and second-semester calculus that it was forbidden to put a variable factor into an integrand in order to integrate. We are allowed to do this here in third semester calculus, provided the "variable" we are putting in as a factor is considered fixed, that is, constant, with respect to the variable of integration! At the point the integral $\int_1^2 \int_0^{y^2} e^{x/y^2}\,dx\,dy$ becomes:

$$\int_1^2 y^2 \int_0^{y^2} \frac{1}{y^2} e^{x/y^2}\,dx\,dy = \int_1^2 y^2 \left[\int e^u\,du\right]dy = \int_1^2 y^2 \left[e^u\right]dy = \int_1^2 y^2 \left[e^{x/y^2}\right]_0^{y^2}\,dy$$

Next, we evaluate the anti-derivative between the limits:

$$\left[e^{x/y^2}\right]_0^{y^2} = e^{y^2/y^2} - e^0 = e^1 - 1 = e - 1$$

But, this is just a constant factor that we can pull in front of the last integral we still need to evaluate:

$$(e-1)\int_1^2 y^2 dy = (e-1)\left[\frac{y^3}{3}\right]_1^2 = (e-1)\left(\frac{2^3}{3} - \frac{1}{3}\right) = \frac{7(e-1)}{3} \quad (answer)$$

(d) The integral $\int_0^\infty \int_0^\infty xye^{-(x^2+y^2)}dydx$ is considered an "improper integral" because of the infinite limits. This particular improper integral is taken from the study of probability theory, where x and y are considered two random variables, and where the integrand is referred to as the "probability density function" for the random variables.

You probably learned about integrals when you took first-semester calculus. You may also recall that there are two possibilities when evaluating an improper integral: 1) the improper integral converges, that is, the result is a finite value, or 2) the improper integral diverges, that is, it "blows up."

We start by integrating with respect to the variable y, and again use u-substitution with the exponential rule. We identify u as the exponent for the integrand's exponential function:

$$u = -\left(x^2 + y^2\right)$$

Next, we determine du, keeping in mind that the "du" will have us do partial differentiation with respect to the variable y (because we are currently integrating with respect to the y.) So, we have that:

$$du = -2y\,dy$$

The good news is that the variable "y" already exists in the integrand. So, we only need to insert a factor of –2, provided we multiply the front of the integral by the reciprocal of –2, or $-\frac{1}{2}$. So, the integral $\int_0^\infty \int_0^\infty xye^{-(x^2+y^2)}dydx$ now becomes:

$$\int_0^\infty \left(-\frac{1}{2}\right)\int_0^\infty x(-2y)e^{-(x^2+y^2)}dydx = \int_0^\infty \left(-\frac{1}{2}\right)x\left[\int e^u du\right]dx = \int_0^\infty \left(-\frac{1}{2}\right)x\left[e^u\right]dx = \int_0^\infty \left(-\frac{1}{2}\right)x\left[e^{-(x^2+y^2)}\right]_0^\infty dx$$

Notice that we also pulled the variable "x" out of the inner integral in the above calculations, since it is considered a constant with respect to the variable of integration y.

Evaluating the exponential anti-derivative between its limits gives us:

$$\left[e^{-(x^2+y^2)}\right]_0^\infty = e^{-(x^2+\infty)} - e^{-(x^2+0)} = e^{-\infty} - e^{-x^2}$$

From your study of limits in first-semester calculus, we know that $\lim\limits_{x\to\infty} e^{-x} = 0$.

So, that result is simply $-e^{-x^2}$. To complete this problem, we still need to integrate once more. The problem at this point looks like:

$$\int_0^\infty \left(-\frac{1}{2}\right)x\left(-e^{-x^2}\right)dx$$

This is now a simple first-semester calculus problem.

Again, we apply u-substitution with the exponential rule. We identify u as the exponent for the integrand's exponential function $u = -x^2$. Next, we determine du, keeping in mind that the "du" will have us do (regular) differentiation with respect to the variable x (because we are currently integrating with respect to the x.)

So, we have that $du = -2xdx$. Continuing on to solve this problem:

$$\int_0^\infty \left(-\frac{1}{2}\right)x\left(-e^{-x^2}\right)dx \;=\; \left(-\frac{1}{2}\right)\left(\frac{1}{2}\right)\int_0^\infty -2xe^{-x^2}dx \;=\; \left(-\frac{1}{4}\right)\int e^u du \;=\; \left(-\frac{1}{4}\right)e^u = \left(-\frac{1}{4}\right)\left[e^{-x^2}\right]_0^\infty$$

We now finish the problem:

$$-\left(\frac{1}{4}\right)\left[e^{-x^2}\right]_0^\infty \;=\; -\left(\frac{1}{4}\right)\left(e^{-\infty} - e^0\right) \;=\; -\left(\frac{1}{4}\right)(-1) \;=\; \frac{1}{4}$$

Thus, the improper double integral converges. This result makes sense in probability theory, where probabilities for a particular event must be a value in the range of 0 to 1 (inclusive).

EXERCISES: # 6–11 all from **Quiz 9**, #4 from **Quiz 10**, #1, 2, 3, 4, 5, 7, 18a, 20 from **Practice Midterm 4**

4.2 DOUBLE INTEGRALS AND VOLUME

In this section, you learn that a double integral can be used to represent volume, as I mentioned above. Then, you can use Fubini's Theorem to write the double integral as an iterated integral. Using notation, we can write:

$$\textbf{\textit{Volume}} = \iint_R f(x,y)dA = \int_a^b \int_{g_1(x)}^{g_2(x)} f(x,y)dydx$$

Where the region R is a plane region in the xy-plane and the volume is that of a solid bounded above by the surface $z = f(x,y)$ and below by the region R.

As we discussed in the last section, the challenge that many students face when doing these types of problems is setting up the iterated integral with the correct limits.

EXAMPLE. Set up an iterated integral for *both* orders of integration for the integral $\iint\limits_{R}(8y+x)dA$, where the region R is the semi-circle bounded by $y=\sqrt{4-x^2}$ and $y=0$. Then evaluate the one you feel is most convenient.

SOLUTION. First, sketch the graphs for the equations: $y=\sqrt{4-x^2}$ and $y=0$.

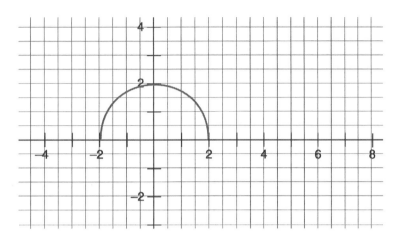

Using the order *dydx*, the limits of x need to be constants—and we see from the graph that those will be –2 and 2. The variable y will have a lower limit of 0 and an upper limit of $y=\sqrt{4-x^2}$. This integral can then be written:

$$\int\limits_{-2}^{2}\int\limits_{0}^{\sqrt{4-x^2}}(8y+x)dydx$$

Using the order *dxdy*, the limits of y will be 0 and 2. The variable x ranges from a low on the left side of the semi-circle, and the upper limit will be the right-hand side of the semi-circle. We need to rewrite the equation $y=\sqrt{4-x^2}$ as a function of y (i.e., solve for x):

$$(y)^2 = \left(\sqrt{4-x^2}\right)^2$$
$$y^2 = 4-x^2$$
$$x^2 = 4-y^2$$
$$\sqrt{x^2} = \pm\sqrt{4-y^2}$$
$$x = \pm\sqrt{4-y^2}$$

So, the lower limit for x will be $-\sqrt{4-y^2}$, and the upper limit for x will be $\sqrt{4-y^2}$. This integral can then be written:

$$\int\limits_{0}^{2}\int\limits_{-\sqrt{4-y^2}}^{\sqrt{4-y^2}}(8y+x)dxdy$$

It turns out that both of these integrals are about the same challenge level. (This may not always be the case, as we have seen!) So we will evaluate this last integral:

$$\int_0^2 \int_{-\sqrt{4-y^2}}^{\sqrt{4-y^2}} (8y+x)\,dx\,dy = \int_0^2 \left[8yx + \frac{x^2}{2} \right]_{-\sqrt{4-y^2}}^{\sqrt{4-y}} dy$$

$$= \int_0^2 \left\{ \left(8y\sqrt{4-y^2} + \frac{\left(\sqrt{4-y^2}\right)^2}{2} \right) - \left(8y\left(-\sqrt{4-y^2}\right) + \frac{\left(-\sqrt{4-y^2}\right)^2}{2} \right) \right\} dy$$

$$= \int_0^2 \left\{ \left(8y\sqrt{4-y^2} + 8y\sqrt{4-y^2} \right) + \left(\frac{4-y^2}{2} - \frac{4-y^2}{2} \right) \right\} dy$$

$$= \int_0^2 16y\sqrt{4-y^2}\,dy \quad = \quad 16\int_0^2 y\sqrt{4-y^2}\,dy$$

This integral is easy to do using u-substitution and the general power rule, where $u - 4 - y^2$ and $du = -2y$:

$$= -\frac{1}{2}(16)\int_0^2 (-2)y\sqrt{4-y^2}\,dy$$

$$= -8\int u^{1/2}\,du$$

$$= -8\left[\frac{2u^{3/2}}{3} \right] \quad = \quad -\frac{16}{3}\left[\left(4-y^2\right)^{3/2} \right]_0^2$$

$$= -\frac{16}{3}\left[\left(4-2^2\right)^{3/2} - \left(4-0^2\right)^{3/2} \right]$$

$$= -\frac{16}{3}\left[0 - 4^{3/2} \right]$$

$$= -\frac{16}{3}\left(-\left(\sqrt{4}\right)^3 \right)$$

$$= +\frac{16}{3}(2)^3 \quad = \quad \frac{16}{3}(8) = \quad \frac{128}{3}$$

The other challenge students face in this section is setting up a double integral "from scratch." The next examples are similar to homework exercises and exam questions demonstrating the strategy needed.

EXAMPLE. Find the volume of the wedge cut from the first octant by the cylinder $z = 12 - 3y^2$ and the plane $x + y = 2$.

SOLUTION. Always remember that the phrase, "first octant" means the location where $x \geq 0$, $y \geq 0$, and $z \geq 0$. The top of the surface will be the graph of the function $z = 12 - 3y^2$. This means that this will be the integrand for our double integral representing volume. The base of the solid can be found by examining the graphs in the *xy*-plane. Recall that points in the *xy*-plane have $z = 0$. So, we let $z = 0$ in the equation $z = 12 - 3y^2$, we have that $0 = 12 - 3y^2$. Solving for y, we have that $y = \pm 2$. We discard the -2, since $y = -2$ does not lie in the first octant. We graph the line $y = 2$, as well as the graph of $x + y = 2$ in the *xy*-plane, which is a line. See diagram:

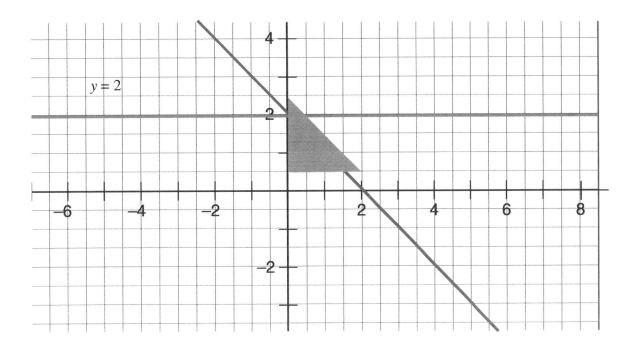

The triangle-shaped region in the first quadrant *(only)* below the line $x + y = 2$ is the base of our solid, where the variable x ranges from 0 to 2. So, the double integral is:

$$\int_0^2 \int_0^{2-x} (12 - 3y^2)\, dy\, dx$$

This is easy to evaluate; I will anti-differentiate with respect to y first:

$$\int_0^2 \left[12y - y^3\right]_0^{2-x} dx \quad = \quad \int_0^2 \left[12(2-x) - (2-x)^3\right] dx$$

I will expand and multiply before evaluating, but you could use *u*-substitution instead if you like and get the same result.

$$= \int_0^2 \left[24 - 12x - (2-x)(2-x)(2-x)\right]dx$$

$$= \int_0^2 \left[24 - 12x - (2-x)(4-4x+x^2)\right]dx$$

$$= \int_0^2 \left[24 - 12x - (8 - 8x + 2x^2 - 4x + 4x^2 - x^3)\right]dx$$

$$= \int_0^2 \left[24 - 8 - 12x + 8x + 4x - 2x^2 - 4x^2 + x^3\right]dx$$

$$= \int_0^2 \left[16 - 6x^2 + x^3\right]dx$$

$$= \left[16x - 2x^3 + \frac{x^4}{4}\right]_0^2$$

$$= 16(2) - 2(2)^3 + \frac{(2)^4}{4} = 32 - 16 + 4 = 20$$

EXAMPLE. Find the volume of the solid bounded by the coordinate planes, and the graphs of $x = 4$ and $z = x^2 - y + 4$.

SOLUTION. It is important that you are able to understand and visualize the coordinate planes. These are the three planes that are the *xy*-plane, the *xz*-plane, and the *yz*-plane. The *xy*-plane has the equation $z = 0$; the *xz*-plane has the equation $y = 0$; and the *yz*-plane has the equation $x = 0$. The graph of $x = 4$ is a plane that is parallel to the *yz*-plane. The base of the solid is a region in the *xy*-plane bounded by the graphs of $x = 0$, $y = 0$, $x = 4$, and the equation $z = x^2 - y + 4$, where we let $z = 0$. We have to graph:

$$0 = x^2 - y + 4$$
$$y = x^2 + 4$$

This graph is a parabola opening upward and intersecting the *y*-axis at 4. A graph of the base of our solid can be found from the diagram:

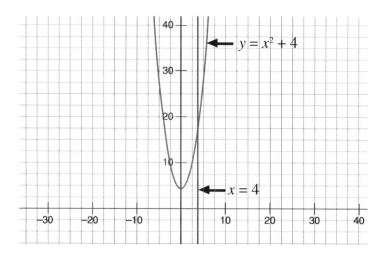

The region representing the base of our solid is vertically simple (that is, we can write using the order *dydx*) and is the region *below* the parabola and to the left of the line $x = 4$ in the first quadrant. The limits will be:

$$0 \le x \le 4 \ \text{ and } \ 0 \le y \le x^2 + 4$$

Now, we can set up the double integral representing the volume of the solid:

$$\text{Volume} = \int_0^4 \int_0^{x^2+4} \left(x^2 - y + 4\right) dy\,dx$$

This is easy to evaluate:

$$= \int_0^4 \left[x^2 y - \frac{y^2}{2} + 4y \right]_0^{x^2+4} dx$$

$$= \int_0^4 \left[x^2\left(x^2 + 4\right) - \frac{\left(x^2 + 4\right)^2}{2} + 4\left(x^2 + 4\right) \right] dx$$

I will expand, multiply, and then collect like terms before evaluating the last integral:

$$= \int_0^4 \left[x^4 + 4x^2 - \frac{1}{2}\left(x^2 + 4\right)\left(x^2 + 4\right) + 4x^2 + 16 \right] dx$$

$$= \int_0^4 \left[x^4 + 8x^2 + 16 - \frac{1}{2}\left(x^4 + 8x^2 + 16\right) \right] dx$$

$$= \int_0^4 \left(\frac{1}{2} x^4 + 4x^2 + 8 \right) dx$$

$$= \left[\frac{x^5}{10} + \frac{4}{3} x^3 + 8x \right]_0^4$$

$$= \frac{(4)^5}{10} + \frac{4}{3}(4)^3 + 8(4)$$

$$= \frac{3296}{15}$$

This concludes the section!

EXERCISES: #12–17 all from **Quiz 9**, #1, 5, from **Quiz 10**, #6, 18b from **Practice Midterm 4**

THIRD SEMESTER CALCULUS STUDENT SUPPLEMENT, 4TH EDITION

4.3   CHANGE OF VARIABLES: POLAR COORDINATES

Some double integrals are much easier to evaluate in polar form than in rectangular form. This is especially true for regions that are circles, cardioids, petal curves, sectors of circles, etc. Remember that a point (x, y) in the rectangular coordinate system can be represented in polar coordinates by the ordered pair (r, θ). The important conversion formulas to remember are the ones you learned in second semester calculus:

$$r^2 = x^2 + y^2, \quad \theta = \tan^{-1}\left(\frac{y}{x}\right), \quad x = r\cos\theta, \, y = r\sin\theta$$

We will use these to help us convert integrals between the two coordinate systems. The other key thing to remember when converting a rectangular double integral into a polar one is the following:

$$\iint_R f(x, y)\,dy\,dx = \iint_R f(r\cos\theta, \, r\sin\theta)\,r\,dr\,d\theta$$

Please notice an extra factor of "r" in the integrand for the integral on the right. You will *always* need to remember to put this in, OK?

EXAMPLE. Change the Cartesian integral into an equivalent polar integral. Then, evaluate the integral.

$$\int_0^{\ln 2} \int_0^{\sqrt{(\ln 2)^2 - y^2}} e^{\sqrt{x^2 + y^2}}\,dx\,dy$$

SOLUTION. First, always sketch the region in the xy-plane. We will need to graph the four equations: $y = 0$, $y = \ln 2$, $x = 0$, $x = \sqrt{(\ln 2)^2 - y^2}$. The last one is a little tricky, but if you rewrite it you will see it is simply a circle of radius ln2:

$$(x)^2 = \left(\sqrt{(\ln 2)^2 - y^2}\right)^2$$
$$x^2 = (\ln 2)^2 - y^2$$
$$x^2 + y^2 = (\ln 2)^2$$

After graphing all four equations, I hope you can see that the region is the quarter of the circle of radius ln2 in the *first quadrant*. See diagram at right: ⟹

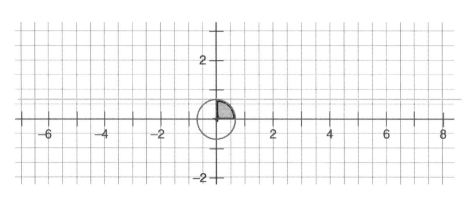

The limits for the variables r and θ are:

$$0 \le r \le \ln 2 \quad \text{and} \quad 0 \le \theta \le \frac{\pi}{2}$$

Now we can set up the corresponding double integral using polar coordinates, keeping in mind the extra "r" you need in the integrand—as well as the fact that $r = \sqrt{x^2 + y^2}$ (substitute in the power of the exponential function):

$$\int_0^{\pi/2} \int_0^{\ln 2} e^r \, r \, dr \, d\theta$$

We need integration by parts to evaluate this integral. Write out the choices for "u," "du," "v," and "dv" to get full credit on your exams in my class! For this problem, we have:

$$u = r, \quad dv = e^r$$
$$du = dr, \quad v = e^r$$

Then, using the Integration By Parts formula, which is $\int u \, dv = uv - \int v \, du$, we have that:

$$\int_0^{\pi/2} \int_0^{\ln 2} e^r \, r \, dr \, d\theta = \int_0^{\pi/2} \left[re^r \Big]_0^{\ln 2} - \int_0^{\ln 2} e^r \, dr \right] d\theta$$

$$= \int_0^{\pi/2} \left\{ \left[\ln 2 e^{\ln 2} - 0 \right] - \left[e^r \Big]_0^{\ln 2} \right\} d\theta$$

$$= \int_0^{\pi/2} \left(\ln 2(2) - \left(e^{\ln 2} - e^0 \right) \right) d\theta = \int_0^{\pi/2} \left(2 \ln 2 - (2 - 1) \right) d\theta$$

$$= \left(2 \ln 2 - 1 \right) \theta \Big]_0^{\pi/2} \quad = \quad \left(2 \ln 2 - 1 \right) \frac{\pi}{2}$$

EXAMPLE. Find the volume of the solid bounded above by the graph of the surface:

$$z = \frac{\ln(x^2 + y^2)}{\sqrt{x^2 + y^2}} \quad \text{over the region } R = \left\{ (x, y) \Big| \ 1 \le x^2 + y^2 \le e \right\}$$

SOLUTION. The region in the xy-plane is the area between two circles that have the equations:

$$x^2 + y^2 = 1^2 \quad \text{and} \quad x^2 + y^2 = \left(\sqrt{e} \right)^2$$

The smaller circle is of radius 1; the larger circle is of radius $\sqrt{e}$. These will be our limits for the variable r in our double integral using polar coordinates. The angle θ will range between 0 and 2π. The volume is represented by:

$$\iint_R \frac{\ln(x^2 + y^2)}{\sqrt{x^2 + y^2}} \, dA = \int_0^{2\pi} \int_1^{\sqrt{e}} \frac{\ln(r^2)}{r} \, r \, dr \, d\theta$$

Reducing the quotient in the integrand to lowest terms gives us an integrand of $\ln(r^2)$. We can simplify this using a property of logarithms: $\ln(r^2) = 2\ln r$. We will integrate using Integration By Parts again, where the choices for *u, du, v, dv* are:

$$u = \ln(r), \quad dv = 1 \cdot dr$$

$$du = \frac{1}{r}dr, \quad v = r$$

Applying the Integration By Parts formula, we have:

$$\int_0^{2\pi}\int_1^{\sqrt{e}} \frac{\ln(r^2)}{r}\,r\,dr\,d\theta \;=\; 2\int_0^{2\pi}\int_1^{\sqrt{e}} \ln r\,dr\,d\theta$$

$$= \; 2\int_0^{2\pi}\left[r\ln r\Big|_1^{\sqrt{e}} - \int_1^{\sqrt{e}} r\cdot\frac{1}{r}\,dr \right]d\theta$$

$$= \; 2\int_0^{2\pi}\left\{ \left[\sqrt{e}\ln\sqrt{e} - 1\ln(1)\right] - [r]_1^{\sqrt{e}} \right\}d\theta$$

We use the property of logarithms again to simplify. We can write that:

$$\ln(\sqrt{e}) = \ln e^{1/2} = \frac{1}{2}\ln e = \frac{1}{2}(1) = \frac{1}{2}$$

Also, please note that $\ln(1) = 0$. Continuing on with the problem:

$$= \; 2\int_0^{2\pi}\left\{ \left[\sqrt{e}\left(\frac{1}{2}\right)\right] - \left[\sqrt{e}-1\right] \right\}d\theta$$

$$= \; 2\int_0^{2\pi}\left(1 - \frac{\sqrt{e}}{2}\right)d\theta$$

$$= \; 2\left(1 - \frac{\sqrt{e}}{2}\right)[\theta]_0^{2\pi}$$

$$= 2\left(1 - \frac{\sqrt{e}}{2}\right)(2\pi) = \; 4\pi - 2\pi\sqrt{e}$$

EXAMPLE. Change the Cartesian integral into an equivalent polar integral. Then, evaluate the polar integral:

$$\int_0^2 \int_{-\sqrt{1-(x-1)^2}}^0 \frac{x+y}{x^2+y^2}\,dy\,dx$$

SOLUTION. First, sketch out the region in the xy-plane. We must sketch the graphs of the equations:

$$x = 0,\ x = 2,\ y = 0,\ \text{and}\ y = -\sqrt{1-(x-1)^2}\ .$$

This last one can be rewritten:

$$(y)^2 = \left(-\sqrt{1-(x-1)^2}\right)^2$$
$$y^2 = 1-(x-1)^2$$
$$(x-1)^2 + y^2 = 1$$

This is a circle centered at the point $(1, 0)$ with a radius of 1. The region is sketched below; keep in mind that the region shaded is *below* the x-axis. Why is that so? Well, the limits for y found from the inside integral range from $-\sqrt{1-(x-1)^2} \le y \le 0$, and this means that the *upper* limit for y is 0.

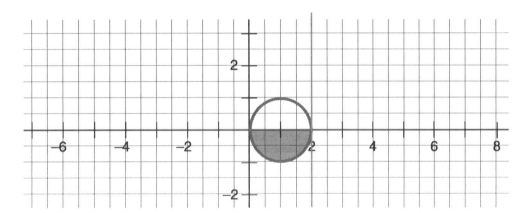

Since the limits for y are $-\sqrt{1-(x-1)^2} \le y \le 0$, this means that y starts at the bottom half of the circle and ends at $y = 0$. Because the circle is *not* centered at the origin, we cannot say that the variable r varies between 0 and 1. Instead, we need to come up with the corresponding polar equation for the rectangular equation of $(x-1)^2 + y^2 = 1$. You did this type of problem back in second semester calculus, remember? Using the conversion formulas $x = r\cos\theta,\ y = r\sin\theta$, we rewrite this equation:

$$(r\cos\theta - 1)^2 + (r\sin\theta)^2 = 1$$
$$r^2\cos^2\theta - 2r\cos\theta + 1 + r^2\sin^2\theta = 1$$
$$r^2(\cos^2\theta + \sin^2\theta) - 2r\cos\theta = 0$$
$$r^2(1) = 2r\cos\theta$$

Now we divide both sides by r to get $r = 2\cos\theta$. Therefore the limits for the variable r in the polar double integral will be $0 \le r \le 2\cos\theta$. Use your graphing calculator (in *polar* mode) to graph the polar equation $r = 2\cos\theta$. What should the range of the angle θ be to sketch out the bottom half of this circle *only*? We need to know the answer to this question to determine the limits for θ.

If we let $0 \leq \theta \leq \pi$ (do this on your calculator by going into "WIN" and changing the range for θ), this sketches out the *entire* circle, so those are *not* the limits. Instead, if we let $\frac{\pi}{2} \leq \theta \leq \pi$, this gives us the bottom half of the circle, so these are indeed the limits. Now we are ready to set up the polar integral:

$$\int_{\pi/2}^{\pi} \int_{0}^{2\cos\theta} \frac{r\cos\theta + r\sin\theta}{r^2} \, r \, dr \, d\theta$$

We reduce the quotient to lowest terms before evaluating. Notice that the factor of r^2 is common to both numerator and denominator, and therefore "cancels" to get:

$$\int_{\pi/2}^{\pi} \int_{0}^{2\cos\theta} (\cos\theta + \sin\theta) \, dr \, d\theta$$

Now we integrate with respect to the variable r. We treat the $(\cos\theta + \sin\theta)$ expression as a constant to do this. We get:

$$= \int_{\pi/2}^{\pi} (\cos\theta + \sin\theta) \left[r \right]_0^{2\cos\theta} d\theta$$

$$= \int_{\pi/2}^{\pi} (\cos\theta + \sin\theta)(2\cos\theta - 0) d\theta$$

$$= 2 \int_{\pi/2}^{\pi} (\cos^2\theta + \sin\theta\cos\theta) d\theta$$

In order to integrate the term $\cos^2\theta$, we use the power reduction formula:

$$\cos^2\theta = \frac{1 + \cos(2\theta)}{2} = \frac{1}{2}(1 + \cos(2\theta))$$

The term $\sin\theta\cos\theta$ will be easy to integrate using u-substitution, where we let $u = \sin\theta$, and $du = \cos\theta \, d\theta$.

Continuing on with the problem, I will split up the integral of the sum into the sum of two separate integrals so that you can see the two separate integration strategies at work:

$$= 2 \int_{\pi/2}^{\pi} \left[\frac{1}{2}(1 + \cos 2\theta) \right] d\theta + 2 \int_{\pi/2}^{\pi} \sin\theta\cos\theta \, d\theta$$

$$= 2 \left[\frac{1}{2}\left(\theta + \frac{1}{2}\sin 2\theta \right) \right]_{\pi/2}^{\pi} + 2 \int u \, du$$

$$= (2)\frac{1}{2}\left[\pi + \frac{1}{2}\sin(2 \cdot \pi) - \left(\frac{\pi}{2} + \frac{1}{2}\sin\left(2 \cdot \frac{\pi}{2} \right) \right) \right] + (2)\frac{u^2}{2}$$

$$= \left[\pi + 0 - \left(\frac{\pi}{2} + 0 \right) \right] + \left[\sin^2\theta \right]_{\pi/2}^{\pi}$$

$$= \left[\frac{\pi}{2} \right] + \left[\sin^2\pi - \sin^2\left(\frac{\pi}{2} \right) \right]$$

$$= \frac{\pi}{2} + [0 - 1]$$

$$= \frac{\pi}{2} - 1$$

EXAMPLE. Change the Cartesian integral into an equivalent polar integral. Then, evaluate the polar integral:

$$\iint_R \frac{\sqrt{x^2 + y^2} + 1}{\left(x^2 + y^2\right)^{3/2}\left(\sqrt{x^2 + y^2} - 1\right)}\, dA,$$

...where the region R is defined as:

$$R = \left\{(x, y)\, \middle|\, 4 \le x^2 + y^2 \le e^2\right\}$$

SOLUTION. The region in the xy-plane is the area between two circles that have the equations:

$$x^2 + y^2 = 2^2 \quad and \quad x^2 + y^2 = \left(e\right)^2.$$

The smaller circle is of radius 2; the larger circle is of radius e. These will be our limits for the variable r in our double integral using polar coordinates. The angle θ will range between 0 and 2π. To convert the integrand into polar coordinates, notice that we can write that:

$$(x^2 + y^2)^{3/2} = \left(r^2\right)^{3/2} = r^{(2)\left(\frac{3}{2}\right)} = r^3$$

So, we have the polar integral:

$$\int_0^{2\pi}\int_2^e \frac{r+1}{r^3(r-1)}\, r\, dr\, d\theta$$

After reducing the quotient in the integrand to lowest terms, we continue on using the method of Partial Fraction Decomposition. *Remember that technique from last semester's calculus class?*

$$\int_0^{2\pi}\int_2^e \frac{r+1}{r^2(r-1)}\, dr\, d\theta$$

The factor of r^2 may be treated in one of two ways:

    1)   a irreducible quadratic factor

    2)   a repeated linear factor

It is entirely up to you, as the result will be the same. I will treat the factor of r^2 as an irreducible quadratic factor:

$$\int_0^{2\pi}\int_2^e \frac{r+1}{r^2(r-1)}\, dr\, d\theta = \int_0^{2\pi}\int_2^e \left[\frac{Ar+B}{r^2} + \frac{C}{r-1}\right] dr\, d\theta$$

The *Basic Equation* can be found by writing out the equality:

$$\frac{r+1}{r^2(r-1)} = \frac{Ar+B}{r^2} + \frac{C}{r-1}$$

Multiply both sides of the equation by the LCD to clear fractions:

$$r^2(r-1)\left[\frac{r+1}{r^2(r-1)}\right] = \left[\frac{Ar+B}{r^2}+\frac{C}{r-1}\right]r^2(r-1)$$

$$r+1 = (Ar+B)(r-1)+Cr^2$$

The last equation obtained is known as the "Basic Equation." We use this to help us determine the unknowns A, B, and C. If we let $r = 1$, we have that:

$$(1)+1 = (A(1)+B)((1)-1)+C(1)^2$$

$$2 = C$$

To obtain the unknowns A and B, I will use the method of equating coefficients, which you studied last semester (I hope). Let's multiply the basic equation all the way out and collect like terms:

$$r+1 = Ar^2 - Ar + Br - B + Cr^2$$

$$r+1 = (A+C)r^2 + (-A+B)r - B$$

We need the polynomial on the left to be equal to the polynomial on the right. Let us equate the second degree terms. Since the polynomial on the left contains no second degree term, we say that the coefficient for that term is 0. To be succinct:

$$0 = (A+C)$$

But we already found that $C = 2$. Substitute this in, and we find that $A = -2$. We equate the coefficients for the first-degree terms:

$$1 = -A + B$$

Since we found that $A = -2$, we then can conclude that $B = -1$. Substituting all of these into our quotients, we then can write that:

$$\int_0^{2\pi}\int_2^e \frac{r+1}{r^2(r-1)}\,dr d\theta = \int_0^{2\pi}\int_2^e \left[\frac{-2r-1}{r^2}+\frac{2}{r-1}\right]dr d\theta$$

I will use the "split-up trick" on the first quotient, then reduce before integrating. The last quotient in the integrand requires use of the Log Rule to integrate. See next page.

$$= \int_{0}^{2\pi}\int_{2}^{e}\left[\frac{-2r-1}{r^2}+\frac{2}{r-1}\right]dr\,d\theta$$

$$= \int_{0}^{2\pi}\int_{2}^{e}\left[\frac{-2r}{r^2}+\frac{-1}{r^2}+\frac{2}{r-1}\right]dr\,d\theta$$

$$= \int_{0}^{2\pi}\int_{2}^{e}\left[-\frac{2}{r}-\frac{1}{r^2}+\frac{2}{r-1}\right]dr\,d\theta$$

$$= \int_{0}^{2\pi}\left[-2\ln|r|-\left(\frac{r^{-1}}{-1}\right)+2\ln|r-1|\right]_{2}^{e}d\theta = \int_{0}^{2\pi}\left[-2\ln|r|+\frac{1}{r}+2\ln|r-1|\right]_{2}^{e}d\theta$$

$$= \int_{0}^{2\pi}\left\{\left[-2\ln e+\frac{1}{e}+2\ln(e-1)\right]-\left[-2\ln 2+\frac{1}{2}+2\ln(2-1)\right]\right\}d\theta$$

$$= \int_{0}^{2\pi}\left(-2+\frac{1}{e}+2\ln(e-1)+2\ln 2-\frac{1}{2}\right)d\theta = \int_{0}^{2\pi}\left(-2-\frac{1}{2}+\frac{1}{e}+\ln(e-1)^2+\ln 2^2\right)d\theta$$

$$= \int_{0}^{2\pi}\left(-\frac{5}{2}+\frac{1}{e}+\ln 4(e-1)^2\right)d\theta$$

I used log properties in the last step to condense the two logs into one. Now we integrate with respect to the angle θ to get:

$$= \left(-\frac{5}{2}+\frac{1}{e}+\ln 4(e-1)^2\right)\left[\theta\right]_{0}^{2\pi}$$

$$= \left(-\frac{5}{2}+\frac{1}{e}+\ln 4(e-1)^2\right)2\pi$$

Whew! That was a tough problem! This concludes the section.

EXERCISES: #18, 19, 20 from **Quiz 9**, #6, 7 from **Quiz 10**, #7, 21 from **Practice Midterm 4**

4.4 CENTER OF MASS

I only cover the concept of finding the mass of a planar lamina in this section. So far, you should understand that we have studied two uses for double integrals, and they are:

1) To find area in the plane:

$$Area = \iint\limits_{R} dA$$

Note that the integrand for finding area will have only "*dydx*" for rectangular double integrals or "*rdrdq*" for polar double integrals, and nothing more.

2) To find volume in space for a solid bounded above by a surface given by $z = f(x, y)$ and bounded below by a region R in the *xy*-plane:

$$Volume = \iint\limits_{R} f(x, y) dA$$

Note that the integrand here has something besides just the "*dydx*."

There are other uses for double integrals that we will look at in this section and the next section on surface area. This third use covered in this section will involve finding the mass of a planar lamina with a density function given by $\rho = f(x, y)$. Here is the formula:

3) Mass of a planar lamina region R (in the *xy*-plane) with density of...

$$Mass = \iint\limits_{R} \rho \, dA = \iint\limits_{R} f(x, y) \, dA$$

This looks exactly like the double integral formula for volume, doesn't it? (It is!)

EXAMPLE. Find the mass of the planar lamina described by the region bounded by the graphs:

$$x = 3, x = 5, y = 0, \text{ and } xy = 1, \text{ given that its density is } \rho = \sqrt{x^2 - 9}$$

SOLUTION. First, sketch the region in the *xy*-plane to determine the limits for the iterated integral. Note that it is easier to graph the equation $xy = 1$ if you first solve for y and then graph $y = \dfrac{1}{x}$ instead.

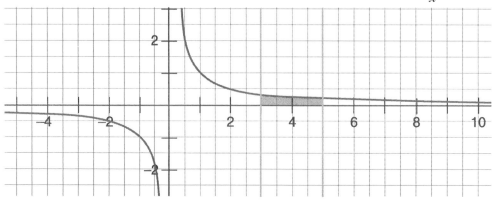

QUESTION: What region should you shade to represent the region that we are integrating over?

ANSWER: The region below the graph of $y = \dfrac{1}{x}$ between the vertical lines represented by $x = 3$ and $x = 5$, but above the *x*-axis.

After examining the region, we can then set up the double integral representing mass:

$$\text{Mass} = \int\limits_{3}^{5} \int\limits_{0}^{1/x} \sqrt{x^2 - 9}\; dy dx$$

First we integrate with respect to *y*, and then run into an integral that requires the use of trigonometric substitution that you studied in last semester's calculus class, remember? Recall that trigonometric substitution is a technique of integration that works quite nicely with radicands of the form:

$$\sqrt{a^2 \pm u^2}\quad or \quad \sqrt{u^2 \pm a^2}$$

$$= \int\limits_{3}^{5} \sqrt{x^2 - 9}\,\big[y\big]_{0}^{1/x}\; dx$$

$$= \int\limits_{3}^{5} \sqrt{x^2 - 9}\left(\frac{1}{x} - 0\right) dx$$

$$= \int\limits_{3}^{5} \frac{\sqrt{x^2 - 9}}{x}\; dx$$

Now, we will draw a corresponding right triangle, and then use change of variables to convert the integral from the variable *x* to θ. The point of doing this is to make the integral easier to evaluate!

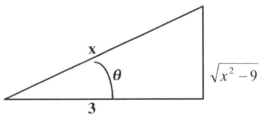

Next, we use this triangle to obtain the expressions *x, dx,* and the radicand $\sqrt{x^2 - 9}$ all in terms of the variable θ.

$$\sec\theta = \frac{hyp}{adj} = \frac{x}{3},\quad x = 3\sec\theta$$

$$dx = 3\sec\theta \tan\theta\, d\theta$$

$$\tan\theta = \frac{opp}{adj} = \frac{\sqrt{x^2 - 9}}{3},\quad \sqrt{x^2 - 9} = 3\tan\theta$$

Substitute all of these into the integrand:

$$= \int_{3}^{5} \frac{\sqrt{x^2-9}}{x} dx$$

$$= \int \frac{3\tan\theta}{3\sec\theta} 3\sec\theta\tan\theta d\theta$$

$$= 3\int \tan^2\theta d\theta$$

Now what? Well, recall from the section on powers of trigonometric functions from second-semester calculus class that when the power of a tangent is positive and even, we convert a tangent-squared factor into a secant-squared factor (using the trig identity $\tan^2\theta = \sec^2\theta - 1$), expand, then repeat as necessary. Luckily, we only have one tangent-squared factor to worry about here. So, we go ahead and use the trig identity just mentioned, and then we can integrate easily using the fact that $\int \sec^2 x dx = \tan x + C$. Continuing on with the problem:

$$= 3\int (\sec^2\theta - 1) d\theta$$

$$= 3[\tan\theta - \theta]$$

I did not put a constant of integration of "C" here, because remember that we are working on a definite integral, not an indefinite one. We cannot use the limits of $x = 3$ and $x = 5$ yet, because those limits are valid only for the variable of integration "x" and *not* for the variable "θ," which we have at this moment. Instead, I must first "back-substitute." I must convert this result back into the variable x. I will do this by using my triangle diagram. Looking back at my triangle diagram above, I see that I can back-substitute the "$\tan\theta$" with:

$$\frac{\sqrt{x^2-9}}{3}$$

The back-substitution for just the "θ" is a bit more tricky, since I have not "solved" any equation for θ up to this point. Using the fact that $\sec\theta = \frac{x}{3}$, I will solve this equation for θ by taking the inverse secant of both sides of the equation, thusly:

$$\sec^{-1}(\sec\theta) = \sec^{-1}\left(\frac{x}{3}\right)$$

$$\theta = \sec^{-1}\left(\frac{x}{3}\right)$$

Therefore, we have the anti-derivative, so we can apply the Fundamental Theorem of Calculus (i.e., evaluate at the limits)...

$$= 3\left[\frac{\sqrt{x^2-9}}{3} - \sec^{-1}\left(\frac{x}{3}\right)\right]_3^5$$

$$= 3\left\{\left[\frac{\sqrt{5^2-9}}{3} - \sec^{-1}\left(\frac{5}{3}\right)\right] - \left[\frac{\sqrt{3^2-9}}{3} - \sec^{-1}\left(\frac{3}{3}\right)\right]\right\}$$

$$= 3\left\{\left[\frac{\sqrt{16}}{3} - \cos^{-1}\left(\frac{3}{5}\right)\right] - \left[0 - \cos^{-1}(1)\right]\right\}$$

To evaluate $\cos^{-1}\left(\frac{3}{5}\right)$, I will use my calculator (be sure to use RADIANS mode). To evaluate $\cos^{-1}(1)$, I can use my brain! That is, what is the angle whose cosine value is equal to 1? That would be 0 degrees (also 0 radians)! Therefore, my final answer for this problem is that the mass of the lamina will be $3\left(\frac{4}{3} - 0.93\right) \approx 1.22$.

EXERCISES: #21 from **Quiz 9**, #8 from **Quiz 10**

4.5 SURFACE AREA

The definition for surface area states:

"If f and its partial derivatives are continuous on the closed region R in the xy-plane, then the area of the surface S given by $z = f(x, y)$ over R is given by:

$$\iint\limits_R dS = \iint\limits_R \sqrt{1 + \left[f_x(x, y)\right]^2 + \left[f_y(x, y)\right]^2}\, dA \text{ ,"}$$

Note the similarity of this integral to arc length. This integral is often very difficult to integrate. We will need to memorize this formula, as it will come in handy for the next chapter on "surface integrals." Surface integrals will look almost exactly like this surface area formula, except that there will be a bit more in the integrand. Don't worry so much about this now, but put it in your thinking cap for later.

Sometimes, you may have to do a bit of algebra to simplify the radicand. Sometimes, it helps to convert the double integral to polar coordinates to make it easier to integrate. You need to be able to distinguish what strategy is needed, and when.

This first example is a demonstration of simplifying the radicand first.

EXAMPLE. Find the surface area of the portion of the cone $z = \sqrt{x^2 + y^2}$ over the region bounded by the graphs of $x = 16 - y^2$ and $x = 0$.

SOLUTION. First, sketch the region in the xy-plane to determine the limits. That is, graph the equations $x = 16 - y^2$ and $x = 0$.

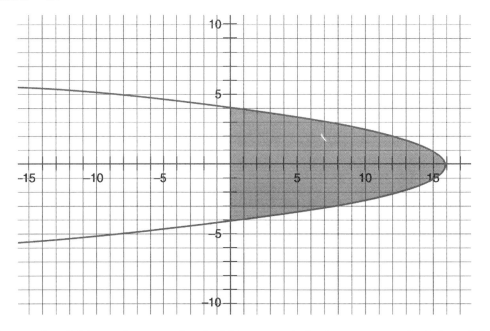

Notice that the graphs yield a region contained within a parabola that opens to the left and the y-axis. The partial derivatives are found using the Chain Rule and are as follows:

$$f_x = \frac{1}{2}\left(x^2 + y^2\right)^{-1/2} \cdot (2x) = \frac{x}{\sqrt{x^2 + y^2}}$$

$$f_y = \frac{1}{2}\left(x^2 + y^2\right)^{-1/2} \cdot (2y) = \frac{y}{\sqrt{x^2 + y^2}}$$

The double integral representing the surface area of the cone is:

$$\int_{-4}^{4} \int_{0}^{16-y^2} \sqrt{1 + \left[\frac{x}{\sqrt{x^2 + y^2}}\right]^2 + \left[\frac{y}{\sqrt{x^2 + y^2}}\right]^2}\ dxdy$$

This looks more awful than it actually is. We need to simplify the radicand. It turns out that the radicand is a constant! Take a look:

$$= \int_{-4}^{4} \int_{0}^{16-y^2} \sqrt{1 + \frac{x^2}{\left(\sqrt{x^2+y^2}\right)^2} + \frac{y^2}{\left(\sqrt{x^2+y^2}\right)^2}} \, dxdy$$

$$= \int_{-4}^{4} \int_{0}^{16-y^2} \sqrt{1 + \frac{x^2}{x^2+y^2} + \frac{y^2}{x^2+y^2}} \, dxdy$$

$$= \int_{-4}^{4} \int_{0}^{16-y^2} \sqrt{1 \cdot \left(\frac{x^2+y^2}{x^2+y^2}\right) + \frac{x^2}{x^2+y^2} + \frac{y^2}{x^2+y^2}} \, dxdy$$

$$= \int_{-4}^{4} \int_{0}^{16-y^2} \sqrt{\frac{x^2+y^2+x^2+y^2}{x^2+y^2}} \, dxdy$$

$$= \int_{-4}^{4} \int_{0}^{16-y^2} \sqrt{\frac{2x^2+2y^2}{x^2+y^2}} \, dxdy$$

$$= \int_{-4}^{4} \int_{0}^{16-y^2} \sqrt{\frac{2(x^2+y^2)}{x^2+y^2}} \, dxdy$$

$$= \int_{-4}^{4} \int_{0}^{16-y^2} \sqrt{2} \, dxdy$$

$$= \sqrt{2} \int_{-4}^{4} [x]_{0}^{16-y^2} \, dy$$

$$= \sqrt{2} \int_{-4}^{4} \left(16 - y^2\right) dy$$

$$= \sqrt{2} \left[16y - \frac{y^3}{3}\right]_{-4}^{4}$$

$$= \sqrt{2} \left[\left(16(4) - \frac{(4)^3}{3}\right) - \left(16(-4) - \frac{(-4)^3}{3}\right)\right]$$

$$= \sqrt{2} \left[\left(64 - \frac{64}{3}\right) - \left(-64 - \frac{(-64)}{3}\right)\right]$$

$$= \sqrt{2} \left[64 + 64 - \frac{64}{3} - \frac{64}{3}\right]$$

$$= \sqrt{2} \left[128 - \frac{128}{3}\right]$$

$$= \sqrt{2} \left(\frac{256}{3}\right)$$

This next example shows the advantage of recognizing when a double integral can be evaluated only if you reverse the order of integration.

EXAMPLE. Find the surface area of $f(x,y) = \dfrac{2}{5}y^{5/2}$ over the region bounded by the graphs of $y = \sqrt{x}$, $x = 4$, $x = 0$, and $y = 0$.

SOLUTION. After sketching the region, the limits for x and y are $0 \le y \le \sqrt{x}$ and $0 \le x \le 4$. See diagram below:

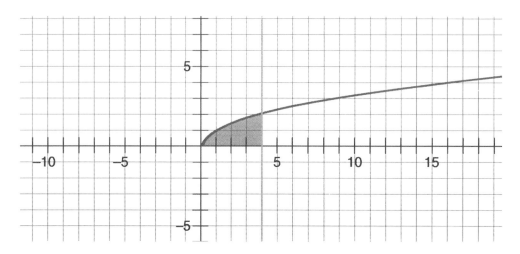

The partial derivatives for the function are:

$$f_x = 0 \quad \text{and} \quad f_y = \left(\frac{2}{5}\right)\left(\frac{5}{2}\right)y^{\frac{5}{2}-1} = y^{3/2}$$

We set up the double integral representing the surface area:

$$\int_0^4 \int_0^{\sqrt{x}} \sqrt{1 + \left[y^{3/2}\right]^2}\, dy\, dx \quad = \quad \int_0^4 \int_0^{\sqrt{x}} \sqrt{1 + y^3}\, dy\, dx$$

This integral is not integrable right now, even knowing all of our advanced integration techniques. However, if we reverse the order of integration, it can then be easily evaluated using the General Power Rule, as we shall see. Since we want the order of integration to be "*dxdy*," the limits for the variable y will need to be constants. From our sketch of the region, I hope you can see that the new limits for y will be $0 \le y \le 2$. To get the limits for x, we need to solve the equation $y = \sqrt{x}$ for x. After squaring both sides, we have that $x = y^2$. The lower limit for x is y^2. The new double integral representing surface area is:

$$\int_0^2 \int_{y^2}^4 \sqrt{1 + y^3}\, dx\, dy$$

We can integrate with respect to x easily. Continuing on:

$$= \int_0^2 \sqrt{1+y^3} \left[x \right]_{y^2}^4 dy$$

$$= \int_0^2 \sqrt{1+y^3} \left(4 - y^2 \right) dy$$

The integral $= \int_0^2 \sqrt{1+y^3} \left(4 - y^2 \right) dy$ is impossible to integrate using basic intehration techniques. We can use a calculator with an integration utility, for example the TI-84. Use the FNINT command to evaluate this definite integral.

The final answer is: 7.1875

This third example demonstrates the benefit of knowing how to convert to polar coordinates to make the integral easier to evaluate.

EXAMPLE. Find the surface area of the hemisphere $z = \sqrt{9 - x^2 - y^2}$ over the circle:

$$R = \left\{ (x, y) \mid x^2 + y^2 \le 4 \right\}$$

SOLUTION. The region has the limits $-\sqrt{4 - x^2} \le y \le \sqrt{4 - x^2}$ and $-2 \le x \le 2$. The partial derivatives are easier to see once you rewrite the function using an exponent instead of the radical. That is, we write $z = \left(9 - x^2 - y^2 \right)^{1/2}$. Then we use the Chain Rule to differentiate:

$$z_x = \frac{1}{2} \left(9 - x^2 - y^2 \right)^{-1/2} \cdot (-2x) = \frac{-x}{\sqrt{9 - x^2 - y^2}}$$

$$z_y = \frac{1}{2} \left(9 - x^2 - y^2 \right)^{-1/2} \cdot (-2y) = \frac{-y}{\sqrt{9 - x^2 - y^2}}$$

Now we can set up the double integral representing surface area:

$$\int_{-2}^{2} \int_{-\sqrt{4-x^2}}^{\sqrt{4-x^2}} \sqrt{1 + \left[\frac{-x}{\sqrt{9 - x^2 - y^2}} \right]^2 + \left[\frac{-y}{\sqrt{9 - x^2 - y^2}} \right]^2} \, dy dx$$

This is yet another instance where your algebra skills come in handy to simplify this messy integrand:

$$= \int_{-2}^{2} \int_{-\sqrt{4-x^2}}^{\sqrt{4-x^2}} \sqrt{1 + \frac{x^2}{\left(\sqrt{9-x^2-y^2}\right)^2} + \frac{y^2}{\left(\sqrt{9-x^2-y^2}\right)^2}} \, dydx$$

$$= \int_{-2}^{2} \int_{-\sqrt{4-x^2}}^{\sqrt{4-x^2}} \sqrt{\frac{9-x^2-y^2}{9-x^2-y^2} + \frac{x^2}{9-x^2-y^2} + \frac{y^2}{9-x^2-y^2}} \, dydx$$

$$= \int_{-2}^{2} \int_{-\sqrt{4-x^2}}^{\sqrt{4-x^2}} \sqrt{\frac{9-x^2-y^2+x^2+y^2}{9-x^2-y^2}} \, dydx$$

$$= \int_{-2}^{2} \int_{-\sqrt{4-x^2}}^{\sqrt{4-x^2}} \sqrt{\frac{9}{9-x^2-y^2}} \, dydx$$

$$= \int_{-2}^{2} \int_{-\sqrt{4-x^2}}^{\sqrt{4-x^2}} \frac{\sqrt{9}}{\sqrt{9-x^2-y^2}} \, dydx$$

$$= \int_{-2}^{2} \int_{-\sqrt{4-x^2}}^{\sqrt{4-x^2}} \frac{3}{\sqrt{9-x^2-y^2}} \, dydx$$

This integral will be awfully hard to integrate unless we convert to polar coordinates. Using the fact that $9-x^2-y^2 = 9-\left(x^2+y^2\right) = 9-r^2$, and the limits for r and θ over the region (a circle of radius 2) will be $0 \le r \le 2$ and $0 \le \theta \le 2\pi$. Now we can represent the surface area as a double integral in polar coordinates:

$$= \int_{0}^{2\pi} \int_{0}^{2} \frac{3}{\sqrt{9-r^2}} \, rdrd\theta$$

The good news is that this integral is easy to evaluate using u-substitution. We let $u = 9-r^2$ and $du = -2rdr$. This means we need to introduce a factor of -2 into the integrand (provided we divide it out in front):

$$= 3\left(-\frac{1}{2}\right) \int_{0}^{2\pi} \int_{0}^{2} \frac{1}{\sqrt{9-r^2}} (2r) drd\theta$$

$$= -\frac{3}{2} \int \int \frac{1}{u^{1/2}} \, du \quad = \quad -\frac{3}{2} \int \int u^{-1/2} \, du$$

$$= -\frac{3}{2} \int \left[2u^{1/2}\right] \quad = \quad -\frac{3}{2} \int_{0}^{2\pi} 2\left[\left(9-r^2\right)^{1/2}\right]_{0}^{2} \, d\theta$$

$$= -3\int_0^{2\pi}\left[\sqrt{9-r^2}\right]_0^2 d\theta$$

$$= -3\int_0^{2\pi}\left(\sqrt{9-(2)^2} - \sqrt{9-(0)^2}\right)d\theta$$

$$= -3\int_0^{2\pi}\left(\sqrt{5} - 3\right)d\theta$$

$$= -3(\sqrt{5}-3)\left[\theta\right]_0^{2\pi}$$

$$= +3(3-\sqrt{5})(2\pi)$$

$$= 6\pi\left(3-\sqrt{5}\right) \approx 14.4\, sq\, units$$

This last example will show that sometimes there are no "outs." Even if you were to reverse the order of integration or convert to polar coordinates, you will *still* be unable to evaluate the integral. In these instances, I will merely ask you to set up the double integral representing the surface area, and not to evaluate it.

EXAMPLE. Set up the double integral (but no need to evaluate) that represents the surface area of the graph of $f(x,y) = e^{xy}$ over the region $R = \left\{(x,y)\,\middle|\, 0 \le x \le 8,\ 0 \le y \le x\right\}$

SOLUTION. The limits for the region in the xy-plane are already given. So, let's go ahead and find the partial derivatives:

$$f_x = ye^{xy} \quad \text{and} \quad f_y = xe^{xy}\ .$$

We set up the double integral representing surface area:

$$\int_0^8\int_0^x \sqrt{1+\left[ye^{xy}\right]^2 + \left[xe^{xy}\right]^2}\,dydx \quad = \quad \int_0^8\int_0^x \sqrt{1+y^2 e^{2xy} + x^2 e^{2xy}}\,dydx$$

$$= \int_0^8\int_0^x \sqrt{1+e^{2xy}\left(x^2+y^2\right)}\,dydx$$

Now, it is tempting to try to convert to polar coordinates, especially since the expression $x^2 + y^2$ is easily replaced with r^2. However, remember that all instances of the variables x and y need to be converted to polar, and this integral will still be too hard after doing all of the conversions to polar! We would need a computer algebra system to do the evaluation.

This concludes the section.

EXERCISES: #10c, 13 from **Quiz 10**, #13b, 14, 19 from **Practice Midterm 4**

4.6 TRIPLE INTEGRALS

This section covers triple integrals using rectangular coordinates. The next section will cover triple integrals in two other coordinate systems: cylindrical and spherical (that you studied back in Chapter 1). A triple integral can be used in one of two ways:

1) To represent the volume of a solid region, Q. The notation is:

$$Volume = \iiint\limits_{Q} dV$$

Note that the integrand will not contain anything except one of the following:

$$dzdydx \qquad dzdxdy \qquad dydzdx \qquad dydxdz \qquad dxdzdy \qquad dxdydz$$

2) To represent the mass of a solid region Q, where the density function $\rho = f(x, y, z)$ is continuous over the solid region. The notation is:

$$Mass = \iiint\limits_{Q} f(x, y, z)dV$$

Here, the integral will have something in the integrand.

Key point: For both integrals, you are integrating over a solid region Q instead of a region R in the plane (as you did for double integrals).

So, now you have *two* ways to represent the volume of a solid: one using a double integral and one using a triple integral. It is easy to convert between the two:

$$Volume = \int_a^b \int_{g_1(x)}^{g_2(x)} f(x, y)dydx = \int_a^b \int_{g_1(x)}^{g_2(x)} \int_0^{f(x,y)} dzdydx$$

I will do five examples using triple integrals. The first one will be simply to evaluate the given triple integral. The second and third examples will be exercises in changing the order of integration. The fourth one will be an exercise in setting up a triple integral to represent volume. The last example will involve an improper integral.

EXAMPLE 1. Evaluate the triple integral $\displaystyle\int_0^2 \int_0^{\sqrt{3}/2} \int_0^{1/(4y^2+3)} dzdydx$.

SOLUTION. We integrate with respect to z first. We will then use the Arctan Rule to integrate:

$$= \int_0^2 \int_0^{\sqrt{3}/2} \Big[z\Big]_0^{1/(4y^2+3)} dydx$$

$$= \int_0^2 \int_0^{\sqrt{3}/2} \left[\frac{1}{4y^2+3}\right]dydx$$

Recall that the Arctan Rule formula for integration says:

$$\int \frac{1}{u^2 + a^2} du = \frac{1}{a} \arctan\left(\frac{u}{a}\right) + C$$

We will let $u = 2y$, $a = \sqrt{3}$, and $du = 2dy$. This means we are missing a factor of 2 in the integrand that we will put in (provided we also divide it out in front):

$$= \int_0^2 \left\{ \frac{1}{2} \int \frac{2}{(2y)^2 + \left(\sqrt{3}\right)^2} du \right\} dx$$

$$= \frac{1}{2} \int_0^2 \left[\frac{1}{\sqrt{3}} \arctan\left(\frac{2y}{\sqrt{3}}\right) \right]_0^{\sqrt{3}/2} dx$$

$$= \frac{1}{2\sqrt{3}} \int_0^2 \left[\arctan\left(\frac{2}{\sqrt{3}} \cdot \frac{\sqrt{3}}{2}\right) - \arctan(0) \right] dx$$

$$= \frac{1}{2\sqrt{3}} \int_0^2 \left[\arctan(1) - \arctan(0) \right] dx$$

We can use either our brains or our calculators to get the values for these arctangents. Remember, I will always take points off of your exams for decimal values (unless otherwise specified). To get arctan(1), read it as "the angle whose tangent is equal to 1." *Or*…using your graphing calculator (in "DEGREES" mode), you can see that arctan(1) = 45°. However, you must convert this to radians to get the correct answer; $\arctan(1) = \frac{\pi}{4}$. To get arctan(0), read it as, "the angle whose tangent is equal to 0." This gives 0 degrees (i.e., 0 radians). Continuing with the problem:

$$= \frac{1}{2\sqrt{3}} \int_0^2 \left[\frac{\pi}{4} \right] dx$$

$$= \frac{1}{2\sqrt{3}} \left(\frac{\pi}{4}\right) [x]_0^2$$

$$= \frac{\pi}{8\sqrt{3}} (2)$$

$$= \frac{\pi}{4\sqrt{3}}$$

EXAMPLE 2. Rewrite the triple integral $\int_0^2 \int_{2x}^4 \int_0^{\sqrt{y^2-4x^2}} dz\,dy\,dx$ using the order of integration "*dxdydz*." (No need to evaluate it, just set it up.)

SOLUTION. First, sketch what is going on in the *xy*-plane by sketching the graphs of the equations $x=0$, $x=2$, $y=2x$, and $y=4$. Notice that the region is a triangle-shaped area above the line $y=2x$, but below the line $y=4$ and to the right of the *y*-axis. This represents the base of the solid region for which we are finding the volume.

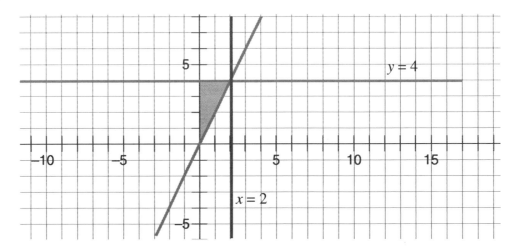

Then, solve for *x* with the equation $z=\sqrt{y^2-4x^2}$. The reason why we are solving for *x* is because the inside integral will be using the *x* for its variable of integration. The equation $z=\sqrt{y^2-4x^2}$ represents the top of the surface of the solid region for which we are finding the volume.

Reminder

Recall from Chapter 1 that the equation $z=\sqrt{y^2-4x^2}$ can be rewritten as $z^2=y^2-4x^2$, which is an elliptic cone, one of the quadric surfaces we studied.) After squaring both sides, etc., we have:

$$(z)^2 = \left(\sqrt{y^2-4x^2}\right)^2$$
$$z^2 = y^2 - 4x^2$$
$$4x^2 = y^2 - z^2$$
$$x^2 = \frac{y^2-z^2}{4}$$
$$\sqrt{x^2} = \pm\sqrt{\frac{y^2-z^2}{4}}$$
$$x = \ +\frac{\sqrt{y^2-z^2}}{2}$$

We chose the positive version because we are in the first octant. Therefore the limits for the variable x are:

$$0 \leq x \leq \frac{\sqrt{y^2 - z^2}}{2}$$

To see what the limits are for y and z, we need to take a look at what is going on in the yz-plane. To do this, we know that $x = 0$ in the yz-plane. So, we set $x = 0$ in the equation $z^2 = y^2 - 4x^2$. We get that:

$$z^2 = y^2 - 4(0)^2$$
$$z^2 = y^2$$
$$z = y$$

Therefore, we sketch the line $z = y$ in the yz-plane. We also know that the limits for y used to be $2x \leq y \leq 4$. Since the limits for x used to be $0 \leq x \leq 2$, we know that the maximum value for y will be 4. Thus, the limits for y will be $z \leq y \leq 4$. Then since $y = z$, we have that the lower limit for z will be when $x = 0$, so $2(x) = 2(0) = 0$. And since $z = y$, the upper limit for z will be 4. The final answer can be written:

$$\int_0^4 \int_z^4 \int_0^{\sqrt{y^2-z^2}/2} dx\,dy\,dz$$

In this third example, we will reverse the order of integration again. However, this time, you are to choose the appropriate order of integration to make the integral "do-able." In its current state, there is no way to evaluate the integral even using advanced techniques of integration!

EXAMPLE 3. Evaluate the triple integral $\displaystyle\int_0^2 \int_0^{4-x^2} \int_0^x \frac{\sin(2z)}{4-z} dy\,dz\,dx$ by changing the order of integration in an appropriate way.

SOLUTION. First, let's pretend we did not know we need to reverse the order of integration and start doing the problem by integrating the inside integral (with respect to the variable y). This first step is actually easy; it will be the middle integral that we cannot integrate.

$$= \int_0^2 \int_0^{4-x^2} \frac{\sin(2z)}{4-z} \left[y\right]_0^x dz\,dx$$

$$= \int_0^2 \int_0^{4-x^2} \frac{\sin(2z)}{4-z}(x-0) dz\,dx$$

$$= \int_0^2 \int_0^{4-x^2} \frac{\sin(2z)}{4-z}(x) dz\,dx$$

OUCH! HELP! Notice that there is no way we can evaluate this integrand with respect to z. Instead, let us examine what is going on in the xz-plane and reverse the order of integration. We will let the inside integral remain unchanged. We will let y continue to be the variable of integration for the innermost integral. The order of integration to which we are changing, will be "*dydxdz*."

We sketch the graphs of the equations $x = 0$, $x = 2$, $z = 0$, and $z = 4 - x^2$ to examine what is going on in the xz-plane. Notice that the region is the portion of a downward facing parabola in the xz-plane, with a z-intercept of 4. See diagram below:

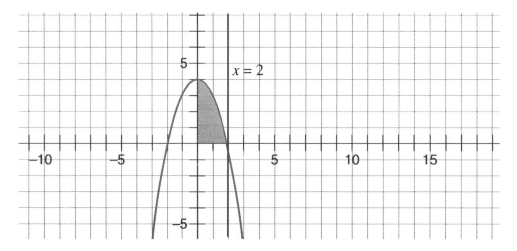

Since z will be the variable of integration for the outside integral, the limits need to be constants. Thus, the limits for z are $0 \leq z \leq 4$. The lower limit for x will be 0. To determine the upper limit for x, we need to solve the equation $z = 4 - x^2$ for x:

$$x^2 = 4 - z$$
$$\sqrt{x^2} = \pm\sqrt{4 - z}$$
$$x = +\sqrt{4 - z}$$

We chose the positive portion since we are in the first octant. Now we set up the triple integral and evaluate:

$$\int_0^4 \int_0^{\sqrt{4-z}} \int_0^x \frac{\sin(2z)}{4 - z} \, dy \, dx \, dz$$

$$= \int_0^4 \int_0^{\sqrt{4-z}} \frac{\sin(2z)}{4 - z} [y]_0^x \, dx \, dz$$

$$= \int_0^4 \int_0^{\sqrt{4-z}} \frac{\sin(2z)}{4 - z} [x] \, dx \, dz$$

$$= \int_0^4 \frac{\sin(2z)}{4 - z} \left[\frac{x^2}{2} \right]_0^{\sqrt{4-z}} \, dz$$

$$= \int_0^4 \frac{\sin(2z)}{4 - z} \left[\frac{\left(\sqrt{4 - z}\right)^2}{2} \right] dz$$

$$= \int_0^4 \left(\frac{\sin(2z)}{4 - z} \right) \cdot \frac{(4 - z)}{2} \, dz$$

We can "cancel" the expressions "$4 - z$" to reduce the fraction to lowest terms! We continue on:

$$= \frac{1}{2}\int_0^4 \sin(2z)dz$$

$$= \left(\frac{1}{2}\right)\left(\frac{1}{2}\right)\left[-\cos(2z)\right]_0^4$$

$$= -\frac{1}{4}\left[\cos(2\cdot 4) - \cos(0)\right]$$

$$= -\frac{1}{4}\left[\cos 8 - 1\right]$$

$$\approx -0.29$$

The next example will involve the set up of a triple integral to represent the volume of a solid region, where the only info given will be the graphs of the boundary of the solid region. The biggest challenge for one of these types of problems is, as usual, setting up the limits of integration.

EXAMPLE 4. Find the volume of the solid region in the first octant bounded by the coordinate planes, the plane $y + z = 2$, and the cylindrical surface $x = 4 - y^2$

SOLUTION. Since no order of integration is specified, you can choose whatever order is most convenient. If we examine what is going on in the xy-plane first, this may help us to determine which order of integration to choose. We will first sketch the graph of the equation $x = 4 - y^2$, a parabola opening to the left with an x-intercept of 4. Then since $z = 0$ in the xy-plane, substitute $z = 0$ into the equation $y + z = 2$, and we have that $y = 2$. So the graph of what is going on in the xy-plane requires that we sketch the graphs for the equations $y = 2$, $x = 0$, $y = 0$, and $x = 4 - y^2$. Why did I add the equations $x = 0$ and $y = 0$? Well, the problem stated that we want the volume of the solid in the *first octant,* which means that we are automatically bound by the coordinate planes (and therefore the axes). A diagram of the graphs are below:

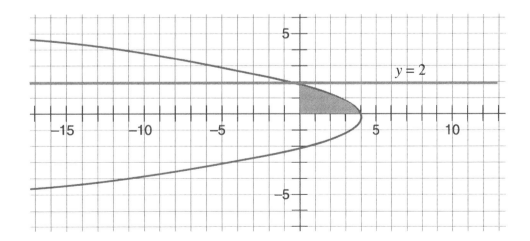

The order of integration in the xy-plane that is most convenient is "*dxdy*," where the limits of y will need to be constants ($0 \le y \le 2$). We will choose the overall order of integration to be "*dzdxdy*." The lower limit for z will be 0, since we are in the first octant. To find the upper limit for z, we solve the equation $y + z = 2$ for z, and get that $z = 2 - y$.

The triple integral representing the volume of the solid will be:

$$Volume = \int_0^2 \int_0^{4-y^2} \int_0^{2-y} dz\,dx\,dy$$

$$= \int_0^2 \int_0^{4-y^2} [z]_0^{2-y} \, dx\,dy$$

$$= \int_0^2 \int_0^{4-y^2} (2-y)\,dx\,dy$$

$$= \int_0^2 (2-y)[x]_0^{4-y^2} \, dy$$

$$= \int_0^2 (2-y)(4-y^2)\,dy$$

$$= \int_0^2 \left(8 - 2y^2 - 4y + y^3\right)dy$$

$$= \left[8y - \frac{2}{3}y^3 - 2y^2 + \frac{1}{4}y^4 \right]_0^2$$

$$= 8(2) - \frac{2}{3}(2)^3 - 2(2)^2 + \frac{1}{4}(2)^4$$

$$= 16 - \frac{16}{3} - 8 + \frac{16}{4}$$

$$= 8 - \frac{16}{3} + 4$$

$$= \frac{20}{3} \; cubic \; units$$

Note: When finding a volume, mass, or a surface area, the answer must always be a positive value! If not, you need to go back and re-do the problem, OK? *However,* it is possible for the value of a triple integral to be a negative number. This just means that the surface dips below the *xy*-plane more often than it is above the *xy*-plane. You remember from first semester calculus that it was possible for the value of a single integral to be a negative number, right? It just meant that the function dipped below the *x*-axis (i.e., was negative more than positive over the given interval). This gave a negative value for "area." Similarly, you can get a "negative" volume if the surface is "negative" over the given region.

This last example will involve an improper integral. This improper integral will be the type that has infinite limits. You studied improper integrals in your second semester calculus class. If you need to review, review your main text in your course.

EXAMPLE 5. Evaluate the triple integral $\int_0^{\pi/4} \int_0^{\ln(\sec v)} \int_{-\infty}^{2t} e^x \, dx \, dt \, dv$.

Note: This is an integral in *tvx*-space instead of *xyz*-space.

SOLUTION. We begin by integrating the inside integral, then evaluating the anti-derivative at the limits.

Note that the lower limit of the inside integral is infinite, that is $-\infty$. This is what makes it an "improper integral." Recall also that $e^{-\infty} = \dfrac{1}{e^{\infty}} = \dfrac{1}{\infty} = 0$

$$\int_0^{\pi/4} \int_0^{\ln(\sec v)} \int_{-\infty}^{2t} e^x \, dx \, dt \, dv = \int_0^{\pi/4} \int_0^{\ln(\sec v)} \left[e^x \right]_{-\infty}^{2t} dt \, dv$$

$$= \int_0^{\pi/4} \int_0^{\ln(\sec v)} \left[e^{2t} - e^{-\infty} \right] dt \, dv$$

$$= \int_0^{\pi/4} \int_0^{\ln(\sec v)} \left[e^{2t} \right] dt \, dv$$

$$= \int_0^{\pi/4} \left[\frac{1}{2} e^{2t} \right]_0^{\ln(\sec v)} dv$$

Before I go on, I want to remind you of the fact that $e^{\ln x} = x$ from your old pre-calculus class. Therefore, $e^{2\ln x} = \left(e^{\ln x} \right)^2 = x^2$. I will be using these facts to help me simplify the integrand in the next couple of steps:

$$= \frac{1}{2} \int_0^{\pi/4} \left[e^{2\ln(\sec v)} - e^0 \right] dv \qquad\qquad = \frac{1}{2} \left[\tan v - v \right]_0^{\pi/4}$$

$$= \frac{1}{2} \int_0^{\pi/4} \left[(\sec v)^2 - 1 \right] dv \qquad\qquad = \frac{1}{2} \left\{ \left[\tan \frac{\pi}{4} - \frac{\pi}{4} \right] - \left[\tan 0 - 0 \right] \right\}$$

Now recall an old integration formula that tells us that $\int \sec^2 u \, du = \tan u + C$. We use that in the next step: $\Longrightarrow$

$$= \frac{1}{2} \left[1 - \frac{\pi}{4} - 0 \right]$$

$$= \frac{1}{2} - \frac{\pi}{8}$$

This concludes the section.

EXERCISES: #9, 10b, 11a from **Quiz 10**, #8, 9, 10, 11a, 13a, 18c from **Practice Midterm 4**

4.7 TRIPLE INTEGRALS IN CYLINDRICAL AND SPHERICAL COORDINATES

Many triple integral problems are far easier done in either cylindrical or spherical coordinates than rectangular coordinates. We have studied both of these coordinate systems back in Chapter 1. A point described by the (rectangular) ordered triple (x, y, z), will be described in cylindrical coordinates as (r, θ, z), and in spherical coordinates by (ρ, θ, ϕ). I will repeat the tables from that section below containing the conversion formulas.

Conversion Formulas

FROM CYLINDRICAL TO RECTANGULAR	FROM RECTANGULAR TO CYLINDRICAL
$x = r\cos\theta$	$r^2 = x^2 + y^2$ or $r = \sqrt{x^2 + y^2}$
$y = r\sin\theta$	$\tan\theta = \dfrac{y}{x}$ or $\theta = \tan^{-1}\left(\dfrac{y}{x}\right)$
$z = z$	$z = z$

Conversion Formulas

FROM SPHERICAL TO RECTANGULAR	FROM RECTANGULAR TO SPHERICAL
$x = \rho\sin\phi\cos\theta$	$\rho^2 = x^2 + y^2 + z^2$ or $\rho = \sqrt{x^2 + y^2 + z^2}$
$y = \rho\sin\phi\sin\theta$	$\tan\theta = \dfrac{y}{x}$ or $\theta = \tan^{-1}\left(\dfrac{y}{x}\right)$
$z = \rho\cos\phi$	$\phi = \arccos\left(\dfrac{z}{\sqrt{x^2 + y^2 + z^2}}\right)$

Conversion Formulas

FROM SPHERICAL TO CYLINDRICAL	FROM CYLINDRICAL TO SPHERICAL
$r^2 = \rho^2\sin^2\phi$	$\rho = \sqrt{r^2 + z^2}$
$\theta = \theta$	$\theta = \theta$
$z = \rho\cos\phi$	$\phi = \arccos\left(\dfrac{z}{\sqrt{r^2 + z^2}}\right)$

To convert a Cartesian (rectangular) triple integral to cylindrical coordinates, use the definition:

$$\iiint_Q f(x,y,z)dV \;=\; \int_{\theta_1}^{\theta_2}\int_{g_1(\theta)}^{g_2(\theta)}\int_{h_1(r\cos\theta,\,r\sin\theta)}^{h_2(r\cos\theta,\,r\sin\theta)} f(r\cos\theta,\,r\sin\theta,\,z)\,rdzdrd\theta$$

Please note that you will replace "*dzdydx*" (or any permutation of this) by "*rdzdrd θ*" when converting the integral to cylindrical coordinates.

To convert a Cartesian triple integral to spherical coordinates, use the definition:

$$\iiint_Q f(x,y,z)dV \;=\; \int_{\theta_1}^{\theta_2}\int_{\phi_1}^{\phi_2}\int_{\rho_1}^{\rho_2} f(\rho\sin\phi\cos\theta,\,\rho\sin\phi\sin\theta,\,\rho\cos\phi)\,\rho^2\sin\phi\,d\rho\,d\phi\,d\theta$$

Please note that you will replace "*dzdydx*" (or any permutation of this) by " $\rho^2\sin\phi\,d\rho\,d\phi\,d\theta$ " when converting the integral to spherical coordinates.

EXAMPLE. Use cylindrical coordinates to find the mass of the solid region Q, where $Q = \left\{(x,y,z)\,\middle|\, 0\le z\le x^2,\;\; -1\le x\le 0,\;\; 0\le y\le\sqrt{1-x^2}\,\right\}$ and the density is given by $\rho = x^2 + y^2$.

SOLUTION. Recall that the mass of a solid region Q is given by:

$$\textbf{\textit{Mass}} = \iiint_Q \rho(x,y,z)dV$$

I will first set up the triple integral using rectangular coordinates:

$$\int_{-1}^{0}\int_{0}^{\sqrt{1-x^2}}\int_{0}^{x^2}\left(x^2 + y^2\right)dzdydx$$

This integral would be somewhat of a challenge to evaluate using these rectangular coordinates. It will be easier to evaluate using cylindrical coordinates anyway. Cylindrical coordinates are ideal for triple integrals where the region in the xy-plane is a circle, a sector of a circle, a cardioid, or a rose curve. After sketching the region in the xy-plane, I hope you can see that it is an area of a quarter of a circle in the second quadrant. If you are still having trouble seeing this, then you need to sketch the graph of the equations.

$x = -1,\; x = 0,\; y = 0,\;$ and $\; y = \sqrt{1-x^2}$.

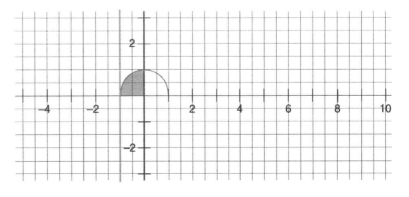

The good news about cylindrical coordinates is that the z variable stays the same. The limits for r and θ will be $0 \le r \le 1$, $\frac{\pi}{2} \le \theta \le \pi$. We convert the triple integral to cylindrical coordinates using the conversion formulas above:

$$\int_{\pi/2}^{\pi} \int_0^1 \int_0^{r^2 \cos^2 \theta} (r^2) r \, dz \, dr \, d\theta$$

$$= \int_{\pi/2}^{\pi} \int_0^1 \int_0^{r^2 \cos^2 \theta} (r^3) \, dz \, dr \, d\theta$$

$$= \int_{\pi/2}^{\pi} \int_0^1 r^3 [z]_0^{r^2 \cos^2 \theta} \, dr \, d\theta$$

$$= \int_{\pi/2}^{\pi} \int_0^1 r^3 (r^2 \cos^2 \theta) \, dr \, d\theta$$

$$= \int_{\pi/2}^{\pi} \int_0^1 r^5 \cos^2 \theta \, dr \, d\theta$$

$$= \int_{\pi/2}^{\pi} \cos^2 \theta \left[\frac{r^6}{6}\right]_0^1 \, d\theta$$

$$= \frac{1}{6} \int_{\pi/2}^{\pi} \cos^2 \theta \, d\theta$$

Do you recall the guidelines from last semester's calculus class on integration techniques for powers of trigonometric functions? Well, when the power of cosine is even and positive, you need to use the power reduction formula $\cos^2 \theta = \frac{1 + \cos(2\theta)}{2}$.

$$= \frac{1}{6} \int_{\pi/2}^{\pi} \frac{1 + \cos(2\theta)}{2} \, d\theta$$

$$= \frac{1}{6}\left(\frac{1}{2}\right) \int_{\pi/2}^{\pi} [(1 + \cos(2\theta)] \, d\theta$$

$$= \frac{1}{12}\left[\theta + \frac{1}{2}\sin(2\theta)\right]_{\pi/2}^{\pi}$$

$$= \frac{1}{12}\left[\pi + \frac{1}{2}\sin(2\pi)\right] - \frac{1}{12}\left[\frac{\pi}{2} + \frac{1}{2}\sin\left(2 \cdot \frac{\pi}{2}\right)\right]$$

$$= \frac{1}{12}[\pi + 0] - \frac{1}{12}\left[\frac{\pi}{2} + 0\right]$$

$$= \frac{\pi}{12} - \frac{\pi}{24}$$

$$= \frac{\pi}{24}$$

The next example will involve converting a rectangular triple integral to *both* cylindrical and spherical integrals, then choosing the one that is easiest to evaluate.

EXAMPLE. Convert the integrals from rectangular coordinates to *both* cylindrical and spherical coordinates, and evaluate the simplest iterated integral.

$$\textbf{(a)} \int_{-2}^{2} \int_{-\sqrt{4-x^2}}^{\sqrt{4-x^2}} \int_{x^2+y^2}^{4} x\,dz\,dy\,dx$$

$$\textbf{(b)} \int_{0}^{1} \int_{0}^{\sqrt{1-x^2}} \int_{0}^{\sqrt{1-x^2-y^2}} \sqrt{x^2+y^2+z^2}\,dz\,dy\,dx$$

SOLUTION.

(a) It is fairly easy to convert to cylindrical coordinates. We graph the region in the *xy*-plane by sketching the graphs of $x = -2$, $x = 2$, $y = -\sqrt{4-x^2}$ and $y = \sqrt{4-x^2}$ (an entire circle of radius 2).

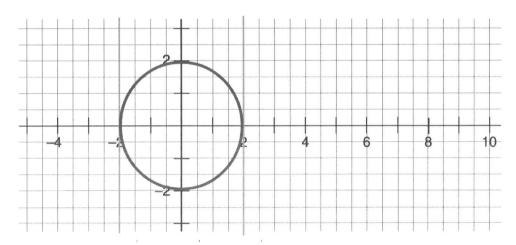

Then, we can convert easily to cylindrical coordinates and write:

$$\int_{0}^{2\pi} \int_{0}^{2} \int_{r^2}^{4} (r\cos\theta)r\,dz\,dr\,d\theta \;=\; \int_{0}^{2\pi} \int_{0}^{2} \int_{r^2}^{4} (r^2\cos\theta)dz\,dr\,d\theta$$

This one is tricky to convert to spherical coordinates. It turns out that we will need to take the sum of two separate integrals to convert this single integral to spherical coordinates.

The angle θ will still range between 0 and 2π. The variables ρ and ϕ will be a bit challenging. We first need to take a look at the equation that we obtain from the limits of the variable z that are $z = 4$ and $z = x^2 + y^2$.

These equations' graphs represent a plane and an elliptic paraboloid, respectively. We need to convert these rectangular equations to corresponding equations in spherical coordinates. Recall that we did a lot of this back in Chapter 1. We use the conversion formulas listed on page 159:

$$\rho\cos\phi = 4 \qquad \text{and} \qquad \rho\cos\phi = (\rho\sin\phi\cos\theta)^2 + (\rho\sin\phi\sin\theta)^2$$

$$\rho = \frac{4}{\cos\phi} \qquad\qquad\qquad \rho\cos\phi = \rho^2\sin^2\phi\cos^2\theta + \rho^2\sin^2\phi\sin^2\theta$$

$$\rho = 4\sec\phi \qquad\qquad\qquad \rho\cos\phi = \rho^2\sin^2\phi(\cos^2\theta + \sin^2\theta)$$

$$\cos\phi = \rho\sin^2\phi$$

$$\frac{\cos\phi}{\sin^2\phi} = \rho$$

$$\frac{\cos\phi}{\sin\phi}\cdot\frac{1}{\sin\phi} = \rho$$

$$\cot\phi\csc\phi = \rho$$

Now we need to figure out at what angle of ϕ (that I refer to as "the drop-down angle") do the plane and paraboloid intersect? When do the two equations found above for ρ equal each other? I will set them equal, and then solve the resulting equation for ϕ.

$$\rho = \rho$$

$$4\sec\phi = \cot\phi\csc\phi$$

$$\frac{4}{\cos\phi} = \frac{\cos\phi}{\sin\phi}\cdot\frac{1}{\sin\phi}$$

$$4 = \frac{\cos^2\phi}{\sin^2\phi}$$

$$\sqrt{4} = \sqrt{\frac{\cos^2\phi}{\sin^2\phi}}$$

$$2 = \frac{\cos\phi}{\sin\phi}$$

Note: We *do not* use the Square Root Property that tells us to remember to take "±" of one side of the equation when we square-root both sides because the value of ρ is always understood to be positive (unlike the "*r*" from cylindrical or polar coordinates which *can* be positive or negative).

$$\frac{1}{2} = \frac{\sin\phi}{\cos\phi}$$

$$\frac{1}{2} = \tan\phi$$

$$\tan^{-1}\left(\frac{1}{2}\right) = \tan^{-1}(\tan\phi)$$

$$\tan^{-1}\left(\frac{1}{2}\right) = \phi$$

So, the angle ϕ will start "drop-down" from the z-axis, and it will drop down to $\tan^{-1}\left(\dfrac{1}{2}\right)$ as it "sweeps out" the plane $z = 4$, then it will go from $\tan^{-1}\left(\dfrac{1}{2}\right)$ to $\dfrac{\pi}{2}$ when we "sweep out" to integrate the portion of the solid that is the elliptic paraboloid.

We can finally write the triple integrals for the conversion to spherical coordinates. The first integral represents the portion of the solid as we "sweep out" the plane $z = 4$, and the second integral represents the portion of the solid as we "sweep out" the elliptic paraboloid, with the equation $z^2 = x^2 + y^2$.

$$\int_0^{2\pi} \int_0^{\tan^{-1}(1/2)} \int_0^{4\sec\phi} (\rho\sin\phi\cos\theta)\,\rho^2\sin\phi\,d\rho\,d\phi\,d\theta \;+\; \int_0^{2\pi} \int_{\tan^{-1}(1/2)}^{\pi/2} \int_0^{\cot\phi\csc\phi} (\rho\sin\phi\cos\theta)\,\rho^2\sin\phi\,d\rho\,d\phi\,d\theta$$

Finally, we choose which one of these conversions is easiest to evaluate. I believe that the triple integral in cylindrical coordinates is easiest. We evaluate that here:

$$\int_0^{2\pi} \int_0^2 \int_{r^2}^4 (r^2\cos\theta)\,dz\,dr\,d\theta$$

$$= \int_0^{2\pi} \int_0^2 (r^2\cos\theta)[z]_{r^2}^4\,dr\,d\theta$$

$$= \int_0^{2\pi} \int_0^2 (r^2\cos\theta)(4 - r^2)\,dr\,d\theta$$

$$= \int_0^{2\pi} \int_0^2 (4r^2\cos\theta - r^4\cos\theta)\,dr\,d\theta$$

$$= \int_0^{2\pi} \left[\frac{4}{3}r^3\cos\theta - \frac{r^5}{5}\cos\theta\right]_0^2 d\theta$$

$$= \int_0^{2\pi} \left[\frac{4}{3}(2)^3\cos\theta - \frac{(2)^5}{5}\cos\theta\right] d\theta$$

$$= \int_0^{2\pi} \left[\frac{32}{3}\cos\theta - \frac{32}{5}\cos\theta\right] d\theta$$

$$= \frac{64}{15}\int_0^{2\pi}\cos\theta\,d\theta \qquad\Longrightarrow\qquad = \frac{64}{15}\left[\sin 2\pi - \sin 0\right]$$

$$= \frac{64}{15}\left[\sin\theta\right]_0^{2\pi} \qquad\qquad\qquad\qquad = 0$$

(b) Again, converting the triple integral $\int\limits_0^1 \int\limits_0^{\sqrt{1-x^2}} \int\limits_0^{\sqrt{1-x^2-y^2}} \sqrt{x^2+y^2+z^2}\,dzdydx$ to cylindrical coordinates is easy. The region in the *xy*-plane is a quarter of a circle of radius 1 centered at the origin in the first quadrant. We find this out by sketching the graphs of $x=0$, $x=1$, $y=0$ and $y=\sqrt{1-x^2}$:

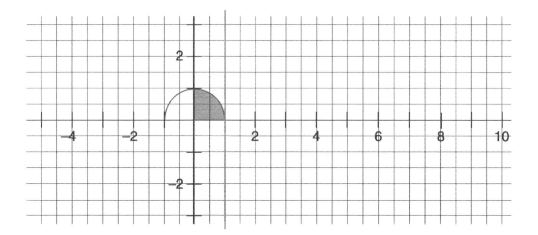

Convert to cylindrical coordinates, and our triple integral becomes:

$$\int\limits_0^{\pi/2} \int\limits_0^1 \int\limits_0^{\sqrt{1-r^2}} \sqrt{r^2+z^2}\ rdzdrd\theta$$

To convert to spherical coordinates, we note that the equation:

$$z=\sqrt{1-x^2-y^2}$$
$$z^2=1-x^2-y^2$$
$$x^2+y^2+z^2=1$$

...is a sphere of radius of 1 centered at the origin. Since this integral represents the portion of the sphere in the first octant only, the "drop-down" angle of ϕ will be equal to $\dfrac{\pi}{2}$. We convert to spherical coordinates:

$$\int\limits_0^{\pi/2} \int\limits_0^{\pi/2} \int\limits_0^1 \sqrt{\rho^2}\ \rho^2 \sin\phi\,d\rho d\theta d\phi \ = \ \int\limits_0^{\pi/2} \int\limits_0^{\pi/2} \int\limits_0^1 \rho^3 \sin\phi\,d\rho d\theta d\phi$$

Now we choose which of the two triple integrals is easiest to evaluate. This time, the triple integral in spherical coordinates is easiest to evaluate. The reason why the integral in cylindrical coordinates is challenging is because to integrate with respect to *z* we would need to use trigonometric substitution (which is not fun, remember?) to tackle that radical of $\sqrt{r^2+z^2}$.

We evaluate the triple integral in spherical coordinates:

$$= \int_0^{\pi/2} \int_0^{\pi/2} \sin\phi \left[\frac{\rho^4}{4}\right]_0^1 d\theta d\phi$$

$$= \int_0^{\pi/2} \int_0^{\pi/2} \sin\phi \left[\frac{1^4}{4}\right] d\theta d\phi$$

$$= \frac{1}{4} \int_0^{\pi/2} \sin\phi [\theta]_0^{\pi/2} \, d\phi$$

$$= \frac{1}{4} \int_0^{\pi/2} \sin\phi \left(\frac{\pi}{2}\right) d\phi$$

$$= \frac{1}{4}\left(\frac{\pi}{2}\right)[-\cos\phi]_0^{\pi/2}$$

$$= -\frac{\pi}{8}\left[\cos\frac{\pi}{2} - \cos 0\right]$$

$$= -\frac{\pi}{8}[0-1]$$

$$= \frac{\pi}{8}$$

EXAMPLE. Evaluate the triple integral:

$$\int_{\pi/6}^{\pi/2} \int_{-\pi/2}^{\pi/2} \int_{\csc\phi}^{2} 5\rho^4 \sin^3\phi \, d\rho \, d\theta \, d\phi$$

SOLUTION. First, integrate with respect to the variable ρ:

$$= \int_{\pi/6}^{\pi/2} \int_{-\pi/2}^{\pi/2} \sin^3\phi \left[\rho^5\right]_{\csc\phi}^{2} d\theta \, d\phi$$

$$= \int_{\pi/6}^{\pi/2} \int_{-\pi/2}^{\pi/2} \sin^3\phi \left[2^5 - \csc^5\phi\right] d\theta \, d\phi$$

$$= \int_{\pi/6}^{\pi/2} \int_{-\pi/2}^{\pi/2} \left[32\sin^3\phi - (\sin^3\phi)(\csc^5\phi)\right] d\theta \, d\phi$$

$$= \int_{\pi/6}^{\pi/2} \int_{-\pi/2}^{\pi/2} \left[32\sin^3\phi - (\sin^3\phi)\left(\frac{1}{\sin^5\phi}\right)\right] d\theta \, d\phi$$

$$= \int_{\pi/6}^{\pi/2} \int_{-\pi/2}^{\pi/2} \left[32\sin^3\phi - \csc^2\phi)\right] d\theta \, d\phi$$

Next, we integrate with respect to the variable θ, which is easy since the entire integrand is a constant with respect to θ.

$$= \int_{\pi/6}^{\pi/2} (32\sin^3\phi - \csc^2\phi)[\theta]_{-\pi/2}^{\pi/2}\, d\phi$$

$$= \int_{\pi/6}^{\pi/2} (32\sin^3\phi - \csc^2\phi)\left(\frac{\pi}{2} - \frac{-\pi}{2}\right) d\phi$$

$$= \int_{\pi/6}^{\pi/2} (32\sin^3\phi - \csc^2\phi)(\pi)\, d\phi$$

Finally, we integrate with respect to the variable ϕ. The first term is a power of the sine function. Since the power is odd and positive, we save a sine factor, and convert the remaining sines to cosines:

$$= \pi\int_{\pi/6}^{\pi/2} (32\sin^2\phi\sin\phi - \csc^2\phi)d\phi$$

$$= \pi\int_{\pi/6}^{\pi/2} \left[32(1 - \cos^2\phi)(\sin\phi) - \csc^2\phi\right]d\phi$$

$$= \pi\int_{\pi/6}^{\pi/2} \left[32\sin\phi - 32\cos^2\phi\sin\phi - \csc^2\phi\right]d\phi$$

Now, recall a formula from first semester calculus that states $\int(-\csc^2 u)du = \cot u + C$. We will use that formula to integrate the last term in the integrand. The second term that will require u-substitution, where $u = \cos^2\phi$ and $du = -\sin\phi$. Then we say that $\int u^2\, du = \dfrac{u^3}{3} = \dfrac{\cos^3\phi}{3}$.

Continuing on:

$$= \pi\left[-32\cos\phi + \frac{32}{3}\cos^3\phi + \cot\phi\right]_{\pi/6}^{\pi/2}$$

$$= \pi\left\{\left[-32\cos\frac{\pi}{2} + \frac{32}{3}\cos^3\frac{\pi}{2} + \cot\frac{\pi}{2}\right] - \left[-32\cos\frac{\pi}{6} + \frac{32}{3}\cos^3\frac{\pi}{6} + \cot\frac{\pi}{6}\right]\right\}$$

To evaluate the expression $\cot\dfrac{\pi}{2}$, we use a trig ID:

$$\cot\frac{\pi}{2} = \frac{\cos(\pi/2)}{\sin(\pi/2)} = \frac{0}{1} = 0$$

Similarly, we have that:

$$\cot\frac{\pi}{6} = \frac{\cos(\pi/6)}{\sin(\pi/6)} = \frac{\sqrt{3}/2}{1/2} = \frac{\sqrt{3}}{2} \cdot \frac{2}{1} = \sqrt{3}.$$

We finish this problem:

$$= \pi\left\{\left[-32(0) + \frac{32}{3}(0) + 0\right] - \left[-32\left(\frac{\sqrt{3}}{2}\right) + \frac{32}{3}\left(\frac{\sqrt{3}}{2}\right)^3 + \sqrt{3}\right]\right\}$$

$$= \pi\left(16\sqrt{3} - \frac{32}{3} \cdot \frac{3\sqrt{3}}{8} - \sqrt{3}\right)$$

$$= \pi\left(16\sqrt{3} - 4\sqrt{3} - \sqrt{3}\right)$$

$$= 11\pi\sqrt{3}$$

EXAMPLE. Use spherical coordinates to *set up* the triple integral that represents the volume of the solid between the spheres $x^2 + y^2 + z^2 = 4$ and $x^2 + y^2 + z^2 = 49$, and inside the cone $z^2 = x^2 + y^2$. (No need to evaluate.)

SOLUTION. The variable ρ will range between the radii of the spheres: $2 \leq \rho \leq 7$.

We need to convert the equation for the cone into an equation in spherical coordinates so that we can obtain the "drop-down" angle, or ϕ. Using the conversion formulas:

$$z^2 = x^2 + y^2$$
$$\rho^2\cos^2\phi = \rho^2\sin^2\phi\cos^2\theta + \rho^2\sin^2\phi\sin^2\theta$$
$$\rho^2\cos^2\phi = \rho^2\sin^2\phi(\cos^2\theta + \sin^2\theta)$$
$$\rho^2\cos^2\phi = \rho^2\sin^2\phi$$
$$1 = \frac{\sin^2\phi}{\cos^2\phi}$$
$$\sqrt{1} = \sqrt{\frac{\sin^2\phi}{\cos^2\phi}}$$
$$1 = \frac{\sin\phi}{\cos\phi}$$

$$1 = \tan\phi$$
$$\tan^{-1}(1) = \tan^{-1}(\tan\phi)$$
$$\frac{\pi}{4} = \phi$$

the "drop-down" angle, or ϕ. Using the conversion formulas:

$$z^2 = x^2 + y^2$$
$$\rho^2 \cos^2 \phi = \rho^2 \sin^2 \phi \cos^2 \theta + \rho^2 \sin^2 \phi \sin^2 \theta$$
$$\rho^2 \cos^2 \phi = \rho^2 \sin^2 \phi (\cos^2 \theta + \sin^2 \theta)$$
$$\rho^2 \cos^2 \phi = \rho^2 \sin^2 \phi$$
$$1 = \frac{\sin^2 \phi}{\cos^2 \phi}$$
$$\sqrt{1} = \sqrt{\frac{\sin^2 \phi}{\cos^2 \phi}}$$
$$1 = \frac{\sin \phi}{\cos \phi}$$

So, the "drop down" angle of ϕ will have the range $0 \le \phi \le \dfrac{\pi}{4}$. The other angle θ, will vary between 0 and 2π, since the projection of the two spheres onto the xy-plane will be two concentric circles that cover all four quadrants. We write the triple integral representing the volume of this solid:

$$\textbf{\textit{Volume}} = \int_0^{\pi/4} \int_0^{2\pi} \int_2^7 \rho^2 \sin \phi \, d\rho \, d\theta \, d\phi$$

Due to symmetry in all eight octants, we can write the triple integral that represents the volume in the first octant only and then multiply by a factor of 8. In a nutshell:

$$\textbf{\textit{Volume}} = 8 \int_0^{\pi/4} \int_0^{\pi/2} \int_2^7 \rho^2 \sin \phi \, d\rho \, d\theta \, d\phi$$

Either answer is OK.

This concludes the chapter!

Exercises: #11b, 14–18 all, Extra Credit Problem from Quiz 10; #11b, 12, 15, 16, 22, 23 from Practice Midterm 4; #27b from **Practice Final**

Chapter 5

VECTOR ANALYSIS

Chapter Comments

This chapter will take everything you've gained over the course of the semester and bring it all together wrapped up in one nice and shiny package! The material here will be challenging, as it contains many new ideas and terminology.

The first four sections of this chapter consider situations in which the integration is done over a plane region bounded by curves. You will be introduced to a brand new type of function called "vector fields." You will learn what a line integral is and how to evaluate one. You will also learn an application from physics to determine the amount of work done using line integrals. The two big theorems in the first part of Chapter 5 are the Fundamental Theorem of Line Integrals and Green's Theorem. The concepts of "conservative vector fields" and "potential functions" are also covered here, which we will be useful to students moving on to take a course in Differential Equations.

The last four sections of Chapter 5 consider situations in which the integration is done over regions in space bounded by surfaces. I will skip the section on parametric surfaces, due to lack of time. We will not need this concept to understand the last three big theorems for this course 1) The Divergence Theorem, 2) Flux and the Divergence Theorem, and 3) Stokes' Theorem. You will learn how to do both surface integrals and flux integrals. It also turns out that both the Divergence Theorem and Stokes' Theorem are simply higher-level analogues for Green's Theorem.

5.1 VECTOR FIELDS

Recall from Chapter 2 the concept of vector-valued functions. A vector-valued function assigned a vector to a real number. Well, a vector *field* is a function that assigns a vector to a given ordered pair (or point) in the plane. Here are two ("function machine") diagrams to help you see the difference:

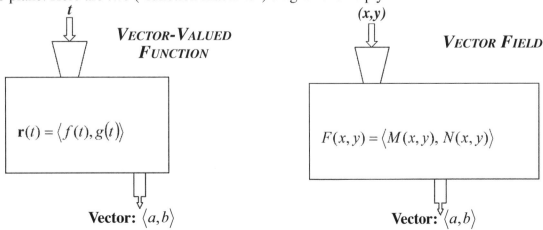

Vector fields are useful in representing various types of force fields (magnetic, electric, gravitational) and velocity fields. Velocity fields describe motions of systems of particles in the plane or space. Believe it or not, you are already familiar with one type of vector field—*gradients!* We just have not told you that a gradient is a vector field yet! But, you know. Can you see why? As an example, suppose you are given the function of two variables: $f(x, y) = x^2 + y^2$. Then the gradient of this function is given by:

$$\nabla f(x, y) = \langle f_x, f_y \rangle = \langle 2x, 2y \rangle$$

This is a vector field. You "feed" the function an input of a point (x,y) and you receive a vector as a result!

Sketching vector fields in the plane is fairly easy. Sketching vector fields in space is a bit more challenging. The Test Bank in this book has problems involve sketching only vector fields in the plane. Please note that because vector fields consist of infinitely many vectors, it is only desirable to sketch enough vectors to help you visualize the field. Usually this means sketching at most, ten vectors.

Procedure to Sketch a Vector Field in the Plane

1) Although you could just plot vectors located at several random points in the plane, it is more "enlightening" to plot vectors of equal magnitude. This corresponds to finding level curves in scalar fields. So, set the magnitude of the vector field function equal to a constant, say k, and take a look at the resulting equation. (Is it an equation of a line? A circle? A parabola? An ellipse?)

2) Make a table with two columns. The first column is for the points (x, y). The second column is for the vector result (output) of the vector field you are attempting to sketch. Fill out the table with enough vectors to get a good picture for the field. Again usually ten rows is enough.

3) Plot the vectors. However, do *not* plot the vectors "sprouting" out of the origin, even though they are in component form. Each vector corresponds to a point, so you want to have the vector "sprouting" out of its corresponding point. You may *start* by sketching it sprouting out of the origin, but then you need to *translate* the vector to its corresponding point, while maintaining both its magnitude and direction.

EXAMPLE. Find the "level curves" for the following vector fields. Set up a table and plot a few of the vectors:

(a) $\mathbf{F}(x, y) = 6\mathbf{i} + 8\mathbf{j}$

(b) $\mathbf{F}(x, y) = x\mathbf{i} - y\mathbf{j}$

SOLUTION.

(a) First, determine where the vectors of equal magnitude will be by finding the "level curves" for the vector field. We do this by setting the magnitude of the force equal to a scalar, k:

$$\|\mathbf{F}(x, y)\| = k$$
$$\sqrt{6^2 + 8^2} = k$$
$$\sqrt{36 + 64} = k$$
$$\sqrt{100} = k$$
$$10 = k$$

Unfortunately, this is *not* an equation. This process just yielded a constant...but this is actually good news! What this means is that all vectors will have a magnitude of 10 at every single point in the plane. So, a table would look something like this:

Point (x, y) (input)	F = <6, 8> (output)
(0, 0)	<6, 8>
(−1, 0)	<6, 8>
(1, 0)	<6, 8>
(1, 1)	<6, 8>
(0, 1)	<6, 8>

The sketch of the vector field will look like a collection of vectors all pointing slightly up and to the right.

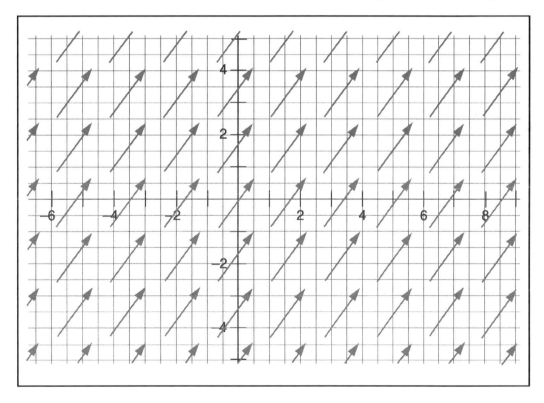

(b) Again, set the magnitude of the force equal to a constant:

$$\|\mathbf{F}(x, y)\| = k$$
$$\sqrt{(x)^2 + (-y)^2} = k$$
$$\sqrt{x^2 + y^2} = k$$
$$x^2 + y^2 = k^2$$

This equation is the equation for a circle of radius k. What this means is that when you set up a table of points is that the points should all be clustered around circles of radius k, for different values of k. For example, if we let $k = 1$, then choose four (or so) points that "live" on a circle of radius 1 to put into the table. If we let $k = 2$, then choose four (or so) points that "live" on a circle of radius 2 to put into the table.

POINT (X, Y)(INPUT)	$F = <X, -Y>$(*OUTPUT*)
(0, 0)	<0, 0> ($k = 0$)
(−1, 0)	<−1, 0> ($k = 1$)
(1, 0)	<1, 0> ($k = 1$)
(0, −1)	<0, 1> ($k = 1$)
(0, 1)	<0, −1> ($k = 1$)
(−2, 0)	<−2, 0> ($k = 2$)
(2, 0)	<2, 0> ($k = 2$)
(0, −2)	<0, 2> ($k = 2$)
(0, 2)	<0, −2> ($k = 2$)

The sketch of the vector field will be a collection of arrows, as shown.

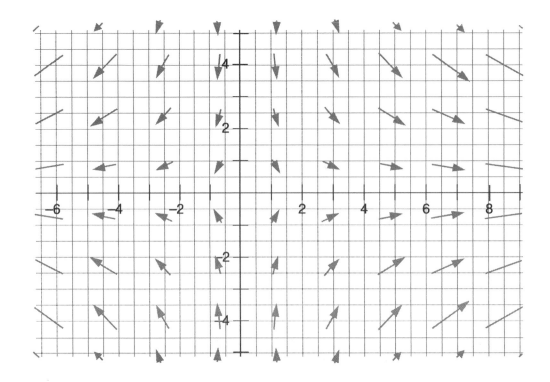

Conservative Vector Fields and Potential Functions.

Fact: *some* vector fields *are* in fact gradients for differentiable functions. (Note that I used the word *some,* not *all.*) These are "special" vector fields, and are referred to as "conservative vector fields." There is a quick test to determine if a vector field is conservative or not: Use the Theorem entitled "Test for Conservative Vector Field in the Plane." (We will find out later how to determine if a vector field in *space* is conservative or not.)

The theorem states that if you are given a vector field $\mathbf{F}(x, y) = \langle M(x, y), N(x, y) \rangle$, it will be conservative if and only if $\dfrac{\partial N}{\partial x} = \dfrac{\partial M}{\partial y}$. If it is in fact conservative, this means you can "recover" the function for which this vector field is a gradient. This "recovered" function is referred to as the "potential function." To find this potential function, we need to anti-differentiate *twice*: once with respect to x, and then again with respect to y.

EXAMPLE. Find the gradient vector field for the scalar function. (***Note:*** You are finding the conservative vector field for the potential function here.):

$$f(x, y, z) = \ln \sqrt{x^2 + y^2 + z^2}$$

SOLUTION. All we need to do is to find the gradient for this function $\nabla f = \langle f_x, f_y, f_z \rangle$. I will first use a property of logarithms to simplify this potential function before we differentiate.

$$f(x, y, z) = \ln \sqrt{x^2 + y^2 + z^2} = \ln\left(x^2 + y^2 + z^2\right)^{1/2} = \frac{1}{2}\ln\left(x^2 + y^2 + z^2\right)$$

Now we will find the gradient:

$$\nabla f = \langle f_x, f_y, f_z \rangle = \left\langle \frac{1}{2}\left(\frac{1}{x^2 + y^2 + z^2}\right)\cdot(2x),\ \frac{1}{2}\left(\frac{1}{x^2 + y^2 + z^2}\right)\cdot(2y),\ \frac{1}{2}\left(\frac{1}{x^2 + y^2 + z^2}\right)\cdot(2z) \right\rangle$$

$$= \left\langle \frac{x}{x^2 + y^2 + z^2},\ \frac{y}{x^2 + y^2 + z^2},\ \frac{z}{x^2 + y^2 + z^2} \right\rangle$$

EXAMPLE. Determine whether the vector fields are conservative. If they are, find the corresponding potential function.

(a) $\mathbf{F}(x, y) = \dfrac{1}{y}\mathbf{i} - \dfrac{y}{x^2}\mathbf{j}$

(b) $\mathbf{F}(x, y) = \left\langle -\dfrac{y}{x^2}, \dfrac{1}{x} \right\rangle$

SOLUTION.

(a) We need to ask: Does $\dfrac{\partial N}{\partial x} = \dfrac{\partial M}{\partial y}$?

Let's check:

$$\frac{\partial N}{\partial x} = -y(-2x^{-3}) = \frac{2y}{x^3} \quad \text{and} \quad \frac{\partial M}{\partial y} = -y^{-2} = -\frac{1}{y^2}$$

No, they are not equal. Therefore, the vector field (a.k.a. the gradient) is not conservative, so no corresponding potential function exists.

(b) Again, we need to check to see if $\dfrac{\partial N}{\partial x} = \dfrac{\partial M}{\partial y}$:

$$\frac{\partial N}{\partial x} = -x^{-2} = -\frac{1}{x^2} \quad \text{and} \quad \frac{\partial M}{\partial y} = -\frac{1}{x^2}$$

Yes! They are equal! Therefore, the vector field is a conservative one. Thus, we can find the potential function. We need to integrate twice. We integrate $M(x, y)$ with respect to x, then we integrate $N(x,y)$ with respect to y. Then, sum up the results. If any "repeats" occur, take just one of them.

$$f(x,y) = \int M(x,y)dx = \int \left(-\frac{y}{x^2}\right)dx = \int (-y)(x^{-2})dx = (-y)\left(\frac{x^{-1}}{-1}\right) + g(y) = \frac{y}{x} + g(y)$$

$$f(x,y) = \int N(x,y)dy = \int \left(\frac{1}{x}\right)dy = \left(\frac{1}{x}\right)y + h(x) = \frac{y}{x} + h(x)$$

We have a repeat here, so just "take it once."

Final answer: $f(x,y) = \dfrac{y}{x} + C$

There are a couple more concepts involving vector fields known as "divergence" and "curl" that we will look at a little later (section 5.4, Green's Theorem).

EXERCISES: #1, 2, 3, 4, 20 from **Quiz 11**

5.2 LINE INTEGRALS

In this section, we will learn what a line integral is and how to evaluate one. Before we do, it is important that you know how to parameterize a curve. The reason why is because a line integral is an integral where you evaluate over a curve (instead of an area in the plane or a solid in space). You may have learned a little bit about this back in Chapter 2 (as well as in second-semester calculus), but it may get a bit challenging. For curves in two dimensions, the third variable is automatically set at 0. The best thing you can do is to get an equation for the curve relating the two variables to each other (e.g., $y = 2x - 1$). Then you can set one of the variables equal to the parameter "t" and the other variable will be easily described with this parameter as well, using the equation.

EXAMPLE. Parameterize the line segment starting from the point $(2, -1, 2)$ to the point $(0, 3, -2)$.

SOLUTION. First, we need to decide what variable to use as our parameter (traditionally, it is "t") and also the starting and ending values for this parameter. Let's use "t" and have t start at $t = 0$ and end at $t = 1$. Remember, these are arbitrary. We are the "boss" and we get to decide. We *could* have had t start at $t = -17$ and ended at $t = 100$ if we wanted to. The result would have been just as valid. Recall from second semester calculus that parameterizations for curves are *not* unique! Now, let us start with parameterizing x as it relates to t. Note that x starts at 2 and ends at 0, where t starts at 0 and ends at 1. We need to come up with an equation describing the relationship between t and x. To do this, we can start by drawing up a chart and then plot these points from the chart on a tx-plane:

t	x
0	2
1	0

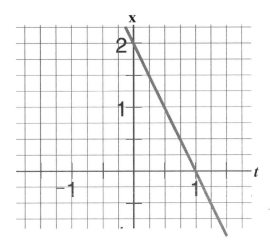

Now, find the equation of the line represented in the diagram. This will give us the relationship between t and x that we require. We'd like to get the equation of this line in slope-intercept form $x = mt + b$:

$$\textbf{\textit{Slope}} = m = \frac{x_2 - x_1}{t_2 - t_1} = \frac{2 - 0}{0 - 1} = \frac{2}{-1} = -2$$

The x-intercept is clearly 2, so we have that $x = -2t + 2$.

Thus, we have one variable parameterized! Next, we parameterize the variable y. Look at only the x and y coordinates. (Pretend z doesn't exist—temporarily!) We have the curve starting at $(2, -1)$ and ending at $(0, 3)$. We sketch these points on the xy-plane and again we need to obtain the corresponding linear equation:

$$\textbf{\textit{Slope}} = m = \frac{y_2 - y_1}{x_2 - x_1} = \frac{3 - (-1)}{0 - 2} = \frac{4}{-2} = -2$$

The y-intercept is clearly equal to 3. So, the linear equation in slope-intercept form is $y = -2x + 3$. However, remember that we have $x = -2t + 2$, which we need to substitute into the equation for y:

$$y = -2x + 3$$
$$y = -2(-2t + 2) + 3$$
$$y = 4t - 4 + 3$$
$$y = 4t - 1$$

Next, we parameterize the variable z by examining only the y and z coordinates. (Pretend x doesn't exist this time.) We have $(-1, 2)$ going to $(3, -2)$. If you graph these points in the yz-plane you will see a line with negative slope:

$$m = \frac{z_2 - z_1}{y_2 - y_1} = \frac{-2 - 2}{3 - (-1)} = \frac{-4}{4} = -1$$

Using the point-slope form for this equation:

$$z - z_1 = m(y - y_1)$$
$$z - 2 = -1(y - (-1))$$
$$z - 2 = -y - 1$$
$$z = -y + 1$$

Now, we substitute the parameterization $y = 4t - 1$ into this equation for z:

$$z = -y + 1$$
$$z = -(4t - 1) + 1$$
$$z = -4t + 1 + 1$$
$$z = -4t + 2$$

The final answer can be written as a vector-valued function (and frequently is):

$$\mathbf{r}(t) = \langle -2t + 2, \, 4t - 1, \, -4t + 2 \rangle \text{ where } 0 \le t \le 1$$

The graph of this vector-valued function is a "curve" (line segment) in space:

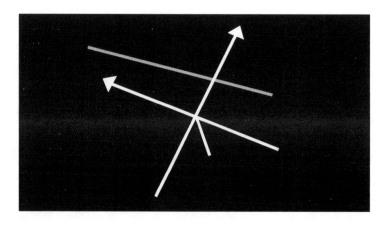

Line Integrals

There are three types of line integrals covered in this section:

1) Line integrals of the form $\int_C f(x,y,z)\,ds$ (used to represent the mass of a wire, where $f(x, y, z)$ is the density function)

   The curve C must be parameterized first, if it isn't already; you need $\mathbf{r}(t)$. The "ds" portion of the integrand is the magnitude of the derivative of $\mathbf{r}(t)$, i.e., $\|\mathbf{r}'(t)\|$. You must substitute the components for $\mathbf{r}(t)$ in place of x, y, and z for the function in the integrand.

2) Line integrals of the form $\int_C \mathbf{F} \cdot d\mathbf{r}$ (used to represent either the work done by an object as it moves along the curve subject to the force given, *or* the flow of a fluid along a curve where $\mathbf{F}$ represents a velocity field instead of a force field). Again, the curve C must be parameterized first, if it isn't already. You will take the dot product between the given force and $\mathbf{r}'(t)$, since "$d\mathbf{r}$" means the derivative of $\mathbf{r}(t)$. Before doing so, you will again substitute the components for $\mathbf{r}(t)$ in place of x, y, and z for the force function in the integrand.

3) Line integrals of the form $\int_C M\,dx + N\,dy$ (known as the "differential form" for a line integral; also used to represent the work done by a field or the flow of a fluid along a curve).

   Again, the curve C must be parameterized first, if it isn't already. The differential form is simply the work form of a line integral where the dot product has been taken already. You will again substitute the components for $\mathbf{r}(t)$ in place of x, y, and z for the force function components, M and N (and P, if given) in the integrand. You also need to find "dx," "dy," (and "dz," if given) by finding the derivative of each of the components of $\mathbf{r}(t)$.

Procedure to Evaluate Type #1 Line Integrals

EXAMPLE. Evaluate the line integral $\int_C \left(3\sqrt[5]{x} - y^2 + xz\right)ds$, where C is the line starting from the point $(2, -1, 2)$ and ending at the point $(0, 3, -2)$.

SOLUTION. We already parameterized this curve from the example above. We will use this result that is $\mathbf{r}(t) = \left\langle -2t + 2,\, 4t - 1,\, -4t + 2 \right\rangle$, $0 \leq t \leq 1$. We evaluate the expression "ds" $= \|\mathbf{r}'(t)\|\,dt$: $\Longrightarrow$

$$\|\mathbf{r}'(t)\| = \|\langle -2, 4, -4 \rangle\|$$
$$= \sqrt{(-2)^2 + (4)^2 + (-4)^2}$$
$$= \sqrt{4 + 16 + 16}$$
$$= \sqrt{36}$$
$$= 6$$

Now, we substitute the components for **r**(*t*) into the integrand and obtain:

$$\int_C \left(3\sqrt[5]{x} - y^2 + xz\right)ds = \int_0^1 \left[3(-2t+2)^{1/5} - (4t-1)^2 + (-2t+2)(-4t+2)\right](6)dt$$

Before evaluating, I will take the factor of 6 outside the integrand, rewrite the problem as a sum of three separate integrals, as well as expand and multiply the second and third terms:

$$= \int_0^1 \left[3(-2t+2)^{1/5} - (4t-1)^2 + (-2t+2)(-4t+2)\right](6)dt$$

$$= 6\int_0^1 3(-2t+2)^{1/5}\,dt - 6\int_0^1 (4t-1)(4t-1)dt + 6\int_0^1 \left(8t^2 - 4t - 8t + 4\right)dt$$

$$= 6(3)\int_0^1 (-2t+2)^{1/5}\,dt - 6\int_0^1 (16t^2 - 8t + 1)dt + 6\int_0^1 \left(8t^2 - 12t + 4\right)dt$$

The first integral will require *u*-substitution, where $u = -2t + 2$, $du = -2dt$. The second and third integrals are easy to integrate, since the integrands are just polynomials.

$$= 18\left(-\frac{1}{2}\right)\int u^{1/5}du - 6\left[\frac{16}{3}t^3 - 4t^2 + t\right]_0^1 + 6\left[\frac{8}{3}t^3 - 6t^2 + 4t\right]_0^1$$

$$= -9\left(\frac{5}{6}\right)u^{6/5} - 6\left[\frac{16}{3}(1)^3 - 4(1)^2 + (1)\right] + 6\left[\frac{8}{3}(1)^3 - 6(1)^2 + 4(1)\right]$$

$$= -\frac{45}{6}\left[(-2t+2)^{6/5}\right]_0^1 - 6\left[\frac{7}{3}\right] + 6\left[\frac{2}{3}\right]$$

$$= -\frac{45}{6}\left[(-2(1)+2)^{6/5} - (-2(0)+2)^{6/5}\right] - 14 + 4$$

$$= -\frac{45}{6}\left[0 - 2^{6/5}\right] - 10$$

$$= \frac{45}{6}\left(2^{6/5}\right) - 10$$

$$\approx 7.23$$

Procedure to Evaluate Type #2 Line Integrals: $\displaystyle\int_C \mathbf{F} \cdot d\mathbf{r}$

1) Parameterize the curve if it is not already. Be sure to determine the range for the parameter t. That is, determine $a \le t \le b$.

2) Evaluate **F** on the curve as a function of the parameter t.

3) Find d**r**/dt.

4) Dot **F** with d**r**/dt.

5) Integrate from $t = a$ to $t = b$.

EXAMPLE. Find the work done by the force field $\mathbf{F}(x, y) = x^2\mathbf{i} - xy\mathbf{j}$ on an object along the path given by the curve C: $\ x = \cos^3 t, \quad y = \sin^3 t$ from $(1, 0)$ to $(0, 1)$.

SOLUTION. Work uses the form of the line integral above $\displaystyle\int_C \mathbf{F} \cdot d\mathbf{r}$. So, we use the procedure given above. Since the curve is already parameterized for us, we simply need to determine the range for t. To start at the point $(1, 0)$, we can have $t = 0$. (I hope this is obvious to you.) To end at the point $(0, 1)$ we require that $x = 0 = \cos^3 t, \quad y = 1 = \sin^3 t$.

The value for t that satisfies both of these equations simultaneously is $t = \dfrac{\pi}{2}$.

Thus, the range for t is $0 \le t \le \dfrac{\pi}{2}$.

Next, we evaluate the force field **F** as a function of t. This means we substitute $x = \cos^3 t \quad and \quad y = \sin^3 t$ into **F** to get that:

$$\mathbf{F}(x, y) = x^2\mathbf{i} - xy\mathbf{j} = (\cos^3 t)^2\mathbf{i} - (\cos^3 t)(\sin^3 t)\mathbf{j} = \left\langle \cos^6 t, \ -\cos^3 t \sin^3 t \right\rangle$$

Next, we find the derivative of $\mathbf{r}(t)$. That is, we find d**r**/dt. After applying the Chain Rule to both components for the curve given we get that:

$$\mathbf{r}'(t) = \left\langle -3\cos^2 t \sin t, \ \ 3\sin^2 t \cos t \right\rangle$$

Then, we take the dot product of the force field with this derivative:

$$\begin{aligned}
\mathbf{F} \cdot d\mathbf{r} &= \left\langle \cos^6 t, -\cos^3 t \sin^3 t \right\rangle \cdot \left\langle -3\cos^2 t \sin t, 3\sin^2 t \cos t \right\rangle \\
&= (\cos^6 t)(-3\cos^2 t \sin t) + (-\cos^3 t \sin^3 t)(3\sin^2 t \cos t) \\
&= -3\cos^8 t \sin t - 3\sin^5 t \cos^4 t
\end{aligned}$$

Finally we integrate on the interval $0 \le t \le \dfrac{\pi}{2}$: $\quad \boldsymbol{Work} = \displaystyle\int_C \mathbf{F} \cdot d\mathbf{r} = \int_0^{\pi/2} \left(-3\cos^8 t \sin t - 3\sin^5 t \cos^4 t \right) dt$

The first term of the integral will be easy to evaluate using u-substitution and the General Power Rule, where $u = \cos t$ and $du = -\sin t$. Then we will simply have:

$$3\int u^8 \, du = 3\left[\frac{u^9}{9}\right] = \frac{1}{3}\left[\cos^9 t\right]_0^{\pi/2} = \frac{1}{3}\left[\cos^9 \frac{\pi}{2} - \cos^9 0\right] = \frac{1}{3}\left[0 - 1^9\right] = -\frac{1}{3}$$

The second term is challenging, as we have to use the technique of integration that we learned last semester on handling powers of trigonometric functions. Since the power of sine is odd and positive, we save a sine factor and convert the remaining sines to cosines using the trig ID $\sin^2 \theta = 1 - \cos^2 \theta$. I will do the integration of this term step-by-step:

$$-3\int_0^{\pi/2} \sin^5 t \cos^4 t \, dt$$

$$= -3\int_0^{\pi/2} \sin^4 t \sin t \cos^4 t \, dt$$

$$= -3\int_0^{\pi/2} \left(\sin^2 t\right)^2 \sin t \cos^4 t \, dt$$

$$= -3\int_0^{\pi/2} \left(1 - \cos^2 t\right)^2 \sin t \cos^4 t \, dt$$

At this point, we need to expand and multiply out the expression $\left(1 - \cos^2 t\right)^2$.

We have that $\left(1 - \cos^2 t\right)^2 = \left(1 - \cos^2 t\right) \cdot \left(1 - \cos^2 t\right) = 1 - 2\cos^2 t + \cos^4 t$.

Let's substitute this into the integrand above and continue evaluating the integral:

$$= -3\int_0^{\pi/2} \left(1 - 2\cos^2 t + \cos^4 t\right) \sin t \cos^4 t \, dt$$

The next step will involve distributing the factor of $\cos^4 t \sin t$ over parentheses. After that, I will split the integral into three separate integrals and integrate term-by-term:

$$= -3\int_0^{\pi/2} \left(\sin t \cos^4 t - 2\cos^6 t \sin t + \cos^8 t \sin t\right) dt$$

$$= -3\int_0^{\pi/2} \sin t \cos^4 t \, dt + (-3)(-2)\int_0^{\pi/2} \cos^6 t \sin t \, dt + (-3)\int_0^{\pi/2} \cos^8 t \sin t \, dt$$

Next, we simply use u-substitution again with the General Power Rule, where $u = \cos t$ and $du = -\sin t$ for each of these three integrals:

$$= 3\int u^4 \, du - 6\int u^6 \, du + 3\int u^8 \, du$$

$$= 3\left[\frac{u^5}{5}\right] - 6\left[\frac{u^7}{7}\right] + 3\left[\frac{u^9}{9}\right]$$

$$= \left[\frac{3}{5}\cos^5 t - \frac{6}{7}\cos^7 t + \frac{1}{3}\cos^9 t\right]_0^{\pi/2}$$

Since $\cos\dfrac{\pi}{2} = 0$, we have that:

$$= 0 - \left[\frac{3}{5}\cos^5 0 - \frac{6}{7}\cos^7 0 + \frac{1}{3}\cos^9 0\right]$$

$$= -\frac{3}{5} + \frac{6}{7} - \frac{1}{3}$$

$$= -\frac{8}{105}$$

Remember, this is not the final answer, as we need to *add* this to the first term we integrated, that where the result was $-\dfrac{1}{3}$.

Thus, the work done was $= -\dfrac{8}{105} - \dfrac{1}{3} = -\dfrac{43}{105}$ *(final answer)*

For the record, it *is possible* for work done to be a negative result.

EXAMPLE. Find the work done by the force $\mathbf{F} = \sqrt{z}\,\mathbf{i} - 2x\mathbf{j} + \sqrt{y}\,\mathbf{k}$ over the path $C_1 \cup C_2$, which consists of the line segment from the point $(0, 0, 0)$ to $(1, 1, 0)$ followed by the segment from $(1, 1, 0)$ to $(1, 1, 1)$.

SOLUTION. We have to separately parameterize the two curves. Then, the work done will be the sum of two separate line integrals. Let's start by parameterizing the first curve. Notice that $z = 0$ for both of the points. So, we only need to worry about the other two coordinates. We have the curve starting at $(0, 0)$ and ending at $(1, 1)$ in the xy-plane. If we declare that the parameter t will start at $t = 0$ and end at $t = 1$, then we can have that $x = t$ and $y = t$ as well. Therefore, we can write that:

$$C_1 : \quad \mathbf{r}_1(t) = \langle t, t, 0 \rangle, \, 0 \le t \le 1$$

For the second curve, we have both the x and y coordinates remaining constant at 1. Only z changes from 0 to 1. Thus, let us again declare that the parameter t will start at $t = 0$ and end at $t = 1$. We write:

$$C_2 : \quad \mathbf{r}_2(t) = \langle 1, 1, t \rangle, \, 0 \le t \le 1$$

Next, we write the sum of the two integrals:

$$\textbf{\textit{Work}} = \int_{C_1} \textbf{F} \cdot d\textbf{r}_1 \quad + \quad \int_{C_2} \textbf{F} \cdot d\textbf{r}_2$$

The force **F** needs to be separately evaluated for each curve. On the first curve, we have that:

$$\textbf{F} = \sqrt{z}\,\textbf{i} - 2x\textbf{j} + \sqrt{y}\,\textbf{k} = \sqrt{0}\,\textbf{i} - 2t\,\textbf{j} + \sqrt{t}\,\textbf{k} = \left\langle 0, -2t, \sqrt{t} \right\rangle \text{ (on the first curve)}$$

For the second curve, the force **F** will be:

$$\textbf{F} = \sqrt{z}\,\textbf{i} - 2x\textbf{j} + \sqrt{y}\,\textbf{k} = \sqrt{t}\,\textbf{i} - 2(1)\,\textbf{j} + \sqrt{1}\,\textbf{k} = \left\langle \sqrt{t}, -2, 1 \right\rangle \text{ (on the second curve)}$$

Next, we find d**r**/d*t* for both curves:

$$C_1 : \quad \textbf{r}_1'(t) = \left\langle 1, 1, 0 \right\rangle , \, 0 \leq t \leq 1$$

$$C_2 : \quad \textbf{r}_2'(t) = \left\langle 0, 0, 1 \right\rangle , \, 0 \leq t \leq 1$$

Finally, we set up the integrals:

$$\textbf{\textit{Work}} = \int_0^1 \left\langle 0, -2t, \sqrt{t} \right\rangle \cdot \left\langle 1, 1, 0 \right\rangle dt \quad + \quad \int_0^1 \left\langle \sqrt{t}, -2, 1 \right\rangle \cdot \left\langle 0, 0, 1 \right\rangle dt$$

$$= \quad \int_0^1 (-2t)\,dt \quad + \quad \int_0^1 (1)\,dt$$

$$= \quad \left[-t^2 \right]_0^1 \quad + \quad \left[t \right]_0^1$$

$$= \quad -1 + 1$$

$$= \quad 0$$

EXAMPLE. Evaluate the integral $\int_C \textbf{F} \cdot d\textbf{r}$, where the force is $\textbf{F} = \dfrac{x\textbf{i} + y\textbf{j} + z\textbf{k}}{\sqrt{x^2 + y^2 + z^2}}$, and the curve C is represented by $\textbf{r}(t) = t\textbf{i} + t\textbf{j} + e^t\textbf{k}, \, 0 \leq t \leq 2$. Use the command "F$_{NINT}$" on your graphing calculator to evaluate the definite integral after you set it up.

SOLUTION. We find d**r**/d*t*: $\mathbf{r}'(t) = 1\mathbf{i} + 1\mathbf{j} + e^t\mathbf{k} = \langle 1, 1, e^t \rangle$. We set up the integral and evaluate the dot product:

$$\int_C \mathbf{F} \cdot d\mathbf{r} = \int_0^2 \left\langle \frac{t}{\sqrt{t^2 + t^2 + \left(e^t\right)^2}}, \frac{t}{\sqrt{t^2 + t^2 + \left(e^t\right)^2}}, \frac{e^t}{\sqrt{t^2 + t^2 + \left(e^t\right)^2}} \right\rangle \cdot \langle 1, 1, e^t \rangle \, dt$$

$$= \int_0^2 \left(\frac{t}{\sqrt{2t^2 + e^{2t}}} + \frac{t}{\sqrt{2t^2 + e^{2t}}} + \frac{e^{2t}}{\sqrt{2t^2 + e^{2t}}} \right) dt$$

$$= \int_0^2 \left(\frac{2t + e^{2t}}{\sqrt{2t^2 + e^{2t}}} \right) dt$$

This integral is not integrable even with advanced integration techniques—that is why we will use "FNINT" on our calculators. The "FNINT" command requires four entries, separated by commas and usually is found under the "CALC" set of commands:

FNINT (integrand, variable integrating with respect to, lower limit of integration, upper limit)

We enter the following:

$$\text{fn int}\left(\frac{\left(2x + e^{2x}\right)}{\sqrt{\left(2x^2 + e^{(2x)}\right)}}, \quad x, \quad 0, \quad 2 \right)$$

Please use parentheses as noted when typing into your calculator.

The answer is approximately 6.9 *(final answer)*.

The last example of this section will be evaluating the differential form for a line integral, that looks like $\int_C M dx + N dy$.

EXAMPLE. Evaluate $\int_C (x - y)dx + (x + y)dy$ along the curve $y = x^2$ from $(-1, 1)$ to $(2, 4)$.

Then, evaluate the line integral again along the same curve but going in the opposite direction.

SOLUTION. First, we need to parameterize the curve. Let's simply let $x = t$, where $-1 \leq t \leq 2$. Then we will have that $y = x^2 = t^2$.

So we have the curve represented by the vector-valued function $\mathbf{r}(t) = \langle t, t^2 \rangle, -1 \leq t \leq 2$.

Then we have that $dx = 1dt$ and $dy = 2tdt$. Substituting everything into the line integral we can write:

$$\int_C (x-y)dx + (x+y)dy = \int_{-1}^{2} \left(t - t^2\right)dt + \left(t + t^2\right)2tdt$$

$$= \int_{-1}^{2} \left(t - t^2 + 2t^2 + 2t^3\right)dt$$

$$= \int_{-1}^{2} \left(t + t^2 + 2t^3\right)dt$$

$$= \left[\frac{t^2}{2} + \frac{t^3}{3} + \frac{t^4}{2}\right]_{-1}^{2}$$

$$= \left(\frac{2^2}{2} + \frac{2^3}{3} + \frac{2^4}{2}\right) - \left(\frac{(-1)^2}{2} + \frac{(-1)^3}{3} + \frac{(-1)^4}{2}\right)$$

$$= \left(\frac{4}{2} + \frac{8}{3} + \frac{16}{2}\right) - \left(\frac{1}{2} + \frac{-1}{3} + \frac{1}{2}\right)$$

$$= 2 + \frac{8}{3} + 8 - \frac{1}{2} + \frac{1}{3} - \frac{1}{2}$$

$$= 12$$

To go in the opposite direction, we could leave the parameterization the same except have the parameter t start at 2 and end at -1. Recall a property of integral from your first semester of calculus which states that:

$$\int_a^b f(x)dx = -\int_b^a f(x)dx$$

Therefore we will have that:

$$\int_{-C} (x-y)dx + (x+y)dy = \int_{2}^{-1} \left(t - t^2\right)dt + \left(t + t^2\right)2tdt$$

$$= -\int_{-1}^{2} \left(t - t^2\right)dt + \left(t + t^2\right)2tdt$$

$$= -12$$

The last thought I would like to leave you with is that integrals are *not* always used to calculate an area or are done over an area. This section has us integrating over lines (a.k.a. curves), and the result is not an area (a few of the examples above had the result measuring the amount of work being done). We integrate *over* a line, which is unlike any other integration you've done up to this point. Please do not confuse line integrals with the types of integrals you did in your first and second semesters of calculus, OK?

This concludes the section.

EXERCISES: #5–11 all from **Quiz 11**, #19, 20 from **Practice Final**

5.3   CONSERVATIVE VECTOR FIELDS AND INDEPENDENCE OF PATH

The main concept in this whole section is that if a vector field is conservative, then evaluating the line integral $\int_C \mathbf{F} \cdot d\mathbf{r}$ (or the line integral $\int_C M dx + N dy$) is independent of path and you may evaluate it using the (easier) Fundamental Theorem of Line Integrals.

Let's review once more what it means for a vector field to be conservative.

1) For a vector field in the plane that has the form $\mathbf{F}(x, y) = \langle M(x, y),\ N(x, y) \rangle$, it will be conservative if $\dfrac{\partial M}{\partial y} = \dfrac{\partial N}{\partial x}$.

2) For a vector field in space that has the form $\mathbf{F}(x, y, z) = \langle M(x, y, z),\ N(x, y, z),\ P(x, y, z) \rangle$, it will be conservative if $\dfrac{\partial M}{\partial y} = \dfrac{\partial N}{\partial x}$ and $\dfrac{\partial P}{\partial y} = \dfrac{\partial N}{\partial z}$ and $\dfrac{\partial M}{\partial z} = \dfrac{\partial P}{\partial x}$ (All three of these conditions must be met simultaneously.)

Further, if a vector field is conservative, then you can find its corresponding potential function. For a conservative vector field in the plane, you can recover the potential function by performing two separate integrations and adding the results (provided you delete the "repeats") in the following way:

$$f(x, y) = \begin{cases} \int M(x, y)\,dx \\ \int N(x, y)\,dy \end{cases}$$

For a conservative vector field in space, you can recover the potential function by performing three separate integrations and adding the results (provided you delete the "repeats") in the following way:

$$f(x, y, z) = \begin{cases} \int M(x, y, z)\,dx \\ \int N(x, y, z)\,dy \\ \int P(x, y, z)\,dz \end{cases}$$

"Independence of path" means that as long as you know the starting point and ending point of a path and you are trying to evaluate a line integral, you won't care about *how* one gets from the starting to the ending point—you will not have to parameterize the curve at all! The line integral will have the same value if one gets from the starting to the ending point along a straight line or if one follows a "curvier" path. It won't matter.

For quick review, the Fundamental Theorem of Calculus that you studied in your first semester of calculus says the following:

$$\int_a^b f(x)dx = \left[F(x)\right]_a^b = F(b) - F(a), \text{ where } F(x) \text{ is the anti-derivative of } f(x)$$

The Fundamental Theorem of Line Integrals says that if you have a conservative vector field $\mathbf{F}(x,y)$ or $\mathbf{F}(x, y, z)$, then you can evaluate the line integrals $\int_C \mathbf{F} \cdot d\mathbf{r}$, or $\int_C Mdx + Ndy$ for a vector field in the plane, or $\int_C Mdx + Ndy + Pdz$ for a vector field in space, by recovering the potential function corresponding to the vector field, and evaluating that potential function from the lower limit (which is the starting point) to the upper limit (which is the ending point) of the curve:

1) $\int_C \mathbf{F} \cdot d\mathbf{r}$ or $\int_C Mdx + Ndy = \int_{(x_1,y_1)}^{(x_2,y_2)} \mathbf{F} \cdot d\mathbf{r} = \left[f(x,y)\right]_{(x_1,y_1)}^{(x_2,y_2)} = f(x_2,y_2) - f(x_1,y_1),$

where $f(x, y)$ is the potential function, (x_1, y_1) is the starting point and (x_2, y_2) is the ending point.

2) $\int_C \mathbf{F} \cdot d\mathbf{r}$ or $\int_C Mdx + Ndy + Pdz = \int_{(x_1,y_1,z_1)}^{(x_2,y_2,z_2)} \mathbf{F} \cdot d\mathbf{r} = \left[f(x,y,z)\right]_{(x_1,y_1,z_1)}^{(x_2,y_2,z_2)} = f(x_2,y_2,z_2) - f(x_1,y_1,z_1)$

where $f(x, y, z)$ is the potential function, (x_1, y_1, z_1) is the starting point and (x_2, y_2, z_2) is the ending point.

Do you see the similarities between the Fundamental Theorem of Line Integrals and the Fundamental Theorem of Calculus that you studied in your first semester of calculus?

Finally, if the curve is *closed* (starting point is same as ending point) *and* the field is conservative, then the value of the line integral will be equal to zero! (So, the work done along that closed path will be equal to zero as well!)

The line integrals in the forms of $\int_C Mdx + Ndy$ or $\int_C Mdx + Ndy + Pdz$ are called "exact differential form."

EXAMPLE. Find the value of the given line integral.

> *Hint:* if the vector field is conservative, then you may use the Fundamental Theorem of Line Integrals. If there is not enough information given to do the problem, so state.

(a) $\displaystyle\int_C \mathbf{F} \cdot d\mathbf{r}$ where $\mathbf{F}(x, y, z) = \left\langle 2x \ln y - yz, \dfrac{x^2}{y} - xz, -xy \right\rangle$ and the path along the curve C starts from the point $(1, 2, 1)$ and ends at the point $(2, 1, 1)$.

(b) $\displaystyle\int_C \mathbf{F} \cdot d\mathbf{r}$ where $\mathbf{F}(x, y, z) = \left\langle -y, x, 3xz^2 \right\rangle$ and the path along the curve C starts from the point $(1, 0, 2)$ and ends at the point $(-1, 0, 1)$.

(c) $\displaystyle\int_C \mathbf{F} \cdot d\mathbf{r}$ where $\mathbf{F}(x, y, z) = \left\langle -y, x, 3xz^2 \right\rangle$ and the path along the curve C is a straight line segment that starts from the point $(1, 0, 2)$ and ends at the point $(-1, 0, 1)$.

(d) $\displaystyle\int_C \dfrac{x\,dx + y\,dy + z\,dz}{\sqrt{x^2 + y^2 + z^2}}$ where the path along the curve C is a straight line segment that starts from the point $(-1, -1, -1)$ and ends at the point $(2, 2, 2)$.

(e) $\displaystyle\int_C \mathbf{F} \cdot d\mathbf{r}$ where $\mathbf{F}(x, y, z) = \left\langle xz \sin(yz), xz \cos(xy), \sin(3yz) \right\rangle$ and the path along the curve C is represented by $\mathbf{r}(t) = \left\langle t, t, 1 \right\rangle$ and $0 \le t \le \dfrac{\pi}{2}$.

SOLUTION.

(a) We need to find out if the vector field given by $\mathbf{F}(x, y, z) = \left\langle 2x \ln y - yz, \dfrac{x^2}{y} - xz, -xy \right\rangle$ is conservative.

$$\text{Does}\quad \frac{\partial M}{\partial y} = \frac{\partial N}{\partial x} \quad \text{and} \quad \frac{\partial P}{\partial y} = \frac{\partial N}{\partial z} \quad \text{and} \quad \frac{\partial M}{\partial z} = \frac{\partial P}{\partial x}\ ?$$

Let's check:

$$\frac{\partial M}{\partial y} = 2x \cdot \frac{1}{y} - z = \frac{\partial N}{\partial x} \quad \text{and} \quad \frac{\partial P}{\partial y} = -x = \frac{\partial N}{\partial z} \quad \text{and} \quad \frac{\partial M}{\partial z} = -y = \frac{\partial P}{\partial x}$$

So, it is conservative! Therefore, we can use the Fundamental Theorem of Line Integrals, and we do not need to know the path or even need to parameterize the path. So, let's find the potential function:

$$\int M(x, y, z)\,dx = \int \left(2x \ln y - yz \right)dx = x^2 \ln y - xyz + g_1(y, z)$$

$$\int N(x, y, z)\,dy = \int \left(\frac{x^2}{y} - xz \right)dy = x^2 \ln y - xyz + g_2(x, z)$$

$$\int P(x, y, z)\,dz = \int \left(-xy \right)dz = -xyz + g_3(x, y)$$

So, we sum these all up (but do not add "duplicates") and we have that the potential function is $f(x, y, z) = x^2 \ln y - xyz + C$. So, we can evaluate the line integral using the Fundamental Theorem of Line Integrals:

$$\int_C \mathbf{F} \cdot d\mathbf{r} = \left[f(x,y,z) \right]_{(1,2,1)}^{(2,1,1)} = \left[x^2 \ln y - xyz \right]_{(1,2,1)}^{(2,1,1)} = \left((2)^2 \ln(1) - (2)(1)(1) \right) - \left((1)^2 \ln(2) - (1)(2)(1) \right)$$

$$= 0 - 2 - \ln 2 + 2 = -\ln 2$$

(b) We will first determine if the vector field $\mathbf{F}(x, y, z) = \left\langle -y, x, 3xz^2 \right\rangle$ is conservative or not. Does...

$$\frac{\partial M}{\partial y} = \frac{\partial N}{\partial x} \quad \text{and} \quad \frac{\partial P}{\partial y} = \frac{\partial N}{\partial z} \quad \text{and} \quad \frac{\partial M}{\partial z} = \frac{\partial P}{\partial x} \text{ ...?}$$

$$\frac{\partial M}{\partial y} = -1 \neq \frac{\partial N}{\partial x} = 1 \quad \text{and} \quad \frac{\partial P}{\partial y} = 0 = \frac{\partial N}{\partial z} \quad \text{and} \quad \frac{\partial M}{\partial z} = 0 \neq \frac{\partial P}{\partial x} = 3z^2 \text{ ...?}$$

So, *no,* the vector field is *not* conservative, so we cannot use the Fundamental Theorem of Line Integrals here. Therefore, we need to know the exact path, which isn't given. So, there is not enough information to do this problem (final answer).

(c) This problem is exactly like the previous one, but the path is given., we still *cannot* use the Fundamental Theorem of Line Integrals here, since the vector field is *not* conservative. This means the value of the line integral *is* dependent upon the path, and since the fact that we have a "straight line segment" is given, we do in fact have enough information to find the parameterization of the path. After we get the path parameterized, we will evaluate the line integral using the dot product, etc., etc.

We need to parameterize the straight-line path between the points $(1, 0, 2)$ and $(-1, 0, 1)$. Note that the y-coordinate is always equal to zero. If we let $0 \leq t \leq 1$ (this is an arbitrary call here; *you* may decide to let the variable t vary between two other values instead), we need to determine the relationship between t and x. I like to make a chart and establish a linear relationship between the two variables. I want to get a linear relationship (in slope-intercept form) between t and x where $x = mt + b$:

t	x
0	1
1	-1

If we want to get the "slope" of this line, we use the slope formula:

$$m = \frac{x_2 - x_1}{t_2 - t_1} = \frac{1 - (-1)}{0 - 1} = \frac{2}{-1} = -2$$

Use the point-slope formula to get the linear equation:

$$x - x_1 = m(t - t_1)$$
$$x - 1 = -2(t - 0)$$
$$x = -2t + 1$$

Let's get the relationship between x and z. We want to get the linear relationship in slope-intercept form $z = mx + b$. We can also make a chart to help us to do this:

x	z
1	2
−1	1

$$m = \frac{z_2 - z_1}{x_2 - x_1} = \frac{1 - 2}{-1 - 1} = \frac{-1}{-2} = \frac{1}{2}$$

Use the point-slope formula to get the linear equation:

$$z - z_1 = m(x - x_1)$$
$$z - 2 = \frac{1}{2}(x - 1)$$
$$z = \frac{1}{2}x - \frac{1}{2} + 2$$
$$z = \frac{1}{2}x + \frac{3}{2}$$

Next, substitute the fact that $x = -2t + 1$ into this equation:

$$z = \frac{1}{2}(-2t + 1) + \frac{3}{2}$$
$$z = -t + \frac{1}{2} + \frac{3}{2}$$
$$z = -t + 2$$

So, our parameterization gives us the vector-valued function:

$$\mathbf{r}(t) = \langle -2t + 1, \, 0, \, -t + 2 \rangle, \text{ where } 0 \le t \le 1$$

Next, we find $\mathbf{dr}$ by taking the derivative of this vector-valued function:

$$\mathbf{r}'(t) = \langle -2, \, 0, \, -1 \rangle$$

Next, we evaluate the vector field using this parameterization:

$$\mathbf{F}(x,y,z) = \left\langle -y, x, 3xz^2 \right\rangle = \left\langle 0, -2t+1, 3(-2t+1)(-t+2)^2 \right\rangle = \left\langle 0, -2t+1, (-6t+3)(t^2-4t+4) \right\rangle$$

$$= \left\langle 0, -2t+1, -6t^3+24t^2-24t+3t^2-12t+12 \right\rangle$$

$$= \left\langle 0, -2t+1, -6t^3+27t^2-36t+12 \right\rangle$$

Finally, we will evaluate the line integral using the dot product:

$$\int_C \mathbf{F}\cdot d\mathbf{r} = \int_0^1 \left\langle 0, -2t+1, -6t^3+27t^2-36t+12 \right\rangle \cdot \left\langle -2, 0, 1 \right\rangle dt$$

$$= \int_0^1 \left(-6t^3+27t^2-36t+12 \right) dt$$

$$= \left[-\frac{3}{2}t^4+9t^3-18t^2+12t \right]_0^1$$

$$= -\frac{3}{2}(1)^4-9\,(1)^2-18\,(1)+12\,(1)$$

$$= -\frac{3}{2}$$

(d) To evaluate the exact differential form line integral $\displaystyle \int_C \frac{xdx+ydy+zdz}{\sqrt{x^2+y^2+z^2}}$, we need to determine if the vector field is conservative or not. This line integral is in the exact differential form which looks like $\displaystyle \int_C Mdx+Ndy+Pdz$. The vector field is equal to:

$$\mathbf{F}(x,y,z)=\left\langle M,N,P\right\rangle = \left\langle \frac{x}{\sqrt{x^2+y^2+z^2}}, \frac{y}{\sqrt{x^2+y^2+z^2}}, \frac{z}{\sqrt{x^2+y^2+z^2}} \right\rangle$$

Does $\dfrac{\partial M}{\partial y}=\dfrac{\partial N}{\partial x}$ and $\dfrac{\partial P}{\partial y}=\dfrac{\partial N}{\partial z}$ and $\dfrac{\partial M}{\partial z}=\dfrac{\partial P}{\partial x}$ ...?

Before we check, let's rewrite the force field using exponents instead of radicals so it will be easier to differentiate. Then our force field will look like:

$$\mathbf{F}(x,y,z)=\left\langle M,N,P\right\rangle = \left\langle \frac{x}{\left(x^2+y^2+z^2\right)^{1/2}}, \frac{y}{\left(x^2+y^2+z^2\right)^{1/2}}, \frac{z}{\left(x^2+y^2+z^2\right)^{1/2}} \right\rangle$$

Does $\dfrac{\partial M}{\partial y}=\dfrac{\partial N}{\partial x}$ and $\dfrac{\partial P}{\partial y}=\dfrac{\partial N}{\partial z}$ and $\dfrac{\partial M}{\partial z}=\dfrac{\partial P}{\partial x}$ ...?

To differentiate each of these components, we need to use the Chain Rule:

$$\frac{\partial M}{\partial y} = x\left(\frac{-1}{2}\right)(x^2 + y^2 + z^2)^{-3/2}(2y) = \frac{-xy}{(x^2 + y^2 + z^2)^{3/2}}$$

This is in fact equal to $\frac{\partial N}{\partial x}$, which will be left to the reader.

Similarly, we also have that $\frac{\partial P}{\partial y} = \frac{\partial N}{\partial z}$ and $\frac{\partial M}{\partial z} = \frac{\partial P}{\partial x}$ as well, so the force field is indeed conservative.

Therefore, we can apply the Fundamental Theorem of Line Integrals to evaluate this line integral. We need to find the potential function, by performing three integrations and summing the results—making sure to eliminate any duplicates. The potential function *f(x, y, z)* will be:

$$f(x,y,z) = \begin{cases} \int M(x,y,z)dx \\ \int N(x,y,z)dy \\ \int P(x,y,z)dz \end{cases}$$

We will do the integration using the Power Rule with *u*-substitution for each of these integrals:

$$\int M(x,y,z)dx \;=\; \int \frac{x}{\sqrt{x^2 + y^2 + z^2}}dx$$

Let $u = x^2 + y^2 + z^2$ *and* $du = 2xdx$. (Remember that when we find "*du*," we need to differentiate with respect to *x*.) We need a factor of 2 in the integrand. We put one in and divide it out:

$$= \;\frac{1}{2}\int \frac{2x}{\sqrt{x^2 + y^2 + z^2}}dx \;=\; \frac{1}{2}\int \frac{1}{\sqrt{u}}du \;=\; \frac{1}{2}\int u^{-1/2}du = \;\frac{1}{2}\left[2u^{1/2}\right]$$

Substitute $u = x^2 + y^2 + z^2$ in, to get that:

$$= \;(x^2 + y^2 + z^2)^{1/2} = \sqrt{x^2 + y^2 + z^2}$$

When we evaluate the integrals $\int N(x,y,z)dy$ *and* $\int P(x,y,z)dz$, we get the exact same thing. I will do one of these integrals and leave the other one to the reader.

$$\int N(x,y,z)dy \;=\; \int \frac{y}{\sqrt{x^2 + y^2 + z^2}}dy$$

Let $u = x^2 + y^2 + z^2$ and $du = 2y\,dy$. (Remember that when we find "*du*," we need to differentiate with respect to *y*.) We need a factor of 2 in the integrand. We put one in and divide it out:

$$= \frac{1}{2}\int \frac{2y}{\sqrt{x^2 + y^2 + z^2}}dy = \frac{1}{2}\int \frac{1}{\sqrt{u}}du = \frac{1}{2}\int u^{-1/2}du = \frac{1}{2}\left[2u^{1/2}\right]$$

Substitute $u = x^2 + y^2 + z^2$ in, to get that:

$$= \left(x^2 + y^2 + z^2\right)^{1/2} = \sqrt{x^2 + y^2 + z^2}$$

Therefore, the potential function is:

$$f(x, y, z) = \sqrt{x^2 + y^2 + z^2}$$

Finally, we evaluate this potential function between the starting and ending points to obtain the value of the line integral:

$$\int_C \frac{x\,dx + y\,dy + z\,dz}{\sqrt{x^2 + y^2 + z^2}} = \left[f(x,y,z)\right]_{(-1,-1,-1)}^{(2,2,2)} = \sqrt{2^2 + 2^2 + 2^2} - \sqrt{(-1)^2 + (-1)^2 + (-1)^2}$$

$$= \sqrt{12} - \sqrt{3} = 2\sqrt{3} - \sqrt{3} = \sqrt{3}$$

(e) Let's start by determining whether the field $\mathbf{F}(x, y, z) = \langle xz\sin(yz), xz\cos(xy), \sin(3yz)\rangle$ is conservative or not.

Does $\dfrac{\partial M}{\partial y} = \dfrac{\partial N}{\partial x}$ and $\dfrac{\partial P}{\partial y} = \dfrac{\partial N}{\partial z}$ and $\dfrac{\partial M}{\partial z} = \dfrac{\partial P}{\partial x}$ ...?

Well, we have that $\dfrac{\partial M}{\partial y} = xz^2\cos(yz)$.

To find $\dfrac{\partial N}{\partial x}$, we need to use the Product Rule that tells us $(fg)' = f'g + g'f$.

Since we have that $N = xz\cos(xy)$, we will let $f = xz$ and $g = \cos(xy)$.

Then, $\dfrac{\partial N}{\partial x} = z\cos(xy) - y\sin(xy)\cdot(xz)$. This is *not* the same as $\dfrac{\partial M}{\partial y} = xz^2\cos(yz)$.

Therefore, the force field is *not* conservative, and we *cannot* use the Fundamental Theorem of Line Integrals to evaluate the line integral. Instead, we will evaluate the line integral the "old-fashioned way"—by brute force using the dot product. Fortunately, the curve is already parameterized for us. This is provided:

$$\mathbf{r}(t) = \langle t, t, 1\rangle \quad \text{and} \quad 0 \le t \le \frac{\pi}{2}$$

We substitute these components into our force field to get our force field in terms of the parameter t:

$$\mathbf{F}(x(t), y(t), z(t)) = \left\langle t\sin(t),\, t\cos(t^2),\, \sin(3t) \right\rangle$$

We find $d\mathbf{r}$ by differentiating the curve $\mathbf{r}(t) = \langle t, t, 1 \rangle$:

$$\mathbf{r}'(t) = \langle 1, 1, 0 \rangle$$

Next, we take the dot product $\mathbf{F} \cdot d\mathbf{r}$ and evaluate the integral:

$$\int_C \mathbf{F} \cdot d\mathbf{r} = \int_0^{\pi/2} \left\langle t\sin t,\, t\cos t^2,\, \sin(3t) \right\rangle \cdot \langle 1, 1, 0 \rangle dt$$

$$= \int_0^{\pi/2} \left(t\sin t + t\cos t^2 \right) dt$$

$$= \int_0^{\pi/2} \left(t\sin t \right) dt \;+\; \int_0^{\pi/2} \left(t\cos t^2 \right) dt$$

I purposely split the integral of the sum into the sum of two integrals because I want to drive home a point: We need to use Integration by Parts on the first integral and u-substitution on the second one. I will do the integrals separately. For the first integral, we need to use the Integration By Parts Formula that says:

$$\int u\,dv = uv - \int v\,du$$

We need to identify u, du, v, and dv for the integral $\int_0^{\pi/2} \left(t\sin t \right) dt$.

We will let… $u = t, \qquad dv = \sin t\,dt$

$$du = 1dt, \quad v = -\cos t$$

Substituting these all into the Integration By Parts Formula, we have that:

$$\int_0^{\pi/2} \left(t\sin t \right) dt = \left[(t)(-\cos t) \right]_0^{\pi/2} - \int_0^{\pi/2} -\cos t\,dt$$

$$= \left[-t\cos t \right]_0^{\pi/2} + \int_0^{\pi/2} \cos t\,dt$$

$$= \left[-t\cos t \right]_0^{\pi/2} + \left[\sin t \right]_0^{\pi/2}$$

$$= \left[-t\cos t + \sin t \right]_0^{\pi/2}$$

$$= \left(-\frac{\pi}{2}\cos\left(\frac{\pi}{2}\right) + \sin\left(\frac{\pi}{2}\right) \right) - \left(0\cos 0 + \sin 0 \right)$$

$$= -\frac{\pi}{2}\cdot 0 + 1 - 0$$

$$= 1$$

Now we turn our attention to the other integral, that is $\int_{0}^{\pi/2} \left(t\cos t^2\right)dt$.

We do not use Integration By Parts here. Instead, we simply use *u*-substitution with the rule for integrating cosine:

$$\int \cos u\, du = \sin u + C \text{ . We let } u = t^2 \quad \text{and} \quad du = 2t\, dt \text{ .}$$

We need a factor of 2 in the integrand. We put one in and divide it out:

$$\int_{0}^{\pi/2} \left(t\cos t^2\right)dt = \frac{1}{2}\int_{0}^{\pi/2}\left(2t\cos t^2\right)dt = \frac{1}{2}\int \cos u\, du = \frac{1}{2}\sin u = \frac{1}{2}\left[\sin t^2\right]_{0}^{\pi/2}$$

$$= \frac{1}{2}\left[\sin\left(\frac{\pi}{2}\right)^2 - \sin(0^2)\right]$$

$$= \frac{1}{2}\sin\left(\frac{\pi^2}{4}\right)$$

$$\approx 0.31$$

(Be sure you are in "RADIANS" mode when typing the answer into your calculator.)

Our final answer for this problem is:

$$\int_{C} \mathbf{F}\cdot d\mathbf{r} = 1 + \frac{1}{2}\sin\left(\frac{\pi^2}{4}\right)$$

Note that I did *not* use the approximation of 0.31 here. I prefer exact answers on exams!

The last two examples has us using line integrals to calculate work done on an object by a force field as it moves along a curve. Remember that we use the fact that:

$$\boldsymbol{Work} = \int_{C} \mathbf{F}\cdot d\mathbf{r}$$

EXAMPLE. Find the work done by $\mathbf{F} = \left(x^2 + y\right)\mathbf{i} + \left(y^2 + x\right)\mathbf{j} + ze^z\mathbf{k}$ over the following paths from $(1, 0, 0)$ to $(1, 0, 1)$.

    (a)   The line segment $x = 1$, $y = 0$, $0 \leq z \leq 1$.

    (b)   The helix $\mathbf{r}(t) = (\cos t)\mathbf{i} + (\sin t)\mathbf{j} + \dfrac{t}{2\pi}\mathbf{k}$, $0 \leq t \leq 2\pi$

    (c)   The x-axis from $(1, 0, 0)$ to $(0, 0, 0)$ followed by the parabola $z = x^2$, $y = 0$ from $(0, 0, 0)$ to $(1, 0, 1)$.

SOLUTION. First, we determine whether the field is conservative or not. If it is, the work done by the field will be the same for all of the paths given, since conservative fields are independent of path!

$$\text{Does } \frac{\partial M}{\partial y} = \frac{\partial N}{\partial x} \quad \text{and} \quad \frac{\partial P}{\partial y} = \frac{\partial N}{\partial z} \quad \text{and} \quad \frac{\partial M}{\partial z} = \frac{\partial P}{\partial x} \text{...?}$$

Let's check:

$$\frac{\partial M}{\partial y} = 1 = \frac{\partial N}{\partial x} \quad \text{and} \quad \frac{\partial P}{\partial y} = 0 = \frac{\partial N}{\partial z} \quad \text{and} \quad \frac{\partial M}{\partial z} = 0 = \frac{\partial P}{\partial x}$$

Yes! The field is conservative. Therefore, we can apply the Fundamental Theorem of Line Integrals to evaluate this line integral. We need to find the potential function by performing three integrations and summing the results, making sure to eliminate any duplicates. The potential function *f(x, y, z)* will be:

$$f(x,y,z) = \begin{cases} \int M(x,y,z)dx \\ \int N(x,y,z)dy \\ \int P(x,y,z)dz \end{cases}$$

I will do each integral separately:

$$\int M(x,y,z)dx = \int (x^2 + y)dx = \frac{1}{3}x^3 + xy + g_1(y,z)$$

$$\int N(x,y,z)dy = \int \left(y^2 + x\right)dy = \frac{1}{3}y^3 + xy + g_2(x,z)$$

The last integral requires Integration By Parts. I will identify *u, du, v,* and *dv*:

$$u = z, \quad dv = e^z dz$$
$$du = 1dz, \quad v = e^z$$

Applying the Integration By Parts formula, we have:

$$\int P(x,y,z)dz = \int ze^z dz = ze^z - \int e^z dz = ze^z - e^z + g_3(x,y)$$

Now, I sum up the results, deleting the "repeats" to get the potential function:

$$f(x,y,z) = \frac{1}{3}x^3 + \frac{1}{3}y^3 + xy + ze^z - e^z$$

Finally, we apply the Fundamental Theorem of Line Integrals:

$$\textbf{\textit{Work}} = \int_C \textbf{F} \cdot d\textbf{r} \quad = \quad \left[f(x,y,z) \right]_{(1,0,0)}^{(1,0,1)} \quad = \quad \left[\frac{1}{3}x^3 + \frac{1}{3}y^3 + xy + ze^z - e^z \right]_{(1,0,0)}^{(1,0,1)}$$

$$= \quad \left[\frac{1}{3}(1)^3 + \frac{1}{3}(0)^3 + (1)(0) + (1)e^1 - e^1 \right] \quad - \quad \left[\frac{1}{3}(1)^3 + \frac{1}{3}(0)^3 + (1)(0) + (0)e^0 - e^0 \right]$$

$$= \quad \frac{1}{3} + 0 - \frac{1}{3} + 1 = 1$$

This answer is valid for all three paths, since the field is conservative.

EXAMPLE. Find the work done by $\textbf{F} = \left(x^2 + y \right)\textbf{i} + \left(y^2 + x \right)\textbf{j} + ze^z\textbf{k}$ over the piece-wise continuous path from the point (1, 0, 1) to (0, 0, 0) by a straight-line segment followed by the parabola $z = x^2$, $y = 0$ from (0, 0, 0) to (1, 0, 1).

SOLUTION. Notice that the force field is the same as the previous example, where we found that it was a conservative field. Also, notice that the curve is closed; it has the same starting and ending point, which is the point (1, 0, 1). Since we already found the potential function, we can say that the work done is:

$$\textbf{\textit{Work}} = \int_C \textbf{F} \cdot d\textbf{r} \quad = \quad \left[f(x,y,z) \right]_{(1,0,1)}^{(1,0,1)} \quad = \quad \left[\frac{1}{3}x^3 + \frac{1}{3}y^3 + xy + ze^z - e^z \right]_{(1,0,1)}^{(1,0,1)} \quad = \quad 0$$

This concludes the section!

EXERCISES: #12, 13, 14 from **Quiz 11**, #21, 22 from **Practice Final**

5.4 Green's Theorem in the Plane

In the last section, we learned how to evaluate line integrals (also known as "flow integrals") for conservative fields. We found a potential function for the field, evaluated it at the path endpoints, and calculated the integral as the appropriate difference of those values.

In this section, we learn how to evaluate both line and flux integrals (a new kind of integral) across closed plane curves when the vector field is not conservative. We do this by using a theorem known as "Green's Theorem," which allows us to convert line integrals to double integrals.

Green's Theorem is one of the great theorems of calculus. In pure mathematics, it ranks in importance with the Fundamental Theorem of Calculus. In applied mathematics, the generalizations of Green's Theorem to three dimensions (a.k.a. Stokes' Theorem, covered later) provide the basis for theorems about electricity, magnetism, and fluid flow.

We talk in terms of velocity fields of fluid flows because fluid flows are easy to picture. However, Green's Theorem applies to any vector field satisfying certain conditions. It does not depend for its validity on the field's having a particular physical interpretation.

Before we go over the two forms for Green's Theorem, we need to backtrack and cover the concepts of divergence and curl, as well as the definition of a simple closed curve. I will now address the concepts of divergence and curl that I skipped until now.

Definition:
Flux Density, a.k.a. Divergence

The flux density or divergence of a vector field $\mathbf{F} = M\mathbf{i} + N\mathbf{j}$ at the point (x, y) is:

$$\operatorname{div} \mathbf{F} = \frac{\partial M}{\partial x} + \frac{\partial N}{\partial y} \quad \textbf{\textit{Note:}} \text{ Divergence is a scalar, } not \text{ a vector!}$$

The physical interpretation for this is: If water were flowing into a region through a small hole at the point (x, y), the lines of flow would diverge there (hence the name) and, since water would be flowing out of a small rectangle about this point, the divergence of $\mathbf{F}$ at that point would be positive. Conversely, if the water were draining *out* of the hole instead of flowing in, the divergence would be negative.

We can extend the definition to find divergence for vector fields in space:

$$\operatorname{div} \mathbf{F} = \frac{\partial M}{\partial x} + \frac{\partial N}{\partial y} + \frac{\partial P}{\partial z}$$

EXAMPLE. Find the divergence of $\mathbf{F} = \ln\!\left(x^2 + y^2\right)\mathbf{i} + xy\mathbf{j} + \ln(y^2 + z^2)\mathbf{k}$

SOLUTION. Using the definition, we have that:

$$\text{div } \mathbf{F} = \frac{\partial M}{\partial x} + \frac{\partial N}{\partial y} + \frac{\partial P}{\partial z} \quad = \quad \frac{2x}{x^2 + y^2} + x + \frac{2z}{y^2 + z^2}$$

Recall that a field in the plane is conservative if $\dfrac{\partial N}{\partial x} = \dfrac{\partial M}{\partial y}$.

Recall that I gave you the condition to determine when a field in space is conservative. We needed:

$$\frac{\partial M}{\partial y} = \frac{\partial N}{\partial x} \quad \text{and} \quad \frac{\partial P}{\partial y} = \frac{\partial N}{\partial z} \quad \text{and} \quad \frac{\partial M}{\partial z} = \frac{\partial P}{\partial x}$$

To see where this last fact comes from, we need to define the curl of a vector field in space:

DEFINITION:
Curl of a Vector Field in Space
(This definition is not valid for fields in the plane.)

The curl of $\mathbf{F}(x, y, z) = M\mathbf{i} + N\mathbf{j} + P\mathbf{k}$ is:

$$\text{curl } \mathbf{F}(x, y, z) \quad = \quad \nabla \times \mathbf{F}(x, y, z) = \left(\frac{\partial P}{\partial y} - \frac{\partial N}{\partial z} \right)\mathbf{i} \quad - \quad \left(\frac{\partial P}{\partial x} - \frac{\partial M}{\partial z} \right)\mathbf{j} \quad + \quad \left(\frac{\partial N}{\partial x} - \frac{\partial M}{\partial y} \right)\mathbf{k}$$

Note: The curl of $\mathbf{F}$ is a vector, *not* a scalar. (Compare and contrast with the concept of divergence!)

Do not try to memorize this! Instead, notice that we are taking a cross-product here of ∇, known as the differential operator, with the field. The differential operator ∇ can be thought of as a vector with the form:

$$\nabla = \left\langle \frac{\partial}{\partial x}, \frac{\partial}{\partial y}, \frac{\partial}{\partial z} \right\rangle$$

You have seen the symbol "∇" from a previous chapter on our discussion of gradients. I am hoping you recall that when we defined the gradient of a function back in Chapter 3, denoted and defined as:

$$\nabla f(x, y, z) = \left\langle \frac{\partial f}{\partial x}, \frac{\partial f}{\partial y}, \frac{\partial f}{\partial z} \right\rangle \dots \text{where we had the differential operator "act" on the function } f.$$

So, when we find the curl of a three-dimensional vector field, we use the method of finding the determinant of a 3×3 matrix, that we learned back in Chapter 1 when we studied cross products. We find the curl of $\mathbf{F}$ by taking the cross product: $\implies$

$$\text{curl } \mathbf{F}(x, y, z) \quad = \quad \nabla \times \mathbf{F}(x, y, z)$$

$$= \begin{vmatrix} \mathbf{i} & \mathbf{j} & \mathbf{k} \\ \dfrac{\partial}{\partial x} & \dfrac{\partial}{\partial y} & \dfrac{\partial}{\partial z} \\ M & N & P \end{vmatrix}$$

$$= \left(\frac{\partial P}{\partial y} - \frac{\partial N}{\partial z} \right)\mathbf{i} \quad - \quad \left(\frac{\partial P}{\partial x} - \frac{\partial M}{\partial z} \right)\mathbf{j} \quad + \quad \left(\frac{\partial N}{\partial x} - \frac{\partial M}{\partial y} \right)\mathbf{k}$$

Note: The curl of a field in space is a vector (not a scalar).

Next, we re-define what it takes for a three-dimensional vector field to be conservative.

Test for a Conservative Vector Field in Space

Suppose *M, N,* and *P* have continuous first partial derivatives in an open sphere *Q* in space. The vector field $\mathbf{F}(x, y, z) = M\mathbf{i} + N\mathbf{j} + P\mathbf{k}$ is conservative if and only if:

$$\text{curl } \mathbf{F}(x, y, z) \quad = \quad 0$$

That is, **F** is conservative if and only if:

$$\frac{\partial M}{\partial y} = \frac{\partial N}{\partial x} \quad \text{and} \quad \frac{\partial P}{\partial y} = \frac{\partial N}{\partial z} \quad \text{and} \quad \frac{\partial M}{\partial z} = \frac{\partial P}{\partial x}$$

This is what I gave you in the last section.

So, have you figured out why we cannot take the curl of a two-dimensional vector field yet? If you are still stumped, I will give you the answer: In order to take a cross product of two vectors, both vectors need to have three components each, but a vector field in the plane has only two components!

EXAMPLE. Find the curl of:

$$\mathbf{F}(x, y, z) = x^2 z\mathbf{i} - 2xz\mathbf{j} + yz\mathbf{k} \text{ at the point } (2, -1, 3)$$

SOLUTION. Just for fun, I wanted to sketch the graph of the (three-dimensional) vector field:

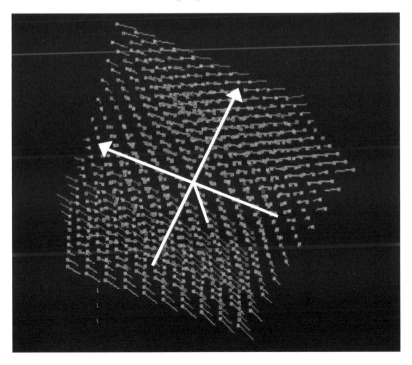

Do not panic, as I would never ask you to sketch a three-dimensional vector field on any of my exams. It would be too challenging!

Let us find the curl for this field. We first find the cross-product of the differential operator ∇ and the field:

$$\text{curl } \mathbf{F}(x,y,z) \;=\; \nabla \times \mathbf{F}(x,y,z)$$

$$= \begin{vmatrix} \mathbf{i} & \mathbf{j} & \mathbf{k} \\ \dfrac{\partial}{\partial x} & \dfrac{\partial}{\partial y} & \dfrac{\partial}{\partial z} \\ M & N & P \end{vmatrix}$$

$$= \begin{vmatrix} \mathbf{i} & \mathbf{j} & \mathbf{k} \\ \dfrac{\partial}{\partial x} & \dfrac{\partial}{\partial y} & \dfrac{\partial}{\partial z} \\ x^2 z & -2xz & yz \end{vmatrix}$$

$$= \left(\frac{\partial[yz]}{\partial y} - \frac{\partial[-2xz]}{\partial z} \right)\mathbf{i} \;\; - \;\; \left(\frac{\partial[yz]}{\partial x} - \frac{\partial[x^2 z]}{\partial z} \right)\mathbf{j} \;\; + \;\; \left(\frac{\partial[-2xz]}{\partial x} - \frac{\partial[x^2 z]}{\partial y} \right)\mathbf{k}$$

$$= \left(z - (-2x) \right)\mathbf{i} \;\; - \;\; \left(0 - x^2 \right)\mathbf{j} \;\; + \;\; \left(-2z - 0 \right)\mathbf{k}$$

$$= \left\langle z + 2x,\; x^2,\; -2z \right\rangle$$

Now we evaluate this curl at the point (2, –1, 3):

$$\text{curl } \mathbf{F}(2,-1,3) = \left\langle 3 + 2(2),\; (2)^2,\; -2(3) \right\rangle \; = \left\langle 7,\, 4,\, -6 \right\rangle$$

Let us go back and discuss vector fields in the plane again. Again, we cannot discuss curl for these (2-D) vector fields, however, we *can* discuss a concept known as *circulation density*.

The circulation density for a vector field $\mathbf{F}$ in the plane is the scalar $\dfrac{\partial N}{\partial x} - \dfrac{\partial M}{\partial y}$.

The positive orientation for this circulation density is the counterclockwise rotation of the fluid around an imaginary vertical axis. You may visualize this imaginary vertical axis as the $\mathbf{k}$ vector on the z-axis, looking downward on the xy-plane from this vantage point. Notice that this circulation density value $\dfrac{\partial N}{\partial x} - \dfrac{\partial M}{\partial y}$ is actually the $\mathbf{k}$-component for the curl of a vector field $\mathbf{F}$ in space, defined above. For Green's Theorem, we need only this $\mathbf{k}$-component. So, we can finally define the concept of curl for a vector in two dimensions.

DEFINITION:
k-Component of Circulation Density, or Curl

The **k**-component of the circulation density, or curl, of a vector field $\mathbf{F} = M\mathbf{i} + N\mathbf{j}$ at the point (x, y) is the scalar:

$$(\text{curl } \mathbf{F}) \cdot \mathbf{k} = \frac{\partial N}{\partial x} - \frac{\partial M}{\partial y}$$

In summary, the curl of a field in the plane is the dot product of the standard vector **k** with its curl, where the vector field's third component is assumed to be zero.

The key difference between curl of a vector field in the plane and that of a vector field in space is that the curl for a vector in the plane is a scalar, not a vector.

EXAMPLE. Find the **k**-component of the curl for the vector field $\mathbf{F}(x, y) = \left(x^2 - y\right)\mathbf{i} + \left(xy - y^2\right)\mathbf{j}$.

SOLUTION. Again, just for fun, I will sketch the graph of this vector field. Fortunately, this vector field is in 2 dimensions only:

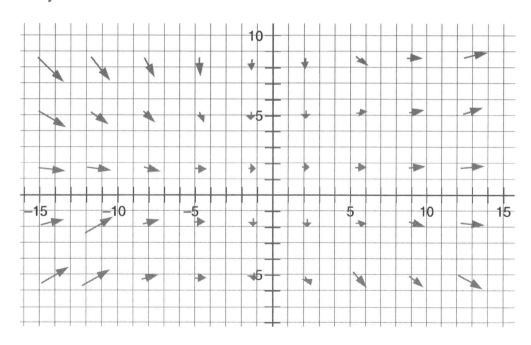

Let's go ahead and find the curl. Use the formula in the preceding definition:

$$(\text{curl } \mathbf{F}) \cdot \mathbf{k} = \frac{\partial N}{\partial x} - \frac{\partial M}{\partial y} = \frac{\partial}{\partial x}\left(xy - y^2\right) - \frac{\partial}{\partial y}\left(x^2 - y\right) = y + 1$$

The physical interpretation for this discussion of curl of a 2-D vector field can be visualized in the following way: If water is moving about a region in the *xy*-plane in a thin layer (where the velocity at a point can be visualized by my diagram above), then the **k**-component of circulation, or curl, at a point (x_0, y_0) gives a way to measure how fast and in what direction a small paddle wheel will spin if it is put into the water at (x_0, y_0) with its axis perpendicular to the plane, parallel to $\mathbf{k}$. For example, if the curl is a positive value, say 10, then the paddle wheel will spin in the counterclockwise direction with a velocity of 10. If the curl is -10, then the wheel will spin in the clockwise direction with a velocity of 10.

We are almost ready for Green's Theorem, except we need to define what a simple curve is. Simple closed curves are curves that are closed (same starting and ending point), yet they do not cross themselves.

We are ready for Green's Theorem. Green's Theorem basically says that if certain conditions are met, the outward flux of a vector field across a simple closed curve in the plane equals the double integral of the divergence of the field over the region enclosed by the curve. In plain language, you can turn a line integral into a double integral (that's way easier to evaluate) under most circumstances!

There are two versions of Green's Theorem:

    1)   The Flux-Divergence or Normal Form (this will be extended to 3-D in section 5.6)

    2)   The Circulation-Curl or Tangential Form (this is emphasized most in many textbooks)

Each of these will typically give a different result. I will give both versions as well as examples for each below.

GREEN'S THEOREM (The Flux-Divergence or Normal Form)

Consider the field $\mathbf{F} = M\mathbf{i} + N\mathbf{j}$. Let C be a simple closed curve oriented counterclockwise (that is, C is traversed once so that the region R always lies to the left). Further, suppose that M and N have continuous partial derivatives in an open region containing R. Then, the outward flux of the field $\mathbf{F}$ across the curve C equals the double integral of the divergence of $\mathbf{F}$ (i.e., div $\mathbf{F}$) over the region R enclosed by C:

$$\oint_C \mathbf{F} \cdot \mathbf{n}\, ds \;\;=\;\; \oint_C M\,dy - N\,dx \;\;=\;\; \iint_R (\text{div } \mathbf{F})\,dA \;\;=\;\; \iint_R \left(\frac{\partial M}{\partial x} + \frac{\partial N}{\partial y} \right) dx\,dy$$

               (OUTWARD FLUX)   *(DIVERGENCE INTEGRAL)*

In the next form, Green's Theorem says that the counterclockwise circulation of a vector field around a simple closed curve is the double integral of the $\mathbf{k}$-component of the curl of the field enclosed by the curve.

Green's Theorem (Circulation-Curl or Tangential Form)

Consider the field $\mathbf{F} = M\mathbf{i} + N\mathbf{j}$. Let C be a simple closed curve oriented counterclockwise (that is, C is traversed once so that the region R always lies to the left). Further, suppose that M and N have continuous partial derivatives in an open region containing R. Then, the counterclockwise circulation of the field $\mathbf{F}$ around the curve C equals the double integral of (curl $\mathbf{F}$) $\mathbf{k}$ over the region R enclosed by C.

$$\oint_C \mathbf{F} \cdot \mathbf{T}\, ds \;\;=\;\; \oint_C M\,dx + N\,dy \;\;=\;\; \iint_R (\text{curl } \mathbf{F}) \cdot \mathbf{k}\,dA \;\;=\;\; \iint_R \left(\frac{\partial N}{\partial x} - \frac{\partial M}{\partial y} \right) dx\,dy$$

          (COUNTERCLOCKWISE CIRCULATION)   *(CURL INTEGRAL)*

I hope the next few examples help you understand how to set up and evaluate integrals using both versions of Green's Theorem.

EXAMPLE 1. Verify Green's Theorem by evaluating *both* sides of the equation:

$$\oint_C y^2\,dx + x^2\,dy = \iint_R \left(\frac{\partial N}{\partial x} - \frac{\partial M}{\partial y} \right) dx\,dy$$

...for the path *C* triangle with vertices (0, 0), (4, 0), and (4, 4)

SOLUTION. To evaluate the left-hand side of the equation *only*, we need to parameterize the curve. Unfortunately, this curve is piece-wise smooth with three different boundaries (one for each side of the triangle). I will parameterize each line segment separately. So we will have the curve *C* equal to the sum of all three separate curves, or $C = C_1 + C_2 + C_3$. Remember, parameterizations are *not* unique, so you may come up with a different set of parameterizations. However, the line integral will still have the same value regardless of the parameterization used. Green's Theorem requires that the path be traversed *counterclockwise* around the region, which I will do.

For the line segment C_1 lying on the *x*-axis from (0, 0) to (4, 0), the *y*-coordinate is always 0, so $y = 0$. I will let $x = t$, where $0 \le t \le 4$. To boil it down:

$$r_1(t) = \langle t, 0 \rangle \quad 0 \le t \le 4$$

For the vertical line segment C_2 connecting the point (4, 0) to (4, 4), we have the *x*-coordinate staying constant at $x = 4$. Let us have $y = t$, where again we'll have the range of *t* be $0 \le t \le 4$. In simpler form:

$$r_2(t) = \langle 4, t \rangle \quad 0 \le t \le 4$$

For the line segment C_3 starting at the point (4, 4) and ending at the point (0,0), I hope you can see that this line segment has a slope of 1, and its corresponding linear equation will be $y = x$. If we again let our parameter of *t* have a range of $0 \le t \le 4$, then we need to find a relationship between *x* and *t*, since *x* *starts* at 4 and *ends* at 0 (which is the reverse of what *t* is doing), we cannot say that $x = t$. Definitely, *x* is not equal to the parameter *t* here. As we have done before, let's make a chart. The two variables will have a linear relationship, which we will put in slope intercept form $x = mt + b$.

t	x
0	4
4	0

$$\text{slope} = m = \frac{x_2 - x_1}{t_2 - t_1} = \frac{0 - 4}{4 - 0} = \frac{-4}{4} = -1$$

If we plot the points in the chart above on a *tx*-plane, the vertical axis (which is *x* here) has an intercept of 4, so $b = 4$. The linear relationship we want is $x = -t + 4$. We know that $y = x$ on this line segment, so we also have that $y = -t + 4$. Restated:

$$r_3(t) = \langle -t + 4, -t + 4 \rangle \quad 0 \le t \le 4$$

Before we evaluate the left-hand side of the equation, let us find the values for "dx" and "dy" for each of these parameterizations. We need these because we will use them to substitute into the left-hand side integral. For C_1, we found that $r_1(t) = \langle t, 0 \rangle$ $0 \le t \le 4$. Since $x = t$ and $y = 0$, we will have that $dx = 1dt = dt$, and $dy = (0)dt = 0$. For C_2, we found that $r_2(t) = \langle 4, t \rangle$ $0 \le t \le 4$. Since $x = 4$ and $y = t$, we will have that $dx = (0)dt = 0$, and $dy = (1)dt = dt$. For C_3, we found that $r_3(t) = \langle -t + 4, -t + 4 \rangle$ $0 \le t \le 4$. Since $x = -t + 4$ and $y = -t + 4$, we will have that $dx = -1dt = -dt$, and $dy = -1dt = -dt$.

We are ready to evaluate the left-hand side of the equation, that is, $\oint_C y^2 dx + x^2 dy$. We can split this up into a sum of three separate integrals, one for each of the curves:

$$\oint_C y^2 dx + x^2 dy = \oint_{C_1} y^2 dx + x^2 dy + \oint_{C_2} y^2 dx + x^2 dy + \oint_{C_3} y^2 dx + x^2 dy$$

$$= \int_0^4 \left[0^2 (1)dt + t^2 (0) \right] + \int_0^4 \left[t^2 (0) + 4^2 \, dt \right] + \int_0^4 \left[(-t + 4)^2 (-1)dt + (-t + 4)^2 (-1)dt \right]$$

$$= \quad 0 \qquad + \quad \int_0^4 16 dt \qquad + \quad -2\int_0^4 (-t + 4)^2 \, dt$$

$$= \left[16t \right]_0^4 \qquad -2\int_0^4 \left(t^2 - 8t + 16 \right) dt$$

$$= 16(4) \quad - \quad 2\left[\frac{1}{3}t^3 - 4t^2 + 16t \right]_0^4$$

$$= \quad 64 \quad - \quad 2\left[\frac{1}{3}(4)^3 - 4(4)^2 + 16(4) \right]$$

$$= \quad 64 \quad - \quad \frac{128}{3} + 128 - 128$$

$$= \quad \frac{64}{3}$$

And we are done with the left-hand side!

The right-hand side of the equation needs to be evaluated. The right hand side has the form:

$$\iint_R \left(\frac{\partial N}{\partial x} - \frac{\partial M}{\partial y} \right) dxdy$$

Although they did not give us M and N explicitly, we can "read" them right off of the left-hand side of the equation, that is:

$$\oint_C y^2 dx + x^2 dy \quad = \quad \oint_C M dx + N dy$$

Since $M = y^2$ and $N = x^2$, we can easily find the appropriate partial derivatives:

$$\frac{\partial N}{\partial x} = 2x \quad \text{and} \quad \frac{\partial M}{\partial y} = 2y$$

Plug these into the right-hand side of the equation:

$$\iint\limits_{R} \left(\frac{\partial N}{\partial x} - \frac{\partial M}{\partial y} \right) dxdy = \iint\limits_{R} (2x - 2y)dxdy$$

We need to determine the limits for the two integrals, meaning, we need to iterate the integral. I prefer the order "*dydx*" instead of "*dxdy*" for this particular problem. We sketch the graph of the region (which is a triangle) by sketching the graphs for the equations $x = 0$, $x = 4$, $y = 0$, and $y = x$:

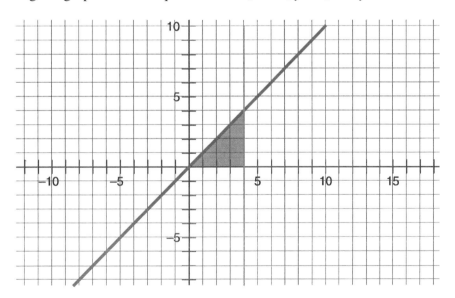

The region is the triangle bounded above by the line $y = x$ and below by the x-axis, and to the right by the vertical line $x = 4$. We can see that the constant limits for x will be $0 \le x \le 4$, and the limits for y will be $0 \le y \le x$. Now I have an iterated integral and I can evaluate the double integral easily:

$$
\begin{aligned}
\iint\limits_{R} (2x - 2y)dxdy &= \int_{0}^{4}\int_{0}^{x} (2x - 2y)dydx \\
&= \int_{0}^{4} \left[2xy - y^2\right]_{0}^{x} dx \\
&= \int_{0}^{4} \left[2x(x) - x^2\right] dx \\
&= \int_{0}^{4} x^2 dx \\
&= \left[\frac{1}{3}x^3\right]_{0}^{4} \\
&= \frac{64}{3}
\end{aligned}
$$

The answers for the left-hand side and right-hand side must match! (Which they do.)

The moral of the story for the last example is that evaluating the double integral (right-hand side of the equation) is *way* easier than evaluating the line integral. If you can use Green's Theorem, you make life much easier!

EXAMPLE 2 Use Green's Theorem to find *both* the counterclockwise circulation and the outward flux for the field $\mathbf{F} = (x^2 + 4y)\mathbf{i} - (xy + y^2)\mathbf{j}$ using the curve given by C: the square bounded by the graphs of $x = 0$, $x = 2$, $y = 0$, and $y = 2$.

SOLUTION. Here we have that $M = x^2 + 4y$ and $N = -(xy + y^2) = -xy - y^2$. I will need all four partial derivatives $\dfrac{\partial M}{\partial x}, \dfrac{\partial M}{\partial y}, \dfrac{\partial N}{\partial x}$ and $\dfrac{\partial N}{\partial y}$ to use *both* versions of Green's Theorem. I will find them all:

$$\frac{\partial M}{\partial x} = 2x, \frac{\partial M}{\partial y} = 4, \frac{\partial N}{\partial x} = -y \text{ and } \frac{\partial N}{\partial y} = -x - 2y$$

To find the counterclockwise circulation, we will use the formula:

$$\oint_C M dx + N dy = \iint_R \left(\frac{\partial N}{\partial x} - \frac{\partial M}{\partial y} \right) dx dy$$

$$= \iint_R (-y - 4) dx dy$$

A square is an easy region since both variables will have constant ranges $0 \le x \le 2$ and $0 \le y \le 2$. We iterate the integral and evaluate:

$$\iint_R (-y - 4) dx dy = \int_0^2 \int_0^2 (-y - 4) dx dy \text{ (circulation)}$$

$$= \int_0^2 \left[-xy - 4x \right]_0^2 dy$$

$$= \int_0^2 (-2y - 4(2)) dy$$

$$= \left[-y^2 - 8y \right]_0^2$$

$$= -(2)^2 - 8(2) = -4 - 16$$

$$= -20 \qquad (circulation)$$

To find the outward flux, we will use the formula:

$$\oint_C M dy - N dx = \iint_R \left(\frac{\partial M}{\partial x} + \frac{\partial N}{\partial y} \right) dx dy$$

Recall that the expression $\dfrac{\partial M}{\partial x} + \dfrac{\partial N}{\partial y}$ is the *divergence* for the field **F,** or div **F**. We already know the iteration of the integral, so we substitute in the appropriate partial derivatives and evaluate:

$$\iint\limits_{R}\left(\frac{\partial M}{\partial x} + \frac{\partial N}{\partial y}\right)dxdy \;=\; \int_{0}^{2}\int_{0}^{2}(2x - x - 2y)dxdy \;=\; \int_{0}^{2}\int_{0}^{2}(x - 2y)dxdy$$

$$= \int_{0}^{2}\left[\frac{x^2}{2} - 2xy\right]_{0}^{2} dy$$

$$= \int_{0}^{2}\left(\frac{2^2}{2} - 2(2)y\right)dy \;=\; \int_{0}^{2}(2 - 4y)dy$$

$$= \left[2y - 2y^2\right]_{0}^{2}$$

$$= 2(2) - 2(2)^2 \;=\; 4 - 8 \;=\; -4 \qquad \text{(Flux)}$$

EXAMPLE 3. Use Green's Theorem to find *both* the counterclockwise circulation and the outward flux for the field $\mathbf{F} = (-x^2 y)\mathbf{i} + (xy^2)\mathbf{j}$, using the curve given by C: the circle $x^2 + y^2 = 25$.

SOLUTION. I will get all four partial derivatives for the field **F**:

$$\frac{\partial M}{\partial x} = -2xy, \frac{\partial M}{\partial y} = -x^2, \frac{\partial N}{\partial x} = y^2 \text{ and } \frac{\partial N}{\partial y} = 2xy$$

To find the counterclockwise circulation, we will use the formula:

$$\oint\limits_{C} Mdx + Ndy \;=\; \iint\limits_{R}\left(\frac{\partial N}{\partial x} - \frac{\partial M}{\partial y}\right)dxdy \;=\iint\limits_{R}(y^2 - (-x^2))dxdy = \iint\limits_{R}\left(y^2 + x^2\right)dxdy$$

We need to iterate the double integral. If we use rectangular coordinates, we can solve for y with the circle equation to get $y = \pm\sqrt{25 - x^2}$. The constant limits for x would then be $-5 \le x \le 5$ and the variable limits for y will be $-\sqrt{25 - x^2} \le y \le \sqrt{25 - x^2}$. We would need to evaluate:

$$\iint\limits_{R}\left(y^2 + x^2\right)dydx = \int_{-5}^{5}\int_{-\sqrt{25-x^2}}^{\sqrt{25-x^2}}\left(y^2 + x^2\right)dydx$$

We learned from the last chapter that integrating over regions that are circular is easier when we convert to polar coordinates. We will do that and evaluate. Remember that in place of "*dydx*" we need to put "*rdrdθ*" and that $x^2 + y^2 = r^2$. A circle of radius 5 will have the limits $0 \le r \le 5$ and $0 \le \theta \le 2\pi$.

$$\int_{-5}^{5}\int_{-\sqrt{25-x^2}}^{\sqrt{25-x^2}}\left(y^2+x^2\right)dydx = \int_{0}^{2\pi}\int_{0}^{5}\left(r^2\right)rdrd\theta$$

$$= \int_{0}^{2\pi}\int_{0}^{5}\left(r^3\right)drd\theta$$

$$= \int_{0}^{2\pi}\left[\frac{1}{4}r^4\right]_{0}^{5}d\theta$$

$$= \frac{1}{4}\int_{0}^{2\pi}5^4 d\theta$$

$$= \frac{625}{4}\left[\theta\right]_{0}^{2\pi}$$

$$= \frac{625}{4}(2\pi)$$

$$= \frac{625\pi}{2} \qquad (Circulation)$$

To find the outward flux, we will use the formula:

$$\text{Flux} = \oint_{C} Mdy - Ndx = \iint_{R}\left(\frac{\partial M}{\partial x}+\frac{\partial N}{\partial y}\right)dxdy = \iint_{R}\left(-2xy+2xy\right)dxdy = \iint_{R}0dxdy = 0$$

EXAMPLE 4. Find the outward flux for the field $\mathbf{F} = \left(3xy - \dfrac{x}{1+y^2}\right)\mathbf{i} + \left(e^x + \tan^{-1}y\right)\mathbf{j}$ across the cardioid $r = 2(1+\cos\theta)$.

SOLUTION. We need to determine the range for the parameter θ in which the cardioid will be sketched out only *once*. It turns out that this range will be $0 \le \theta \le 2\pi$. You can find this out by trial and error and using your graphing calculator in "POLAR" mode.

To find outward flux, we use the formula:

$$\text{Flux} = \oint_{C} Mdy - Ndx = \iint_{R}\left(\frac{\partial M}{\partial x}+\frac{\partial N}{\partial y}\right)dxdy$$

So, we only need to find two of the partial derivatives for the field:

$$\iint_{R}\left(\frac{\partial M}{\partial x}+\frac{\partial N}{\partial y}\right)dxdy = \iint_{R}\left(3y - \frac{1}{1+y^2}+\frac{1}{1+y^2}\right)dxdy = \iint_{R}3ydxdy$$

Now, we need to convert everything to polar coordinates. Recall that $y = r\sin\theta$, and "$dydx$" is the same as "$rdrd\theta$." The limits for the variable r use the equation for the curve given where $0 \le r \le 2(1+\cos\theta)$.

$$\iint\limits_R 3y\,dxdy = \int_0^{2\pi} \int_0^{2(1+\cos\theta)} (3r\sin\theta)r\,drd\theta$$

$$= \int_0^{2\pi} \int_0^{2(1+\cos\theta)} (3r^2\sin\theta)drd\theta$$

$$= \int_0^{2\pi} 3\sin\theta\left[\frac{1}{3}r^3\right]_0^{2(1+\cos\theta)} d\theta$$

$$= \int_0^{2\pi} 3\left(\frac{1}{3}\right)\sin\theta[2(1+\cos\theta)]^3\,d\theta$$

$$= 2^3\int_0^{2\pi}\sin\theta(1+\cos\theta)^3\,d\theta$$

In order to integrate this, we will use u-substitution and the General Power Rule. We let $u = 1 + \cos\theta$ and $du = -\sin\theta\,d\theta$. Of course, we need to put in a minus sign inside and outside the integral to do this:

$$= -8\int_0^{2\pi}(-\sin\theta)(1+\cos\theta)^3\,d\theta$$

$$= -8\int u^3\,du = -8\left[\frac{1}{4}u^4\right]$$

$$= -2\left[(1+\cos\theta)^4\right]_0^{2\pi}$$

$$= -2\left[(1+\cos 2\pi)^4 - (1+\cos 0)^4\right] = 0$$

EXAMPLE 5. Use Green's Theorem to find the work done by the field $\mathbf{F} = \left(y + e^x\ln y\right)\mathbf{i} + \left(\dfrac{e^x}{y}\right)\mathbf{j}$ in moving a particle once counterclockwise around the curve C that forms the boundary of the region R, where R is a region bounded by the graphs of $y = 3 - x^2$ and $y = x^4 + 1$.

SOLUTION. We use the circulation-curl version of Green's Theorem to find the work done by a field. That is:

$$Work = \oint_C M\,dx + N\,dy = \iint\limits_R \left(\frac{\partial N}{\partial x} - \frac{\partial M}{\partial y}\right)dxdy$$

We find the appropriate partial derivatives:

$$\frac{\partial N}{\partial x} = \frac{e^x}{y} \quad \text{and} \quad \frac{\partial M}{\partial y} = 1 + \frac{e^x}{y}$$

We need to graph the equations $y = 3 - x^2$ and $y = x^4 + 1$ to get a picture of what our region looks like to iterate the double integral properly. The region is bounded above by $y = 3 - x^2$ and below by $y = x^4 + 1$.

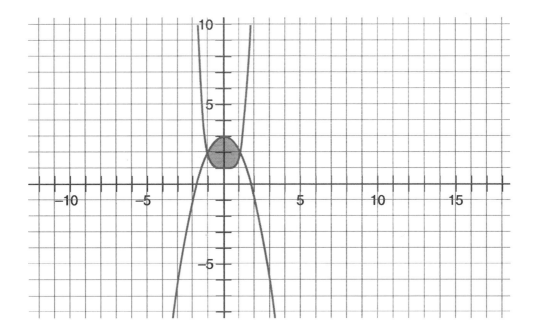

These two graphs intersect twice, and to find these exact intersection points, we set the two equations equal to each other and solve for x:

$$\begin{aligned}
3 - x^2 &= x^4 + 1 \\
0 &= x^4 + 1 - \left(3 - x^2\right) \\
0 &= x^4 + x^2 - 2
\end{aligned}$$

To solve this polynomial equation, I will use u-substitution to convert it to a quadratic equation. Then I can factor it, and I will "back-substitute" to get the correct factorization for the 4ᵗʰ degree polynomial. Finally, I use the Zero Product Property (ZPP) to solve. I will let $u = x^2$. After squaring both sides of this equation, I get that $u^2 = x^4$. Substitute *both* of these into the 4ᵗʰ degree polynomial and we have that:

$$x^4 + x^2 - 2 = u^2 + u - 2 = (u + 2)(u - 1) = (x^2 + 2)(x^2 - 1) = (x^2 + 2)(x + 1)(x - 1)$$

I set this equal to zero and apply the ZPP to solve:

$$\begin{aligned}
0 &= (x^2 + 2)(x + 1)(x - 1) \\
x^2 + 2 &= 0 \quad \text{or} \quad x + 1 = 0 \quad \text{or} \quad x - 1 = 0 \\
x &= \pm\sqrt{-2} = \pm\sqrt{2}i \quad \text{or} \quad x = -1 \quad \text{or} \quad x = 1
\end{aligned}$$

We disregard the imaginary solutions, and we have the intersection points have both $x = 1$ and $x = -1$ as x-coordinates. We can set up the iterated integral:

$$\iint_R \left(\frac{\partial N}{\partial x} - \frac{\partial M}{\partial y} \right) dxdy = \int_{-1}^{1} \int_{x^4+1}^{3-x^2} \left(\frac{e^x}{y} - \left(1 + \frac{e^x}{y} \right) \right) dydx$$

$$= \int_{-1}^{1} \int_{x^4+1}^{3-x^2} -1 \, dydx$$

$$= -\int_{-1}^{1} [y]_{x^4+1}^{3-x^2} \, dx$$

$$= -\int_{-1}^{1} [(3 - x^2) - (x^4 + 1)] dx$$

$$= -\int_{-1}^{1} (2 - x^2 - x^4) dx$$

$$= -\left[2x - \frac{x^3}{3} - \frac{x^5}{5} \right]_{-1}^{1}$$

$$= -\left[\left(2(1) - \frac{(1)^3}{3} - \frac{(1)^5}{5} \right) - \left(2(-1) - \frac{(-1)^3}{3} - \frac{(-1)^5}{5} \right) \right]$$

$$= -\left[\left(2 - \frac{1}{3} - \frac{1}{5} \right) \right] - \left(-2 - \frac{-1}{3} - \frac{-1}{5} \right)$$

$$= -\left[2 - \frac{1}{3} - \frac{1}{5} + 2 - \frac{1}{3} - \frac{1}{5} \right]$$

$$= -4 + \frac{2}{3} + \frac{2}{5} = -\frac{44}{15}$$

EXAMPLE 6. Let C be the boundary of a region on which Green's Theorem holds. Use Green's Theorem to calculate:

$$\oint_C f(x)dx + g(y)dy$$

SOLUTION. Since we have that $M = f(x)$ and $N = g(y)$, then the partial derivatives we will use for the formula:

$$\oint_C Mdx + Ndy = \iint_R \left(\frac{\partial N}{\partial x} - \frac{\partial M}{\partial y} \right) dxdy \text{ ...will be:}$$

$$\frac{\partial N}{\partial x} = \frac{\partial}{\partial x}[g(y)] = 0 = \frac{\partial M}{\partial y} = \frac{\partial}{\partial y}[f(x)]$$

Then the double integral will have a value of 0 as well.

EXAMPLE 7. Show that the value of $\oint_C xy^2\,dx + (x^2y + 2x)\,dy$ around any square depends only on the area of the square and not its location on the plane.

SOLUTION. Using Green's Theorem, we have that:

$$
\begin{aligned}
\oint_C xy^2\,dx + (x^2y + 2x)\,dy &= \iint_R \left(\frac{\partial N}{\partial x} - \frac{\partial M}{\partial y} \right) dx\,dy \\
&= \iint_R (2xy + 2 - 2xy)\,dA = \iint_R 2\,dx\,dy
\end{aligned}
$$

$$= 2 \text{ times the area of the square}$$

EXAMPLE 8. What is so special about the integral $\oint_C 4x^3 y\,dx + x^4\,dy\ \dots$?

Give reasons for your answer.

SOLUTION. Since $M = 4x^3 y$ and $N = x^4$, we have that $\dfrac{\partial N}{\partial x} = 4x^3 = \dfrac{\partial M}{\partial y}$, so that after applying Green's Theorem we have that:

$$
\oint_C 4x^3 y\,dx + x^4\,dy = \iint_R \left(\frac{\partial N}{\partial x} - \frac{\partial M}{\partial y} \right) dx\,dy = \iint_R (4x^3 - 4x^3)\,dx\,dy = 0
$$

Conclusion: the integral is 0 for any simple closed curve C.

This concludes the section!

EXERCISES: #15–19 all, extra credit I, II, and III, 20, divergence and curl problems 21, 22, Extra Credit from **Quiz 11**, #23 from **Practice Final**

SUMMARY FLOWCHART

Below is a summary of the preceding four sections in a flowchart that gives a path (no pun intended!) on how to tackle any given line integral. My students have found this chart to be most helpful on clearing up any confusion on when the Fundamental Theorem of Line Integrals and Green's Theorem may or may not be used, and also where parameterization for the curve is required or not.

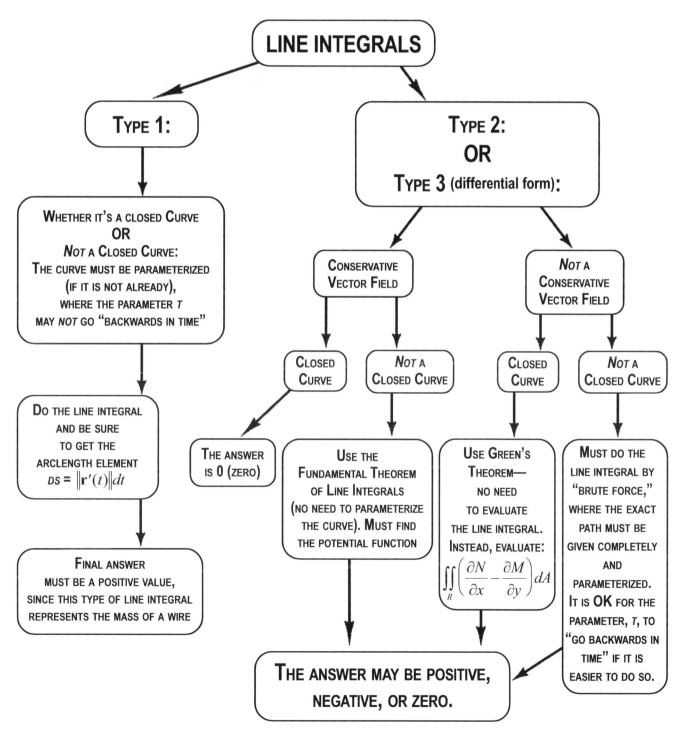

5.5 SURFACE INTEGRALS AND FLUX INTEGRALS

We are going to study what a surface integral is. But *first*, I would like to summarize for you the three new types of integrals you've learned about in the last couple of months:

1)   Line Integrals (integrate over a line, or curve *C*) have the form $\int_C \mathbf{F} \cdot d\mathbf{r}$

2)   Double Integrals (integrate over a region *R* in the plane) have the form $\iint_R f(x,y)dA$

3)   Triple Integrals (integrate over a solid *Q* in space) have the form $\iiint_Q f(x,y,z)dV$

Let's review the formula for surface area we learned back in the last chapter. The reason I am doing this is because surface integrals look very similar to this formula.

Definition: Surface Area

If *f* and its partial derivatives are continuous on the closed region *R* in the *xy*-plane, then the area of the surface *S* given by $z = f(x,y)$ over *R* is given by:

$$\textbf{\textit{Surface Area}} = \iint_R dS = \iint_R \sqrt{1 + [f_x(x,y)]^2 + [f_y(x,y)]^2}\, dA$$

Note that the "*dS*" gets "replaced" by a big expression $\sqrt{1 + [f_x(x,y)]^2 + [f_y(x,y)]^2}\, dA$.

Here's the Theorem for the formula and definition of a surface integral. You will be integrating over a *surface S* now!

Theorem—Evaluating a Surface Integral

Let *S* be a surface with the corresponding equation $z = g(x,y)$ and let *R* be the projection of this surface onto the *xy*-plane. If *g*, g_x, and g_y are all continuous on *R* and *f*(*x, y, z*) is a function that is continuous on *S*, then the surface integral of *f* over *S* is:

$$\iint_S f(x,y,z)dS = \iint_R f(x,y,g(x,y)) \sqrt{1 + [g_x(x,y)]^2 + [g_y(x,y)]^2}\, dA$$

Although the right-hand side of this formula looks scary, it is simply just a *big* ugly double integral that you already know how to handle from the last chapter. Summary: A surface integral can be easily converted into a *big* ugly regular double integral, where you are integrating over a region *R*.

The differences between this and the regular surface area formula you learned back in section 4.5 on surface area are the following:

1)   The integrand has a function in it, not just "*dS*." Concurrently, you have *f*(*x, y, z*) to deal with.

2)   The surface is given by $z = g(x,y)$, instead of $z = f(x,y)$. This means that "*dS*" gets "replaced" by the big expression $\sqrt{1 + [g_x(x,y)]^2 + [g_y(x,y)]^2}\, dA$ instead.

A common student pitfall is neglecting to replace the "*z*" from the function *f(x, y, z)* on the left-hand side of the formula with "$z = g(x, y)$" to obtain the $f(x, y, g(x, y))$ function in the right-hand side of the formula. Since you are supposed to convert the surface integral into a regular double integral, you only want the variables *x* and *y* in the integrand. The variable *z* should not be anywhere!

Another challenge is to convert the surface integral into a regular double integral and finding the limits of integration to iterate the regular double integral. The region *R* you are finding the limits for is the projection of the surface *S* onto the *xy*-plane, which is sometimes tough to visualize properly!

The last challenge may be the integration itself. Because of the huge (and ugly) radical in the integrand, the integral *might* not be integrable using basic techniques of integration.

EXAMPLE. Integrate $f(x, y, z) = \sqrt{x^2 + y^2 + z^2}$ over the surface bound by the graphs of $z = \sqrt{x^2 + y^2}$ (which is an elliptic cone) and the cylinder $(x - 1)^2 + y^2 \leq 1$.

SOLUTION. We are being asked to integrate $\iint\limits_{S} f(x, y, z)dS = \iint\limits_{S} \sqrt{x^2 + y^2 + z^2}\, dS$.

We need to first find the "*dS*" part, which means we need to get the partial derivatives for the function $z = \sqrt{x^2 + y^2}$, and put them into the radical $\sqrt{1 + [z_x]^2 + [z_y]^2}$.

To differentiate $z = \sqrt{x^2 + y^2}$, I will rewrite it using exponents first so that I can easily apply the Chain Rule for derivatives:

$$z = \sqrt{x^2 + y^2} = \left(x^2 + y^2\right)^{1/2}$$

Now, I can find the partials:

$$z_x = \frac{1}{2}\left(x^2 + y^2\right)^{-1/2} \cdot (2x) = \frac{x}{\left(x^2 + y^2\right)^{1/2}} = \frac{x}{\sqrt{x^2 + y^2}}$$

$$z_y = \frac{1}{2}\left(x^2 + y^2\right)^{-1/2} \cdot (2y) = \frac{y}{\left(x^2 + y^2\right)^{1/2}} = \frac{y}{\sqrt{x^2 + y^2}}$$

I substitute these into the radical and simplify:

$$\sqrt{1+\left[z_x\right]^2+\left[z_y\right]^2} = \sqrt{1+\left[\frac{x}{\sqrt{x^2+y^2}}\right]^2+\left[\frac{y}{\sqrt{x^2+y^2}}\right]^2}$$

$$= \sqrt{1+\frac{x^2}{x^2+y^2}+\frac{y^2}{x^2+y^2}}$$

$$= \sqrt{1\left(\frac{x^2+y^2}{x^2+y^2}\right)+\frac{x^2}{x^2+y^2}+\frac{y^2}{x^2+y^2}}$$

$$= \sqrt{\frac{2x^2+2y^2}{x^2+y^2}}$$

$$= \sqrt{\frac{2(x^2+y^2)}{x^2+y^2}}$$

$$= \sqrt{2}$$

Next, I need to rewrite $f(x,y,z)=\sqrt{x^2+y^2+z^2}$ as a function of two variables, where I substitute the function $z=\sqrt{x^2+y^2}$ in place of "z" for the function f:

$$f(x,y,g(x,y))=\sqrt{x^2+y^2+\left(\sqrt{x^2+y^2}\right)^2} = \sqrt{x^2+y^2+x^2+y^2} = \sqrt{2(x^2+y^2)} \quad \sqrt{2}\sqrt{x^2+y^2}$$

I need to figure out what the projection of the surface looks like on the *xy*-plane to get the region *R* we are integrating over. The inequality $(x-1)^2+y^2\le 1$ is a circle of radius 1 with its center point at $(1,0)$. This is actually what the projection of the surface will be. Because this is a circle, it's best to convert this entire problem to polar coordinates. So, I need to convert the equation $(x-1)^2+y^2=1$ to polar. You did a problem like this (I hope) at the end of your second semester calculus class. I will first expand and multiply the first term in the equation:

$$(x-1)(x-1)+y^2=1$$
$$x^2-2x+1+y^2=1$$
$$x^2+y^2-2x=0$$
$$r^2\qquad -2r\cos\theta=0$$
$$r^2 = 2r\cos\theta$$

I will now solve for r by dividing both sides of this equation by r. Ordinarily, it is illegal to divide both sides of an equation by a variable due to potential division by zero. However, if we assume r positive, we are not losing any info since the final result that is $r = 2\cos\theta$, contains the value or $r = 0$ within it. If we let the range of the angle θ be $0 \le \theta \le 2\pi$, then we will have the graph of a circle of radius 1 centered at $(1, 0)$. We can iterate our integral and evaluate:

$$\iint_S f(x,y,z)dS = \iint_R \sqrt{2}\sqrt{x^2 + y^2}\sqrt{2}dA = 2\int_0^{2\pi}\int_0^{2\cos\theta}\sqrt{r^2}\,rdrd\theta = 2\int_0^{2\pi}\int_0^{2\cos\theta}r^2drd\theta$$

$$= 2\int_0^{2\pi}\left[\frac{r^3}{3}\right]_0^{2\cos\theta}d\theta$$

$$= \frac{2}{3}\int_0^{2\pi}(2\cos\theta)^3 d\theta$$

I hope you remember how to integrate using the techniques involving powers of trig functions from your last semester of calculus. The power of cosine is odd and positive, so we save a cosine factor and convert the remaining cosines to sines:

$$= \frac{2}{3}(8)\int_0^{2\pi}(\cos\theta)^3 d\theta$$

$$= \frac{16}{3}\int_0^{2\pi}\cos^2\theta\cdot\cos\theta d\theta$$

$$= \frac{16}{3}\int_0^{2\pi}(1-\sin^2\theta)\cos\theta d\theta$$

$$= \frac{16}{3}\int_0^{2\pi}(\cos\theta - \sin^2\theta\cos\theta)d\theta$$

The first term in the integrand is easy to integrate; it'll simply be $\sin\theta$. The second term will involve u-substitution, where $u = \sin\theta$, and $du = \cos\theta\,d\theta$. So then $\int u^2 du = \frac{1}{3}u^3 = \frac{\sin^3\theta}{3}$.

Finishing the problem:
$$= \frac{16}{3}\left[\sin\theta - \frac{\sin^3\theta}{3}\right]_0^{2\pi}$$

$$= 0$$

EXAMPLE. Integrate $f(x,y,z) = x^2 - 2yz$ over the surface bound by the graphs of:

$$z = \cos x, \text{ where } 0 \le x \le \frac{\pi}{2} \quad \text{and} \quad 0 \le y \le \frac{x}{2}$$

Use your calculator ("FNINT" command) to evaluate the outer integral after you evaluate the inner one.

SOLUTION. We need to integrate $\iint_S f(x,y,z)dS = \iint_S (x^2 - 2yz)dS$.

We must first find the "dS" part, which means we need to get the partial derivatives for the function $z = \cos x$, and put them into the radical $\sqrt{1+\left[z_x\right]^2+\left[z_y\right]^2}$. We do that now:

$$\sqrt{1+\left[z_x\right]^2+\left[z_y\right]^2} = \sqrt{1+\left[-\sin x\right]^2+\left[0\right]^2} = \sqrt{1+\sin^2 x}$$

We need to rewrite the function $f(x,y,z) = x^2 - 2yz$ by replacing the "z" with "$\cos x$."

$$f(x,y,g(x,y)) = x^2 - 2y\cos x$$

We have the limits for x and y, so we can set up the double integral:

$$
\begin{aligned}
\iint_S f(x,y,z)\,dS &= \iint_S \left(x^2 - 2yz\right)dS \\
&= \int_0^{\pi/2} \int_0^{x/2} \left(x^2 - 2y\cos x\right)\sqrt{1+\sin^2 x}\,dy\,dx \\
&= \int_0^{\pi/2} \sqrt{1+\sin^2 x}\left[x^2 y - y^2\cos x\right]_0^{x/2} dx \\
&= \int_0^{\pi/2} \sqrt{1+\sin^2 x}\left[x^2\left(\frac{x}{2}\right) - \left(\frac{x}{2}\right)^2\cos x\right]dx \\
&= \int_0^{\pi/2} \sqrt{1+\sin^2 x}\left[\frac{x^3}{2} - \frac{x^2\cos x}{4}\right]dx
\end{aligned}
$$

OUCH! This integral is too hard to do by hand, so we use the "FNINT" command on our calculators:

$$\text{FNINT}\left(\sqrt{\left(1+(\sin x)^2\right)}\left(\frac{x^3}{2} - \frac{x^2\cos x}{4}\right), x, 0, \frac{\pi}{2}\right) \approx 0.887$$

Flux Integrals

Flux Integrals are surface integrals with a little bit more stuff in the integrand...and a very nice physical interpretation.

DEFINITION:
Orientable Surface

A surface S is *orientable* if a unit normal vector **N** can be defined at every non-boundary point of S in such a way that the normal vectors vary continuously over the surface S. If this is possible, S is called an *oriented surface*.

An orientable surface has two distinct sides. You may think that *all* surfaces have two distinct sides. However, a Möbius strip does *not* have two distinct sides! Therefore, a Möbius strip is not an orientable surface.

The unit normal vector **N** of a surface is a vector that juts out of the surface and is orthogonal to that surface. Recall from Chapter 4 of this book, that for a function of two variables, its gradient is a vector in the plane that is perpendicular (normal) to all of its level curves. For a function of three variables, its gradient is a vector in space that is perpendicular (normal) to all of its level surfaces. Also recall that we cannot graph a function of three variables, since we need 4 dimensions to do so. Where am I going with this discussion? Well, for a surface $z = g(x, y)$, we can find a normal vector by first forming a function of three variables:

$$G(x, y, z) = z - g(x, y)$$

Note: This is the opposite of what you did back in Chapter 3 when you found gradients to get the normal vector to a level surface to find equations of tangent lines and normal lines. We obtained $G(x, y, z) = g(x, y) - z$ in that chapter.

Next, the gradient for the function of three variables will be a vector normal to the level surfaces for this function and is:

$$\nabla \mathbf{G}(x, y, z) = \left\langle \frac{\partial G}{\partial x}, \frac{\partial G}{\partial y}, \frac{\partial G}{\partial z} \right\rangle = \left\langle -g_x(x, y), -g_y(x, y), 1 \right\rangle$$

To find the *unit* normal vector, we simply divide this gradient by its magnitude:

$$\mathbf{N} = \frac{\nabla G(x, y, z)}{\|\nabla G(x, y, z)\|} = \frac{\left\langle -g_x(x, y), -g_y(x, y), 1 \right\rangle}{\sqrt{\left[-g_x(x, y) \right]^2 + \left[-g_y(x, y) \right]^2 + 1^2}}$$

We are ready for the definition of a *flux integral,* which is simply a special case of a surface integral. The physical interpretation for the flux integral relates to the amount of fluid that flows across a surface with a given velocity field over a certain unit of time.

DEFINITION:
Flux Integral

Let $\mathbf{F}(x, y, z) = \left\langle M, N, P \right\rangle$ be a vector field where *M*, *N*, and *P* all have continuous first partial derivatives on a surface *S* oriented by a unit normal vector **N.** The *flux integral* of **F** across *S* is given by:

$$\iint\limits_S \mathbf{F} \cdot \mathbf{N} dS$$

How to evaluate this? Well, we simply substitute all of the appropriate expressions for each quantity and/or vector and we get what the Theorem tells us: how to go about evaluating these flux integrals.

THEOREM:
Evaluating a Flux Integral

Let S be an oriented surface given by $z = g(x, y)$ and let R be its projection on the xy-plane. Then:

$$\iint_S \mathbf{F} \cdot \mathbf{N} dS = \iint_R \mathbf{F} \cdot \frac{\langle -g_x(x, y), -g_y(x, y), 1 \rangle}{\sqrt{[-g_x(x, y)]^2 + [-g_y(x, y)]^2 + 1^2}} \sqrt{[g_x(x, y)]^2 + [g_y(x, y)]^2 + 1^2} \, dA$$

$$= \iint_R \mathbf{F} \cdot \langle -g_x(x, y), -g_y(x, y), 1 \rangle \, dA$$

If a flux integral is positive, the flow *out* of the surface exceeds the flow into S, and we say that there is a *source* of $\mathbf{F}$ within S.

If a flux integral is negative, the flow *into* the surface exceeds the flow out of S, and we say that there is a *sink* of $\mathbf{F}$ within S.

EXAMPLE. Find the flux of the field $\mathbf{F}(x, y, z) = \langle zx, zy, z \rangle$ across the portion of the sphere $x^2 + y^2 + z^2 = 9$ in the first octant in the direction away from the origin (i.e., oriented outward). Is it a source, a sink, or neither?

SOLUTION. First, we need to rewrite the equation for the surface to get that function of three variables discussed earlier: $G(x, y, z) = z - g(x, y)$. To do this, we first need to solve the sphere equation for z:

$$x^2 + y^2 + z^2 = 9$$
$$z^2 = 9 - x^2 - y^2$$
$$z = \pm\sqrt{9 - x^2 - y^2}$$

Since we are only interested in the first octant, we take the positive version only for z. We can find $G(x, y, z) = z - g(x, y)$. We do this now:

$$G(x, y, z) = z - \sqrt{9 - x^2 - y^2} = z - \left(9 - x^2 - y^2\right)^{1/2}$$

I purposely rewrote this using exponents to make it easier for me to differentiate. Next, we find the gradient for this new function:

$$\nabla \mathbf{G}(x, y, z) = \left\langle \frac{\partial G}{\partial x}, \frac{\partial G}{\partial y}, \frac{\partial G}{\partial z} \right\rangle = \left\langle -g_x(x, y), -g_y(x, y), 1 \right\rangle$$

$$= \left\langle \frac{1}{2}\left(9 - x^2 - y^2\right)^{-1/2} \cdot (2x), \frac{1}{2}\left(9 - x^2 - y^2\right)^{-1/2} \cdot (2y), 1 \right\rangle$$

$$= \left\langle \frac{x}{\sqrt{9 - x^2 - y^2}}, \frac{y}{\sqrt{9 - x^2 - y^2}}, 1 \right\rangle$$

Now, we are ready to set up and evaluate the flux integral:

$$\iint_R \mathbf{F} \cdot \langle -g_x(x, y), -g_y(x, y), 1 \rangle dA \; = \; \iint_R \langle zx, zy, z \rangle \cdot \left\langle \frac{x}{\sqrt{9 - x^2 - y^2}}, \; \frac{y}{\sqrt{9 - x^2 - y^2}}, 1 \right\rangle dA$$

$$= \; \iint_R \left(\frac{zx^2}{\sqrt{9 - x^2 - y^2}} + \frac{zy^2}{\sqrt{9 - x^2 - y^2}} + z \right) dA$$

$$= \; \iint_R \left(\frac{z(x^2 + y^2)}{\sqrt{9 - x^2 - y^2}} \right) dA \; + \; \iint_R z \, dA$$

I purposely split this into a sum of two separate integrals to make the point that it'll be easier to evaluate this way. You shall soon see this for yourself! I need to do a couple of things before I can evaluate either of these integrals.

1) I need to determine the projection of the surface (i.e., the sphere) onto the xy-plane for me to iterate these double integrals. This one's easy; the projection of a sphere onto the xy-plane will always be a circle. This projection is a portion of the circle, since we are interested in the first octant only. Remember we prefer to evaluate double integrals with circular regions of integration using polar coordinates. This means we will definitely be converting these double integrals to polar coordinates. The ranges for r and θ for the portion of the circle in the first octant will be:

$$0 \leq r \leq 3 \quad \text{and} \quad 0 \leq \theta \leq \frac{\pi}{2}$$

2) I need to replace the variable "z" in both of these double integrals, since a double integral is allowed to contain only two variables x and y. I replace z with the equation for the sphere that is:

$$z = \sqrt{9 - x^2 - y^2}$$

Of course, we need to convert our integrands to polar also:

$$\iint_R \left(\frac{z(x^2 + y^2)}{\sqrt{9 - x^2 - y^2}} \right) dA \; + \; \iint_R z \, dA \; = \; \int_0^{\pi/2} \int_0^3 \frac{\sqrt{9 - x^2 - y^2}\,(r^2)}{\sqrt{9 - x^2 - y^2}} (r \, dr \, d\theta) \; + \; \int_0^{\pi/2} \int_0^3 \left(\sqrt{9 - x^2 - y^2} \right)(r \, dr \, d\theta)$$

I simplify these integrands by reducing quotients to lowest terms and converting completely to polar coordinates:

$$= \; \int_0^{\pi/2} \int_0^3 r^3 \, dr \, d\theta \; + \; \int_0^{\pi/2} \int_0^3 \sqrt{9 - r^2} \, r \, dr \, d\theta$$

The first integral is very easy to evaluate. The second integral requires the General Power Rule, where $u = 9 - r^2$ and $du = -2rdr$. Of course, we'll have to put a factor of -2 into the integrand, as well as divide it out before evaluating:

$$= \int_0^{\pi/2} \left[\frac{r^4}{4} \right]_0^3 d\theta \; + \; \int_0^{\pi/2} \left(\frac{1}{-2} \right) \int_0^3 \sqrt{9 - r^2} (-2)r dr d\theta$$

$$= \int_0^{\pi/2} \frac{3^4}{4} d\theta \; + \; \frac{-1}{2} \int_0^{\pi/2} \left[\int \sqrt{u} \, du \right] d\theta$$

$$= \frac{81}{4} \int_0^{\pi/2} d\theta \; + \; \frac{-1}{2} \int_0^{\pi/2} \left[\frac{2}{3} u^{3/2} \right] d\theta$$

$$= \frac{81}{4} [\theta]_0^{\pi/2} \; + \; \frac{-1}{3} \int_0^{\pi/2} \left[(9 - r^2)^{3/2} \right]_0^3 d\theta$$

$$= \frac{81}{4} \left(\frac{\pi}{2} \right) \; + \; \frac{-1}{3} \int_0^{\pi/2} \left[(9 - 3^2)^{3/2} - (9 - 0^2)^{3/2} \right] d\theta$$

$$= \frac{81\pi}{8} \; + \; \frac{-1}{3} \int_0^{\pi/2} \left[(0)^{3/2} - (9)^{3/2} \right] d\theta$$

$$= \frac{81\pi}{8} \; + \; \frac{-1}{3} (-27) \int_0^{\pi/2} d\theta$$

$$= \frac{81\pi}{8} \; + \; 9 \left(\frac{\pi}{2} \right) \; = \; \frac{81\pi}{8} \; + \; \frac{36\pi}{8} \; = \; \frac{117\pi}{8}$$

This positive value for the flux integral tells us that we have a *source*.

EXAMPLE. Set up and evaluate the integral(s) that would calculate the flux of the field $\mathbf{F}(x, y, z) = \langle 2xy, 2yz, 2xz \rangle$ across the surface of the cube cut from the first octant by the planes $x = 5$, $y = 5$, and $z = 5$. Source, sink, or neither?

SOLUTION. Because a cube has six sides, we will need to calculate the total flux by taking the *sum* of six separate flux integrals. Further, the region we will be integrating over for each integral will vary. We will determine the projection of each surface onto an appropriate plane—not necessarily the *xy*-plane as we have in previous examples. Finally, we can identify the normal vector **N** for each surface without differentiating, as we shall soon see. Let us name each surface:

S_1: The top of the cube, where the surface is the plane $z = 5$. The unit normal vector **N** will be the standard unit vector **k**. We project this plane onto the xy-plane and we have a rectangle: $0 \le x \le 5, 0 \le y \le 5$.

S_2: The bottom of the cube, where the surface is the plane $z = 0$. The unit normal vector **N** will be the standard unit vector **-k**. We project this plane onto the xy-plane and we have a rectangle: $0 \le x \le 5, 0 \le y \le 5$.

S_3: The front of the cube, where the surface is the plane $x = 5$. The unit normal vector **N** will be the standard unit vector **i**. We project this plane onto the yz-plane and we have a rectangle: $0 \le y \le 5, 0 \le z \le 5$.

S_4: The back of the cube, where the surface is the plane $x = 0$. The unit normal vector **N** will be the standard unit vector **-i**. We project this plane onto the yz-plane and we have a rectangle: $0 \le y \le 5, 0 \le z \le 5$.

S_5: The right side of the cube, where the surface is the plane $y = 5$. The unit normal vector **N** will be the standard unit vector **j**. We project this plane onto the xz-plane and we have a rectangle: $0 \le x \le 5, 0 \le z \le 5$.

S_6: The left side of the cube, where the surface is the plane $y = 0$. The unit normal vector **N** will be the standard unit vector **-j**. We project this plane onto the xz-plane and we have a rectangle: $0 \le x \le 5, 0 \le z \le 5$.

Again, we must sum each of the six flux integrals (one for each surface of the cube) to get our answer:

$$\iint_S \mathbf{F} \cdot \mathbf{N} dS = \iint_{S_1} \mathbf{F} \cdot \mathbf{N} dS + \iint_{S_2} \mathbf{F} \cdot \mathbf{N} dS + \iint_{S_3} \mathbf{F} \cdot \mathbf{N} dS + \iint_{S_4} \mathbf{F} \cdot \mathbf{N} dS + \iint_{S_5} \mathbf{F} \cdot \mathbf{N} dS + \iint_{S_6} \mathbf{F} \cdot \mathbf{N} dS$$

I will do each flux integral separately:

$$\iint_{S_1} \mathbf{F} \cdot \mathbf{N} dS = \int_0^5 \int_0^5 \langle 2xy, 2yz, 2xz \rangle \cdot \mathbf{k} \, dy dx = \int_0^5 \int_0^5 2xz \, dy dx = \int_0^5 \int_0^5 2x(5) \, dy dx = 10 \int_0^5 x[y]_0^5 \, dx = 50 \left[\frac{x^2}{2}\right]_0^5 = 625$$

$$\iint_{S_2} \mathbf{F} \cdot \mathbf{N} dS = \int_0^5 \int_0^5 \langle 2xy, 2yz, 2xz \rangle \cdot -\mathbf{k} \, dy dx = -\int_0^5 \int_0^5 2xz \, dy dx = -\int_0^5 \int_0^5 2x(0) \, dy dx = 0$$

Be careful! This next integral has us working with the plane given by $x = 5$. The region we consider is in the yz-plane, and we replace any instance of "x" in the double integral with "5."

$$\iint_{S_3} \mathbf{F} \cdot \mathbf{N} dS = \int_0^5 \int_0^5 \langle 2xy, 2yz, 2xz \rangle \cdot \mathbf{i} \, dy dz = \int_0^5 \int_0^5 2xy \, dy dz = \int_0^5 \int_0^5 2(5)y \, dy dz = 10 \int_0^5 \left[\frac{y^2}{2}\right]_0^5 dz = 125[z]_0^5 = 625$$

The next integral is similar to the last one, except that $x = 0$:

$$\iint_{S_4} \mathbf{F} \cdot \mathbf{N} dS = \int_0^5 \int_0^5 \langle 2xy, 2yz, 2xz \rangle \cdot -\mathbf{i} \, dy dz = \int_0^5 \int_0^5 -2xy \, dy dz = \int_0^5 \int_0^5 -2(0)y \, dy dz = 0$$

The last two integrals have us working with the planes $y = 5$ and $y = 0$. Both will have the regions be the projections of these planes onto the xz-plane. Any instance of "y" appearing in the integrand must be replaced with either "5" or "0" depending on which surface we are using:

$$\iint\limits_{S_5} \mathbf{F} \cdot \mathbf{N} dS = \int_0^5\int_0^5 \langle 2xy, 2yz, 2xz \rangle \cdot \mathbf{j} \, dxdz = \int_0^5\int_0^5 2yzdxdz = \int_0^5\int_0^5 2(5)zdxdz = 10\int_0^5 [x]_0^5 zdz = 50\left[\frac{z^2}{2}\right]_0^5 = 625$$

$$\iint\limits_{S_6} \mathbf{F} \cdot \mathbf{N} dS = \int_0^5\int_0^5 \langle 2xy, 2yz, 2xz \rangle \cdot -\mathbf{j} \, dxdz = \int_0^5\int_0^5 -2yzdxdz = \int_0^5\int_0^5 -2(0)zdxdz = 0$$

For the final answer, we take the sum of all of these results: $625 + 625 + 625 = 1875$. Because this result is positive, we have a *source!*

One last note: Since the last example had us calculating the flux on a *closed surface,* we will be able to use a Theorem on it covered in the next section that will make the problem much easier. We won't have to break it down into six separate integrals…so stay tuned!

This concludes the section.

EXERCISES: #24, 25, from **Practice Final**

5.6 THE DIVERGENCE THEOREM

As long as you have a closed and oriented surface, the Divergence Theorem gives a relationship between a flux integral over a surface S (that contains a solid region) and a triple integral over the solid Q.

Recall that one of the versions we studied for Green's Theorem (section 5.4) stated that the flux of a vector field across a simple closed curve can be calculated by integrating the divergence of the field over the region enclosed by the curve. Green's Theorem gave us a relationship between a line integral and a double integral over a region R (bounded by the curve). Let's review Green's Theorem.

GREEN'S THEOREM: (Flux Version)

$$\oint_C \mathbf{F} \cdot \mathbf{n} \, ds \quad = \quad \oint_C Mdy - Ndx \quad = \quad \iint\limits_R (\text{div } \mathbf{F})dA \quad = \quad \iint\limits_R \left(\frac{\partial M}{\partial x} + \frac{\partial N}{\partial y}\right)dxdy$$

(*OUTWARD FLUX*)   (*DIVERGENCE INTEGRAL*)

DIVERGENCE THEOREM

The Divergence Theorem is the corresponding three-dimensional analog for Green's Theorem. It basically states that the net outward flux of a vector field across a closed surface in space can be calculated by integrating the divergence of the field over the solid region enclosed by the surface:

$$(\text{\textit{OUTWARD FLUX}}) \quad \iint_S \mathbf{F} \cdot \mathbf{N} dS \quad = \quad \iiint_Q \text{div} \mathbf{F} dV \quad (\text{\textit{DIVERGENCE INTEGRAL}})$$

For review, recall that divergence for a vector field is found by:

$$\text{div} F = \nabla \cdot \mathbf{F} = \frac{\partial M}{\partial x} + \frac{\partial N}{\partial y} + \frac{\partial P}{\partial z}$$

EXAMPLE. Do the example all over again from the previous section:

Set up and evaluate the integral(s) that would calculate the flux of the field $\mathbf{F}(x, y, z) = \langle 2xy, 2yz, 2xz \rangle$ across the surface of the cube cut from the first octant by the planes $x = 5$, $y = 5$, and $z = 5$.

Source, sink, or neither?

SOLUTION. Because we know the Divergence Theorem, we only require a single integral instead of having to take the sum of six separate integrals!

$$\iint_S \mathbf{F} \cdot \mathbf{N} dS = \iiint_Q \text{div} \mathbf{F} dV = \int_0^5\int_0^5\int_0^5 (2y + 2z + 2x) dz dy dx \qquad = \int_0^5 \left[5y^2 + 25y + 10xy \right]_0^5 dx$$

$$= \int_0^5\int_0^5 \left[2yz + z^2 + 2xz \right]_0^5 dy dx \qquad = \int_0^5 \left[5(5)^2 + 25(5) + 10x(5) \right] dx$$

$$= \int_0^5\int_0^5 \left[2y(5) + (5)^2 + 2x(5) \right] dy dx \qquad\Longrightarrow\qquad = \int_0^5 (125 + 125 + 50x) dx$$

$$= \int_0^5\int_0^5 (10y + 25 + 10x) dy dx \qquad = \left[250x + 25x^2 \right]_0^5$$

$$= 250(5) + 25(5)^2$$

This is (of course) the same result we got the last time!

$$= 1250 + 625$$

$$= 1875$$

EXAMPLE. Calculate the outward flux of the gravitational field...

$$\mathbf{F} = -\frac{GM(x\mathbf{i} + y\mathbf{j} + z\mathbf{k})}{\left(x^2 + y^2 + z^2\right)^{3/2}}$$

...where G is the constant for gravity and M is the mass of the object (also a constant) across the surface enclosed by the sphere $x^2 + y^2 + z^2 = 1$.

SOLUTION. We first find the divergence of **F**. We need to use the Quotient Rule three times—to differentiate each of the three components of **F**.

$$\text{div}\mathbf{F} = \frac{\partial M}{\partial x} + \frac{\partial N}{\partial y} + \frac{\partial P}{\partial z}$$

$$\frac{\partial M}{\partial x} = -GM \frac{(1)\left(x^2 + y^2 + z^2\right)^{3/2} - \frac{3}{2}\left(x^2 + y^2 + z^2\right)^{1/2} \cdot (2x)(x)}{\left[\left(x^2 + y^2 + z^2\right)^{3/2}\right]^2} = \frac{\left(x^2 + y^2 + z^2\right)^{3/2} - 3x^2\left(x^2 + y^2 + z^2\right)^{1/2}}{\left(x^2 + y^2 + z^2\right)^3}$$

$$\frac{\partial N}{\partial y} = -GM \frac{(1)\left(x^2 + y^2 + z^2\right)^{3/2} - \frac{3}{2}\left(x^2 + y^2 + z^2\right)^{1/2} \cdot (2y)(y)}{\left[\left(x^2 + y^2 + z^2\right)^{3/2}\right]^2} = \frac{\left(x^2 + y^2 + z^2\right)^{3/2} - 3y^2\left(x^2 + y^2 + z^2\right)^{1/2}}{\left(x^2 + y^2 + z^2\right)^3}$$

$$\frac{\partial P}{\partial z} = -GM \frac{(1)\left(x^2 + y^2 + z^2\right)^{3/2} - \frac{3}{2}\left(x^2 + y^2 + z^2\right)^{1/2} \cdot (2z)(z)}{\left[\left(x^2 + y^2 + z^2\right)^{3/2}\right]^2} = \frac{\left(x^2 + y^2 + z^2\right)^{3/2} - 3z^2\left(x^2 + y^2 + z^2\right)^{1/2}}{\left(x^2 + y^2 + z^2\right)^3}$$

Now, I take the sum. Notice that all three quotients have the same denominator. Also notice that all three quotients have the same first term in the numerator's sum, so I can add all three fractions and collect like terms in the numerator:

$$\frac{\partial M}{\partial x} + \frac{\partial N}{\partial y} + \frac{\partial P}{\partial z} = \frac{3\left(x^2 + y^2 + z^2\right)^{3/2} - 3x^2\left(x^2 + y^2 + z^2\right)^{1/2} - 3y^2\left(x^2 + y^2 + z^2\right)^{1/2} - 3z^2\left(x^2 + y^2 + z^2\right)^{1/2}}{\left(x^2 + y^2 + z^2\right)^3}$$

Notice that the last three terms in the numerator have a common factor, that is $-3\left(x^2 + y^2 + z^2\right)^{1/2}$. I will factor this out, and continue to simplify the numerator:

$$= \frac{3\left(x^2 + y^2 + z^2\right)^{3/2} - 3\left(x^2 + y^2 + z^2\right)^{1/2}\left(x^2 + y^2 + z^2\right)}{\left(x^2 + y^2 + z^2\right)^3}$$

$$= \frac{3\left(x^2 + y^2 + z^2\right)^{3/2} - 3\left(x^2 + y^2 + z^2\right)^{3/2}}{\left(x^2 + y^2 + z^2\right)^3}$$

$$= 0$$

Since the divergence is equal to 0, this means that the triple integral of this divergence will also have a value of 0. Therefore, the outward flux of this field over *any* closed surface will be equal to zero! So I don't even need to evaluate any integrals to figure this out! *WOW!*

EXAMPLE. Determine the outward flux of the field $\mathbf{F}(x,y,z) = \left\langle xy^2, x^2y, e^z \right\rangle$ through the surface of the solid bounded below by the cone $z = \sqrt{x^2 + y^2}$ and above by the plane $z = 4$. See diagram below:

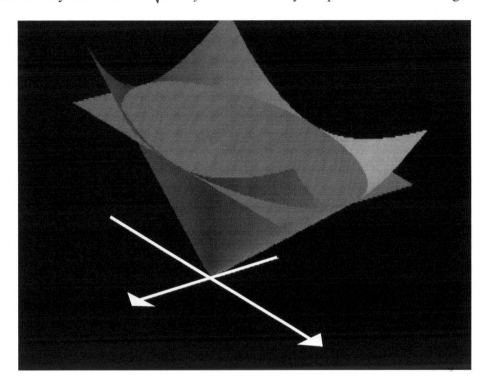

SOLUTION. From the diagram, we see that this is indeed an orientable, closed surface. We use the Divergence Theorem:

$$\textit{Outward flux} = \iint\limits_S \mathbf{F} \cdot \mathbf{N}\,dS \;\; = \;\; \iiint\limits_Q \text{div}\mathbf{F}\,dV \;\; = \;\; \iiint\limits_Q \left(y^2 + x^2 + e^z\right)dV$$

We need to iterate this triple integral now. To figure out what is going on in the xy-plane, we project the surface onto the xy-plane. It will be a circle of radius 4. How do we know this? Set the two equations for the surfaces equal to each other:

$$z = \sqrt{x^2 + y^2} = 4$$
$$\left(\sqrt{x^2 + y^2}\right)^2 = 4^2$$
$$x^2 + y^2 = 4^2$$

This means the ranges for x and y will be:

$$-4 \le x \le 4$$
$$-\sqrt{16 - x^2} \le y \le \sqrt{16 - x^2}$$

The range for z will be $\sqrt{x^2 + y^2} \le z \le 4$. We iterate the triple integral, but then make it easier on everyone by converting to *cylindrical* coordinates. (Remember those?)

$$\iiint\limits_{Q} \left(y^2 + x^2 + e^z\right) dV = \int_{-4}^{4} \int_{-\sqrt{16-x^2}}^{\sqrt{16-x^2}} \int_{\sqrt{x^2+y^2}}^{4} \left(y^2 + x^2 + e^z\right) dz\,dy\,dx$$

$$= \int_{0}^{2\pi} \int_{0}^{4} \int_{\sqrt{r^2}}^{4} \left(r^2 + e^z\right) dz\,r\,dr\,d\theta$$

$$= \int_{0}^{2\pi} \int_{0}^{4} \int_{r}^{4} \left(r^3 + re^z\right) dz\,dr\,d\theta$$

$$= \int_{0}^{2\pi} \int_{0}^{4} \left[r^3 z + re^z\right]_{r}^{4} dr\,d\theta$$

$$= \int_{0}^{2\pi} \int_{0}^{4} \left[\left(r^3(4) + re^4\right) - \left(r^3(r) + re^r\right)\right]_{r}^{4} dr\,d\theta$$

$$= \int_{0}^{2\pi} \int_{0}^{4} \left(4r^3 + re^4 - r^4 - re^r\right) dr\,d\theta$$

The last term in the integrand requires integration by parts where:

$$u = r \qquad dv = e^r\,dr$$
$$du = dr \qquad v = e^r$$

So, we have the anti-derivative as $\int re^r = re^r - \int e^r\,dr = re^r - e^r$. Substitute this in and keep going:

$$= \int_{0}^{2\pi} \left[r^4 + \frac{r^2}{2}e^4 - \frac{r^5}{5} - \left(re^r - e^r\right)\right]_{0}^{4} d\theta$$

$$= \int_{0}^{2\pi} \left[\left((4)^4 + \frac{(4)^2}{2}e^4 - \frac{(4)^5}{5} - (4)e^4 + e^4\right) - \left(0 + e^0\right)\right] d\theta$$

$$= \int_{0}^{2\pi} \left(256 + 6e^4 - \frac{1024}{5} - 1\right) d\theta$$

$$= \left[255\theta + 6e^4\theta - \frac{1024}{5}\theta\right]_{0}^{2\pi}$$

$$= 255(2\pi) + 6e^4(2\pi) - \frac{1024}{5}(2\pi)$$

$$= \left(\frac{502}{5} + 12e^4\right)\pi$$

EXAMPLE. Determine the outward flux of the field:

$$\mathbf{F}(x,y,z) = \left(5x^3 + 12xy^2\right)\mathbf{i} + \left(y^3 + e^y \sin z\right)\mathbf{j} + \left(5z^3 + e^y \cos z\right)\mathbf{k}$$

…through the surface of the solid between the spheres $x^2 + y^2 + z^2 = 1$ and $x^2 + y^2 + z^2 = 2$.

SOLUTION. First, we determine the divergence of **F**:
$$\frac{\partial M}{\partial x} = 15x^2 + 12y^2$$

$$\frac{\partial N}{\partial y} = 3y^2 + e^y \sin z$$

$$\frac{\partial P}{\partial z} = 15z^2 - e^y \sin z$$

Taking the sum now:

$$\mathrm{div}F = \frac{\partial M}{\partial x} + \frac{\partial N}{\partial y} + \frac{\partial P}{\partial z} = 15x^2 + 12y^2 + 3y^2 + e^y \sin z + 15z^2 - e^y \sin z$$

$$= 15x^2 + 15y^2 + 15z^2$$

$$= 15(x^2 + y^2 + z^2)$$

Next, we use the Divergence Theorem. To iterate the triple integral, we note that the projection of the two spheres onto the *xy*-plane will give us two concentric circles: one with a radius of 1 and the other with a radius of $\sqrt{2}$. It will be much easier if we convert to *spherical* coordinates. We recall that spherical coordinates are useful when we are integrating over solids that are spheres. The ranges for each of the variables (that are ρ, θ, and ϕ) will be:

$$1 \le \rho \le \sqrt{2}$$
$$0 \le \theta \le 2\pi$$
$$0 \le \phi \le \pi$$

Remember we also must convert the integrand to spherical coordinates, where we replace "*dzdydx*" with "$\rho^2 \sin\phi\, d\rho\, d\phi\, d\theta$." We have that outward flux will be:

$$\iint_S \mathbf{F} \cdot \mathbf{N}dS = \iiint_Q \mathrm{div}\mathbf{F}dV = \iiint_Q 15\left(x^2 + y^2 + z^2\right)dV$$

$$= 15\int_0^{2\pi}\int_0^{\pi}\int_1^{\sqrt{2}} \left(\rho^2\right)\left(\rho^2 \sin\phi\right)d\rho\, d\phi\, d\theta$$

$$= 15\int_0^{2\pi}\int_0^{\pi}\int_1^{\sqrt{2}} \left(\rho^4\right)\left(\sin\phi\right)d\rho\, d\phi\, d\theta$$

$$= 15\int_0^{2\pi}\int_0^{\pi} \sin\phi\left[\frac{\rho^5}{5}\right]_1^{\sqrt{2}} d\phi\, d\theta$$

$$= 15\int_0^{2\pi}\int_0^{\pi} \sin\phi\left[\frac{\left(\sqrt{2}\right)^5}{5} - \frac{1}{5}\right]d\phi\, d\theta$$

$$= 15\left[\frac{2^{5/2}}{5} - \frac{1}{5}\right]\int_0^{2\pi}\left[-\cos\phi\right]_0^{\pi} d\theta$$

$$= 15\left[\frac{2^{5/2}}{5} - \frac{1}{5}\right]_0^{2\pi} \int [-\cos\pi - (-\cos 0)]d\theta \qquad = (15)(2)\left(\frac{2^{5/2} - 1}{5}\right)[\theta]_0^{2\pi}$$

$$= 15\left[\frac{2^{5/2}}{5} - \frac{1}{5}\right]_0^{2\pi} \int [-(-1) - (-1)]d\theta \qquad = (30)\left(\frac{2^{5/2} - 1}{5}\right)(2\pi)$$

$$= 15\left[\frac{2^{5/2}}{5} - \frac{1}{5}\right]_0^{2\pi} \int 2d\theta \qquad = (6)(2\pi)(4\sqrt{2} - 1)$$

$$= 12(4\sqrt{2} - 1)\pi$$

This concludes this section.

EXERCISES: #26, 27, 28 from **Practice Final**

5.7 STOKES' THEOREM

Stokes' Theorem gives us a way to generalize the circulation-curl of Green's Theorem in the plane to velocity fields in space. In 3-D, the circulation around a point P in a plane is described with a vector. This vector is not only normal to the plane of circulation, but it points in the direction that gives it a right-hand relation to the circulation line. You will use the right-hand rule to determine this direction. The length of the vector gives the rate of the fluid's rotation, which usually varies as the circulation plane is tilted about P. It turns out that the vector of greatest circulation in a flow with velocity field $\mathbf{F}(x,y,z) = \langle M, N, P \rangle$ is:

$$\text{curl } \mathbf{F}(x,y,z) = \left\langle \frac{\partial P}{\partial y} - \frac{\partial N}{\partial z}, \frac{\partial M}{\partial z} - \frac{\partial P}{\partial x}, \frac{\partial N}{\partial x} - \frac{\partial M}{\partial y} \right\rangle$$

Remember that we found this curl by taking the cross product of the operator ∇ with the field $\mathbf{F}$, which requires the method for finding the determinant of a 3×3 matrix. To explain further:

$$
\begin{aligned}
\text{curl } \mathbf{F}(x,y,z) &= \nabla \times \mathbf{F}(x,y,z) \\[2mm]
&= \begin{vmatrix} \mathbf{i} & \mathbf{j} & \mathbf{k} \\ \dfrac{\partial}{\partial x} & \dfrac{\partial}{\partial y} & \dfrac{\partial}{\partial z} \\ M & N & P \end{vmatrix} \\[2mm]
&= \left(\frac{\partial P}{\partial y} - \frac{\partial N}{\partial z}\right)\mathbf{i} - \left(\frac{\partial P}{\partial x} - \frac{\partial M}{\partial z}\right)\mathbf{j} + \left(\frac{\partial N}{\partial x} - \frac{\partial M}{\partial y}\right)\mathbf{k}
\end{aligned}
$$

Recall that $(\text{curl } \mathbf{F}) \cdot \mathbf{k} = \dfrac{\partial N}{\partial x} - \dfrac{\partial M}{\partial y}$ for a field that has the form $\mathbf{F}(x,y) = \langle M, N \rangle$.

Stokes' Theorem tells us that (under certain conditions) the circulation of a vector field around the boundary of an oriented surface in space in the direction counterclockwise with respect to the surface's unit normal vector **N,** equals the integral of the normal component of the curl of the field over the surface. It enables one to convert a line integral over a closed curve to a surface integral.

The circulation of a vector field $\mathbf{F}(x, y, z) = \langle M, N, P \rangle$ around the boundary C of an oriented surface S in the direction counterclockwise with respect to the surface's unit normal vector **N,** equals the integral of (curl F) **N** over S. That is:

$$\oint_C \mathbf{F} \cdot d\mathbf{r} = \iint_S (\text{curl } \mathbf{F}) \cdot \mathbf{N} \, dS = \iint_S (\nabla \times \mathbf{F}) \cdot \mathbf{N} dS$$

Note: Recall that the normal vector **N** is $\langle -g_x, -g_y, 1 \rangle$, where $z = g(x, y)$ is the equation for the surface. If the curve C is in the xy-plane, oriented counterclockwise, and R is the region in the xy-plane bounded by C, then Stokes' equation becomes:

$$\oint_C \mathbf{F} \cdot d\mathbf{r} = \iint_S (\text{curl } \mathbf{F}) \cdot \mathbf{N} \, dS = \iint_S (\nabla \times \mathbf{F}) \cdot \mathbf{k} dS = \iint_R \left(\frac{\partial N}{\partial x} - \frac{\partial M}{\partial y} \right) dA$$

…which is the circulation-curl form for Green's Theorem!

The "paddle-wheel interpretation" for Stokes' Theorem is the following:

Suppose a small paddle wheel is at a point P in space (where P is in the domain of **F**). Further, suppose the paddle wheel's axis is directed along some vector **u**. The circulation of fluid around a curve C will affect the rate of spin of the paddle wheel. Assume further that the radius of the paddle wheel does not exceed the radius of the "circle" contained by C. The paddle wheel will spin fastest when the axle of the paddle wheel points in the direction of curl $\mathbf{F} = \nabla \times \mathbf{F}$ vector.

EXAMPLE. Use the surface integral in Stokes' Theorem to calculate the circulation of the field $\mathbf{F}(x, y, z) = \langle y, xz, x^2 \rangle$ around the curve C, the boundary of the triangle cut from the plane $x + y + z = 1$ by the first octant, counterclockwise when viewed from above.

SOLUTION. First, we need to find the curl of the field **F**:

$$\text{curl } \mathbf{F}(x, y, z) = \nabla \times \mathbf{F}(x, y, z)$$

$$= \begin{vmatrix} \mathbf{i} & \mathbf{j} & \mathbf{k} \\ \dfrac{\partial}{\partial x} & \dfrac{\partial}{\partial y} & \dfrac{\partial}{\partial z} \\ M & N & P \end{vmatrix}$$

$$= \begin{vmatrix} \mathbf{i} & \mathbf{j} & \mathbf{k} \\ \dfrac{\partial}{\partial x} & \dfrac{\partial}{\partial y} & \dfrac{\partial}{\partial z} \\ y & xz & x^2 \end{vmatrix}$$

$$= \left(\frac{\partial(x^2)}{\partial y} - \frac{\partial(xz)}{\partial z} \right)\mathbf{i} - \left(\frac{\partial(x^2)}{\partial x} - \frac{\partial(y)}{\partial z} \right)\mathbf{j} + \left(\frac{\partial(xz)}{\partial x} - \frac{\partial(y)}{\partial y} \right)\mathbf{k}$$

$$= (0 - x)\mathbf{i} - (2x - 0)\mathbf{j} + (z - 1)\mathbf{k}$$

$$= \langle -x, -2x, z - 1 \rangle$$

We need to determine the normal vector **N** to the surface that is the plane $x + y + z = 1$. First, solve for z to get $z = 1 - x - y$.

Then, we can find **N**:

$$\mathbf{N} = \langle -g_x, -g_y, 1 \rangle = \langle -(-1), -(-1), 1 \rangle = \langle 1, 1, 1 \rangle$$

Now we use Stokes' Theorem:

$$\oint_C \mathbf{F} \cdot d\mathbf{r} = \iint_S (\text{curl } \mathbf{F}) \cdot \mathbf{N} \, dS = \iint_S \langle -x, -2x, z-1 \rangle \cdot \langle 1, 1, 1 \rangle dS$$

$$= \iint_R (-3x + z - 1) dA$$

To iterate this double integral, we need to project the image of the plane $x + y + z = 1$ onto the xy-plane. On the xy-plane, we know that $z = 0$. So substitute $z = 0$ into the equation for the plane, and get that:

$$x + y + 0 = 1$$
$$y = 1 - x$$

This is the graph of a line with y-intercept 1 and a slope of -1. In the first octant only, we have that the x-intercept will be 1 also.

If we choose the order of integration to be "*dydx*," we have the limits for x as constants, and the limits for y will be variable. Keep in mind we are also dealing with the first octant only. The limits for the variables will be $0 \le x \le 1$ and $0 \le y \le 1 - x$.

We also need to be sure that only the variable x and y appear in the double integral. Right now, we have the variable z in the integrand. We replace it with $z = 1 - x - y$. We continue:

$$= \iint_R (-3x + (1 - x - y) - 1) dy dx$$

$$= \int_0^1 \int_0^{1-x} (-4x - y) dy dx$$

$$= \int_0^1 \left[-4xy - \frac{y^2}{2} \right]_0^{1-x} dx$$

$$= \int_0^1 \left[-4x(1-x) - \frac{(1-x)^2}{2} \right] dx$$

Congratulations on making it through the semester!

$$= \int_0^1 \left[-4x - x^2 - \frac{1}{2}(1 - 2x + x^2) \right] dx$$

$$= \left[-2x^2 - \frac{x^3}{3} - \frac{1}{2}x + \frac{x^2}{2} - \frac{1}{6}x^3 \right]_0^1$$

EXERCISES: #30 from **Practice Final**

$$= -2 - \frac{1}{3} - \frac{1}{2} + \frac{1}{2} - \frac{1}{6}$$

$$= -\frac{5}{2}$$

PART II

TEST BANK

THIRD SEMESTER CALCULUS STUDENT SUPPLEMENT, 4TH EDITION

Quiz and Test Table of Contents

$\sqrt{}$ **PART II** TEST BANK

CALCULUS III   ✠ **QUIZ 1** ✠   **NAME:** _____
TOTAL PAGES: 1

Instructions: *Please show all work and circle your answers.*
No notes, books, or scratch paper. Calculators are OK.

---

§ 1.1 AND 1.2

1. A vector **v** has initial point $(3, -1, -5)$ and terminal point $(0, -1, 2)$.

    (a) Write **v** in component form. (that is: <a, b, c> form)

    (b) Write **v** as a linear combination of the standard unit vectors.

    (c) Find the magnitude of **v**.

    (d) Find the unit vector in the direction of **v**.

    (e) Find the unit vector in the opposite direction of **v**.

§ 1.2

2. Find the standard equation of a sphere that has points $(-1, 2, 8)$ and $(-7, -2, -2)$ as endpoints of a diameter.

§ 1.1

3. A vector **v** has magnitude of 2 and direction $\theta = 300°$. Find its component form.

CALCULUS III   ✠ **QUIZ 2** ✠   **NAME:** _____

TOTAL PAGES: 7

Instructions: *Please show all work and circle your answers.*
No notes, books, or scratch paper. Calculators are OK.

---

§ 1.1

1. If the point $(7, -1)$ is the initial point for the vector $\mathbf{v} = \langle 1, 2 \rangle$, find the terminal point.

§ 1.1

2. Find a vector of magnitude 3 in the direction of $\mathbf{v} = \langle 1, 2 \rangle$. *(Hint:* First, come up with the unit vector in the same direction as $\mathbf{v}$. Then multiply that unit vector by a scalar.)

§ 1.1

3. Two forces have an angle of $110°$ between them. $\mathbf{F_1}$ has a magnitude of 390 N. (N is *Newtons,* a unit of force.) $\mathbf{F_2}$ has a magnitude of 260 N. Find *both* the magnitude and direction of $\mathbf{F_1} + \mathbf{F_2}$. *(Hint:* Superimpose an *xy*-coordinate plane such that $\mathbf{F_1}$ is on the positive *x*-axis. Then, resolve the two force vectors into component form. Add the two vectors to get the resultant force. Finally, find both the magnitude and direction or angle of this resultant force.)

§ 1.1

4. Three forces with magnitudes of 50, 20, and 40 pounds act on an object at angles of $60°$, $30°$, and $-90°$ respectively, with the positive *x*-axis. Find *both* the direction and magnitude of the resultant force. *(Hint:* same as for previous problem.)

§ 1.1

5. Given the function, $f(x)=x^3$, and the point that lies on the graph of this function: $(-2, -8)$...

(a) Find the equation of the tangent line at this point. *(**Hint:** First, find the derivative at this point, which is the same as the slope of the tangent line at this point. Now that you have the slope and a point that lies on the line, you have enough info to get the equation of that line.)*

(b) Find a vector parallel to the graph at this point. *(**Hint:** the tangent line is parallel to the graph at this point. So a vector parallel to the graph will simply be a line segment of this very same tangent line. Simply find another point that "lives" on this tangent line, which will be the terminal point of the vector that is parallel to the graph.)*

(c) Find a vector normal to the graph at this point. *(**Hint:** the tangent line is parallel, which you have the equation for already. Now, determine the equation for a line that is perpendicular to the tangent line, using your knowledge of what relationship slopes of perpendicular lines have. Two points on that perpendicular line will be the initial and terminal points for a vector that is normal to the graph.)*

§ 1.1

6. *Static equilibrium.* Two workers carry a 350-pound cylindrical weight which has two short ropes tied to the eyelet on the top center of the cylinder. One rope makes a 35 °-angle with the vertical and the other a 60°-angle with the vertical. Find the tension (force) in each rope. *(**Hint:** The tension is the magnitude of the force!)*

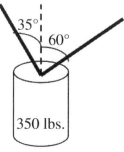

§ 1.2

7. Find the standard equation of a sphere that has points $(4, -3, 5)$ and $(-6, 1, -1)$ as endpoints of a diameter.

§ 1.2

8. Find the center and radius of a sphere given by $x^2 + y^2 + z^2 + 2x - 2y + 6z + 7 = 0$ (**Hint:** you will need to complete the square three times.)

§ 1.2

9. Find the initial point of the vector $\mathbf{v} = 5\mathbf{i} - \mathbf{k}$ if the terminal point is $(5, 5, 0)$.

§ 1.2

10. Determine whether each vector is parallel to the vector $\mathbf{v} = \langle -4, -1, 5 \rangle$. If it is, find c such that $\mathbf{u} = c\mathbf{v}$.

(a) $\mathbf{u} = \langle 8, 4, 10 \rangle$

(b) $\mathbf{u} = \langle -8, -2, 10 \rangle$

(c) $\mathbf{u} = \left\langle 3, \dfrac{3}{4}, -\dfrac{15}{4} \right\rangle$

(d) $\mathbf{u} = \left\langle -2, \dfrac{1}{2}, \dfrac{5}{2} \right\rangle$

§ 1.3

11. Given that $\mathbf{v} = <5, -2>$ and $\mathbf{w} = <-3, 1>$, find $\mathbf{v} \cdot \mathbf{w}$.

§ 1.3

12. Determine whether the vectors $\mathbf{v} = <2, -3>$ and $\mathbf{w} = <1, 1>$ are orthogonal, parallel, or neither.

§ 1.3

13. Given $\mathbf{v} = <3, 2>$ and $\mathbf{w} = <1, -3>$, find the projection of $\mathbf{v}$ onto $\mathbf{w}$.

§ 1.3

14. Find the angle θ between the vectors $\mathbf{v} = <3, 2>$ and $\mathbf{w} = <7, -5>$.

§ 1.3

15. Find the cosine of the angle θ between the vectors $\mathbf{v} = <3, -9>$ and $\mathbf{w} = <2, 1>$.

CALCULUS III **QUIZ 2, CONTINUED** **PAGE 5 OF 7**

§ 1.3

16. Let $\mathbf{u}$ = < 3, –1, –2 > and $\mathbf{v}$ = <–2, -3, 2 >. Calculate $\mathbf{u} \cdot \mathbf{v}$.

§ 1.3

17. Let $\mathbf{u}$ = < 3, –1, –2 > and $\mathbf{v}$ = <–2, –3, 2 >. Calculate $\text{proj}_v \mathbf{u}$.

§ 1.3

18. ***Multiple choice:*** Which of the following statements is true about the vectors $\mathbf{u} = \left\langle -\frac{1}{2}, \frac{1}{3}, -\frac{1}{4} \right\rangle$ and $\mathbf{v} = \left\langle 2, -\frac{4}{3}, 1 \right\rangle$? *(Please circle your choice.)*

(a) $\mathbf{u}$ and $\mathbf{v}$ are orthogonal

(b) $\mathbf{u}$ and $\mathbf{v}$ are parallel

(c) $\mathbf{u}$ is a unit vector of $\mathbf{v}$

(d) the angle between $\mathbf{u}$ and $\mathbf{v}$ is $\frac{\pi}{4}$

(e) none of these

§ 1.3

19. Find $\mathbf{u} \cdot \mathbf{v}$ if $\|\mathbf{u}\| = 7$, $\|\mathbf{v}\| = 12$, and the angle between $\mathbf{u}$ and $\mathbf{v}$ is $\frac{\pi}{4}$.

§ 1.3

20. Find the vector component of **u** orthogonal to **v** for **u** = < 1, 2 > and **v** = <–4, 4 >. *(**Hint:** First, you will need to find the projection of **u** onto **v**. Then you will have to subtract this result from **u**.)*

§ 1.3

21. Find the direction cosines for the vector with initial point (2, –1, 3) and terminal point (4, 3, –5). (Fill in the blanks.)

cos α = _____

cos β = _____

cos γ = _____

§ 1.3

22. Determine a scalar, k, such that the vectors **u** = < 3, 2 > and **v** = <2, k > are orthogonal.

§ 1.4

23. Let **u** = < –1, 2, 2 > and **v** = <3, –2, 1 >.

(a) Find **u** $\times$ **v**.

(b) Show that the vector **u** $\times$ **v** is orthogonal to the vector **v**. *(**Hint:** Use the dot product here!)*

§ 1.4

24.   *Multiple choice:* Let **u** = <3, –1, –2 > and **v** = <–2, –3, 2 >. Which of the following vectors is orthogonal to both **u** and **v**? *(Please circle your choice.)*

   (a)   **k**   **(b)**   **w** = < -8, 2, –11 >

   (c)   **v** × **u**   **(d)**   **w** = <1, 2, 0 >

   (e)   none of these

§ 1.4

25.   Let **u** = <3, –1, –2 > and **v** = <–2, –3, 2 >, and **w** = <1, 2, 0 >. Calculate **u** · (**v** × **w**).

§ 1.3

26.   Find the work done in moving a particle from $P(0, 2, 1)$ to $Q(3, –2, 2)$ if the magnitude and direction of the force is given by **v** = <4, –2, 3>.

§ 1.3

27.   A toy wagon is pulled by exerting a force of 15 pounds on the handle that makes a 30°-angle with the horizontal (see figure). Find the work done in pulling the wagon 50 feet.

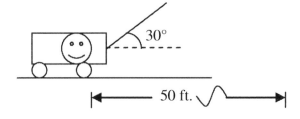

§ 1.3

28.   A 36,000-pound truck is parked on a 12°-slope (see figure). Assume the only force to overcome is that due to gravity. Find the force required to keep the truck from rolling down the hill.

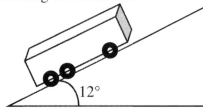

$\sqrt{}$ *PART II* T_EST B_ANK

C_ALCULUS_ **III**   ✠ **QUIZ 3** ✠   **NAME:** _____
T_OTAL_ P_AGES_: **1**

Instructions: *Please show all work and circle your answers.*
No notes, books, or scratch paper. Calculators are OK.

───

§ 1.5

1. Find a set of parametric equations for the line that passes through the points (2, 0, 3) and (4, 3, 3).

§ 1.5

2. Calculate the distance from the point (1, –2, 3) to the plane given by $(x-3) + 2(y+1) - 4z = 0$.

§ 1.5

3. Find an equation of the plane determined by the points (1, 2, –3), (2, 3, 1) and (0, –2, –1). (**Hint:** First, find the two vectors that lie in the plane with a shared initial point using the three points given.)

§ 1.6

4. *Multiple choice:* If the equation of a cylinder is given by $x^2 - z = 0$, then... *(Please circle your choice.)*

   (a) the rulings are parallel to the z-axis   **(b)** the graph is a parabola

   (c) the rulings are parallel to the x-axis   **(d)** the generating curve is a parabola

   (e) both (a) and (d)   **(e)** none of these

CALCULUS III ❈ PRACTICE MIDTERM 1 ❈   NAME: _____
TOTAL PAGES: 5

Instructions: *Please show all work and circle your answers.*
One 3"× 5" index card of formulas (front and back) is permitted which must be turned in with the exam. No books or scratch paper. Calculators OK, except TI-89 , TI-92, or any other symbolic differentiation/integration utility.

1. Let $\mathbf{v} = \overrightarrow{PQ}$ and $\mathbf{w} = \overrightarrow{PR}$ given the points $P(5, 0, 0)$, $Q(4, 4, 0)$, and $R(2, 0, 6)$. Find the following:

(a) component forms of $\mathbf{v}$ and $\mathbf{w}$

(b) $\mathbf{v} \cdot \mathbf{w}$

(c) $\mathbf{v} \times \mathbf{w}$

(d) a unit vector going in the same direction as $\mathbf{w}$

(e) the vector component of $\mathbf{v}$ in the direction of $\mathbf{w}$

(f) the vector component of $\mathbf{v}$ orthogonal to $\mathbf{w}$

(g)   an equation of the plane containing P, Q, and R

(h)   the distance of a point $M(-1, 5, 9)$ to the plane containing P, Q, and R

(i)   a set of parametric equations of the line through the points P and Q

(j)   the distance of the point R to the line through the points P and Q

(k)   the angle θ between the vectors **v** and **w**

2.   Find the component form of **u** if...

(a)   the angle, measured counterclockwise, from the positive *x*-axis is 135° and $\|\mathbf{u}\|$ is 4. (Assume that **u** is in the plane.)

(b)   **u** is a unit vector perpendicular to the lines...

$$x = 4, \quad y = 3 + 2t, \quad z = 1 + 5t$$
$$x = -3 + 7s, \quad y = -2 + s, \quad z = 1 + 2s$$

3.   Determine whether the vectors are orthogonal, parallel, or neither.

(a)   <7, –2, 3>, <–1, 4, 5>

(b)   <4, –1, 5> , <3, 2, 2>

(c)   <–4, 3, –6> , <16, –12, 24>

4.   Find the tension in each of the supporting cables in the figure.

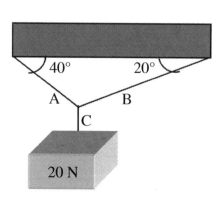

$\sqrt{}$ *PART II* TEST BANK

5. Identify each of the quadric surfaces:

    (a) $3x^2 + 3y^2 + 3z^2 - 2x + 3y - 11 = 0$

    (b) $2x^2 + 5z^2 = y^2$

    (c) $2x^2 - y^2 - z^2 = 1$

6. Complete the chart and convert the coordinates from one system to another among the rectangular, cylindrical, and spherical coordinate systems.

RECTANGULAR	CYLINDRICAL	SPHERICAL
$(6, -2, -3)$		
	$\left(-2, \dfrac{11\pi}{6}, 3\right)$	
		$\left(9, \dfrac{3\pi}{4}, \dfrac{\pi}{2}\right)$

7. Sketch the following planes:

    (a) $2x - y + z = 4$

    (b) $x + 2y = 4$

    (c) $z = 2$

---

8.   Find an equation in rectangular coordinates from cylindrical coordinates.

   (a) $r = \dfrac{1}{2}z$

   (b) $r = 2\cos\theta$

   (c) $\theta = \dfrac{\pi}{3}$

9.   Find an equation in rectangular coordinates from spherical coordinates.

   (a) $\rho = 3$

   (b) $\phi = \dfrac{\pi}{6}$

   (c) $\rho = 4\cos\phi$

10.   Convert the rectangular equation $x^2 + y^2 = z$ to...

   (a)   cylindrical coordinates

   (b)   spherical coordinates

11.   Convert the rectangular equation $x^2 + y^2 = 2x$ to...

   (a)   cylindrical coordinates

   (b)   spherical coordinates

CALCULUS III QUIZ 4 **NAME:** ______________________________

TOTAL PAGES: 5

Instructions: ***Please show all work and circle your answers.***
No notes, books, or scratch paper. Calculators are OK.

---

§ 1.1

1. Given the function $f(x) = \sqrt{2-x}$ ...

(a) Find a unit vector parallel to the graph at the point (–2, 2). ***(Hint:*** Find the derivative to get the slope of the line at that point. Then come up with the equation of the tangent line at that point. Use this equation to get another point on the line.)

(b) Find a unit vector perpendicular to the graph at the point (–2, 2)

§ 1.3

2. A 32,000-pound truck is parked on a 25°-hill. Assume that the only force to overcome is that due to gravity.

(a) Find the force required to keep the truck from rolling down the hill.

(b) Find the force perpendicular to the hill.

§ 1.5

3. Find the distance between the line given by $\begin{cases} x = 2 - t \\ y = -1 + 3t \\ z = -1 + t \end{cases}$ and the point $Q(1, 1, 0)$.

§ 2.1

4. Represent the parabola $x = y^2 + 5$ by a vector-valued function.

§ 2.1

5. Represent the intersection of the surfaces $x^2 + y^2 + z^5 = 3$ and $x = y$ as a vector-valued function. (**Hint:** There's an example in the text.)

§ 2.1

6. Find the limit (if it exists): $\displaystyle\lim_{t \to -2}\left\langle \frac{t^2 - 3t - 10}{t+2},\ \sin\left(\frac{\pi t}{2}\right),\ \frac{t+2}{\sin(t+2)} \right\rangle.$

§ 2.1

7. Find the domain of the vector-valued function $\mathbf{r}(t) = \left\langle e^{-t/2},\ \sqrt{t^2 - 4},\ \dfrac{1}{t+5} \right\rangle$. (**Hint:** You may need to review how to solve a polynomial inequality from your old Pre-Calculus textbook.)

§ 2.3

8. Calculate the speed of the object having the position vector $\mathbf{r}(t) = \left\langle \sin t,\ \cos t,\ -16t^2 \right\rangle$ at the point when $t = 1$.

§ 2.3

9. The vector-valued function $\mathbf{r}(t) = \left\langle \ln\left(\dfrac{t^5}{1+t^2}\right),\ \left(\ln\dfrac{t}{2}\right)^5,\ \sqrt[5]{(1-t)^2} \right\rangle$ describes the position of an object moving in space. Find its acceleration at $t = 2$.

CALCULUS III   **QUIZ 4, CONTINUED**   **PAGE 3 OF 5**

§ 2.3

10.   Determine the intervals on which the vector-valued function is smooth: $\mathbf{r}(t) = \left\langle 2t^3 - \dfrac{7}{2}t^2 - 5t, \ \dfrac{t^3}{3} - \dfrac{t}{4} \right\rangle$.

§ 2.3

11.   An object starts from rest at the point $(0, 1, 1)$ and moves with an acceleration of $\mathbf{a}(t) = \langle 1, 1, 0 \rangle$. Find its position at $t = 4$. (***Hint:*** You will need to anti-differentiate twice.)

§ 2.3

12.   Given the position function $\mathbf{r}(t) = \left\langle t^2, t^3 \right\rangle$ ...

   (a)   Sketch the path of the object for $-2 \le t \le 2$. Use your graphing calculator in parametric mode to help you. Be sure and show the orientation of the curve.

   (b)   Determine the value for t when the position of the object is $(1, -1)$.

   (c)   Determine the velocity function.

   (d)   Sketch the velocity vector at the point $(1, -1)$ on the same graph as in part **(a)**. (***Hint:*** Use the value for time that you got from part **(b)** to get the coordinates for the terminal point of the vector.)

   (e)   Sketch the acceleration vector at the point $(1, -1)$ on the same graph as in part **(a)**.

CALCULUS III **QUIZ 4, CONTINUED** **PAGE 4 OF 5**

§ 2.3

13. ***Projectile Motion.*** A projectile is fired from the ground at an angle of elevation of 60° with an initial velocity of 210 feet per second.

(a) Write the vector-valued function for the path of the projectile.

(b) How high did it go? (That is, what was the projectile's maximum height?)

(c) How long was it in the air?

(d) How far did it go? (In other words, find the range of the projectile.)

§ 2.3

14. ***Projectile Motion.*** The quarterback of a football team releases a pass at a height of 5 feet above the playing field, and the football is caught by a receiver 50 feet directly downfield at a height of 4 feet. The pass is released at an angle of 30° with the horizontal.

(a) Find the speed of the football when it was released. ***(Hint:*** You need to solve two equations with two unknowns.)

(b) Find the maximum height of the football.

(c) Find the time the receiver has to reach the proper position after the quarterback releases the football.

§ 2.4

15. Let $\mathbf{r}(t) = \langle 2\cos t, 2\sin t, 3t \rangle$.

    (a) Calculate the unit tangent vector, a.k.a. $\mathbf{T}(t)$.

    (b) Calculate the principal unit normal vector, a.k.a. $\mathbf{N}(t)$.

§ 2.4

16. Find the principal unit normal vector for the curve represented by $\mathbf{r}(t) = \langle t^2, t \rangle$ when $t = -1$. **(*Hint:*** This involves a lot of work, which includes using the Quotient Rule. Show all steps!)

§ 2.4

17. Find a set of parametric equations for the line tangent to the space curve given by $\mathbf{r}(t) = \langle t, t^2, t^3 \rangle$ at the point (2, 4, 8).

CALCULUS III   ✠ **QUIZ 5** ✠   **NAME:** _____

TOTAL PAGES: 7

Instructions: *Please show all work and circle your answers.*
No notes, books, or scratch paper. Calculators are OK.

---

§ 2.4

1. Find a set of parametric equations for the line tangent to the helix $\mathbf{r}(t) = \langle 2\cos t, 2\sin t, t \rangle$ at the point corresponding to $t = \dfrac{\pi}{6}$.

§ 2.4

2. Find the tangential and normal components of acceleration at $t = 0$ for the position function given by $\mathbf{r}(t) = \left\langle e^{t}, e^{-t}, \sqrt{2}\,t \right\rangle$.

§ 2.4

3. A projectile is fired from the ground with initial speed of 160 feet per second at an angle of $\pi/6$.

    (a) How far did the projectile travel?

    (b) Find the tangential and normal components of acceleration when $t = 1$ second.

§ **2.4**

4. Let $r(t) = \langle t^2, t^3 \rangle$.

    (a) Find $\mathbf{T}(2)$.

    (b) Find $\mathbf{N}(2)$. (***Hint:*** Since this is a plane curve, the unit normal vector is easy to get once you already have the unit tangent vector. That is, if $\mathbf{T}(t) = <x, y>$ then $\mathbf{N}(t) = <-y, x>$ *or* $\mathbf{N}(t) = <y, -x>$, since we must have that these two vectors are orthogonal to each other, or $\mathbf{T} \cdot \mathbf{N} = 0$! So, use your answer from part **(a)**.)

    (c) Write the acceleration vector corresponding to $t = 2$ as a linear combination of $\mathbf{T}$ and $\mathbf{N}$.

    (d) Sketch the graph of the plane curve. At the point when $t = 2$, sketch the vectors $\mathbf{T}(2)$ and $\mathbf{N}(2)$. Be sure and show the orientation of the curve using arrows.

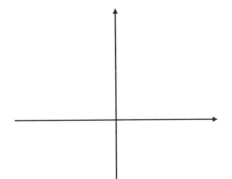

Calculus III **QUIZ 5, continued** **Page 3 of 7**

§ 2.5

5. Find the arc length of the curve C given by $\mathbf{r}(t)=\left\langle t,\frac{\sqrt{6}}{2}t^2,t^3\right\rangle$ from $-1\le t\le 1$.

§ 2.5

6. Consider the curve C given by $\mathbf{r}(t)=\langle 2t,t^2,\ln t\rangle$.

(a) Calculate the curvature, K, of the curve. Use $K=\dfrac{\|\mathbf{T}'(t)\|}{\|\mathbf{r}'(t)\|}$

(b) Find the arc length of the curve from $1\le t\le 2$.

CALCULUS III   QUIZ 5, CONTINUED   PAGE 4 OF 7

§ 2.5

7. Show that the helix given by $\mathbf{r}(t) = \langle \cos t, \sin t, t \rangle$ has constant curvature.

§ 2.5

8. Find the length of the curve $\mathbf{r}(t) = \langle e^t \cos t, e^t \sin t, e^t \rangle$ over the interval [0, 1].

§ 2.5

9. Consider the curve given by $\mathbf{r}(t) = \langle t^2, 1 \rangle$.

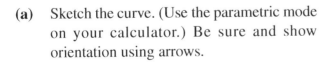

   (a) Sketch the curve. (Use the parametric mode on your calculator.) Be sure and show orientation using arrows.

   (b) Calculate the curvature. Explain the result.

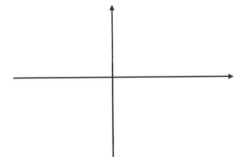

§ 2.5

10. Consider the curve given by $\mathbf{r}(t) = \langle -1, -6t, 8t \rangle$.

   (a) Find the arc length function, s, as a function of t.

   (b) Express $\mathbf{r}$ as a function of the parameter s. In other words, find $\mathbf{r}(s)$. (***Hint:*** Use your answer to part **(a)** and solve the equation for t. Then, substitute that version of t into $\mathbf{r}$.

   (c) Show that $\|\mathbf{r}'(s)\| = 1$, as it should! (***Hint:*** First, use your answer from part **(b)** to obtain $\mathbf{r}'(s)$).

§ 2.5

11. Consider the curve given by $\mathbf{r}(s) = \langle 3 + s, 1 \rangle$. Find the curvature using the formula $K = \|\mathbf{T}'(s)\|$ and explain the result.

§ 2.5

12. 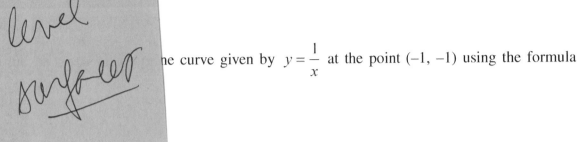 he curve given by $y = \dfrac{1}{x}$ at the point $(-1, -1)$ using the formula

§ 3.1

13. Describe the level surfaces of the function $f(x,y,z) = x^2 + y^2 + z^2 - 2y + 4z + 5$. (That is, are they parabolas? circles? spheres? cones? hyperbolas?)

§ 3.1

14. ***Multiple choice:*** Describe the level surfaces of the function $f(x,y,z) = 4x - 3y$.
     (Please circle your choice.)

     (a) Planes parallel to the xy-plane **(b)** Lines in the xy-plane

     (c) Planes parallel to the z-axis **(d)** Lines parallel to the z-axis

     (e) None of these

§ 3.1

15. ***Multiple choice:*** Find the domain of the function $f(x,y) = \dfrac{2}{x - y^2}$. *(Please circle your choice.)*

     (a) All (x, y) such that $x = y^2$ **(b)** All (x, y) such that $x \neq y$

     (c) All (x, y) such that $y \neq x^2$ **(d)** All (x, y) such that $x \neq y^2$

     (e) None of these

CALCULUS III   **QUIZ 5, CONTINUED**

§ 3.1

16.   Find the domain of the function $f(x,y) = \ln(x^2 + y^2 - 4)$.

§ 3.1

17.   Find $\dfrac{f(x+\Delta x, y) - f(x,y)}{\Delta x}$ for $f(x,y) = -x^2 + y^2 - 2x$.

§ 3.1

18.   *Multiple choice:* Describe the level curves of $f(x,y) = \sqrt{1 + x^2 - y}$.

   (a)   The level curves are circles.

   (b)   The level curves are straight lines passing through the origin.

   (c)   The level curves are parabolas with vertices on the *x*-axis.

   (d)   The level curves are parabolas with vertices on the *y*-axis.

§ 3.1

19.   Let $f(x,y) = x^2 e^{xy^2}$.

   (a)   Find the domain *and* range of *f*.

   (b)   Find $f(1, 0)$.

   (c)   Find $f(-2, 2)$.

CALCULUS III   ✠ **QUIZ 6** ✠   **NAME:** _____
TOTAL PAGES: 2

Instructions: *Please show all work and circle your answers.*
No notes, books, or scratch paper. Calculators are OK.

---

§ **3.2**

1.   Evaluate the limit $\displaystyle\lim_{(x,y)\to(0,e)} e^x \ln y$.

§ **3.2**

2.   Evaluate the limit $\displaystyle\lim_{(x,y)\to(1,1)} \frac{x^2 - y^2}{x - y}$. (***Hint:*** Factor the numerator, then use direct substitution).

§ **3.2**

3.   Where is the function $f(x,y) = \dfrac{x}{y\sqrt{4 - x^2 - y^2}}$ continuous?

§ **3.2**

4.   Prove that the limit does not exist by considering the two paths given below approaching the point $(0, 0)$ for $\displaystyle\lim_{(x,y)\to(0,0)} \frac{xy^2}{x^2 + y^4}$.

Use the path where $x = 0$:

Use the path where $x = y^2$:

CALCULUS **III**   **QUIZ 6,** CONTINUED   PAGE 2 OF 2

§ 3.3

5. Find $f_{xy}(x,y)$ for $f(x,y) = \dfrac{4x^2}{y} + \dfrac{y^2}{2x}$.

§ 3.4

6. Find the total differential of $f(x,y) = 2x^2 y^3$.

§ 3.4

7. Given $z = \ln(x^2 + y^2)$ …

    (a) Use the differential dz to *approximate* the change in z as (x, y) moves from $(0, 1)$ to the point $(0.01, 0.98)$.

    (b) Find the *actual* change in z as (x, y) moves from $(0, 1)$ to the point $(0.01, 0.98)$.

$\sqrt{}$ *PART II*   TEST BANK

CALCULUS III ✠ PRACTICE MIDTERM 2 ✠   NAME: _____
TOTAL PAGES: 7

Instructions: *Please show all work and circle your answers.*
One 3"× 5" index card of formulas (front and back) is permitted which must be turned in with the exam. No books or scratch paper. Calculators OK, except TI-89 , TI-92, or any other symbolic differentiation/integration utility.

1. Complete the chart and convert the coordinates from one system to another among the rectangular, cylindrical, and spherical coordinate systems.

RECTANGULAR	CYLINDRICAL	SPHERICAL
$\left(\dfrac{\sqrt{3}}{4}, \dfrac{3}{4}, \dfrac{3\sqrt{3}}{2}\right)$		
	$\left(81, -\dfrac{5\pi}{6}, 27\sqrt{3}\right)$	
		$\left(12, -\dfrac{\pi}{2}, \dfrac{2\pi}{3}\right)$

2. Convert the equation to rectangular coordinates from the given coordinate system.

   (a) spherical: $\rho = 2\cos\theta$

   (b) spherical: $\theta = \dfrac{\pi}{4}$

   (c) cylindrical: $r = 4\sin\theta$

3. Convert the rectangular equation $x^2 - y^2 = 2z$ to...

   (a) cylindrical coordinates

   (b) spherical coordinates

4. Find the domain of the vector-valued functions. Express your answer in interval notation, please.

(a) $r(t) = e^{-t}\mathbf{i} + \dfrac{1}{t}\mathbf{j} - \sqrt{1-t}\,\mathbf{k}$

(b) $r(t) = |t|\mathbf{i} + \ln t\mathbf{j} + \dfrac{1}{3t^2 - 13t - 10}\mathbf{k}$

5. Sketch the curve represented by the vector-valued function $r(t) = (2-t)\mathbf{i} + t^2\mathbf{j}$ and indicate the orientation of the curve. Then, write the corresponding rectangular equation. (***Hint:*** eliminate the parameter.)

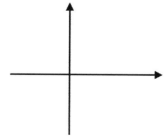

6. Find the limits (if they exist). If they do not exist, say so.

(a) $\displaystyle\lim_{t \to 1}\left(\ln t\mathbf{i} + \dfrac{t^2 + 7t - 8}{t-1}\mathbf{j} + \cos(\pi t)\mathbf{k} \right)$

(b) $\displaystyle\lim_{t \to 0}\left(\dfrac{\sin t}{t}\mathbf{i} + e^t\mathbf{j} + \dfrac{t+3}{t-2}\mathbf{k} \right)$

(c) $\displaystyle\lim_{t \to \infty}\left(\dfrac{t}{t^2 - 2}\mathbf{i} + e^{-t}\mathbf{j} + \dfrac{t+3}{t-2}\mathbf{k} \right)$

(d) $\displaystyle\lim_{t \to 1}\left(\dfrac{t^2 + 1}{t-1}\mathbf{i} + \sin\dfrac{\pi t}{2}\mathbf{j} + t^2\mathbf{k} \right)$

C~ALCULUS~ **III**   **Practice Midterm 2,** C~ONTINUED~   P~AGE~ 3 OF 7

7. Find $\mathbf{r}'(t)$.

   (a) $\mathbf{r}(t) = e^{-t}\mathbf{i} + \dfrac{1}{t}\mathbf{j} - \sqrt{1-t}\,\mathbf{k}$

   (b) $\mathbf{r}(t) = \sin^3 t\mathbf{i} + \ln\sqrt[3]{\dfrac{5}{t^2}}\,\mathbf{j} - \arctan t\,\mathbf{k}$

8. Find the open intervals on which the curve given by the vector-valued function is smooth.

   (a) $\mathbf{r}(t) = (t + \sin t)\mathbf{i} + (1 - \cos t)\mathbf{j}$

   (b) $\mathbf{r}(t) = \sqrt{t}\,\mathbf{i} + (t^2 - 1)\mathbf{j} + \dfrac{1}{4}t\mathbf{k}$

9. Given the functions $\mathbf{r}(t) = (t^2 - 1)\mathbf{i} + (t^3 - 3t^2)\mathbf{j} + 5t\mathbf{k}$ and $\mathbf{u}(t) = \sqrt{2}\,t\mathbf{i} + e^t\mathbf{j} + e^{-t}\mathbf{k} \cdots$
   Find $\dfrac{d}{dt}\big(\mathbf{r}(t)\cdot\mathbf{u}(t)\big)$:

10. Find $\mathbf{r}(t)$ if $\mathbf{r}'(t) = \dfrac{1}{1+t^2}\mathbf{i} + e^{-t}\mathbf{k}$ and $\mathbf{r}(0) = \mathbf{i} - \mathbf{j}$.

11. Find $\mathbf{r}(t)$ given the following conditions. (*Hint:* you will need to anti-differentiate twice.)
$$\mathbf{r}''(t) = -4\cos t\,\mathbf{i} - 3\sin t\,\mathbf{k}, \quad \mathbf{r}'(0) = 2\mathbf{k}, \quad \mathbf{r}(0) = 4\mathbf{j}$$

12. Evaluate the definite integrals…

   (a) $\displaystyle\int_0^1 \left(\frac{t}{\sqrt{t^2+1}}\mathbf{i} + t\mathbf{j} - \mathbf{k} \right) dt$

   (b) $\displaystyle\int_{-1}^1 \left(t\mathbf{i} + t^3\mathbf{j} + \sqrt[3]{t}\ \mathbf{k} \right) dt$

   (c) $\displaystyle\int_0^{\pi/2} \left[(2\cos t)\mathbf{i} + (2\sin t)\mathbf{j} + \mathbf{k} \right] dt$

13. Find a set of parametric equations for the tangent line to the graph of the vector-valued function
$\mathbf{r}(t) = \left\langle \ln(t-3), t^2, \frac{1}{2}t \right\rangle$ at the point $(0, 16, 2)$.

14. Given the vector-valued function $\mathbf{r}(t) = \left\langle t-1, t, \frac{1}{t} \right\rangle \dots$

  (a) Find the velocity, speed, and acceleration of the function at time t.

  (b) Find the unit tangent vector, $\mathbf{T}(t)$, the unit normal vector, $\mathbf{N}(t)$, as well as the tangential and normal components of acceleration, a_T and a_N, at time t.

  (c) Find the curvature of the function at time t.

15. Find the maximum height and the range of a projectile fired from ground level at an angle of elevation of $30°$ if the initial velocity is 76 feet per second. (Use $g = 32 f/s^2$.)

16. Find the arc-length of the space curve given by $\mathbf{r}(t) = \langle 10\cos^3 t, 10\sin^3 t \rangle$ from $0 \le t \le 2\pi$.

17. Given the function $f(x,y) = e^{\frac{x}{y}} \dots$

    (a) Find the domain.

    (b) Find the range.

18. Discuss the continuity of the function $f(x,y,z) = \dfrac{1}{\sqrt{x^2 + y^2 + z^2}}$.

19. Find the limits (if they exist):

(a) $\displaystyle\lim_{(x,y)\to(-2,1)} \frac{x^2 - 4y^2}{x + 2y}$

(b) $\displaystyle\lim_{(x,y)\to(0,0)} \frac{y + xe^{-y^2}}{1 + x^2}$

(c) $\displaystyle\lim_{(x,y)\to(0,0)} \frac{\sin(x - y)}{x - y}$

(*Hint:* Use a theorem we learned in Calculus I that $\displaystyle\lim_{\theta\to0} \frac{\sin\theta}{\theta} = 1$.)

20. Given the function $f(x, y) = x^2 + y^2 \ldots$

(a) Sketch the graph of the function.
(*Hint:* quadric surface!)

(b) Describe the level curves for the function. Then sketch the level curves for the values of $c = 0, 2, 4$.

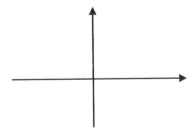

21. Sketch the level surface for the function of three variables $f(x, y, z) = 2y + 3z$ for the value of $c = 6$.

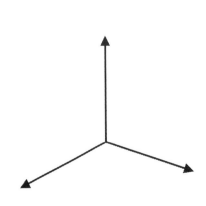

CALCULUS III   ✠ **QUIZ 7** ✠   **NAME:** _____

TOTAL PAGES: 1

Instructions: *Please show all work and circle your answers.*
No notes, books, or scratch paper. Calculators are OK.

§ 3.7

1. Consider the surface given by $f(x,y) = \dfrac{1}{7}\left(4x^2 - 2y^2\right)$.

  (a) Find an equation of the tangent plane (in general form) to the surface at the point $(-2, -1, 2)$.

  (b) Find a set of parametric equations for the normal line to the surface at the point $(-2, -1, 2)$.

  (c) Find the angle of inclination of the tangent plane to the surface at the point $(-2, -1, 2)$.

CALCULUS III   ✠ QUIZ 8 ✠   **NAME:** _____

TOTAL PAGES: 8

Instructions: *Please show all work and circle your answers.*
No notes, books, or scratch paper. Calculators are OK.

§ 3.7

1. Given the equations of two surfaces $z = x^2 + y^2$ and $z = 4 - y$, and a point of intersection of the two surfaces $(2, -1, 5)$, do the following:

   (a) Find a set of parametric equations of the tangent line to the curve of intersection of the surfaces at the given point. (**Hint:** Find the separate gradients for the two surfaces. This gives two normal vectors to each surface. Then take the cross-product of these two gradients. This will yield a vector that is orthogonal to *both* of the gradients. This orthogonal vector happens to be going in the direction of the tangent line to the curve of intersection. So, you may use this vector's components for the direction numbers of your line.)

   (b) Find the cosine of the angle between the two gradient vectors at the given point.

   (c) State whether or not the surfaces are orthogonal at the given point.

§ 3.7

2. Repeat problem #1 for the two surfaces $z = \sqrt{x^2 + y^2}$ and $2x + y + 2z = 20$ at the point of intersection $(3, 4, 5)$.

§ 3.7

3. Find the angle of inclination of the tangent plane to the elliptic paraboloid $4x^2 + y^2 - 16z = 0$ at the point (2, 4, 2).

§ 3.6 AND 3.7

4. Find the path of a heat-seeking particle in the temperature field $T(x,y) = 100 - 2x^2 - y^2$ placed on the point (3, 4) on a hot metal plate. (***Hint:*** You need differential equations.)

§ 3.8

5. Determine the relative extrema and saddle points (if any) for the function $f(x,y) = 120x + 120y - xy - x^2 - y^2$. Give answers as ordered triples.

§ 3.8

6. Determine the relative extrema and saddle points (if any) for the function $f(x,y) = y^5 + x^4 - 5y - 32x - 8$. Give answers as ordered triples.

§ 3.8

7. Determine the relative extrema and saddle points (if any) for the function $f(x,y) = 4xy - x^4 - y^4$. Give answers as ordered triples.

§ 3.8

8. Find the **absolute** extrema of $f(x,y) = y^2 - 3x^2 - 2y + 6x$ on the square with vertices $(0, 0)$, $(2, 0)$, $(2, 2)$, and $(0, 2)$.

§ 3.8

9. Find the critical points and test for relative extrema. List the critical points for which the Second Partials Test fails $f(x,y) = x^3 + y^3 - 3x^2 + 6y^2 + 3x + 12y + 7$.

§ 3.9

10. Use Lagrange Multipliers to minimize the distance from the point $P(2, 2)$ to the unit circle centered at the origin given by $x^2 + y^2 = 1$.

§ 3.9

11. Lagrange Multipliers are often used in economics. A commonly used function in economics, known as the Cobb-Douglas production model has the form $f(x, y) = Cx^p y^{1-p}$, where $0 < p < 1$, C is a positive constant, x represents the units of labor (i.e., the number of "man-hours" available), and y represents the units of capital (i.e., the number of dollars available).

Suppose we want to maximize the Cobb-Douglas production function for a software company which is given by $f(x, y) = 100x^{3/4} y^{1/4}$ where each labor unit, x, costs the company \$150 and each capital unit, y, costs the company \$250. The company is subject to the constraint that together labor and capital cannot exceed \$50,000.

C*ALCULUS* **III**   **QUIZ 8,** C*ONTINUED*   P*AGE* **5** O*F* **8**

§ 3.8

12. An open box (a box that has no top) has a volume of 64 cubic feet. Find the dimensions that minimize the surface area.

   (a)   Use the techniques of section 3.8—the Second Partials test—to prove it's a minimum.

   (b)   Do the problem all over again, only now use the Method of Lagrange Multipliers to find the minimum.

§ 3.8

13.   Find three positive numbers x, y, and z, such that the sum is 54 and the product is a maximum.

   (a)   Use the techniques of section 3.8—the Second Partials Test—to prove it's a maximum.

   (b)   Do the problem all over again, only now use the Method of Lagrange Multipliers to find the maximum.

§ 3.8

14. Help Elmer Fudd, the marketing analyst at ACME Company, to maximize his company's profits. The cost of producing x widgets at location #1 and y widgets at location #2 are given by $C_1(x) = 0.01x^2 + 2x + 1000$ and $C_2(y) = 0.03y^2 + 2y + 300$, respectively. If widgets sell for $14 each, and the revenues obtained from selling the widgets is given by $R(x, y) = 14(x + y)$, find the quantity that must be produced at each location in order to maximize profit. Remember: Profit = Revenue - Costs, so $P(x, y) = R(x, y) - C_1 - C_2$. Finally, calculate this maximum profit. Why is the Method of Lagrange Multipliers not appropriate for this problem at all?

§ 4.1

15. Evaluate the iterated integral $\displaystyle\int_0^\pi \int_0^{\cos y} x \sin y \, dx \, dy$.

§ 4.1

16. Evaluate the integral $\displaystyle\iint_R (x^2 + 4y)\, dA$, where R is the region bounded by the graphs of $y = 2x$ and $y = x^2$.

§ 4.1

17. Sketch the region R whose area is given by the iterated integral $\displaystyle\int_0^4 \int_{\sqrt{x}}^2 dy \, dx$. (Please shade in the region.) Then, evaluate the iterated integral. Finally, switch the order of integration and show that both orders yield the same area.

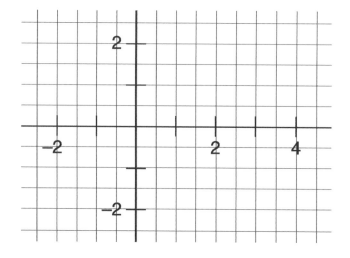

CALCULUS III   ✖ QUIZ 9 ✖   NAME: _____

TOTAL PAGES: 9

Instructions: *Please show all work and circle your answers.*
No notes, books, or scratch paper. Calculators are OK.

§ 3.8

1. Determine the relative extrema and saddle points (if any) for the function $f(x,y) = 2x^4 + y^2 - x^2 - 2y$. Give answer(s) as ordered triples.

§ 3.8

2. Find the ***absolute*** extrema for $f(x, y) = x^2 + y^2$ on the disc $x^2 + y^2 \le 1$.

§ 3.9

A rectangular box has its base resting on the xy-plane and a vertex at the origin. Find the dimensions of the box that would yield the maximum volume if another vertex of the box was in the plane $2x + y + z = 6$. Then, calculate this maximum volume.

(a) Use the techniques of section 3.8—the Second Partials Test—to prove it's a maximum.

(b) Do the problem all over again, only now use the Method of Lagrange Multipliers to find the maximum.

§ 3.9

4. *Minimizing the cost of a container*. A rectangular box with no top is to be constructed having a volume of 12 cubic feet. The cost per square foot of the material to be used is $4 for the bottom, $3 for two of the opposite sides, and $2 for the remaining pair of sides. Use the Method of Lagrange Multipliers to find the dimensions of the box that will minimize the cost.

§ 3.8

5. Help Elmer Fudd, the marketing analyst at ACME Company, to maximize his company's revenue. The company produces two competitive products, the Widget-2000 and the Widget-EX, the prices (in dollars) of which are p_1 and p_2, respectively. The total revenue obtained from selling both products is given by $R(p_1, p_2) = 300p_1 + 900p_2 + 1.8p_1p_2 - 1.5p_1^2 - p_2^2$.

    (a) Determine the maximum revenue.

    (b) What prices should Mr. Fudd charge for each product? (i.e., what do the values of p_1 and p_2 need to be in order to maximize revenue?)

§ 4.1

6. Evaluate the iterated integral $\displaystyle\int_0^{\sqrt{\pi}} \int_{\pi/6}^{y^2} 2y \cos x \; dxdy$.

§ 4.1

7. Evaluate the integral $\displaystyle\iint xy \, dA$, where R is the region bounded by the graphs of $y = \sqrt{x}$, $y = \dfrac{1}{2}x$, $x = 2$, and $x = 4$.

§ 4.1

8. Evaluate the integral $\displaystyle\int_0^1 \int_{2x}^2 e^{y^2} \; dydx$ by reversing the order of integration. (*Note: without* reversing the order; this integral is not integrable!)

$$\sqrt{} \quad \textit{PART II} \quad \textsc{Test Bank}$$

§ 4.1

9. Evaluate the integral $\displaystyle\int_0^1 \int_{\sqrt{y}}^1 \sin(x^3)\,dxdy$ by reversing the order of integration. *(**Note:** without reversing the order; this integral is not integrable!)*

§ 4.1

10. ***Multiple choice:*** Reverse the order of integration and rewrite the iterated integral $\displaystyle\int_{-2}^0 \int_0^{\sqrt{4-x^2}} f(x,y)\,dydx$. *(Please circle your choice.)* Show your work!

(a) $\displaystyle\int_0^2 \int_0^{\sqrt{4-y^2}} f(x,y)\,dxdy$ **(b)** $\displaystyle\int_2^0 \int_0^{\sqrt{4-y^2}} f(x,y)\,dxdy$

(c) $\displaystyle\int_2^0 \int_{\sqrt{4-y^2}}^0 f(x,y)\,dxdy$ **(d)** $\displaystyle\int_0^2 \int_{-\sqrt{4-y^2}}^0 f(x,y)\,dxdy$

(e)   none of these

§ 4.1

11. Evaluate the improper integral $\displaystyle\int_1^\infty \int_x^{2x} ye^{-x}\,dydx$. *(**Hint:** You will need to do repeated integration by parts!)*

§ 4.1

12. Evaluate the improper integral $\displaystyle\int_0^3 \int_0^\infty \frac{x^2}{1+y^2}\,dy\,dx$. (***Hint:*** The Arctan Rule will be used here!)

§ 4.2

13. Use a double integral to find the volume of the solid bounded by the plane $z = 4 - 2x - 4y$ and the coordinate planes.

§ 4.2

14. Evaluate $\displaystyle\int_0^{\pi/2} \int_0^{\pi/4} \cos x \sin^2 y \, dy\,dx$. (***Hint:*** You will need to use a trig ID before integrating.)

§ 4.2

15. Evaluate $\displaystyle\int_0^2 \int_{y+5}^0 e^{2x+y} \, dx\,dy$.

§ 4.2

16. Evaluate $\displaystyle\int_0^1 \int_{-3}^0 \frac{1}{(x+2)(1-y)} \, dy\,dx$.

§ 4.2

17. Evaluate $\displaystyle\int_0^2 \int_0^x \frac{1}{x+1} \, dy\,dx$. (***Hint:*** *Do not* reverse the order of integration—that will make it too hard. Instead, use long division to simplify the result of the inside integral (after integrating), remembering that the answer to a long division problem looks like $\text{Quotient} + \dfrac{\text{Remainder}}{\text{Divisor}}$. Another way you can do it instead (if you really despise long division) is to use Change of Variables.)

§ 4.2

18. Reverse the order of integration, then evaluate $\displaystyle\int_0^2 \int_{y^2}^4 \frac{y}{1+x^2}\, dx\, dy$.

§ 4.2 AND 4.3

19. Use a double integral to find the volume of the solid bounded by the three coordinate planes and $z = 1 - x^2 - y^2$ in the first octant (as shown). **(*Hint:* Converting to polar coordinates makes this problem *so* much easier!)**

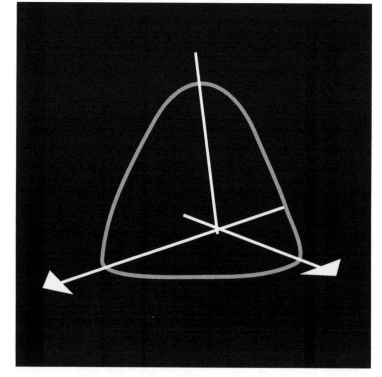

§ **4.3**

20. Evaluate the double integral by converting to polar coordinates: $\displaystyle\int_0^2 \int_0^{\sqrt{4-x^2}} x \, dy\, dx$.

§ **4.3**

21. Use a double integral in polar coordinates to find the volume of the solid bounded by the graphs of the equations $z = 9 - x^2 - y^2$, $z = 0$, $x^2 + y^2 \ge 1$, $x^2 + y^2 \le 4$.

§ **4.4**

22. Find the mass of the triangular lamina with vertices $(0,0)$, $(0,3)$, and $(3,0)$ given that the density at (x, y) is $\rho(x,y) = 10y$.

CALCULUS III ✠ PRACTICE MIDTERM 3 ✠   **NAME:** _____
TOTAL PAGES: 9

Instructions: *Please show all work and circle your answers.*
One 3"× 5" index card of formulas (front and back) is permitted which must be turned in with the exam. No books or scratch paper. Calculators OK, except TI-89 , TI-92, or any other symbolic differentiation/integration utility.

1. Find the limits (if they exist)…

    (a) $\displaystyle \lim_{(x,y)\to(4,5)} \frac{5x-4y}{15x^2-22xy+8y^2}$

    (b) $\displaystyle \lim_{(x,y)\to(\ln 8,\ln 8)} \frac{10e^y \sin(x^2-y^2)}{x^2-y^2}$

2. Give the points at which the function is continuous in set notation:
    $$f(x,y) = \ln(4-x-y) + \arccos\left(\frac{y}{x}\right)$$

3. Find the partial derivatives. Write answers without using negative exponents and simplify all complex fractions.

    (a) Find f_y for $f(x,y) = e^{xy}(\cos x \sin y)$.

    (b) Find f_x for $f(x,y) = \arctan\dfrac{y}{x}$.

4. Show that the limit does not exist by considering two different paths approaching the point (0, 0).
    $$\lim_{(x,y)\to(0,0)} \frac{x-y}{x^2+y^2}$$

    (a) Path #1: along the line $y = x$.

    (b) Path #2: along the y-axis.

5. Use implicit partial differentiation to find $\dfrac{\partial z}{\partial x}$ for $x^2 y - 2yz - xz - z^2 = 0$.

6. Show that the function $z = \dfrac{y}{x^2 + y^2}$ satisfies LaPlace's equation $\dfrac{\partial^2 z}{\partial x^2} + \dfrac{\partial^2 z}{\partial y^2} = 0$.

7. Find the mixed partials and verify that they are equal for the function $h(x,y) = x \sin y + y \cos x$.

8. Given the function $f(x,y) = \dfrac{x}{x^2 + y^2} \ldots$

   (a) Use the total differential to approximate change in the function as (x, y) varies from the point $(1, 2)$ to the point $(0.98, 2.01)$.

   (b) Calculate the actual change in the function as (x, y) varies from the point $(1, 2)$ to the point $(0.98, 2.01)$.

9. The height and radius of a right circular cylinder are approximately 15 centimeters and 8 centimeters, respectively. The maximum error in each measurement is ±0.02 centimeters.

   (a) Find the approximate volume of the cylinder.

   (b) Use the total differential to estimate the propagated error in the calculated volume of the cylinder.

   (c) Use the total differential to estimate the relative error in the calculated volume of the cylinder and express this relative error in percent.

10. The length, width, and height of a rectangular chamber are changing at the rate of 3 feet per minute, 2 feet per minute, and ½-foot per minute, respectively. Find the rate at which the volume is changing at the instant the length is 10 feet, the width is 6 feet, and the height is 4 feet.

11. Use the Chain Rule to find $\dfrac{dw}{dt}$ when $w = \ln(xy) + xy$ and $x = e^{t}$, $y = e^{-t}$. You will not get any credit unless you use the Chain Rule.

12. The surface of a mountain is described by the equation $f(x,y) = 1500 - x^{2} - 3y^{2}$. If a skier is at the point (10, 20, 200), what direction should she move in order to descend at the greatest rate?

CALCULUS III **Practice Midterm 3,** CONTINUED PAGE 4 OF 9

---

13. Given the function $f(x,y,z) = x^2 - 2y^2 + z^2 e^z \dots$

   (a) Find the gradient of f at the point $(-1, 2, 1)$.

   (b) Calculate the directional derivative in the direction of the vector $\mathbf{v} = \langle -2, 7, 4 \rangle$ at the point $(-1, 2, 1)$.

   (c) Find the maximum value of the directional derivative at the point $(-1, 2, 1)$.

14. Given the function $f(x,y) = \ln(x^2 + y^2 + 1) + e^{2xy} \dots$

   (a) Find ∇f at the point $(0, -2)$.

   (b) Calculate the directional derivative in the direction of the vector $\mathbf{v} = \langle 5, -12 \rangle$ at the point $(0, -2)$.

   (c) Find the maximum value of the directional derivative at the point $(0, -2)$.

15. Consider the surface given by $x^2 + 4y^2 = 10z$.

   (a) Find an equation of the tangent plane (in general form) to the surface at the point $(2, -2, 2)$.

   (b) Find the angle of inclination θ (in degrees) of the tangent plane at the point $(2, -2, 2)$. Round your answer to nearest tenths.

   (c) Find a set of parametric equations for the normal line to the surface at the point $(2, -2, 2)$.

16. Test the function $f(x,y) = x^3 - 3xy + y^2 + y - 5$ for relative extrema and saddle points. Give answer(s) using ordered triples to get full credit.

17. Find the point on the surface where the tangent plane is horizontal: $z = 3x^2 + 2y^2 - 3x + 4y - 5$.

18. Find the path of a heat-seeking particle placed at the point in space $(2, 2, 5)$, with a temperature field $T(x, y, z) = 100 - 3x - y - z^2$. (**Hint:** Use differential equations.)

19. Determine the relative extrema and saddle points (if any) for the function $f(x, y) = 3x^2 + 6xy + 7y^2 - 2x + 4y$. Give answer(s) as ordered triples.

20. Suppose you are *not* given the original function, but you are given the first partial derivatives $f_x(x,y) = x^2 - 121$ and $f_y(x,y) = 8y^3 + 14y^2 - 15y$. Determine the relative extrema and saddle points (if any) for the function. Give answer(s) as ordered *pairs,* not ordered triples, since you do not have access to the original function. Do not attempt to find the original function, please. You do not require it to complete this problem!

21. Determine the relative extrema and saddle points (if any) for the function $f(x,y) = \dfrac{1}{x^2 + y^2 - 1}$. Give answer(s) as ordered triples.

22. Find the *absolute* extrema of $f(x,y) = (4x - x^2)\cos y$ on the rectangular plate $1 \le x \le 3,\ -\pi/4 \le y \le \pi/4$

23. Find the critical points and test for relative extrema. List the critical points for which the Second Partials Test fails $f(x,y) = x^{2/3} + y^{2/3}$.

24. Determine the relative extrema and saddle points (if any) for the function $f(x, y) = 9x^3 + \dfrac{y^3}{3} - 4xy$. Give answer(s) as ordered triples.

25. Find the minimum distance from the point $(0, 0, 0)$ to the plane given by $x + 2y + z = 6$. You are forbidden from using the formula for calculating the distance from a point to a plane from Chapter 1.

   (a) Use the techniques from section 3.8 and the Second Partials Test to find this minimum distance.

   (b) Do the problem all over again, only now use the Method of Lagrange Multipliers to find the minimum distance.

26. Use Lagrange multipliers to locate the maximum for the function $z = 3xy + x$ subject to the constraint $11x + 15y = 215$.

27. Use the Method of Lagrange Multipliers to find the *minimum* temperature on the surface given by $x^2 + y^2 + z^2 = 9$ if the temperature is given by $T(x, y, z) = x + y + z$. (Give exact answer.)

CALCULUS **III**   ✠ **QUIZ 10** ✠   **NAME:** _____

TOTAL PAGES: **8**

Instructions: *Please show all work and circle your answers.*
No notes, books, or scratch paper. Calculators are OK.

---

§ 4.2

1. Set up and evaluate the double integral that would calculate the volume of the solid bounded above by $f(x,y) = y\cos(xy)$ and the region in the xy-plane $0 \le x \le \pi$, $0 \le y \le 1$.

§ 4.1

2. Evaluate $\displaystyle\int_0^8 \int_{\sqrt[3]{x}}^2 \frac{dy\,dx}{1+y^4}$ after reversing the order of integration. (***Hint:*** It might help to first sketch the region.)

§ 4.1

3. Evaluate $\displaystyle\int_0^{1/16} \int_{y^{1/4}}^{1/2} \cos(16\,\pi\,x^5)\,dx\,dy$ after reversing the order of integration. (***Hint:*** It might help to first sketch the region.)

§ 4.1

4.   Given the iterated integral $\displaystyle\int_0^{\ln 2}\int_{e^y}^{2} dx\,dy\,\ldots$

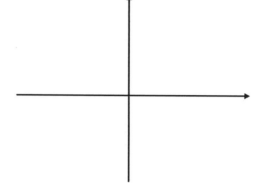

  (a)   Sketch the region.

  (b)   Evaluate the integral.

  (c)   Switch the order of integration and show that the value matches your answer to part **(b)**. (***Hint:*** Integration by parts will be needed.)

§ 4.2

5.   Use a double integral to find the volume of the solid bounded above by the cylinder $z = x^2$ and below by the region enclosed by the parabola $y = 2 - x^2$, and the line $y = x$ in the *xy*-plane. (***Hint:*** The region in the *xy*-plane includes *more* than just the first quadrant!)

§ **4.3**

6.   Convert the double integral to polar, and then evaluate…

(a) $\displaystyle\int_0^1 \int_0^{\sqrt{1-x^2}} e^{-\left(x^2+y^2\right)} \, dy \, dx$

(b) $\displaystyle\int_{-1}^0 \int_{-\sqrt{1-x^2}}^0 \frac{2}{1+\sqrt{x^2+y^2}} \, dy \, dx$

(Double Hint: **1)** This region is tricky to convert to polar. First, sketch the region!

                     2) You will need either change-of-variables or long division to integrate.

(c) $\displaystyle\int_0^2 \int_{-\sqrt{1-(y-1)^2}}^0 xy^2 \, dx \, dy$

CALCULUS III   **QUIZ 10, CONTINUED**   **PAGE 4 OF 8**

§ 4.3

7. ***Converting to a polar integral.*** Integrate $f(x,y) = \dfrac{\ln(x^2 + y^2)}{x^2 + y^2}$ over the region $1 \le x^2 + y^2 \le e^2$.

§ 4.4

8. Find the mass of a thin plate covering the region *outside* the circle $r = 3$ and *inside* the circle $r = 6\sin\theta$ if the plate's density function is $\rho(x,y) = \dfrac{1}{r}$. ***(Hint:*** To find the range of the angle, θ, you will need to find the point(s) of intersection of these two graphs by equating the equations. ***Another Hint:*** It also might help to graph the equations using the POLAR mode on your calculator so you can identify the region.)

§ 4.6

9. Given the triple integral $\displaystyle\int_{0}^{6}\int_{\frac{3x-18}{2}}^{0}\int_{0}^{\frac{18+2y-3x}{6}} dz\,dy\,dx$ …

    Rewrite the order of the integral in the order $dx\,dy\,dz$. *(No need to evaluate!)*

§ 4.2, 4.5, AND 4.6

10. Consider the solid can be found by the bounds of the three coordinate planes and the plane $x + 2y + z = 6$.

   (a)   Set up (but do not evaluate) the *double* integral that would calculate the volume of the solid.

   (b)   Set up (but do not evaluate) the *triple* integral that would calculate the volume of the solid.

   (c)   Set up *and* evaluate the integral that would calculate the surface area of the top of the solid (i.e., the portion of the plane).

§ 4.6 AND 4.7

11. Set up (but do not evaluate) the triple integral to calculate the volume of the solid bounded by $z = 36 - x^2 - y^2$ and $z = 0 \ldots$

   (a)   using rectangular coordinates.

   (b)   using cylindrical coordinates.§ 4.6

§ 4.6

12. Set up (no need to evaluate) the *triple* integral that would calculate the volume of the solid bounded by the three coordinate planes, the plane $y + z = 2$, and the cylindrical surface $x = 4 - y^2$.

§ 4.5

13. Find the surface area for that portion of the surface $x^2 + y^2 - z = 0$ that is inside the cylinder $x^2 + y^2 = 2$. (***Hint:*** Convert to polar.)

§ 4.7

14. Set up (but don't evaluate) the iterated triple integral (in cylindrical coordinates) that would find the volume of a solid Q if Q is the right circular cylinder whose base is the circle $r = 2\sin\theta$ in the xy-plane and whose top lies in the plane $z = 4 - y$ as shown. (***Hint:*** Graph the polar equation using your graphing calculator to determine the limits of θ. It's not what you'd expect!)

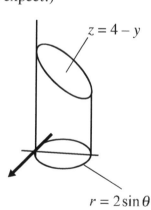

$z = 4 - y$

$r = 2\sin\theta$

§ 4.7

15. Evaluate this triple integral: $\displaystyle\int_{-1}^{1}\int_{-\sqrt{1-x^2}}^{\sqrt{1-x^2}}\int_{-\sqrt{x^2+y^2}}^{\sqrt{x^2+y^2}} \frac{1}{\left[1 + \left[\sqrt{\left(x^2+y^2\right)}\right]^3\right]^2}\,dz\,dy\,dx$. (***Hint:*** Convert to cylindrical coordinates.)

§ 4.7

16. Evaluate the triple integral $\int\limits_{0}^{3\pi/2} \int\limits_{0}^{\pi} \int\limits_{0}^{1} 5\rho^3 \sin^3\phi \, d\rho \, d\phi \, d\theta$. (*Hint:* You will need to review the integration technique back in 2ND semester calculus covering powers of trig functions in the integrand.)

§ 4.7

17. Evaluate $\iiint\limits_{Q} \dfrac{1}{\sqrt{x^2 + y^2 + z^2}} \, dV$ using spherical coordinates if Q is…

(a)   the upper hemisphere of $x^2 + y^2 + z^2 = 25$.

(b)   the lower hemisphere of $x^2 + y^2 + z^2 = 25$.

§ 4.7

18. Let Q be the sphere $x^2 + y^2 + z^2 = 9$.

    (a) Use *cylindrical* coordinates to set up the triple integral to calculate the volume of the upper hemisphere of Q. (No need to evaluate.)

    (b) Use *spherical* coordinates to set up the triple integral to calculate the volume of the upper hemisphere of Q. (No need to evaluate.)

§ 3.8

19. (Review) Test the function $f(x,y) = x^3 + 3x^2 + y^3 - 3y^2$ for relative extrema and saddle points. Give answer(s) as ordered triples.

§ 4.7

EXTRA CREDIT (optional). Convert the triple integral to spherical coordinates (No need to evaluate.):

$$\int_{-1}^{1} \int_{-\sqrt{1-x^2}}^{\sqrt{1-x^2}} \int_{\sqrt{x^2+y^2}}^{1} dz\,dy\,dx$$

CALCULUS III ✠ PRACTICE MIDTERM 4 ✠   **NAME:** _____

TOTAL PAGES: 8

Instructions: *Please show all work and circle your answers.*
One 3"× 5" index card of formulas (front and back) is permitted which must be turned in with the exam. No books or scratch paper. Calculators OK, except TI-89 , TI-92, or any other symbolic differentiation/integration utility.

1. Evaluate $\displaystyle\int_0^{\frac{\pi}{2}} \int_y^{2y} \sin(x+y)\,dx\,dy$.

2. Evaluate $\displaystyle\iint_R \frac{y}{1+x^2}\,dA$, where R is the region bounded by the graphs of $y=0,\ y=\sqrt{x},\ $ and $x=4$.

3. Evaluate the integral $\displaystyle\int_0^3 \int_0^{x^3} e^{y/x^3}\,dy\,dx$

4.   Evaluate the integral $\displaystyle\int_0^\infty\int_0^\infty xye^{-(x^2+3y^2)}\,dxdy$

5.   Evaluate $\displaystyle\int_0^1\int_x^1 \cos(y^2)\,dydx$ by reversing the order of integration.

6.   Use a double integral to find the volume of the solid in the first octant bounded above by the plane $x+y+z=4$ and below by the rectangle on the xy-plane $\{(x,y)\colon 0\le x\le 1,\ 0\le y\le 2\}$.

7.   Evaluate the integral $\int_0^\pi \int_0^{6\sin\theta} r\,dr\,d\theta$.

8.   Find the limits of integration for calculating the volume of the solid Q enclosed by the graphs of $y = x^2$, $z = 0$, and $y + z = 2$ if $V = \iiint\limits_Q dz\,dy\,dx$. Just set up the integral. *No need to evaluate!*

9.   Rewrite the order of the integral $\int_0^4 \int_0^{\frac{4-x}{2}} \int_0^{(12-3x-6y)} dz\,dy\,dx$ in the order $dy\,dx\,dz$. *No need to evaluate the integral—just set it up!*

10. Evaluate the integral $\displaystyle\int_1^3 \int_x^{x^2} \int_0^{\ln x} xe^y \, dy\,dz\,dx$.

11. Set up the triple integral to calculate the volume of the solid bounded by $z = 9 - x^2 - y^2$ and $z = 0$ ...

    (a) using rectangular coordinates.

    (b) using cylindrical coordinates.

12. Consider the triple integral $\displaystyle\int_{-5}^5 \int_0^{\sqrt{25-x^2}} \int_0^{\frac{1}{x^2+y^2}} \sqrt{x^2+y^2} \, dz\,dy\,dx$.

    (a) Set up the integral using cylindrical coordinates.

    (b) Evaluate the integral you set up in part **(a)**.

13. Consider the solid bounded above by the plane $x + y + z = 3$ in the first octant.

(a) Find the volume of the solid. (Use any method you prefer.)

(b) Find the surface area of the top of the solid (i.e., the portion of the plane).

14. Find the surface area for that portion of the surface $z = xy$ that is inside the cylinder $x^2 + y^2 = 1$.

15. Evaluate $\displaystyle\iiint\limits_{Q} \frac{1}{\sqrt{x^2 + y^2 + z^2}}\, dV$ using spherical coordinates if Q is the sphere $x^2 + y^2 + z^2 = 25$.

16. Let Q be the sphere $x^2 + y^2 + z^2 = 9$.

(a) Use cylindrical coordinates to set up the triple integral to calculate the volume of the upper hemisphere of Q. (No need to evaluate.)

(b) Use spherical coordinates to set up the triple integral to calculate the volume of the upper hemisphere of Q. (No need to evaluate.)

17. Evaluate the integral $\displaystyle\int_0^1 \int_0^{e^{-9x}} xy\,dy\,dx$. *(**Hint:** Integration by parts will be involved somehow, somewhere in this problem.)*

18. Given a solid Q enclosed by the graphs of $x = y^4$, $z = 0$, *and* $x + z = 16$. ...

(a) Set up but do not evaluate the double integral that would calculate the *area* of the *base* of the solid region Q. *(**Hint:** Identify what is going on in the xy-plane.)*

(b) Set up, but do not evaluate, the double integral that would calculate the *volume* of the solid region Q.

(c) Set up, but do not evaluate, the *triple* integral that would calculate the *volume* of the solid region Q.

19. Find the surface area for that portion of the surface $z = \dfrac{x^2}{2} - y$ that lies above the triangle in the first quadrant of the *xy*-plane bounded by the lines $y = 3x$, $y = 0$, and $x = 2$.

20. Evaluate $\displaystyle\iint_R \dfrac{x^2}{\sqrt{1+y^2}}\, dA$, where R is the region bounded by the graphs of $y = x^3$, $y = 8$, and $x = 0$.

21. Convert to polar coordinates and then evaluate the double integral $\displaystyle\int_{-6}^{6} \int_{-\sqrt{36-y^2}}^{0} \dfrac{\sqrt{x^2+y^2}}{1+\sqrt{x^2+y^2}}\, dx\, dy$.

22. Set up the triple integral using cylindrical coordinates, and then evaluate $\displaystyle\int_{0}^{1}\int_{-\sqrt{1-x^2}}^{0}\int_{-(x^2+y^2)}^{(x^2+y^2)} 21xy^2\, dz\, dy\, dx$.

23. Set up and evaluate the triple integral using spherical coordinates that would calculate the mass of a sphere with radius 7 if the density function is given by $f(x,y,z)=k\sqrt{x^2+y^2+z^2}$, where k is a constant.

$\sqrt{}$ *PART II* Tᴇsᴛ Bᴀɴᴋ

CᴀʟᴄᴜʟᴜS III   ✠ **QUIZ 11** ✠   **NAME:** _____

Tᴏᴛᴀʟ Pᴀɢᴇs: **8**

Instructions: *Please show all work and circle your answers.*
No notes, books, or scratch paper. Calculators are OK.

§ 5.1

1. Sketch several representative vectors in the vector field.

   (a) $\mathbf{F}(x,y) = \langle 0, -2 \rangle$ 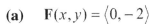   **(b)** $\mathbf{F}(x,y) = \langle x, 0 \rangle$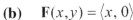

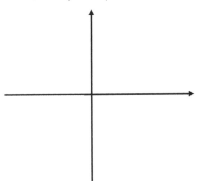 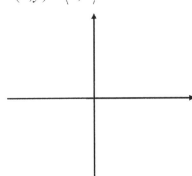

§ 5.1

2. Find a three-dimensional vector field that has the potential function $f(x,y,z) = x^2 e^{yz}$. *(**Hint:** Differentiate.)*

§ 5.1

3. Given the three-dimensional vector field $\mathbf{F}(x,y,z) = \langle e^y, xe^y + y, 0 \rangle$. Find the potential function for **F**. *(**Hint:** Anti-differentiate.)*

§ 5.1

4. ***Multiple choice:*** Determine which of the following vector fields is *not* conservative. *(Please circle your choice.)*

   (a) $\mathbf{F}(x,y) = \left\langle -\dfrac{y}{x^2}, \dfrac{1}{x} \right\rangle$ **(b)** $\mathbf{F}(x,y) = \langle e^y, xe^y + y \rangle$

   (c) $\mathbf{F}(x,y) = \langle -2y^3 \sin 2x, 3y^2(1 + \cos 2x) \rangle$ **(d)** $\mathbf{F}(x,y) = \langle 4x^2 - 4y^2, 8xy - \ln y \rangle$

   (e) $\mathbf{F}(x,y) = \left\langle \dfrac{x}{\sqrt{x^2 + y^2}}, \dfrac{y}{\sqrt{x^2 + y^2}} \right\rangle$

§ 5.2

5.   Let C be the line segment from the point $(0, 0, 0)$ to the point $(1, 3, -2)$. Find $\int_C (x + y^2 - 2z)\,ds$.

§ 5.2

6.   Let $\mathbf{F}(x,y,z) = \langle 2x - y,\ 2z,\ y - z \rangle$. Find the work done by the force $\mathbf{F}$ on an object moving along the straight line from the point $(0, 0, 0)$ to the point $(1, 1, 1)$.

§ 5.2

7.   A particle moves along a path parameterized by $\mathbf{r}(t) = \langle t, t^2, t^3 \rangle$ from the point $(0, 0, 0)$ to the point $(1, 1, 1)$ under a force given by $\mathbf{F}(x,y,z) = \langle 2xz,\ -yz,\ yz^2 \rangle$. Calculate the work done on the particle by the force.

§ 5.2

8. Let $\mathbf{F}(x,y,z) = \langle y, x, z^2 \rangle$ and evaluate $\int_C \mathbf{F} \cdot d\mathbf{r}$ for the curve $r(t) = \langle t, \cos t, \sin t \rangle$ for $0 \le t \le 2\pi$.

   (**Hint:** You neeed integration by parts to do this problem.)

§5.2

9. Let $\int_C y\,ds$, where $C = C_1 \cup C_2$ is the path given in the figure shown.

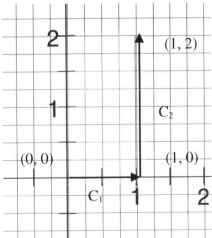

§ 5.2

10. Evaluate $\int_C x\,dx - xy\,dy$ over the path C given by $x = t^2, y = 3t, 0 \le t \le 1$.

§ 5.3

11. Let $\mathbf{F}(x,y) = \left\langle \dfrac{y}{x^2 + y^2}, \dfrac{-x}{x^2 + y^2} \right\rangle$. Calculate $\displaystyle\int_C \mathbf{F} \cdot d\mathbf{r}$, where C is the semi-circle $\mathbf{r}(t) = \langle \cos t, \sin t \rangle$ for $0 \le t \le \pi$.

§ 5.3

12. Use the Fundamental Theorem of Line Integrals to evaluate $\displaystyle\int_C \left(y^2 - 3x^2\right)dx + \left(2xy + 2\right)dy$, where C is a smooth curve from (1, 1) to (−1, 0).

§ **5.3**

13. Evaluate $\int_C e^x \sin y \, dx + e^x \cos y \, dy$, where C is the square with vertices $(0, 0)$, $(1, 1)$, $(0, 2)$, $(-1, 1)$.

§ **5.3**

14. Find the work done by the force field $\mathbf{F}(x, y) = \langle y, x \rangle$ in moving a particle from the point $(0, 4)$ to the point $(3, 1)$ along the following paths:

(a) $C_1 : y = -x + 4$

(b) $C_2 : y = (x - 2)^2$

§ 5.4

15. Let C be the closed curve shown to the right, oriented counter-clockwise. Evaluate $\int_C 2y^3\,dx + \left(x^4 + 6y^2x\right)dy$ using Green's Theorem.

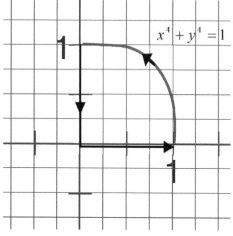

§ 5.4

16. Use Green's Theorem to evaluate the line integral $\int_C \left(x^2 + 2y\right)dx + \left(\frac{1}{2}x^2 - y^3\right)dy$, where C is the path from $(0, 0)$ to $(1, 0)$ along the path $y = 0$, and then from $(1, 0)$ to $(1, 1)$ along $x = 1$, and finally from $(1, 1)$ to $(0, 0)$ along the path $y = \sqrt{x}$.

§ 4.4

17. Use Green's Theorem to evaluate the line integral $\int_C \left(\sin x^2 + y\right)dx + \left(3x - \arctan(e^y)\right)dy$, where C is the square with vertices $(0, 0)$, $(2, 0)$, $(2, 2)$, and $(0, 2)$.

§ **4.4**

18. Show that the work done by a force $\mathbf{F}(x, y) = \left\langle e^x \sin y, e^x \cos y \right\rangle$ around a simple closed curve is zero.

§ **4.4**

19. Use Green's Theorem to evaluate the line integral $\int_C y^3 dx + \left(x^3 + 3xy^2\right) dy$, where C is a circle of radius 3 centered around the origin, oriented counter-clockwise. (**Hint:** Polar coordinates may be helpful.)

EXTRA CREDIT—*(optional).* Use back of page if you need more room to answer these.

  I. Could Green's Theorem be used in Problem #11? (Yes or no.) If yes, use Green's Theorem and show answers are the same. If no, explain why not.

  II. For problem #19: This problem involves a closed-curve path, yet the answer is not zero. *Why not?*

  III. Suppose in problem #11, the path was the entire circle and not just the semi-circle. Then the path is closed *and* the vector field is conservative, yet the answer is not zero. (It will be -2π.) Why isn't the answer zero? (This one is hard.)

§ 5.1

20. Given the vector field $\mathbf{F}(x,y,z) = \left\langle 2xy + z^2,\ x^2,\ 2xz + \pi \cos(\pi z) \right\rangle \dots$

Is this a conservative vector field? (Yes or no.) If yes, find the potential function of $\mathbf{F}$.

§ 5.1

21. Let $\mathbf{F}(x,y,z) = \left\langle x^3 \ln z,\ xe^{-y},\ -\left(y^2 + 2z\right) \right\rangle$.

(a) Calculate the divergence of $\mathbf{F}$.

(b) Evaluate the divergence at the point $(2, \ln 2, 1)$.

(c) Calculate the curl of $\mathbf{F}$.

(d) Evaluate the curl of $\mathbf{F}$ at the point $(1, 1, 1)$.

§ 5.1

22. Let $\mathbf{F}(x,y,z) = \left\langle \cos x,\ \sin y,\ e^{xy} \right\rangle$. Show that div $(\mathbf{curl\ F}) = 0$.

PART II   Test Bank

Calculus III

The Practice

❈ FINAL ❈

EXAM

NAME: _____

INSTRUCTIONS

---

1. This *Final* consists of 11 pages besides this cover. Please be sure that you have all of the pages.

2. *One* 8½" × 11" *handwritten* page of Formulas and Theorems *only* (front and back) is permitted, which must be turned in with the exam. Graphing calculators are required. Calculators with symbolic differentiation and/or integration features are excluded, such as the TI-89 and TI-92. *No swapping* of calculators for obvious reasons.

3. Please show all relevant work on the exam. If you do not show how you arrived at your answer, you shall not receive any credit—even if the answer is correct.

4. Giving or receiving aid on this exam is a definite no-no and will have serious consequences. Cell phones and other electronic devices *are prohibited.*

5. Relax and just do your best!

P.S.   I have enjoyed our brief association and wish you luck in your future studies!

CALCULUS III ✠ PRACTICE FINAL EXAM ✠   **NAME:** _____
TOTAL PAGES: 11

Instructions: *Please show all work and circle your answers.*

---

1. **(a)** The vector **v** has magnitude 8 and direction $q = 120°$. Find its component form.

(b) Suppose this same vector has as its initial point: $\left(-2, 7\sqrt{3}\right)$. Use your answer from part **(a)** to find its terminal point.

2. Determine if the following pairs of vector are orthogonal, parallel, or neither. Show your work.

(a) $\mathbf{v} = \langle 3, -2 \rangle$ and $\mathbf{w} = \langle -1, 2 \rangle$ **(b)** $\mathbf{v} = \langle -2, 0 \rangle$ and $\mathbf{w} = \langle 0, 5 \rangle$

(c) $\mathbf{v} = \langle -1, 2 \rangle$ and $\mathbf{w} = \left\langle 0, -\dfrac{1}{2} \right\rangle$ **(d)** $\mathbf{v} = \langle 2, -3 \rangle$ and $\mathbf{w} = \langle -2, 3 \rangle$

3. Given the vectors $\mathbf{u} = \langle 2, -1, 1 \rangle$ and $\mathbf{w} = \langle -3, 2, 2 \rangle$…

(a) Calculate the angle (in degrees) between the vectors. Round answer to nearest hundredth.

(b) Find $\text{proj}_{\mathbf{u}} \mathbf{w}$.

4. *Orthogonal vectors:*

(a) Find a vector orthogonal to the yz-plane.

(b) Find a vector orthogonal to the two given lines:

line #1: $\begin{cases} x = -1 + 3t \\ y = 3 - 2t \\ z = 1 + t \end{cases}$ **line #2:** $\begin{cases} x = 4 + 5t \\ y = 2 - t \\ z = -1 - 2t \end{cases}$

(c) Find a vector orthogonal to the plane given by $2x - 3y + z = 11$.

5. Determine the parametric equations for the line passing through the points $(-3, 2, 0)$ and $(4, 2, 3)$.

6. Three forces with magnitudes 3, 4, and 5 pounds act on a machine part at angles of $-15°$, $150°$, and $220°$ respectively, with the positive x-axis. Find both the magnitude and direction of the resultant force. Round all answers to the nearest tenths of a unit. Be sure and include units in your final answer to get full credit.

7. Consider the following plane curves. Eliminate the parameter and represent each curve by a vector-valued function $\mathbf{r}(t) = \langle x(t), y(t) \rangle$.

    (a) $x = y^2 - 1$

    (b) $y = x^2 - 1$

8. Find the domain of the vector-valued function $\mathbf{r}(t) = \left\langle \ln(3-t), \ \dfrac{\sqrt[4]{2+t}}{\ln|t|}, \ \dfrac{t+1}{6t^2 - 7t - 3} \right\rangle$.
    Write your answer answer using interval notation to get full credit.

9. Evaluate $\displaystyle\int \left\langle t \ln t, \ \sqrt{1+5t}, \ \dfrac{t}{t+1} \right\rangle dt$.

10. An object starts from rest at the point $(0, 1, 1)$ and moves with an acceleration $\mathbf{a}(t) = \langle 1, \cos t, 0 \rangle$. Find the position, $\mathbf{r}(t)$, at time $t = 4$.

11. Given the vector-valued function $\mathbf{r}(t) = \langle t + 4, 1 - t^2 \rangle$ ...

(a) Sketch the graph, be sure and identify all intercepts.

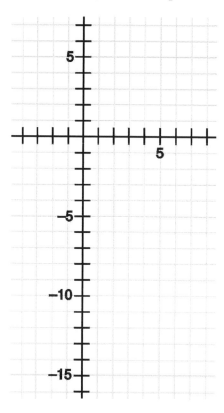

(b) Evaluate the velocity vector when $t = 2$, and sketch it on the same graph at the appropriate position.

12. ***Projectile Motion.*** A projectile is fired at a height of 2 meters above the ground with an initial velocity of 100 meters per second at an angle of 35° with the horizontal. Round each result to the nearest tenths of a unit.

    (a) Find the vector-valued function describing the motion. ***Hint:*** Use $g = 9.8$ meters per second per second.

    (b) Find the maximum height.

    (c) How long was the projectile in the air?

    (d) Find the range.

13. Find the unit tangent vector, $\mathbf{T}(t)$, for the curve given by $\mathbf{r}(t) = \langle 4\cos t, -3\sin t, 1 \rangle$ when $t = \dfrac{3\pi}{2}$.

14. Find the length of the curve $\mathbf{r}(t) = \langle e^t, e^{-t}, \sqrt{2}\,t \rangle$ when $t \in [0, 2]$.

15. ***Domain for a function of 2 variables.*** Find the domain for the given function and write the answer using set notation:
$$f(x, y) = \frac{\ln(x - y)}{e^{1/x}} + \sin(x - y^2) + \sqrt[3]{x + 3y} + \sqrt{x^2 + 2y^3} + \frac{x}{6x - 7y}$$

16. Find the f_{xy} for $f(x,y) = x^2y + 2y^2x^2 + 4x$.

17. Find the first partial derivative with respect to x: $F(x,y,z) = xe^{xyz}$

18. Use the total differential dz to approximate the change in $z = \dfrac{y}{x}$ as (x, y) moves from the point $(2, 1)$ to the point $(2.1, 0.8)$. Then, calculate the actual change Δz.

19. The radius of a right circular cylinder is decreasing at the rate of 4 inches per minute and the height is increasing at the rate of 8 inches per minute. What is the rate of change of the volume when $r = 4$ inches and $h = 8$ inches? (***Hint:*** Use the Chain Rule for function of several variables.)

20. Find the directional derivative of $f(x,y) = x^2y$ at the point $(1, -3)$ in the direction $< -2, 1>$.

21. Given the surface $2x^2 + 3y^2 + 4z^2 = 18\ldots$

    (a) Use implicit differentiation and find the slope in the x-direction, $\dfrac{\partial z}{\partial x}$, at the point $(-1, 2, 1)$.

    (b) Find an equation of the tangent plane (in general form) to the surface at the point $(-1, 2, 1)$.

22. Find a set of parametric equations for the normal line to the surface given by $z = f(x, y) = x^2 y$ at the point $(2, 1, 4)$.

23. Find extrema and saddle point(s), if any, for the function $f(x, y) = x^2 + x - 3xy + y^3 - 5$. Write your answer(s) in ordered triple(s) to get full credit.

24. Use Lagrange Multipliers to find the dimensions of a rectangular box of maximum volume with one vertex at the origin and the opposite vertex lying in the plane given by $6x + 4y + 3z = 24$. Then, give the actual maximum volume.

25. **(a)** Evaluate the integral: $\displaystyle\int_{0}^{\ln 2}\int_{e^{y}}^{2} x\, dx\, dy$

(b) Evaluate the integral $\displaystyle\int_{0}^{1}\int_{y}^{1} x^{2}e^{xy}\, dx\, dy$. (**Hint:** Reverse the order of integration first.)

26. Evaluate the double integral $\displaystyle\int_{-1}^{1}\int_{0}^{\sqrt{1-x^{2}}} e^{x^{2}+y^{2}}\, dy\, dx$ by changing to polar coordinates.

27. Set up the triple integrals (but do not evaluate) that would calculate the volume of the solid bounded by the graphs of $z = 0$, $x^{2} + y^{2} = 16$, and $z = 5 - y$ using…

(a) rectangular coordinates

(b) cylindrical coordinates

28. Find work done by the force $\mathbf{F} = \langle y - x^2, z - y^2, x - z^2 \rangle$ over the curve $r(t) = \langle t, t^2, t^3 \rangle$ from the point $(0, 0, 0)$ to $(1, 1, 1)$.

29. Evaluate , where $C = \displaystyle\int_C \frac{x + y^2}{\sqrt{1 + x^2}}\, ds$. The curve is the straight line segment from the point $(1, 0)$ to the point $\left(1, \dfrac{1}{2}\right)$, and $y = \dfrac{x^2}{2}$ is the curve along the graph of $\left(1, \dfrac{1}{2}\right)$ from the point to $(0, 0)$.

30. Given the field $\mathbf{F} = \left\langle 2x, -y^2, -\dfrac{4}{1 + z^2} \right\rangle \ldots$
    (a) Show that the field is conservative.

    (b) Evaluate $\displaystyle\int_{(0,0,0)}^{(3,3,1)} 2x\, dx - y^2\, dy - \frac{4}{1 + z^2}\, dz$.

31. Find the work done by the field $\mathbf{F} = \left\langle 2\cos y, \dfrac{1}{y} - 2x\sin y \right\rangle$ on the object that follows a path from the point (2, 0), to the point (2, 1), and then to the point $\left(1, \dfrac{\pi}{2}\right)$.

32. Use Green's Theorem to evaluate the line integral $\displaystyle\int_C \left(x - y^3\right)dx + x^3 dy$, where C is the right half of a circle of radius 2, $x^2 + y^2 = 4$.

33. Evaluate the surface integral $\displaystyle\iint_S y \, dS$ if S is the part of the plane $z = 6 - 3x - 2y$ in the first octant.

34. Find the flux integral $\int_S \int \mathbf{F} \cdot \mathbf{N} \, dS$ if $\mathbf{F} = \langle x, y, z \rangle$ where S is the surface $z = 1 - x^2 - y^2$ above the xy-plane.

35. Let Q be the cube bounded by the planes $x = \pm 1$, $y = \pm 1$, and $z = \pm 1$, and let $\mathbf{F}(x, y, z) = \langle x^2, y^2, z^2 \rangle$. Use the Divergence Theorem to evaluate $\iint_S \mathbf{F} \cdot \mathbf{N} dS$. Source, sink, or neither?

36. Let Q be the solid bounded by the cylinder $x^2 + y^2 = 1$ and the planes $z = 0$ and $z = 1$. Use the Divergence Theorem to evaluate $\iint_S \mathbf{F} \cdot \mathbf{N} dS$ and calculate the outward flux of $\mathbf{F}$ through S, where S is the surface of Q and $\mathbf{F}(x, y, z) = \langle x, y, z \rangle$. Source, sink, or neither?

37. Let Q be the region bounded above by the sphere $x^2 + y^2 + z^2 = 9$ and below by the plane $z = 0$ in the first octant. Use the Divergence Theorem to evaluate $\iint\limits_{S} \mathbf{F} \cdot \mathbf{N}\,dS$ and find the outward flux of $\mathbf{F}$ through S, where S is the surface of the solid and $\mathbf{F}(x, y, z) = \langle xy, 4x, 2y \rangle$. Source, sink, or neither?

38. Find the curl of the vector field $\mathbf{F}(x, y, z) = \langle z^2, -x^2, y \rangle$. Is the field conservative?

39. Use Stokes's Theorem to evaluate $\int\limits_{C} \mathbf{F} \cdot d\mathbf{r}$, where $\mathbf{F}(x, y, z) = \langle z, 2x, 2y \rangle$ and S is the surface of the paraboloid (oriented upward) of $z = 4 - x^2 - y^2$, $z \geq 0$, and C is its boundary.

PART III

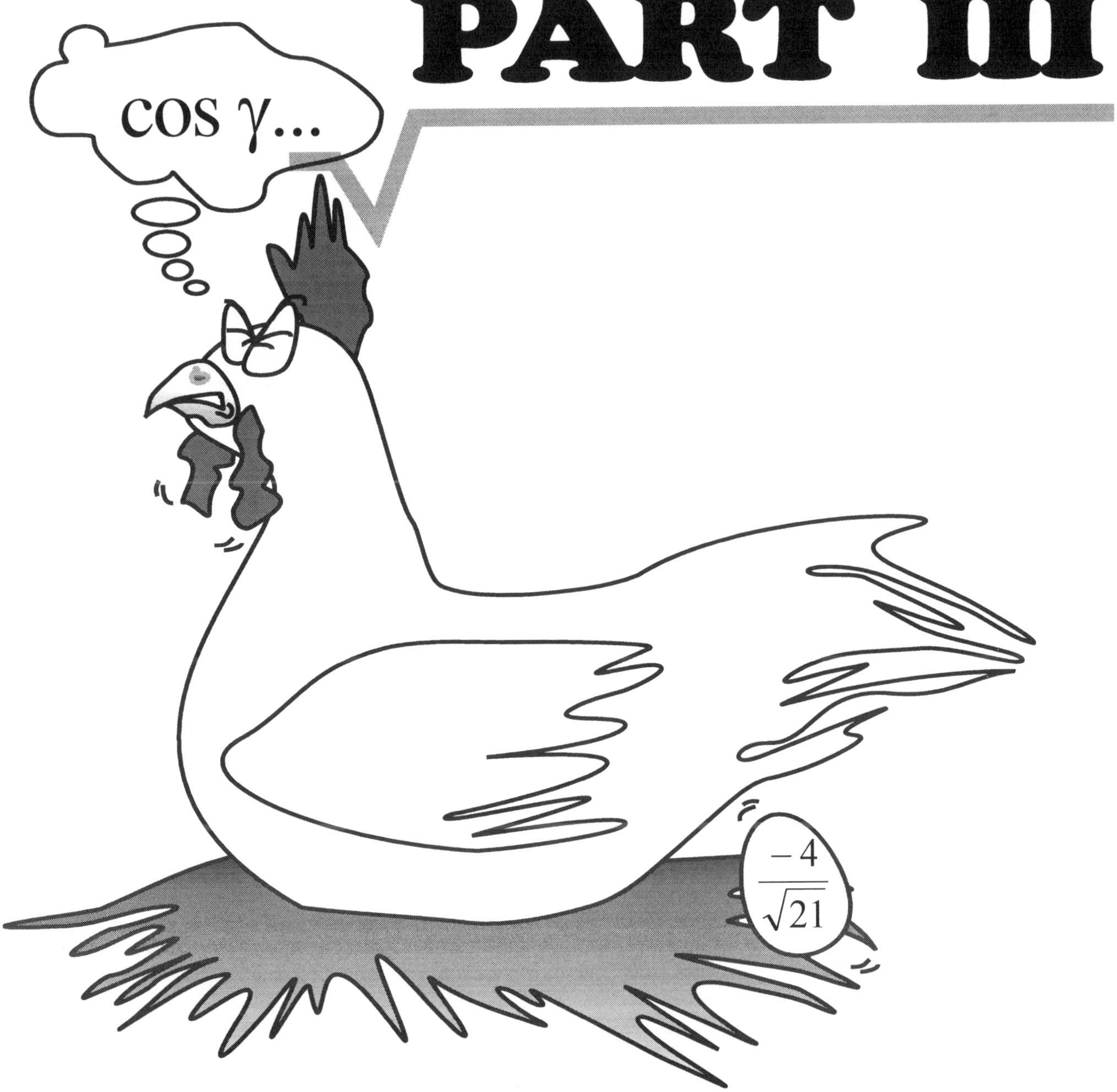

Table of Test KEYS

CALCULUS III   ❈ **QUIZ 1 KEY** ❈

1. **A vector v has initial point (3, –1, –5) and terminal point (0, –1 , 2).**

    (a)   Write v in component form (that is, <a, b, c> form).

    $$v = \left\langle q_1 - p_1, q_2 - p_2, q_3 - p_3 \right\rangle = \left\langle 0 - 3, -1 - (-1), 2 - (-5) \right\rangle = \left\langle -3, 0, 7 \right\rangle$$

    (b)   Write v as a linear combination of the standard unit vectors.

    $$\mathbf{v} = \boxminus 3\mathbf{i} + 0\mathbf{j} + 7\mathbf{k} \quad \text{OR} \quad \mathbf{v} = \boxminus 3\mathbf{i} + 7\mathbf{k}$$

    (c)   Find the magnitude of v.

    $$\|\mathbf{v}\| = \sqrt{(-3)^2 + (0)^2 + (7)^2} = \sqrt{9 + 49} = \sqrt{58}$$

    (d)   Find the unit vector in the direction of v.

    $$\mathbf{u} = \frac{\mathbf{v}}{\|\mathbf{v}\|} = \frac{\left\langle -3, 0, 7 \right\rangle}{\sqrt{58}} = \left\langle \frac{-3}{\sqrt{58}}, 0, \frac{7}{\sqrt{58}} \right\rangle$$

    (e)   Find the unit vector in the opposite direction of v.

    Now we take the *opposite*—that means to take the negative of the vector obtained from part **(d)** above:

    $$-\left\langle \frac{-3}{\sqrt{58}}, 0, \frac{7}{\sqrt{58}} \right\rangle = \left\langle \frac{3}{\sqrt{58}}, 0, \frac{-7}{\sqrt{58}} \right\rangle$$

2. **Find the standard equation of a sphere that has points (–1, 2, 8) and (–7, –2, –2) as endpoints of a diameter.**

    Use the Midpoint Formula to get the center of the sphere:

    $$M = \left(\frac{x_1 + x_2}{2}, \frac{y_1 + y_2}{2}, \frac{z_1 + z_2}{2} \right) = \left(\frac{-1 + -7}{2}, \frac{2 + -2}{2}, \frac{8 + -2}{2} \right) = (-4, 0, 3) = (x_0, y_0, z_0) \text{ (center)}$$

    Using the distance formula, we find the radius for the sphere:

    $$d = \sqrt{(x_1 - x_2)^2 + (y_1 - y_2)^2 + (z_1 - z_2)^2} = \sqrt{(-1 - (-4))^2 + (2 - 0)^2 + (8 - 3)^2} = \sqrt{3^2 + 2^2 + 5^2} = \sqrt{9 + 4 + 25}$$

    $$= \sqrt{38} \quad (radius)$$

    Finally, we use the standard equation for a sphere and "plug in" all of the "ingredients" we found:

    $$(x - x_0)^2 + (y - y_0)^2 + (z - z_0)^2 = r^2$$
    $$(x - (-4))^2 + (y - 0)^2 + (z - 3)^2 = (\sqrt{38})^2$$
    $$(x + 4)^2 + y^2 + (z - 3)^2 = 38$$

3. **A vector v has magnitude of 2 and direction θ = 300°. Find its component form.**

    x-component: $\cos\theta = \dfrac{adj}{hyp} \Rightarrow \cos 60° = \dfrac{x}{2} \Rightarrow x = 2\cos 60° = 2\left(\dfrac{1}{2}\right) = 1$

    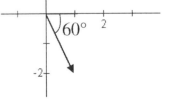

    y-component: $\sin\theta = \dfrac{opp}{hyp} \Rightarrow \sin 60° = \dfrac{y}{2} \Rightarrow y = 2\sin 60° = 2\left(\dfrac{\sqrt{3}}{2}\right) = \sqrt{3}$

    answer: $\mathbf{v} = \left\langle 1, -\sqrt{3} \right\rangle$. The *y*-coordinate is negative since the vector is in Quadrant IV.

CALCULUS III ✠ **QUIZ 2 KEY** ✠

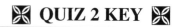

1. **If the point (7, –1) is the initial point for the vector v = <1, 2>, find the terminal point.**

   Let $Q(q_1, q_2)$ be the terminal point.

   Then we need $q_1 - 7 = 1$ *and* $q_2 - (-1) = 2 \Rightarrow q_1 = 8$ *and* $q_2 = 1 \Rightarrow Q(8, 1)$

2. **Find a vector of magnitude 3 in the direction of v = <1, 2>.** (*Hint:* **First, come up with the unit vector in the same direction as v. Then multiply that unit vector by a scalar.**)

$$\mathbf{u} = \frac{\mathbf{v}}{\|\mathbf{v}\|} = \frac{\langle 1, 2 \rangle}{\sqrt{1^2 + 2^2}} = \left\langle \frac{1}{\sqrt{5}}, \frac{2}{\sqrt{5}} \right\rangle$$

   Then, a vector of magnitude 3 would be: $3\mathbf{u} = 3\left\langle \frac{1}{\sqrt{5}}, \frac{2}{\sqrt{5}} \right\rangle = \left\langle \frac{3}{\sqrt{5}}, \frac{6}{\sqrt{5}} \right\rangle$ (*answer*)

3. **Two forces have an angle of 110° between them. F₁ has a magnitude 390 N. (N is *Newtons*, a unit of force.) F₂ has a magnitude of 260 N. Find both the magnitude and direction of F₁ + F₂ .** (*Hint:* **Superimpose an *xy*-coordinate plane such that F₁ is on the positive *x*-axis. Then, resolve the two force vectors into component form. Add the two vectors to get the resultant force. Finally, find both the magnitude and direction or angle of this resultant force.**)

   Component form for $\mathbf{F_1}$: $\mathbf{F_1} = \langle 390, 0 \rangle$

   Component form for $\mathbf{F_2}$: $\mathbf{F_2} = \langle \|\mathbf{F_2}\|\cos\theta, \|\mathbf{F_2}\|\sin\theta \rangle = \langle 260\cos(110°), 260\sin(110°) \rangle \approx \langle -88.9, 244.3 \rangle$

   $\mathbf{F_1} + \mathbf{F_2} = \langle 390 + (-88.9), 0 + 244.3 \rangle = \langle 301.1, 244.3 \rangle$

   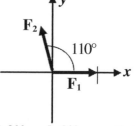

   Magnitude = $\|\mathbf{F_1} + \mathbf{F_2}\| = \sqrt{(301.1)^2 + (244.3)^2} \approx 387.7$ *Newtons*

   Direction (angle) = $\theta = \arctan\left(\frac{y}{x}\right) = \arctan\left(\frac{244.3}{301.1}\right) = 39°$

4. **Three forces with magnitudes of 50, 20, and 40 pounds act on an object at angles of 60°, 30°, and –90° respectively, with the positive x-axis. Find both the direction and magnitude of the resultant force.** (*Hint:* **same as for previous problem.**)

   Find the component forms for all three force vectors:

   $\mathbf{F_1}$: $\mathbf{F_1} = \langle \|\mathbf{F_1}\|\cos\theta, \|\mathbf{F_1}\|\sin\theta \rangle = \langle 50\cos(60°), 50\sin(60°) \rangle \approx \langle 25, 43.3 \rangle$

   $\mathbf{F_2}$: $\mathbf{F_2} = \langle \|\mathbf{F_2}\|\cos\theta, \|\mathbf{F_2}\|\sin\theta \rangle = \langle 20\cos(30°), 20\sin(30°) \rangle \approx \langle 17.3, 10 \rangle$

   $\mathbf{F_3}$: $\mathbf{F_3} = \langle \|\mathbf{F_3}\|\cos\theta, \|\mathbf{F3}\|\sin\theta \rangle = \langle 40\cos(-90°), 40\sin(-90°) \rangle \approx \langle 0, -40 \rangle$

   $\mathbf{F_1} + \mathbf{F_2} + \mathbf{F_3} = \langle 25 + 17.3 + 0, 43.3 + 10 + -40 \rangle = \langle 42.3, 13.3 \rangle$

   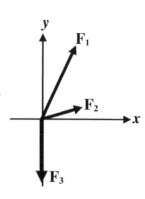

   Magnitude = $\|\mathbf{F_1} + \mathbf{F_2} + \mathbf{F_3}\| = \sqrt{(42.3)^2 + (13.3)^2} \approx 44.3$ *pounds*

   Direction (angle) = $\theta = \arctan\left(\frac{y}{x}\right) = \arctan\left(\frac{13.3}{42.3}\right) \approx 17.4°$

CALCULUS **III**   QUIZ 2 KEY, CONTINUED

5.   Given the function, $f(x) = x^3$, and the point that lies on the graph of this function (–2, –8)...

(a)   **Find the equation of the tangent line at this point.** *(Hint:* **First, find the derivative at this point, which is the same as the slope of the tangent line at this point. Now that you have the slope and a point that lies on the line, you have enough info to get the equation of that line.)**

slope of $f'(x) = 3x^2 \Rightarrow$ at $x = -2$, we have slope = $f'(-2) = 3(-2)^2 = 12$.

We now use the point slope form of the equation of a line that is:

$$(y - y_1) = m(x - x_1) \Rightarrow y - (-8) = 12(x - (-2))$$
$$y + 8 = 12(x + 2) \Rightarrow y + 8 = 12x + 24 \Rightarrow y = 12x + 16$$

(b)   **Find a vector parallel to the graph at this point.** *(Hint:* **The tangent line is parallel to the graph at this point. So, a vector parallel to the graph will simply be a line segment of this very same tangent line. Simply find another point that "lives" on this tangent line, which will be the terminal point of the vector that is parallel to the graph.)**

Let (–2, –8) be the initial point for the vector.

To find a terminal point on the tangent line, pick an arbitrary value of x (say, $x = 1$) and plug it into the equation for the line obtained above: $y = 12(1) + 16 \Rightarrow y = 28$.

Then the point (1, 28) is the terminal point for the vector.

Forming the vector now: $\langle 1 - (-2), 28 - (-8) \rangle = \langle 3, 36 \rangle$

(c)   **Find a vector normal to the graph at this point.** *(Hint:* **The tangent line is parallel, which you have the equation for already. Now, determine the equation for a line that is perpendicular to the tangent line, using your knowledge of what relationship slopes of perpendicular lines have. Two points on that perpendicular line will be the initial and terminal points for a vector that is normal to the graph.)**

A line perpendicular to the tangent line will have a slope that is the negative reciprocal of the tangent line's slope. That is, the slope of the perpendicular line will have a slope that is $m = -\dfrac{1}{12}$

Using the point-slope form again, we have the equation of the perpendicular line as:

$$(y - y_1) = m(x - x_1) \Rightarrow y - (-8) = -\frac{1}{12}(x - (-2)) \Rightarrow y + 8 = -\frac{1}{12}x - \frac{1}{6}$$

To get a vector that is normal to the graph at the point given, we simply need to find a vector that "lives" on this perpendicular line. We'll use the point (–2, –8) as the initial point and find a terminal point the same way we found one above. Pick an arbitrary value of x again (say, $x = 8$). Then we find the corresponding y-coordinate by plugging this into the equation for our perpendicular line:

$$y + 8 = -\frac{1}{12}(8) - \frac{1}{6} \Rightarrow y + 8 = -\frac{4}{6} - \frac{1}{6} \Rightarrow y + 8 = -\frac{5}{6} \Rightarrow y = -\frac{5}{6} - 8 = -\frac{53}{6}$$

So, our terminal point is $\left(8, -\dfrac{53}{6}\right)$.

The vector on the perpendicular line (and hence normal to the graph) is: $\left\langle 8 - (-2), -\dfrac{53}{6} - (-8) \right\rangle = \left\langle 10, -\dfrac{5}{6} \right\rangle$.

CALCULUS III

6. *Static equilibrium.* **Two workers carry a 350-pound cylindrical weight, which has two short ropes tied to the eyelet on the top center of the cylinder. One rope makes a 35 °-angle with the vertical and the other a 60°-angle with the vertical. Find the tension (force) in each rope.** *(Hint: The tension is the magnitude of the force!)*

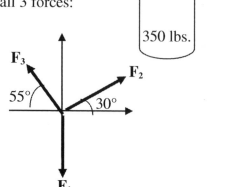

Create a force diagram, then get the component form for all 3 forces:

$\mathbf{F}_1 = \langle 0, -350 \rangle$

$\mathbf{F}_2 = \langle \|\mathbf{F}_2\| \cos 30°, \|\mathbf{F}_2\| \sin 30° \rangle = \langle \|\mathbf{F}_2\|(0.866), \|\mathbf{F}_2\|(0.5) \rangle$

$\mathbf{F}_3 = \langle -\|\mathbf{F}_3\| \cos 55°, \|\mathbf{F}_3\| \sin 55° \rangle = \langle \|\mathbf{F}_3\|(-0.574), \|\mathbf{F}_3\|(0.819) \rangle$

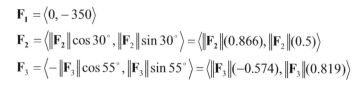

Static equilibrium means the sum of all three forces equals the zero vector: $\mathbf{F}_1 + \mathbf{F}_2 + \mathbf{F}_3 = \mathbf{0}$. Equate all of the components to get a linear system of two equations and two unknowns:

$\begin{cases} 0 + 0.866\|\mathbf{F}_2\| - 0.574\|\mathbf{F}_3\| = 0 \\ -350 + 0.5\|\mathbf{F}_2\| + 0.819\|\mathbf{F}_3\| = 0 \end{cases}$ $\Rightarrow$ solve for $\|\mathbf{F}_2\|$ $\Rightarrow \|\mathbf{F}_2\| = \dfrac{0.574}{0.866}\|\mathbf{F}_3\|$

Plug this into the second equation, and solve for $\|\mathbf{F}_3\|$:

$$0.5\left(\frac{0.574}{0.866}\|\mathbf{F}_3\|\right) + 0.819\|\mathbf{F}_3\| = 350 \Rightarrow \|\mathbf{F}_3\| = 304 \text{lbs and } \|\mathbf{F}_2\| = 201.7 \text{lbs}$$

7. **Find the standard equation of a sphere that has points (4, –3, 5) and (–6, 1, –1) as endpoints of a diameter.**

    Use the Midpoint Formula to get the center of the sphere:

$$M = \left(\frac{x_1 + x_2}{2}, \frac{y_1 + y_2}{2}, \frac{z_1 + z_2}{2} \right) = \left(\frac{4 + -6}{2}, \frac{-3 + 1}{2}, \frac{5 + -1}{2} \right) = (-1, -1, 2) = (x_0, y_0, z_0) \text{ (center)}$$

Using the distance formula, we find the radius for the sphere:

$$d = \sqrt{(x_1 - x_2)^2 + (y_1 - y_2)^2 + (z_1 - z_2)^2} = \sqrt{(4 - (-1))^2 + (-3 - (-1))^2 + (5 - 2)^2} = \sqrt{5^2 + (-2)^2 + 3^2}$$

$$= \sqrt{38} \quad (radius)$$

Finally, we use the standard equation for a sphere and "plug in" all of the "ingredients" we found:

$$(x - x_0)^2 + (y - y_0)^2 + (z - z_0)^2 = r^2$$
$$(x - (-1))^2 + (y - (-1))^2 + (z - 2)^2 = (\sqrt{38})^2$$
$$(x + 1)^2 + (y + 1)^2 + (z - 2)^2 = 38$$

8. **Find the center and radius of a sphere given by** $x^2 + y^2 + z^2 + 2x - 2y + 6z + 7 = 0$. *(Hint: You will need to* **complete the square three times.)**

$$x^2 + 2x + M_1 + y^2 - 2y + M_2 + z^2 + 6z + M_3 = -7 + M_1 + M_2 + M_3$$

CALCULUS III

where $M_1 = \left(\dfrac{b}{2}\right)^2 = \left(\dfrac{2}{2}\right)^2 = 1$ $\quad and \quad$ $M_2 = \left(\dfrac{b}{2}\right)^2 = \left(\dfrac{-2}{2}\right)^2 = 1$ $\quad and \quad$ $M_3 = \left(\dfrac{b}{2}\right)^2 = \left(\dfrac{6}{2}\right)^2 = 9$

$$\left(x^2 + 2x + 1\right) + \left(y^2 - 2y + 1\right) + \left(z^2 + 6z + 9\right) = -7 + 1 + 1 + 1$$
$$(x+1)^2 + (y-1)^2 + (z+3)^2 = 4$$

So, the center is: $(-1, 1, -3)$ and the radius is 2 (answer)

9. **Find the initial point of the vector v = 5i –k if the terminal point is (5, 5, 0).**

First, we write **v** in component form $\mathbf{v} = \langle 5, 0, -1 \rangle$.

Let the initial point be $P(p_1, p_2, p_3)$.

$$5 - p_1 = 5 \Rightarrow p_1 = 0$$
Then we have that $\quad 5 - p_2 = 0 \Rightarrow p_2 = 5$
$$0 - p_3 = -1 \Rightarrow p_3 = 1$$

Thus, our initial point is $(0, 5, 1)$.

10. **Determine whether each vector is parallel to the vector v = <–4, –1, 5>. If it is, find c such that u = cv.**

(a) **u = <8, 4, 10>** $\quad$ (a) $\quad 8 = c(-4) \Rightarrow c = 2$

$\qquad\qquad\qquad\qquad\qquad\quad 4 = c(-1) \Rightarrow c = -4 \Rightarrow$ different values for c means that the vectors are *not* parallel.

(b) **u = <–8, –2, 10>** $\quad$ (b) $\quad -8 = c(-4) \Rightarrow c = 2$

$\qquad\qquad\qquad\qquad\qquad\qquad -2 = c(-1) \Rightarrow c = 2 \Rightarrow$ same values for c means that the vectors *are* parallel.

$\qquad\qquad\qquad\qquad\qquad\qquad 10 = c(5) \Rightarrow c = 2$

(c) $\mathbf{u} = \left\langle 3, \dfrac{3}{4}, -\dfrac{15}{4} \right\rangle$ $\quad$ (c) $\quad 3 = c(-4) \Rightarrow c = -\dfrac{3}{4}$

$\qquad\qquad\qquad\qquad\qquad\qquad \dfrac{3}{4} = c(-1) \Rightarrow c = -\dfrac{3}{4} \Rightarrow$ same values for c means that the vectors *are* parallel.

$\qquad\qquad\qquad\qquad\qquad\qquad -\dfrac{15}{4} = c(5) \Rightarrow c = -\dfrac{3}{4}$

(d) $\mathbf{u} = \left\langle -2, \dfrac{1}{2}, \dfrac{5}{2} \right\rangle$ $\quad$ (d) $\quad -2 = c(-4) \Rightarrow c = -\dfrac{1}{2}$

$\qquad\qquad\qquad\qquad\qquad\qquad \dfrac{1}{2} = c(-1) \Rightarrow c = -\dfrac{1}{2} \Rightarrow$ different values for c means that the vectors *not* parallel!

$\qquad\qquad\qquad\qquad\qquad\qquad \dfrac{5}{2} = c(5) \Rightarrow c = \dfrac{1}{2}$

CALCULUS **III**   **Quiz 2 KEY,** CONTINUED

11.   **Given that v = < 5, –2 > and w = <–3, 1 >, find v · w.**

$$\mathbf{v} \cdot \mathbf{w} = \langle 5, 2 \rangle \cdot \langle -3, 1 \rangle = (5)(-3) + (-2)(1) = -15 + -2 = -17$$

12.   **Determine whether the vectors v = < 2, –3 > and w = <1, 1 > are orthogonal, parallel, or neither.**

If orthogonal, then the angle between them would be 90 °.

This means that the dot product would be equal to 0 since $\cos 90° = \dfrac{\mathbf{u} \cdot \mathbf{w}}{\|\mathbf{u}\|\|\mathbf{w}\|} = 0$.

Check to see if the dot product is equal to zero: $\mathbf{v} \cdot \mathbf{w} = \langle 2, -3 \rangle \cdot \langle 1, 1 \rangle = (2)(1) + (-3)(1) = 2 + -3 = -1$.
No, so the vectors are not orthogonal.

If they are parallel, then one is a multiple of the other. In other words, we can find a scalar *c* such that $\mathbf{v} = c\mathbf{w}$.

Check if this is true: $2 = c(1) \Rightarrow c = 2 \qquad -3 = c(1) \Rightarrow c = -3$.

No, they are not parallel since we are getting different values for *c*.

Therefore, the vectors are *neither* (orthogonal or parallel). *(answer)*

13.   **Given v = < 3, 2 > and w = <1, –3 >, find the projection of v onto w.**

Use the formula in the Theorem for projection of a vector onto another vector:

$$\text{proj}_{\mathbf{w}} \mathbf{v} = \left(\frac{\mathbf{v} \cdot \mathbf{w}}{\|\mathbf{w}\|^2} \right) \mathbf{w} = \left(\frac{(3)(1) + (2)(-3)}{1^2 + (-3)^2} \right) \mathbf{w} = \left(\frac{-3}{10} \right) \langle 1, -3 \rangle = \left\langle -\frac{3}{10}, \frac{9}{10} \right\rangle$$

14.   **Find the angle θ between the vectors v = < 3, 2 > and w = <7, –5 >.**

We know that:
$$\cos \theta = \frac{\mathbf{v} \cdot \mathbf{w}}{\|\mathbf{v}\|\|\mathbf{w}\|} = \frac{(3)(7) + (2)(-5)}{\sqrt{3^2 + 2^2} \sqrt{7^2 + 5^2}} = \frac{11}{\sqrt{13}\sqrt{74}} \approx 0.355$$
$$\theta = \arccos(0.355) \approx 69.2°$$

15.   **Find the cosine of the angle θ between the vectors v = < 3, –9 > and w = <2, 1 >.**

$$\cos \theta = \frac{\mathbf{v} \cdot \mathbf{w}}{\|\mathbf{v}\|\|\mathbf{w}\|} = \frac{(3)(2) + (-9)(1)}{\sqrt{3^2 + 9^2} \sqrt{2^2 + 1^2}} = \frac{-3}{\sqrt{90}\sqrt{5}} = \frac{-3}{3\sqrt{10}\sqrt{5}} = -\frac{1}{5\sqrt{2}} \quad OR \quad -\frac{\sqrt{2}}{10}$$

(after rationalizing the denominator)

16.   **Let u = <3, –1, –2> and v = <–2, –3, 2 >. Calculate u · v.**

$$\mathbf{u} \cdot \mathbf{v} = \langle 3, -1, -2 \rangle \cdot \langle -2, -3, 2 \rangle = (3)(-2) + (-1)(-3) + (-2)(2) = -6 + 3 + -4 = -7$$

17.   **Let u = < 3, –1, –2 > and v = <–2, –3, 2 >. Calculate proj$_v$ u.**

$$\text{proj}_{\mathbf{v}} \mathbf{u} = \left(\frac{\mathbf{u} \cdot \mathbf{v}}{\|\mathbf{v}\|^2} \right) \mathbf{v} = \left(\frac{(3)(-2) + (-1)(-3) + (-2)(2)}{(-2)^2 + (-3)^2 + 2^2} \right) \mathbf{v} = \left(\frac{-7}{17} \right) \langle -2, -3, 2 \rangle = \left\langle \frac{14}{17}, \frac{21}{17}, -\frac{14}{17} \right\rangle$$

$\sqrt{}$ *PART III* TEST ANSWER KEY

CALCULUS **III** $\qquad\qquad$ **Quiz 2 KEY,** CONTINUED

18. *Multiple choice:* **Which of the following statements is true about the vectors...?**

$$\mathbf{u} = \left\langle -\frac{1}{2}, \frac{1}{3}, -\frac{1}{4} \right\rangle \text{ and } \mathbf{v} = \left\langle 2, -\frac{4}{3}, 1 \right\rangle \text{ (Please circle your choice.)}$$

(a) **u and v are orthogonal** $\qquad$ (b) **u and v are parallel**

(c) **u is a unit vector of v** $\qquad$ (d) **the angle between u and v is** $\dfrac{\pi}{4}$

(e) **none of these**

They are not orthogonal since the dot product is not equal to zero. We know that **u** is not a unit vector of **v** because if it were, then...

$$\mathbf{u} = \frac{\mathbf{v}}{\|\mathbf{v}\|} = \frac{\left\langle 2, -\frac{4}{3}, 1 \right\rangle}{\sqrt{2^2 + \left(\frac{4}{3}\right)^2 + 1^2}} \approx \frac{\left\langle 2, -\frac{4}{3}, 1 \right\rangle}{2.6} \neq \left\langle -\frac{1}{2}, \frac{1}{3}, -\frac{1}{4} \right\rangle$$

However, they are parallel because they are indeed multiples of each other. In other words, we can find a scalar c such that $\mathbf{u} = c\mathbf{v}$. Check if this is true:

$$-\frac{1}{2} = c(2) \Rightarrow c = -\frac{1}{4}$$

$$\frac{1}{3} = c\left(-\frac{4}{3}\right) \Rightarrow c = -\frac{1}{4} \Rightarrow \text{ same values for c means that the vectors are parallel!}$$

$$-\frac{1}{4} = c(1) \Rightarrow c = -\frac{1}{4}$$

19. **Find u · v if** $\|\mathbf{u}\| = 7$, $\|\mathbf{v}\| = 12$, **and the angle between u and v is** $\dfrac{\pi}{4}$ **.**

$$\cos\theta = \frac{\mathbf{u}\cdot\mathbf{v}}{\|\mathbf{u}\|\|\mathbf{v}\|} \Rightarrow \mathbf{u}\cdot\mathbf{v} = \cos\theta\|\mathbf{u}\|\|\mathbf{v}\| \Rightarrow \mathbf{u}\cdot\mathbf{v} = \cos\left(\frac{\pi}{4}\right)(7)(12) = \left(\frac{\sqrt{2}}{2}\right)(7)(12) = 42\sqrt{2}$$

20. **Find the vector component of u orthogonal to v for u = <1, 2> and v = <–4, 4>.** *(Hint:* **First, you will need to find the projection of u onto v. Then you will have to subtract this result from u.)**

We first find $\mathbf{w}_1 = \text{proj}_{\mathbf{v}}\mathbf{u}$:

$$\mathbf{w}_1 = \text{proj}_{\mathbf{v}}\mathbf{u} = \left(\frac{\mathbf{u}\cdot\mathbf{v}}{\|\mathbf{v}\|^2}\right)\mathbf{v} = \left(\frac{(1)(-4) + (2)(4)}{(-4)^2 + (4)^2}\right)\mathbf{v} = \left(\frac{4}{32}\right)\langle -4, 4\rangle = \left\langle -\frac{1}{2}, \frac{1}{2} \right\rangle$$

The component of **u** that is orthogonal to **v** is often called $\mathbf{w}_2$, and is found by the formula $\mathbf{w}_2 = \mathbf{u} - \mathbf{w}_1 = \mathbf{u} - \text{proj}_{\mathbf{v}}\mathbf{u}$.

So, we now have $\mathbf{w}_2 = \mathbf{u} - \mathbf{w}_1 = \mathbf{u} - \text{proj}_{\mathbf{v}}\mathbf{u} = \langle 1, 2\rangle - \left\langle -\frac{1}{2}, \frac{1}{2} \right\rangle = \left\langle \frac{3}{2}, \frac{3}{2} \right\rangle$

21. **Find the direction cosines for the vector with initial point (2, –1, 3) and terminal point (4, 3, –5). (Fill in the blanks.)**

$$\cos \alpha = \frac{1}{\sqrt{21}}$$

$$\cos \beta = \frac{2}{\sqrt{21}}$$

$$\cos \gamma = \frac{-4}{\sqrt{21}}$$

We first find the component form for the vector **v** by taking the terminal point minus the initial point:

$$\mathbf{v} = \langle 4 - 2, 3 - (-1), -5 - 3 \rangle = \langle 2, 4, -8 \rangle$$

Then :

$$\cos \alpha = \frac{v_1}{\|\mathbf{v}\|} = \frac{2}{\sqrt{2^2 + 4^2 + 8^2}} = \frac{2}{\sqrt{84}} = \frac{2}{2\sqrt{21}} = \frac{1}{\sqrt{21}}$$

$$\cos \beta = \frac{v_2}{\|\mathbf{v}\|} = \frac{4}{2\sqrt{21}} = \frac{2}{\sqrt{21}}$$

$$\cos \gamma = \frac{v_3}{\|\mathbf{v}\|} = \frac{-8}{2\sqrt{21}} = \frac{-4}{\sqrt{21}}$$

22. **Determine a scalar, *k*, such that the vectors u = < 3, 2 > and v = <2, *k* > are orthogonal.**

We want $\cos \theta = \dfrac{\mathbf{u} \cdot \mathbf{v}}{\|\mathbf{u}\|\|\mathbf{v}\|} = 0$.

This means we need the dot product of **u** and **v** to be equal to 0:

$$(3)(2) + (2)(k) = 0 \text{ means that } k = -3. \text{ *(answer)*}$$

23. **Let u = <–1, 2, 2> and v = <3, –2, 1>.**

   (a) **Find u × v =**
   $$= \begin{vmatrix} \mathbf{i} & \mathbf{j} & \mathbf{k} \\ -1 & 2 & 2 \\ 3 & -2 & 1 \end{vmatrix} = \begin{vmatrix} 2 & 2 \\ -2 & 1 \end{vmatrix} \mathbf{i} - \begin{vmatrix} -1 & 2 \\ 3 & 1 \end{vmatrix} \mathbf{j} + \begin{vmatrix} -1 & 2 \\ 3 & -2 \end{vmatrix} \mathbf{k}$$

   $$= \left[(2)(1) - (2)(-2) \right] \mathbf{i} - \left[(-1)(1) - (2)(3) \right] \mathbf{j} + \left[(-1)(-2) - (3)(2) \right] \mathbf{k}$$

   $$= 6\mathbf{i} + 7\mathbf{j} - 4\mathbf{k} \quad OR: \ \langle 6, 7, -4 \rangle$$

   (b) **Show that the vector u × v is orthogonal to the vector v.** *(Hint:* **use the dot product here!)**

   We need to show that the dot product between the vector we obtained in part (a), that is: **u × v** and the vector **v** is equal to zero in order to show orthogonality:

   $$(\mathbf{u} \times \mathbf{v}) \cdot \mathbf{v} = \langle 6, 7, -4 \rangle \cdot \langle 3, -2, 1 \rangle = (6)(3) + (7)(-2) + (-4)(1) = 18 + -14 + -4 = 0$$

CALCULUS III   **Quiz 2 KEY**, CONTINUED

24. *Multiple choice:* Let u = <3, –1, –2> and v = <–2, –3, 2>. Which of the following vectors is orthogonal to both u and v? (Please circle your choice.)

(a) k   (b) w = <–8, 2, –11>

(c) v × u *By definition!*   (d) w = <1, 2, 0>

(e) none of these

25. *Let u = <3, –1, –2> and v = <–2, –3, 2>, and w = <1, 2, 0>. Calculate u · (v × w).*

By the Theorem:
$$\mathbf{u}\cdot(\mathbf{v}\times\mathbf{w}) = \begin{vmatrix} 3 & -1 & -2 \\ -2 & -3 & 2 \\ 1 & 2 & 0 \end{vmatrix} = \begin{vmatrix} -3 & 2 \\ 2 & 0 \end{vmatrix}3 - \begin{vmatrix} -2 & 2 \\ 1 & 0 \end{vmatrix}(-1) + \begin{vmatrix} -2 & -3 \\ 1 & 2 \end{vmatrix}(-2)$$

$$= (-4)(3) - (-2)(-1) + (-1)(-2) = -12 + -2 + 2 = -12$$

26. **Find the work done in moving a particle from $P(0, 2, 1)$ to $Q(3, –2, 2)$ if the magnitude and direction of the force is given by v = <4, –2, 3>.**

The definition of work is the dot product of the force vector with the distance vector:

$$W = \mathbf{F}\cdot\overrightarrow{PQ} = \langle 4, -2, 3\rangle\cdot\langle 3-0, -2-2, 2-1\rangle = \langle 4, -2, 3\rangle\cdot\langle 3, -4, 1\rangle = (4)(3) + (-2)(-4) + (3)(1) = 23$$

27. **Pulling a toy wagon exerts a force of 15 pounds on the handle that makes a 30°-angle with the horizontal (see figure). Find the work done in pulling the wagon 50 feet.**

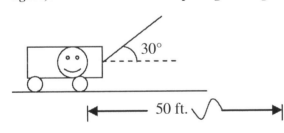

Another definition of work is $W = \cos\theta\|\mathbf{F}\|\|\overrightarrow{PQ}\|$, where the angle θ is the angle between the force vector and the vector along the line of motion.

$$W = \cos(30°)(15)(50) = \frac{\sqrt{3}}{2}(750) = 375\sqrt{3}$$
foot-pounds ≈ 649.5ft-lbs.

28. **A 36,000-pound truck is parked on a 12°-slope (see figure). Assume the only force to overcome is that due to gravity. Find the force required to keep the truck from rolling down the hill.**

Project the force vector (pointing downward with a magnitude of 36,000) onto the ramp.

A *unit* vector on the ramp is given by $\mathbf{v} = \cos(12°)\mathbf{i} + \sin(12°)\mathbf{j}$ » <0.978, 0.208>.

Now we project **F** onto **v**:

$$\text{proj}_{\mathbf{v}}\mathbf{F} = \left(\frac{\mathbf{F}\cdot\mathbf{v}}{\|\mathbf{v}\|^2}\right)\mathbf{v} = \left(\frac{\langle 0, -36000\rangle\cdot\langle 0.978, 0.208\rangle}{1^2}\right)\mathbf{v} \approx (-7484)\langle 0.978, 0.208\rangle$$

Since **v** is already a unit vector, multiplying it by a scalar of –7484 makes it a vector of magnitude 7484. Therefore, the magnitude of this force is 7484 lbs. and this is the force necessary to prevent the truck from rolling down the hill!

CALCULUS III   ✠ QUIZ 3 KEY ✠

1.   **Find a set of parametric equations for the line that passes through the points (2, 0, 3) and (4, 3, 3).**

We first need to form a direction vector for the line, which we will do by forming a vector between the 2 given points (use terminal point minus initial point) $\mathbf{v} = \overrightarrow{PQ} = \langle 4-2, 3-0, 3-3 \rangle = \langle 2, 3, 0 \rangle$.

Therefore, the direction numbers are $a = 2$, $b = 3$, and $c = 0$. Next, we use the formulas for the parametric equations of a line that are:

$$\begin{cases} x = x_1 + at \\ y = y_1 + bt \\ z = z_1 + ct \end{cases} \Rightarrow \begin{cases} x = 2 + 2t \\ y = 0 + 3t \\ z = 3 + 0t \end{cases} \Rightarrow \begin{cases} x = 2 + 2t \\ y = 3t \\ z = 3 \end{cases}$$

2.   **Calculate the distance from the point (1, –2, 3) to the plane given by** $(x-3) + 2(y+1) - 4z = 0$.

We know that a vector normal to the plane is $\mathbf{n} = \langle 1, 2, -4 \rangle$ (We use the coefficients for the variables to get the components for the normal vector.) We now need an arbitrary point P that "lives" on the plane.

Using the fact the standard equation of a plane has the form: $a(x - x_1) + b(y - y_1) + c(z - z_1) = 0$, where the point (x_1, y_1, z_1) is a point in the plane. So, we have the point $P(3, -1, 0)$.

Let the point Q be the point given that is (1, –2, 3) and now form the vector *PQ*:

$$\overrightarrow{PQ} = \langle 1-3, -2-(-1), 3-0 \rangle = \langle -2, -1, 3 \rangle$$

Use the formula for the distance between a point and a plane:

$$D = \frac{\left| \overrightarrow{PQ} \cdot \mathbf{n} \right|}{\|\mathbf{n}\|} = \frac{\left| \langle -2, -1, 3 \rangle \cdot \langle 1, 2, -4 \rangle \right|}{\sqrt{1^2 + 2^2 + 4^2}} = \frac{\left| (-2)(1) + (-1)(2) + (3)(-4) \right|}{\sqrt{21}} = \frac{|-16|}{\sqrt{21}} = \frac{16}{\sqrt{21}} \approx 3.5$$

*(**Note:** Do not forget the absolute value in the numerator of the formula!)*

3.   **Find an equation of the plane determined by the points (1, 2, –3) , (2, 3, 1) and (0, –2, –1). (*Hint:* First, find the two vectors that lie in the plane with a shared initial point using the three points given.)**

Name the points *P, Q,* and *R* respectively. We form two vectors that share the point P as an initial point. Here's a drawing of what I am saying:

We form the two vectors:

$\mathbf{u} = \overrightarrow{PQ} = \langle 2-1, 3-2, 1-(-3) \rangle = \langle 1, 1, 4 \rangle$

$\mathbf{v} = \overrightarrow{PR} = \langle 0-1, -2-2, -1-(-3) \rangle = \langle -1, -4, 2 \rangle$

Next, we take the cross product of these two vectors, $\mathbf{u} \times \mathbf{v}$, to get the normal vector to the plane:

$$= \begin{vmatrix} \mathbf{i} & \mathbf{j} & \mathbf{k} \\ 1 & 1 & 4 \\ -1 & -4 & 2 \end{vmatrix} = \begin{vmatrix} 1 & 4 \\ -4 & 2 \end{vmatrix} \mathbf{i} - \begin{vmatrix} 1 & 4 \\ -1 & 2 \end{vmatrix} \mathbf{j} + \begin{vmatrix} 1 & 4 \\ -1 & 2 \end{vmatrix} \mathbf{k}$$

$$= [(2)(1) - (4)(-4)]\mathbf{i} - [(1)(2) - (-1)(4)]\mathbf{j} + [(1)(2) - (-1)(4)]\mathbf{k}$$

$$= 18\mathbf{i} - 6\mathbf{j} - 3\mathbf{k} \quad OR: \quad \langle 18, -6, -3 \rangle$$

CALCULUS III   **Quiz 3 KEY,** CONTINUED

Finally, we use the equation for a plane $a(x - x_1) + b(y - y_1) + c(z - z_1) = 0 \dots$

$18(x - 1) + -6(y - 2) + -3(z - (-3)) = 0 \Rightarrow 18x - 18 - 6y + 12 - 3z - 9 = 0 \Rightarrow 3(6x - 2y - z) = 15$

$6x - 2y - z = 5$ (*answer*)

4. *Multiple choice:* **If the equation of a cylinder is given by** $x^2 - z = 0$, **then... (Please circle your choice.)**

Since the variable y is missing from the equation, the rulings will be parallel to this missing variable's axis!

(a) **The rulings are parallel to the z-axis**   (b) **The graph is a parabola**

(c) **The rulings are parallel to the x-axis**   (d) **The generating curve is a parabola**

(e) **Both (a) and (d)**   (f) **None of these**

CALCULUS III   ❈ **PRACTICE MIDTERM 1 KEY** ❈

---

1. **Let $\mathbf{v} = \overrightarrow{PQ}$ and $\mathbf{w} = \overrightarrow{PR}$ given the points $P(5, 0, 0)$, $Q(4, 4, 0)$, and $R(2, 0, 6)$. Find the following:**

(a) **Component forms of v and w**

$$\mathbf{v} = \overrightarrow{PQ} = \langle 4 - 5, 4 - 0, 0 - 0 \rangle = \langle -1, 4, 0 \rangle$$

$$\mathbf{w} = \overrightarrow{PR} = \langle 2 - 5, 0 - 0, 6 - 0 \rangle = \langle -3, 0, 6 \rangle$$

(b) $\mathbf{v} \cdot \mathbf{w} = \langle -1, 4, 0 \rangle \cdot \langle -3, 0, 6 \rangle = (-1)(-3) + (4)(0) + (0)(6) = 3$

(c) $\mathbf{v} \times \mathbf{w}$

This is the cross-product of v and w and is found using the 3×3 determinant of a matrix method:

$$= \begin{vmatrix} \mathbf{i} & \mathbf{j} & \mathbf{k} \\ -1 & 4 & 0 \\ -3 & 0 & 6 \end{vmatrix} = \begin{vmatrix} 4 & 0 \\ 0 & 6 \end{vmatrix}\mathbf{i} - \begin{vmatrix} -1 & 0 \\ -3 & 6 \end{vmatrix}\mathbf{j} + \begin{vmatrix} -1 & 4 \\ -3 & 0 \end{vmatrix}\mathbf{k}$$

$$= [(4)(6) - (0)(0)]\mathbf{i} - [(-1)(6) - (-3)(0)]\mathbf{j} + [(-1)(0) - (-3)(4)]\mathbf{k}$$

$$= 24\mathbf{i} - (-6)\mathbf{j} + 12\mathbf{k} \quad OR: \quad \langle 24, 6, 12 \rangle$$

(d) **A unit vector going in the same direction as w**

$$\mathbf{u} = \frac{\mathbf{w}}{\|\mathbf{w}\|} = \frac{\langle -3, 0, 6 \rangle}{\sqrt{3^2 + 0^2 + 6^2}} = \left\langle \frac{-3}{\sqrt{45}}, 0, \frac{6}{\sqrt{45}} \right\rangle = \left\langle \frac{-3}{3\sqrt{5}}, 0, \frac{6}{3\sqrt{5}} \right\rangle = \left\langle \frac{-1}{\sqrt{5}}, 0, \frac{2}{\sqrt{5}} \right\rangle$$

(e) **The vector component of v in the direction of w**

This means finding the projection of **v** onto **w**:

$$\text{proj}_\mathbf{w}\,\mathbf{v} = \left(\frac{\mathbf{v} \cdot \mathbf{w}}{\|\mathbf{w}\|^2} \right)\mathbf{w} = \left(\frac{(-1)(-3) + (4)(0) + (0)(6)}{(-3)^2 + (0)^2 + 6^2} \right)\mathbf{w} = \left(\frac{3}{45} \right)\langle -3, 0, 6 \rangle = \left(\frac{1}{15} \right)\langle -3, 0, 6 \rangle = \left\langle -\frac{1}{5}, 0, \frac{2}{5} \right\rangle$$

CALCULUS III   **Practice Midterm 1 KEY,** CONTINUED

(f)   **The vector component of v orthogonal to w**

The component of $\mathbf{v}$ that is orthogonal to $\mathbf{w}$ is often called $\mathbf{w}_2$, and is found by the formula $\mathbf{w}_2 = \mathbf{v} - \mathbf{w}_1 = \mathbf{v} - \mathrm{proj}_{\mathbf{w}}\,\mathbf{v}$.

So we now have:

$$\mathbf{w}_2 = \mathbf{v} - \mathbf{w}_1 = \mathbf{v} - \mathrm{proj}_{\mathbf{w}}\,\mathbf{v} = \left\langle -1, 4, 0 \right\rangle - \left\langle -\frac{1}{5}, 0, \frac{2}{5} \right\rangle = \left\langle -1 - \left(-\frac{1}{5}\right), 4 - 0, 0 - \frac{2}{5} \right\rangle = \left\langle -\frac{4}{5}, 4, -\frac{2}{5} \right\rangle$$

(g)   **An equation of the plane containing P, Q, and R**

First, we need a vector normal to the plane containing all three of these points, but we found this vector back in part **(c)** of this problem when we took the cross product of $\mathbf{v}$ and $\mathbf{w}$:

$$\mathbf{n} = \left\langle 24, 6, 12 \right\rangle$$

We will use the point $P(5, 0, 0)$ as our point on the plane, and use the standard equation of a plane that is:

$$a(x - x_1) + b(y - y_1) + c(z - z_1) = 0$$

$$24(x - 5) + 6(y - 0) + 12(z - 0) = 0 \quad \Rightarrow \quad 24x - 120 + 6y + 12z = 0 \quad \Rightarrow \quad 24x + 6y + 12z = 120$$

(h)   **The distance of a point M(–1, 5, 9) to the plane containing P, Q, and R**

Using the distance between a point and a plane formula:

$$D = \frac{\left| \overrightarrow{PM} \cdot \mathbf{n} \right|}{\|\mathbf{n}\|} = \frac{\left| \left\langle -1 - 5,\, 5 - 0,\, 9 - 0 \right\rangle \cdot \left\langle 24,\, 6,\, 12 \right\rangle \right|}{\sqrt{24^2 + 6^2 + 12^2}} = \frac{\left| \left\langle -6,\, 5,\, 9 \right\rangle \cdot \left\langle 24,\, 6,\, 12 \right\rangle \right|}{\sqrt{756}}$$

$$= \frac{\left| (-6)(24) + (5)(6) + (9)(12) \right|}{6\sqrt{21}} = \frac{\left| -6 \right|}{6\sqrt{21}} = \frac{6}{6\sqrt{21}} = \frac{1}{\sqrt{21}} \approx 0.218$$

(i)   **A set of parametric equations of the line through the points P and Q**

We use the vector $\mathbf{v} = \overrightarrow{PQ} = \left\langle 4 - 5,\, 4 - 0,\, 0 - 0 \right\rangle = \left\langle -1, 4, 0 \right\rangle$ for the direction vector for the line. Use the point $P(5, 0, 0)$ for the point on the line. Now we use the parametric equations for a line:

$$\begin{cases} x = x_1 + at \\ y = y_1 + bt \\ z = z_1 + ct \end{cases} \Rightarrow \begin{cases} x = 5 - t \\ y = 0 + 4t \\ z = 0 + 0t \end{cases} \Rightarrow \begin{cases} x = 5 - t \\ y = 4t \\ z = 0 \end{cases}$$

(j)   **The distance of the point R to the line through the points P and Q**

The formula for the distance between a point and a line is $D = \dfrac{\left\| \overrightarrow{PR} \times \mathbf{v} \right\|}{\|\mathbf{v}\|}$.

However, we already found the cross-product $\overrightarrow{PR} \times \mathbf{v}$ from part **(c)** above, that was $\mathbf{n} = \left\langle 24, 6, 12 \right\rangle$.

So, we only need the magnitude of this for the numerator in the formula, where the denominator is the magnitude of the vector: $\mathbf{v} = \overrightarrow{PQ} = \left\langle 4 - 5,\, 4 - 0,\, 0 - 0 \right\rangle = \left\langle -1, 4, 0 \right\rangle$.

CALCULUS III   **Practice Midterm 1 KEY, CONTINUED**

$$D = \frac{\left\|\overrightarrow{PR} \times \mathbf{v}\right\|}{\|\mathbf{v}\|} = \frac{\sqrt{24^2 + 6^2 + 12^2}}{\sqrt{1^2 + 4^2 + 0^2}} = \frac{\sqrt{756}}{\sqrt{17}} \approx 6.67$$

(k)   **The angle θ between the vectors v and w**

$$\cos\theta = \frac{\mathbf{v} \cdot \mathbf{w}}{\|\mathbf{v}\|\|\mathbf{w}\|} = \frac{3}{\sqrt{17}\sqrt{45}} = \frac{3}{\sqrt{17}(3)\sqrt{5}} = \frac{1}{\sqrt{85}} \quad \Rightarrow \quad \theta = \arccos\left(\frac{1}{\sqrt{85}}\right) \approx 83.8°$$

2.   **Find the component form of u if...**

(a)   **The angle, measured counterclockwise, from the positive x-axis is 135° and $\|\mathbf{u}\|$ is 4. (Assume that u is in the plane.)**

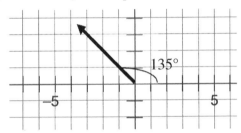

x-component = $\|\mathbf{u}\|\cos\theta = (4)\cos 135° = (4)\left(-\frac{\sqrt{2}}{2}\right) = -2\sqrt{2}$

y-component = $\|\mathbf{u}\|\sin\theta = (4)\sin 135° = (4)\left(\frac{\sqrt{2}}{2}\right) = 2\sqrt{2}$

answer: $\mathbf{u} = \left\langle -2\sqrt{2}, 2\sqrt{2}\right\rangle$. The *x*-coordinate is negative since the vector is in Quadrant II.

(b)   **u is a unit vector perpendicular to the lines:**

$$x = 4, \quad y = 3 + 2t, \quad z = 1 + 5t$$
$$x = -3 + 7s, \quad y = -2 + s, \quad z = 1 + 2s$$

The direction vectors for these lines are $\mathbf{v}_1 = \langle 0, 2, 5\rangle$ *and* $\mathbf{v}_2 = \langle 7, 1, 2\rangle$, respectively. A vector perpendicular to *both* of these vectors will be the cross-product of them:

$$\mathbf{v}_1 \times \mathbf{v}_2 = \begin{vmatrix} \mathbf{i} & \mathbf{j} & \mathbf{k} \\ 0 & 2 & 5 \\ 7 & 1 & 2 \end{vmatrix} = \begin{vmatrix} 2 & 5 \\ 1 & 2 \end{vmatrix}\mathbf{i} - \begin{vmatrix} 0 & 5 \\ 7 & 2 \end{vmatrix}\mathbf{j} + \begin{vmatrix} 0 & 2 \\ 7 & 1 \end{vmatrix}\mathbf{k}$$

$$= [(2)(2) - (5)(1)]\mathbf{i} - [(0)(2) - (5)(7)]\mathbf{j} + [(0)(1) - (2)(7)]\mathbf{k}$$

$$= -1\mathbf{i} - (-35)\mathbf{j} - 14\mathbf{k} \quad OR: \quad \langle -1, 35, -14\rangle$$

Now we just need to make this a *unit* vector:

$$u = \frac{\langle -1, 35, -14\rangle}{\sqrt{1^2 + 35^2 + 14^2}} = \left\langle -\frac{1}{3\sqrt{158}}, \frac{35}{3\sqrt{158}}, -\frac{14}{3\sqrt{158}}\right\rangle$$

CALCULUS **III**   **Practice Midterm 1 KEY, CONTINUED**

3. **Determine whether the vectors are orthogonal, parallel, or neither.**

(a) **<7, –2, 3>, <–1, 4, 5>**

These are orthogonal since the dot product equals zero:

$$\langle 7, -2, 3 \rangle \cdot \langle -1, 4, 5 \rangle = (7)(-1) + (-2)(4) + (3)(5) = -7 - 8 + 15 = 0$$

(b) **<4, –1, 5> , <3, 2, 2>**

These are neither since the dot product is *not* equal to zero and one is not a multiple of the other.

(c) **<–4, 3, –6> , <16, –12, 24>**

These are parallel since one is a multiple of the other:

$$-4 = c(16) \Rightarrow c = -\frac{1}{4}$$

$$3 = c(-12) \Rightarrow c = -\frac{1}{4} \Rightarrow \text{same values for } c \text{ means that the vectors are parallel.}$$

$$-6 = c(24) \Rightarrow c = -\frac{1}{4}$$

4. **Find the tension in each of the supporting cables in the figure.**

We first draw a force diagram:

*(**Note:** To translate the degrees, we used an axiom from geometry that gives us info about angles when a parallel line is cut by a transversal line.)*

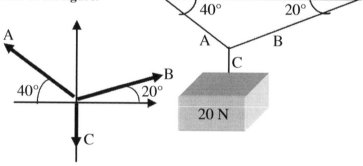

For equilibrium, the forces $\overrightarrow{CA}, \overrightarrow{CB}$ *and* $\overrightarrow{CD} = \langle 0, -20 \rangle$ must sum to zero. The magnitudes of the vectors $\overrightarrow{CA}, \overrightarrow{CB}$ will be the tension (forces) in the cables. We write these forces, where:

$$\overrightarrow{CA} = \left\langle -\left\|\overrightarrow{CA}\right\| \cos 40^\circ, \left\|\overrightarrow{CA}\right\| \sin 40^\circ \right\rangle \ and \ \overrightarrow{CB} = \left\langle \left\|\overrightarrow{CB}\right\| \cos 20^\circ, \left\|\overrightarrow{CB}\right\| \sin 20^\circ \right\rangle$$

$$\overrightarrow{CA} + \overrightarrow{CB} + \overrightarrow{CD} = \vec{0} \Rightarrow gives \ a \ linear \ system:$$

$$\begin{cases} 0 - \left\|\overrightarrow{CA}\right\| \cos 40^\circ + \left\|\overrightarrow{CB}\right\| \cos 20^\circ = 0 \\ -20 + \left\|\overrightarrow{CA}\right\| \sin 40^\circ + \left\|\overrightarrow{CB}\right\| \sin 20^\circ = 0 \end{cases}$$

Solve for $\left\|\overrightarrow{CA}\right\|$ from equation #1: $\left\|\overrightarrow{CA}\right\| = \dfrac{\cos 20^\circ}{\cos 40^\circ} \left\|\overrightarrow{CB}\right\|$ and substitute this into equation #2 to solve for $\left\|\overrightarrow{CB}\right\|$ and get that $\approx 21.7\text{N}, \approx 17.7\text{N}.$

CALCULUS III   **Practice Midterm 1 KEY,** CONTINUED

5. **Identify each of the quadric surfaces:**

    (a) $3x^2 + 3y^2 + 3z^2 - 2x + 3y - 11 = 0$.

    After completing the square three
    times, we have a sphere!

    (b) $2x^2 + 5z^2 = y^2$

    An elliptic cone with the y-axis as its axis.

    (c) $2x^2 - y^2 - z^2 = 1$

    A hyperboloid of two sheets (with the x-axis
    as its axis).

$$\left(\frac{1}{3}\right)\left(3x^2 + 3y^2 + 3z^2 - 2x + 3y - 11\right) = (0)\left(\frac{1}{3}\right)$$

$$x^2 - \frac{2}{3}x + M_1 + y^2 + y + M_2 + z^2 = \frac{11}{3} + M_1 + M_2$$

$$x^2 - \frac{2}{3}x + \frac{1}{9} + y^2 + y + \frac{1}{4} + z^2 = \frac{11}{3} + \frac{1}{9} + \frac{1}{4}$$

$$\left(x - \frac{1}{3}\right)^2 + \left(y + \frac{1}{2}\right)^2 + z^2 = \frac{145}{36}$$

6. **Complete the chart and convert the coordinates from one system to another among the rectangular, cylindrical, and spherical coordinate systems.**

RECTANGULAR	*CYLINDRICAL*	*SPHERICAL*
$(6, -2, -3)$	$\left(2\sqrt{10}, -18.4°, -3\right)$	$(7, -18.4°, 115°)$
$\left(-\sqrt{3}, 1, 3\right)$	$\left(-2, \frac{11\pi}{6}, 3\right)$	$\left(\sqrt{13}, \frac{11\pi}{6}, 33.7°\right)$
$\left(-\frac{9\sqrt{2}}{2}, \frac{9\sqrt{2}}{2}, 0\right)$	$\left(9, \frac{3\pi}{4}, 0\right)$	$\left(9, \frac{3\pi}{4}, \frac{\pi}{2}\right)$

ROW #1: $r = \sqrt{x^2 + y^2} = \sqrt{6^2 + 2^2} = \sqrt{40} = 2\sqrt{10}$

$\theta = \arctan\frac{y}{x} = \arctan\frac{-2}{6} \approx -18.4°, \quad z = z = -3$

$\rho = \sqrt{x^2 + y^2 + z^2} = \sqrt{6^2 + 2^2 + 3^2} = \sqrt{49} = 7$

$\theta = \arctan\frac{y}{x} = \arctan\frac{-2}{6} \approx -18.4°,$

$\phi = \arccos\left(\frac{z}{\rho}\right) = \arccos\left(\frac{-3}{7}\right) \approx 115°$

ROW #2: $x = r\cos\theta = -2\cos(11\pi/6) = (-2)\left(\sqrt{3}/2\right) = -\sqrt{3}$

$y = r\sin\theta = -2\sin(11\pi/6) = (-2)(-1/2) = 1, \quad z = z = 3$

ROW #3: $r = \rho\sin\phi = 9\sin(\pi/2) = 9$

$\theta = \theta = 3\pi/4, \quad z = \rho\cos\phi = 9\cos(\pi/2) = 0$

$x = \rho\sin\phi\cos\theta = 9\sin(\pi/2)\cos(3\pi/4) = (9)(1)\left(-\sqrt{2}/2\right) = -9\sqrt{2}/2$

$y = \rho\sin\phi\sin\theta = 9\sin(\pi/2)\sin(3\pi/4) = (9)(1)\left(\sqrt{2}/2\right) = 9\sqrt{2}/2$

$z = \rho\cos\phi = 9\cos(\pi/2) = 0$

CALCULUS III   **Practice Midterm 1 KEY, CONTINUED**

---

7. **Sketch the following planes:**

   (a) $2x - y + z = 4$

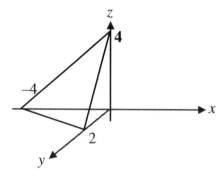

   (b) $x + 2y = 4$

missing the z-coordinate, so the plane is parallel to the z-axis

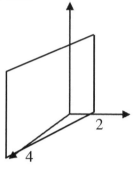

   (c) $z = 2$

(parallel to the xy-plane)

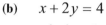

8. **Find an equation in rectangular coordinates from cylindrical coordinates.**

   (a) $r = \dfrac{1}{2}z$

$\sqrt{x^2 + y^2} = \dfrac{1}{2}z \ \Rightarrow \ x^2 + y^2 = \dfrac{z^2}{4}$

   (b) $r = 2\cos\theta$

First, multiply both sides of the equation by r:

$r^2 = 2r\cos\theta \ \Rightarrow \ x^2 + y^2 = 2x$

   (c) $\theta = \dfrac{\pi}{3}$

First, use the conversion formula that says $\tan\theta = \dfrac{y}{x}$.

Then, since $\theta = \dfrac{\pi}{3}$, plug that into the conversion formula to get that:

$$\tan\frac{\pi}{3} = \frac{y}{x} \ \Rightarrow \ \sqrt{3} = \frac{y}{x} \ \Rightarrow \ y = \sqrt{3}x$$

(linear equation, so the graph is a plane—parallel to the z-axis)

9. **Find an equation in rectangular coordinates from spherical coordinates.**

   (a) $\rho = 3$

First, square both sides of the equation:

$$\rho^2 = 3^2 \ \Rightarrow \ x^2 + y^2 + z^2 = 9$$

(the graph is a sphere of radius 3)

   (b) $\phi = \dfrac{\pi}{6}$

First, use the conversion formula that says $\cos\phi = \dfrac{z}{\rho}$.

CALCULUS III   Practice Midterm 1 KEY, CONTINUED

Then, since $\phi = \dfrac{\pi}{6}$, plug that into the conversion formula to get that:

$$\cos\left(\frac{\pi}{6}\right) = \frac{z}{\rho} \quad \Rightarrow \quad \cos\left(\frac{\pi}{6}\right) = \frac{z}{\sqrt{x^2 + y^2 + z^2}}$$

$$\frac{\sqrt{3}}{2} = \frac{z}{\sqrt{x^2 + y^2 + z^2}} \quad \Rightarrow \quad \left(\frac{\sqrt{3}}{2}\right)^2 = \left(\frac{z}{\sqrt{x^2 + y^2 + z^2}}\right)^2 \quad \Rightarrow \quad \frac{3}{4} = \frac{z^2}{x^2 + y^2 + z^2}$$

Now, clear fractions and simplify:

$$(4)(x^2 + y^2 + z^2)\frac{3}{4} = \left(\frac{z^2}{x^2 + y^2 + z^2}\right)(4)(x^2 + y^2 + z^2) \quad \Rightarrow \quad 3x^2 + 3y^2 + 3z^2 = 4z^2 \quad \Rightarrow 3x^2 + 3y^2 - z^2 = 0$$

(c) $\rho = 4\cos\phi$

Using the facts that $\rho = \sqrt{x^2 + y^2 + z^2}$ *and* $\cos\phi = \dfrac{z}{\sqrt{x^2 + y^2 + z^2}}$, we get that the equation

now looks like $\sqrt{x^2 + y^2 + z^2} = 4\left(\dfrac{z}{\sqrt{x^2 + y^2 + z^2}}\right)$.

Multiplying both sides of the equation by $\sqrt{x^2 + y^2 + z^2}$ to clear fractions, we get:

$$x^2 + y^2 + z^2 = 4z$$

10. **Convert the rectangular equation** $x^2 + y^2 = z$ **to...**

(a) **cylindrical coordinates** $r^2 = z$ *(answer)*

(b) **spherical coordinates** $\left(\rho\sin\phi\cos\theta\right)^2 + \left(\rho\sin\phi\sin\theta\right)^2 = \rho\cos\phi$

$$\rho^2 \sin^2\phi\cos^2\theta + \rho^2\sin^2\phi\sin^2\theta = \rho\cos\phi$$

Factor out a common factor from the first two terms on left side:

$$\rho^2\sin^2\phi\left(\cos^2\theta + \sin^2\theta\right) = \rho\cos\phi \quad \Rightarrow \quad \rho^2\sin^2\phi = \rho\cos\phi \text{ (divide both sides by r now)}$$

$$\rho\sin^2\phi = \cos\phi \quad \Rightarrow \quad \rho = \frac{\cos\phi}{\sin^2\phi} = \frac{1}{\sin\phi}\cdot\frac{\cos\phi}{\sin\phi} \quad \Rightarrow \quad \rho = \csc\phi\cot\phi \text{ (answer)}$$

11. **Convert the rectangular equation** $x^2 + y^2 = 2x$ **to...**

(a) **cylindrical coordinates** $r^2 = 2r\cos\theta$

Divide both sides by r now: $r = 2\cos\theta$ *(answer)*

(b) **spherical coordinates** $\left(\rho\sin\phi\cos\theta\right)^2 + \left(\rho\sin\phi\sin\theta\right)^2 = 2\rho\sin\phi\cos\theta$

$$\rho^2\sin^2\phi\left(\cos^2\theta + \sin^2\theta\right) = \rho\cos\phi \quad \Rightarrow \quad \rho^2\sin^2\phi = 2\rho\sin\phi\cos\theta \quad \Rightarrow \quad \rho^2\sin^2\phi - 2\rho\sin\phi\cos\theta = 0$$

$\rho\sin\phi(\rho\sin\phi - 2\cos\theta) = 0 \Rightarrow$ *Apply ZPP* $\Rightarrow \rho\sin\phi = 0$ *(assume can't happen!) or* $\rho\sin\phi - 2\cos\theta = 0 \Rightarrow$

solve for ρ: $\rho = \dfrac{2\cos\theta}{\sin\phi} \quad \Rightarrow \quad \rho = 2\cos\theta\csc\phi$ *(answer)*

CALCULUS III   ❈ QUIZ 4 KEY ❈

1. **Given the function** $f(x) = \sqrt{2-x}$ ...

   (a) **Find a unit vector parallel to the graph at the point (–2, 2).**

   We'll find the equation for the tangent line to the graph at the point *first*. Slope of tangent line is the derivative evaluated at the point:

   $$f'(x) = \frac{1}{2}(2-x)^{-1/2}(-1) = \frac{-1}{2\sqrt{2-x}} \Rightarrow f'(-2) = \frac{-1}{2\sqrt{2-(-2)}} = \frac{-1}{4}$$

   Now, use the point-slope form for the equation of a line:

   $$y - y_1 = m(x - x_1) \Rightarrow y - 2 = -\frac{1}{4}(x-(-2)) \Rightarrow y - 2 = -\frac{1}{4}x - \frac{1}{2}$$

   To get a second point on the line, pick an arbitrary value for *x* (say, *x* = 6) and plug into this equation for a line:

   $$y - 2 = -\frac{1}{4}(6) - \frac{1}{2} \Rightarrow y = -2 + 2 = 0.$$ So, the point (6, 0) is on the line.

   Now, form the vector (use terminal minus initial points) $\mathbf{v} = \langle 6-(-2), 0-2 \rangle = \langle 8, -2 \rangle$.

   Next, normalize this vector (i.e., make it a unit vector):

   $$\mathbf{u} = \frac{\mathbf{v}}{\|\mathbf{v}\|} = \frac{\langle 8, -2 \rangle}{\sqrt{8^2 + 2^2}} = \frac{\langle 8, -2 \rangle}{\sqrt{68}} = \left\langle \frac{8}{2\sqrt{17}}, \frac{-2}{2\sqrt{17}} \right\rangle = \left\langle \frac{4}{\sqrt{17}}, -\frac{1}{\sqrt{17}} \right\rangle$$

   (answer can also be the opposite of this vector, and still be valid!)

   (b) **Find a unit vector perpendicular to the graph at the point (–2, 2).**

   We will get the equation of a line that is perpendicular to the tangent line first. The slope of the perpendicular line will be the negative reciprocal of the slope of the tangent line. So, the slope of the perpendicular line will be +4.

   $$y - y_1 = m(x - x_1) \Rightarrow y - 2 = 4(x-(-2)) \Rightarrow y - 2 = 4x + 8 \Rightarrow y = 4x + 10$$

   Let *x* = 0, then *y* = 10. So, the point (0, 10) is on this perpendicular line.

   Now, form the vector $\mathbf{v} = \langle 0-(-2), 10-2 \rangle = \langle 2, 8 \rangle$. Normalizing this vector, we have:

   $$\mathbf{u} = \frac{\mathbf{v}}{\|\mathbf{v}\|} = \frac{\langle 2, 8 \rangle}{\sqrt{2^2 + 8^2}} = \frac{\langle 2, 8 \rangle}{\sqrt{68}} = \left\langle \frac{2}{2\sqrt{17}}, \frac{8}{2\sqrt{17}} \right\rangle = \left\langle \frac{1}{\sqrt{17}}, \frac{4}{\sqrt{17}} \right\rangle \text{ (or, its opposite)}$$

2. **A 32,000-pound truck is parked on a 25°-hill. Assume that the only force to overcome is that due to gravity.**

   (a) **Find the force required to keep the truck from rolling down the hill.**

   Project the force of the truck onto the ramp. The magnitude of *this* force will be the force required to keep the truck from rolling downhill.

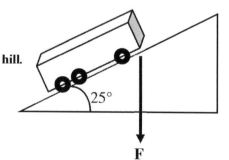

A *unit* vector on the ramp is given by $\mathbf{v} = \cos(25°)\mathbf{i} + \sin(25°)\mathbf{j} \approx <0.91, 0.42>$.

Now we project $\mathbf{F}$ onto $\mathbf{v}$:

$$\text{proj}_\mathbf{v}\mathbf{F} = \left(\frac{\mathbf{F}\cdot\mathbf{v}}{\|\mathbf{v}\|^2}\right)\mathbf{v} = \left(\frac{\langle 0, -32000\rangle\cdot\langle 0.906, 0.423\rangle}{1^2}\right)\mathbf{v} \approx (-13523)\langle 0.906, 0.423\rangle$$

Since $\mathbf{v}$ is already a unit vector, multiplying it by a scalar of –13,523 makes it a vector of magnitude 13523. Therefore, the magnitude of this force is 13,523 lbs. and this is the force necessary to prevent the truck from rolling down the hill!

(b)   Find the force perpendicular to the hill.

This is simply the (magnitude) of the component of the force vector that is orthogonal to the ramp vector:

$$\mathbf{w}_2 = \mathbf{F} - \text{proj}_\mathbf{v}\mathbf{F} = \langle 0, -32000\rangle - (-13523)\langle 0.906, 0.423\rangle \approx \langle -12305, -26321\rangle$$

$$\|w_2\| = \sqrt{12305^2 + 26321^2} \approx 29055 \text{ lbs. } (answer)$$

3.   **Find the distance between the line given by** $\begin{cases} x = 2-t \\ y = -1+3t \\ z = -1+t \end{cases}$ **and the point** $Q(1, 1, 0)$.

The direction vector for the line is $\mathbf{v} = \langle -1, 3, 1\rangle$. A point P on the line is: $P(2, -1, -1)$.

The formula for the distance between a point and a line is: $D = \dfrac{\left\|\overrightarrow{PQ}\times\mathbf{v}\right\|}{\|\mathbf{v}\|}$.

We need to from the vector $\overrightarrow{PQ}$ $\Rightarrow$ $\overrightarrow{PQ} = \langle 1-2, 1-(-1), 0-(-1)\rangle = \langle -1, 2, 1\rangle$

Next, we need to take the cross product: $\overrightarrow{PQ}\times\mathbf{v}$: $\overrightarrow{PQ}\times\mathbf{v} =$

$$= \begin{vmatrix} \mathbf{i} & \mathbf{j} & \mathbf{k} \\ -1 & 2 & 1 \\ -1 & 3 & 1 \end{vmatrix} = \begin{vmatrix} 2 & 1 \\ 3 & 1 \end{vmatrix}\mathbf{i} - \begin{vmatrix} -1 & 1 \\ -1 & 1 \end{vmatrix}\mathbf{j} + \begin{vmatrix} -1 & 2 \\ -1 & 3 \end{vmatrix}\mathbf{k} = [(2)(1)-(3)(1)]\mathbf{i} - [(-1)(1)-(-1)(1)]\mathbf{j} + [(-1)(3)-(-1)(2)]\mathbf{k}$$

$$= -1\mathbf{i} - (0)\mathbf{j} + -1\mathbf{k} \quad OR: \quad \langle -1, 0, -1\rangle$$

Now, we can use the distance formula:

$$D = \frac{\left\|\overrightarrow{PQ}\times\mathbf{v}\right\|}{\|\mathbf{v}\|} = \frac{\sqrt{1^2 + 0^2 + 1^1}}{\sqrt{1^2 + 3^2 + 1^2}} = \frac{\sqrt{2}}{\sqrt{11}} \approx 0.43$$

4.   **Represent the parabola** $x = y^2 + 5$ **by a vector-valued function.**

Let $y = t$, then we have $x = t^2 + 5$ $\Rightarrow$ $\mathbf{r}(t) = \langle t^2 + 5, t\rangle$ (*answer*)

(***Caution!*** Do *not* let $x = t$, because then you will have two different expressions for y. That is, you will have that $y = \pm\sqrt{x-5}$ and that is *not OK*!

CALCULUS III   **Quiz 4 KEY, CONTINUED**

5. **Represent the intersection of the surfaces** $x^2 + y^2 + z^5 = 3$ **and** $x = y$ **as a vector-valued function.**

First, we let $x = t$. Then, since we have that $x = y$, we get that $y = t$ *also!*

Now, solve for z using the first equation:

$$z^5 = 3 - x^2 - y^2 \quad \Rightarrow \quad z = \sqrt[5]{3 - x^2 - y^2} \quad \Rightarrow \quad z = \sqrt[5]{3 - t^2 - t^2} = \sqrt[5]{3 - 2t^2}$$

Finally, we have our vector-valued function: $\mathbf{r}(t) = \left\langle t, t, \sqrt[5]{3 - 2t^2} \right\rangle$ *(answer)*

6. **Find the limit (if it exists):** $\displaystyle\lim_{t \to -2} \left\langle \frac{t^2 - 3t - 10}{t + 2}, \ \sin\left(\frac{\pi t}{2}\right), \ \frac{t + 2}{\sin(t + 2)} \right\rangle.$

I will evaluate each limit separately and then write the final answer as a vector-valued function.

$$\lim_{t \to -2} \frac{t^2 - 3t - 10}{t + 2} = \frac{(-2)^2 - 3(-2) - 10}{-2 + 2} = \frac{0}{0}$$

(indeterminate form; use L'Hopital's Rule)

$$\Rightarrow \lim_{t \to -2} \frac{2t - 3}{1} = 2(-2) - 3 = -7$$

The second component will be $\displaystyle\lim_{t \to -2} \sin\left(\frac{\pi t}{2}\right) = \sin\left(\frac{\pi(-2)}{2}\right) = \sin(-\pi) = 0$.

The third component will be $\displaystyle\lim_{t \to -2} \frac{t + 2}{\sin(t + 2)} = \frac{(-2) + 2}{\sin(-2 + 2)} = \frac{0}{0}$

(indeterminate form; use L'Hopital's Rule)

$$\lim_{t \to -2} \frac{1}{\cos(t + 2)} = \frac{1}{\cos(-2 + 2)} = \frac{1}{\cos 0} = \frac{1}{1} = 1$$

Final answer: $\mathbf{r}(t) = \left\langle -7, 0, 1 \right\rangle$

7. **Find the domain of the vector-valued function** $\mathbf{r}(t) = \left\langle e^{-t/2}, \ \sqrt{t^2 - 4}, \ \dfrac{1}{t + 5} \right\rangle$
    (Hint: **You may need to review how to solve a polynomial inequality from your old Pre-Calculus textbook.)**

The first component has no variables in denominators, logs, or even roots, so there are no restrictions on the domain of $e^{-t/2}$. This means that the domain for $e^{-t/2}$ is $(-\infty, \infty)$.

The second component has an even root, so we need to "worry." We need the radicand to be greater than or equal to zero: $t^2 - 4 \geq 0 \quad \Rightarrow \quad (t + 2)(t - 2) \geq 0$. The critical numbers for this polynomial inequality are ± 2.

Test three arbitrary values in the test intervals created by the critical numbers to see what intervals satisfy the inequality. Try $t = -3$ first: $(-3)^2 - 4 \geq 0$.

This is true, so the interval $(-\infty, -2)$ is part of the solution (a.k.a., *domain*).

$$\sqrt{}\ \textbf{PART III}\quad \textsc{Test Answer Key}$$

CALCULUS III   **Quiz 4 KEY,** CONTINUED

Try $t = 0$ now: $0^2 - 4 \geq 0$. This is a false statement, so we do not include the interval $(-2, 2)$ in the solution.

Now, try $t = 3$: $3^2 - 4 \geq 0$. This is true, so the interval $(2, \infty)$ is part of the solution.

Thus, the domain for the second component is $(-\infty, -2) \cup (2, \infty)$.

The last component will be undefined if $t = -5$. Therefore, the intervals $(-\infty, -5) \cup (-5, \infty)$ give the domain.

Finally, we take the intersection of all three components' domains, and we get the answer $t \in (-\infty, -5) \cup (-5, -2] \cup [2, \infty)$.

8. **Calculate the speed of the object having the position vector $r(t) = \langle \sin t, \cos t, -16t^2 \rangle$ at the point when $t = 1$.**

We use the fact that speed is simply the magnitude of the velocity vector. So, we need to obtain the velocity vector first:

$$\mathbf{r}'(t) = \langle \cos t, -\sin t, -32t \rangle \quad \Rightarrow \quad \|\mathbf{r}'(t)\| = \sqrt{(\cos t)^2 + (-\sin t)^2 + (-32t)^2} = \sqrt{\cos^2 t + \sin^2 t + 1024t^2}$$

$$= \sqrt{1 + 1024t^2}$$

Finally, we evaluate the speed at $t = 1$: $\|\mathbf{r}'(1)\| = \sqrt{1 + 1024(1)^2} = \sqrt{1025} = \sqrt{(25)(41)} = 5\sqrt{41}$

9. **The vector-valued function $r(t) = \left\langle \ln\left(\dfrac{t^5}{1+t^2}\right), \left(\ln\dfrac{t}{2}\right)^5, \sqrt[5]{(1-t)^2} \right\rangle$ describes the position of an object moving in space. Find its acceleration at $t = 2$.**

Acceleration is the second derivative of position. I will first simplify the first component of the position function using log properties. This is in order to make it easier to find its derivative!

$$\ln\left(\frac{t^5}{1+t^2}\right) = \ln t^5 - \ln(1+t^2) = 5\ln t - \ln(1+t^2)$$

Please note that the second component cannot be simplified using log properties! I will rewrite the third component using rational exponents in order to make it easier to differentiate:

$$\sqrt[5]{(1-t)^2} = (1-t)^{2/5}$$

We will take the derivative of position now to find the velocity vector. I will use the Log Rule for the first component, and the Chain Rule for both the second and third components:

$$\mathbf{r}'(t) = \left\langle \frac{5}{t} - \frac{2t}{1+t^2}, 5\left(\ln\frac{t}{2}\right)^4 \frac{1}{t/2}\left(\frac{1}{2}\right), \frac{2}{5}(1-t)^{-3/5}(-1) \right\rangle = \left\langle 5t^{-1} - \frac{2t}{1+t^2}, \frac{5}{t}\left(\ln\frac{t}{2}\right)^4, -\frac{2}{5}(1-t)^{-3/5} \right\rangle$$

To find the second derivative, I will need the Quotient Rule on the second term in the first component. I will need *both* the Product Rule and Chain Rule in order to differentiate the second component. I will use the Chain Rule to differentiate the last component:

$$\mathbf{r}''(t) = \left\langle -5t^{-2} - \frac{2(1+t^2) - 2t(2t)}{(1+t^2)^2}, -5t^{-2}\left(\ln\frac{t}{2}\right)^4 + \frac{5}{t}(4)\left(\ln\frac{t}{2}\right)^3\left(\frac{1}{t/2}\right)\left(\frac{1}{2}\right), \left(-\frac{2}{5}\right)\left(-\frac{3}{5}\right)(1-t)^{-8/5}(-1) \right\rangle$$

Evaluate this at $t = 2$ now:

$$\mathbf{r}''(2) = \left\langle \frac{-5}{2^2} - \frac{2(1+2^2) - 2(2)(2)(2)}{(1+2^2)^2}, -\frac{5}{2^2}\left(\ln\frac{2}{2}\right)^4 + \frac{5}{2}(4)\left(\ln\frac{2}{2}\right)^3\left(\frac{1}{2/2}\right)\left(\frac{1}{2}\right), \left(-\frac{2}{5}\right)\left(-\frac{3}{5}\right)(1-2)^{-8/5}(-1) \right\rangle$$

$$= \left\langle -\frac{101}{100}, 0, -\frac{6}{25} \right\rangle$$

CALCULUS **III**   **Quiz 4 KEY,** CONTINUED

10. **Determine the intervals on which the vector-valued function is smooth.**

The curve is not smooth when $\mathbf{r}'(t)$ is equal to the zero vector. So, first find $\mathbf{r}'(t)$:

$$\mathbf{r}'(t) = \left\langle 6t^2 - 7t - 5, \, t^2 - \frac{1}{4} \right\rangle$$

This will be equal to the zero vector when both components are simultaneously equal to zero:

$6t^2 - 7t - 5 = 0$ $\Rightarrow$ factor and apply the Zero Product Property to solve:

$$6t^2 - 10t + 3t - 5 = 0 \qquad\qquad t^2 - \frac{1}{4} = 0$$
$$2t(3t - 5) + 1(3t - 5) = 0 \qquad\qquad \left(t - \frac{1}{2}\right)\left(t + \frac{1}{2}\right) = 0$$
$$(2t + 1)(3t - 5) = 0$$
$$2t + 1 = 0 \quad and \quad 3t - 5 = 0 \qquad\qquad t - \frac{1}{2} = 0 \quad and \quad t + \frac{1}{2} = 0$$
$$t = -\frac{1}{2} \quad and \quad t = \frac{5}{3} \qquad\qquad t = \frac{1}{2} \quad and \quad t = -\frac{1}{2}$$

The only value that makes both components equal to zero at the same time is $t = -\frac{1}{2}$.

Therefore, the curve is smooth when $t \ne -\frac{1}{2}$.

Using interval notation, we say the curve is smooth on the intervals $t \in \left(-\infty, -\frac{1}{2}\right) \cup \left(-\frac{1}{2}, \infty\right)$.

11. **An object starts from rest at the point (0, 1, 1) and moves with an acceleration of $\mathbf{a}(t) = \langle 1, 1, 0 \rangle$. Find its position at $t = 4$. (*Hint:* You will need to anti-differentiate twice.)**

Starting from rest is a cute way to give the initial condition for velocity. That is, $\mathbf{v}(0) = \mathbf{r}'(0) = \langle 0, 0, 0 \rangle$.

We will start by anti-differentiating acceleration to get velocity:

$$\mathbf{v}(t) = \int \mathbf{a}(t)dt = \int \langle 1,1,0 \rangle dt = \langle t + C_1, \, t + C_2, \, C_3 \rangle$$

To find the constants of integration, we use the initial condition $\mathbf{v}(t) = \mathbf{r}'(t) = \langle 0, 0, 0 \rangle$

$$\mathbf{v}(0) = \mathbf{r}'(0) = \langle 0, 0, 0 \rangle = \langle 0 + C_1, \, 0 + C_2, \, C_3 \rangle \quad \Rightarrow \quad C_1 = 0, C_2 = 0, \, and \, C_3 = 0$$

Therefore the velocity vector is $\mathbf{v}(t) = \langle t, t, 0 \rangle$.

Next, we integrate velocity to get position $\mathbf{r}(t) = \int \mathbf{v}(t)dt = \int \langle t, t, 0 \rangle dt = \left\langle \frac{t^2}{2} + C_{11}, \, \frac{t^2}{2} + C_{22}, \, C_{33} \right\rangle$.

To find the constants of integration, we use the initial condition that says $\mathbf{r}(0) = \langle 0, 1, 1 \rangle$.

$$\mathbf{r}(0) = \left\langle \frac{0^2}{2} + C_{11}, \, \frac{0^2}{2} + C_{22}, \, C_{33} \right\rangle = \langle 0, 1, 1 \rangle \quad \Rightarrow C_{11} = 0, C_{22} = 1, \, and \, C_{33} = 1$$

Therefore the position vector is $\mathbf{r}(t) = \left\langle \frac{t^2}{2}, \, \frac{t^2}{2} + 1, 1 \right\rangle$.

The position at $t = 4$ is: $\mathbf{r}(4) = \left\langle \frac{4^2}{2}, \, \frac{4^2}{2} + 1, 1 \right\rangle = \langle 8, 9, 1 \rangle$

CALCULUS III   **Quiz 4 KEY,** CONTINUED

12. Given the position function $\mathbf{r}(t) = \left\langle t^2, t^3 \right\rangle$

   (a) **Sketch the path of the object for** $-2 \le t \le 2$**. Use your graphing calculator in parametric mode to help you. Be sure to show the orientation of the curve.**

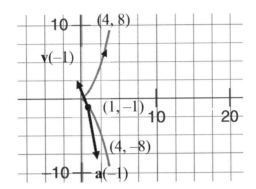

   (b) **Determine the value for t when the position of the object is (1, –1).**

   What value of t gives $\mathbf{r}(t) = \left\langle 1, -1 \right\rangle$? We need $t^2 = 1$ *and* $t^3 = -1$ simultaneously.
   Only t = $\boxdot$1 works *(answer)*

   (c) **Determine the velocity function.**

$$\mathbf{v}(t) = \mathbf{r}'(t) = \left\langle 2t, 3t^2 \right\rangle$$

   (d) **Sketch the velocity vector at the point (1, –1) on the same graph as in part (a).** *(Hint:* **Use the value for time that you got from part (b) to get the coordinates for the terminal point of the vector.**

   As determined above, $t = -1$ at the point $(1, -1)$. So, the velocity at $t = -1$ is:

$$\mathbf{v}(-1) = \mathbf{r}'(-1) = \left\langle 2(-1), 3(-1)^2 \right\rangle = \left\langle -2, 3 \right\rangle$$

   (d) **Sketch the acceleration vector at the point (1, –1) on the same graph as in part (a).**

   First, we find the acceleration vector: $\mathbf{a}(t) = \mathbf{v}'(t) = \left\langle 2, 6t \right\rangle$.

   At $t = -1$, we have $\mathbf{a}(-1) = \left\langle 2, 6(-1) \right\rangle = \left\langle 2, -6 \right\rangle$.

13. *Projectile Motion.* **A projectile is fired from the ground at an angle of elevation of 60° with an initial velocity of 210 feet per second.**

   (a) **Write the vector-valued function for the path of the projectile.**

   Use $g = 32$ feet per second per second. Initial height $h = 0$ (since it was fired from the ground) and the Theorem to get that:

$$\mathbf{r}(t) = \left\langle v_0 \cos\theta \cdot t, \, h + v_0 \sin\theta \cdot t - \frac{1}{2}(32)t^2 \right\rangle = \left\langle 210\cos(60°)t, \, 210\sin(60°)t - 16t^2 \right\rangle$$

$$= \left\langle 210\left(\frac{1}{2}\right)t, \, 210\left(\frac{\sqrt{3}}{2}\right)t - 16t^2 \right\rangle = \left\langle 105t, \, 105\sqrt{3}t - 16t^2 \right\rangle$$

ANSWER 13, CONTINUED...

(b) **How high did it go? (That is, what was the projectile's maximum height?)**

The projectile reaches maximum height when the vertical component of velocity equals zero. Set it equal to zero and solve for time, t. Of course, we need to differentiate the position function first to get the velocity vector:

$$\mathbf{v}(t) = \mathbf{r}'(t) = \left\langle 105,\, 105\sqrt{3} - 32t \right\rangle \;\Rightarrow 105\sqrt{3} - 32t = 0 \;\Rightarrow t = \frac{105\sqrt{3}}{32} \approx 5.68\,\text{sec}$$

The maximum height can be found by substituting this value of time into the vertical component of position:

$$105\sqrt{3}(5.68) - 16(5.68)^2 \approx 516.8\,ft$$

(c) **How long was it in the air?**

First, we need to find when the height is equal to zero, i.e., when the vertical component of position equals zero. So, set it equal to zero and solve for time t:

$$105\sqrt{3}t - 16t^2 = 0$$

This is a quadratic equation; we will factor and apply the ZPP to solve:

$$t(105\sqrt{3} - 16t) = 0 \;\Rightarrow\; t = 0 \;\; or \;\; 105\sqrt{3} - 16t = 0 \;\Rightarrow t = \frac{105\sqrt{3}}{16} \approx 11.37\,\text{sec}$$

(Height did in fact equal zero when time $t = 0$, but this is not the time we are interested in.)

(d) **How far did it go? (In other words, find the range of the projectile.)**

The range of the projectile can be found by substituting the time $t = 11.37$ seconds into the horizontal component of position: $105(11.37) \approx 1{,}193.5$ ft.

14. *Projectile Motion.* **The quarterback of a football team releases a pass at a height of 5 feet above the playing field, and the football is caught by a receiver 50 feet directly downfield at a height of 4 feet. The pass is released at an angle of 30° with the horizontal.**

(a) **Find the speed of the football when it was released.** *(Hint:* **You need to solve two equations with two unknowns.)**

Let v_0 be the speed of the football when it was released. The position function is:

$$\mathbf{r}(t) = \left\langle v_0 \cos\theta \cdot t,\;\; h + v_0 \sin\theta \cdot t - \frac{1}{2}(32)t^2 \right\rangle = \left\langle v_0 \cos(30°)t,\;\; 5 + v_0 \sin(30°)t - 16t^2 \right\rangle$$

$$= \left\langle \frac{\sqrt{3}}{2}v_0 t,\;\; 5 + \frac{1}{2}v_0 t - 16t^2 \right\rangle$$

We learned that the receiver is 50 feet downfield (this is the horizontal component of position function) at a height of 4 feet (this is the vertical component of position). This gives us a (non-linear) system of two equations and two unknowns. The unknowns are time t and the initial velocity.

$$\begin{cases} \dfrac{\sqrt{3}}{2}v_0 t = 50 \\[2mm] 5 + \dfrac{1}{2}v_0 t - 16t^2 = 4 \end{cases}$$

CALCULUS III   **Quiz 4 KEY, CONTINUED**

We will solve for t using the first equation, and then substitute this expression for t into the second equation:

$$t = \frac{50}{v_0}\left(\frac{2}{\sqrt{3}}\right) = \frac{100}{\sqrt{3}v_0} \quad \Rightarrow \quad 5 + \frac{1}{2}v_0\left(\frac{100}{\sqrt{3}v_0}\right) - 16\left(\frac{100}{\sqrt{3}v_0}\right)^2 = 4 \quad \Rightarrow \quad \frac{50}{\sqrt{3}} - 16\left(\frac{10,000}{3v_0^2}\right) = -1$$

Solve for v_0 now:

(answer)

$$\frac{50}{\sqrt{3}} + 1 = 16\left(\frac{10,000}{3v_0^2}\right) \quad \Rightarrow \quad \left(\frac{50}{\sqrt{3}}+1\right)\left(\frac{3}{160000}\right) = \frac{1}{v_0^2} \quad \Rightarrow \quad v_0^2 = \frac{1}{\left(\dfrac{50}{\sqrt{3}}+1\right)\left(\dfrac{3}{160000}\right)}$$

$$v_0 \approx 42\,\frac{ft}{\sec}$$

Note: This means that the position function is now:

$$\mathbf{r}(t) = \left\langle v_0\cos(30°)t,\ 5 + v_0\sin(30°)t - 16t^2 \right\rangle$$

$$= \left\langle \frac{\sqrt{3}}{2}(42)t,\ 5 + \frac{1}{2}(42)t - 16t^2 \right\rangle = \left\langle 21\sqrt{3}t,\ 5 + 21t - 16t^2 \right\rangle$$

(b)   **Find the maximum height of the football.**

Maximum height occurs when the vertical component of velocity equals zero. So, we first need to differentiate the position function to get the velocity function:

$$\mathbf{v}(t) = \mathbf{r}'(t) = \left\langle 21\sqrt{3}, 21 - 32t \right\rangle \quad \Rightarrow 21 - 32t = 0? \quad \Rightarrow t = \frac{21}{32} \approx 0.66\,\sec$$

Plug this value of time t into the vertical component of the position function in order to get the maximum height obtained:

$$5 + 21(0.66) - 16(0.66)^2 \approx 11.89\ \textit{feet}\ \textit{(answer)}$$

(c)   **Find the time the receiver has to reach the proper position after the quarterback releases the football.**

We must find what the value of time t will be when the horizontal component of position is equal to 50 feet. So, set the horizontal component of position equal to 50 and solve for time t:

$$21\sqrt{3}t = 50 \quad \Rightarrow t = \frac{50}{21\sqrt{3}} \approx 1.37\,\sec$$

(answer)

Calculus III   Quiz 4 KEY, continued

15.   Let $\mathbf{r}(t) = \langle 2\cos t, 2\sin t, 3t \rangle$.

   (a)   **Calculate the unit tangent vector, a.k.a., T(*t*).**

   By definition, the unit tangent vector is given by:
   (answer)

$$\mathbf{T}(t) = \frac{\mathbf{r}'(t)}{\|\mathbf{r}'(t)\|} = \frac{\langle -2\sin t, 2\cos t, 3 \rangle}{\sqrt{(2\sin t)^2 + (2\cos t)^2 + (3)^2}} = \frac{\langle -2\sin t, 2\cos t, 3 \rangle}{\sqrt{4\sin^2 t + 4\cos^2 t + 9}} = \frac{\langle -2\sin t, 2\cos t, 3 \rangle}{\sqrt{4(\sin^2 t + \cos^2 t) + 9}}$$

$$= \frac{\langle -2\sin t, 2\cos t, 3 \rangle}{\sqrt{13}} = \left\langle \frac{-2\sin t}{\sqrt{13}}, \frac{2\cos t}{\sqrt{13}}, \frac{3}{\sqrt{13}} \right\rangle$$

   (b)   **Calculate the principal unit normal vector, a.k.a., N(*t*).**

   By definition, the unit normal vector is given by:
   (answer)

$$\mathbf{N}(t) = \frac{\mathbf{T}'(t)}{\|\mathbf{T}'(t)\|} = \frac{\left(1/\sqrt{13}\right)\langle -2\cos t, -2\sin t, 0 \rangle}{\left(1/\sqrt{13}\right)\sqrt{(2\cos t)^2 + (2\sin t)^2 + (0)^2}} = \frac{\langle -2\cos t, -2\sin t, 0 \rangle}{\sqrt{4\sin^2 t + 4\cos^2 t}} = \frac{\langle -2\cos t, -2\sin t, 0 \rangle}{\sqrt{4(\sin^2 t + \cos^2 t)}}$$

$$= \frac{\langle -2\cos t, -2\sin t, 0 \rangle}{\sqrt{4}} = \langle -\cos t, -\sin t, 0 \rangle$$

16.   **Find the principal unit normal vector for the curve represented by** $\mathbf{r}(t) = \langle t^2, t \rangle$ **when** $t = -1$. **(Hint: This involves a lot of work, including using the Quotient Rule.** *Show all steps!*

   First, we find the unit tangent vector:

$$\mathbf{T}(t) = \frac{\mathbf{r}'(t)}{\|\mathbf{r}'(t)\|} = \frac{\langle 2t, 1 \rangle}{\sqrt{(2t)^2 + (1)^2}} = \frac{\langle 2t, 1 \rangle}{\sqrt{4t^2 + 1}} = \left\langle \frac{2t}{\sqrt{4t^2 + 1}}, \frac{1}{\sqrt{4t^2 + 1}} \right\rangle = \left\langle \frac{2t}{(4t^2 + 1)^{1/2}}, (4t^2 + 1)^{-1/2} \right\rangle$$

   Next we will use both the Quotient Rule and the Chain Rule to take derivatives in order to find the principal unit normal vector, **N**(*t*):

$$\mathbf{T}'(t) = \left\langle \frac{2(4t^2 + 1)^{1/2} - \frac{1}{2}(4t^2 + 1)^{-1/2}(8t)(2t)}{\left[(4t^2 + 1)^{1/2}\right]^2}, \; -\frac{1}{2}(4t^2 + 1)^{-3/2}(8t) \right\rangle = \left\langle \frac{2\sqrt{4t^2 + 1} - \frac{8t^2}{\sqrt{4t^2 + 1}}}{4t^2 + 1}, \; -\frac{4t}{(4t^2 + 1)^{3/2}} \right\rangle$$

   The first component of this vector is a complex fraction and we need to simplify it by multiplying both numerator and denominator by $\sqrt{4t^2 + 1}$.

   This is the numerator for our unit normal vector, **N**(t)

$$\mathbf{T}'(t) = \left\langle \frac{2\sqrt{4t^2+1} - \dfrac{8t^2}{\sqrt{4t^2+1}}\left(\dfrac{\sqrt{4t^2+1}}{\sqrt{4t^2+1}}\right)}{4t^2+1}, -\frac{4t}{(4t^2+1)^{3/2}} \right\rangle = \left\langle \frac{2(4t^2+1)-8t^2}{(4t^2+1)^{3/2}}, -\frac{4t}{(4t^2+1)^{3/2}} \right\rangle$$

$$= \left\langle \frac{2}{(4t^2+1)^{3/2}}, -\frac{4t}{(4t^2+1)^{3/2}} \right\rangle$$

Next, we'll find the magnitude of this (that will be the denominator for our unit normal vector):

$$\|\mathbf{T}'(t)\| = \left\| \left\langle \frac{2}{(4t^2+1)^{3/2}}, -\frac{4t}{(4t^2+1)^{3/2}} \right\rangle \right\| = \sqrt{\left(\frac{2}{(4t^2+1)^{3/2}}\right)^2 + \left(-\frac{4t}{(4t^2+1)^{3/2}}\right)^2}$$

$$= \sqrt{\frac{4}{(4t^2+1)^3} + \frac{16t^2}{(4t^2+1)^3}} = \sqrt{\frac{4+16t^2}{(4t^2+1)^3}} = \sqrt{\frac{4(1+4t^2)}{(4t^2+1)^3}} = \sqrt{\frac{4}{(4t^2+1)^2}} = \frac{2}{(4t^2+1)}$$

Now, we can form our unit normal vector:

$$\mathbf{N}(t) = \frac{\mathbf{T}'(t)}{\|\mathbf{T}'(t)\|} = \frac{\left\langle \dfrac{2}{(4t^2+1)^{3/2}}, \dfrac{-4t}{(4t^2+1)^{3/2}} \right\rangle}{\dfrac{2}{(4t^2+1)}} = \frac{4t^2+1}{2} \left\langle \frac{2}{(4t^2+1)^{3/2}}, \frac{-4t}{(4t^2+1)^{3/2}} \right\rangle$$

$$= \left\langle \frac{1}{(4t^2+1)^{1/2}}, \frac{-2t}{(4t^2+1)^{1/2}} \right\rangle$$

Finally, we evaluate this unit normal vector at time $t = -1$:
$$\mathbf{N}(-1) = \left\langle \frac{1}{(4(-1)^2+1)^{1/2}}, \frac{-2(-1)}{(4(-1)^2+1)^{1/2}} \right\rangle = \left\langle \frac{1}{\sqrt{5}}, \frac{2}{\sqrt{5}} \right\rangle$$

17.   **Find a set of parametric equations for the line tangent to the space curve given by** $\mathbf{r}(t) = \left\langle t, t^2, t^3 \right\rangle$ **at the point (2, 4, 8).**

First, we need to determine the value of time t at the point $(2, 4, 8)$. We do this by asking ourselves, "What value of t would make our function $\mathbf{r}(t) = \langle 2, 4, 8 \rangle$?" The only value that makes sense is that $t = 2$. Because velocity vectors are tangent to the graphs of the position vectors, we will find the velocity vector now and use that as our direction vector for the our line:

$$\mathbf{v}(t) = \mathbf{r}'(t) = \left\langle 1, 2t, 3t^2 \right\rangle \implies At \ t = 2 \implies \mathbf{v}(2) = \left\langle 1, 2(2), 3(2)^2 \right\rangle = \left\langle 1, 4, 12 \right\rangle$$

The components of this vector are the direction numbers for our line. That is, $a = 1$, $b = 4$, and $c = 12$.

Now we use the parametric equations for our line that have the form:

$$\begin{cases} x = x_0 + at \\ y = y_0 + bt \\ z = z_0 + ct \end{cases} \implies \begin{cases} x = 2 + t \\ y = 4 + 4t \\ z = 8 + 12t \end{cases}$$

CALCULUS III   ✠ QUIZ 5 KEY ✠

1. **Find a set of parametric equations for the line tangent to the helix** $\mathbf{r}(t) = \langle 2\cos t, 2\sin t, t \rangle$ **at the point corresponding to** $t = \dfrac{\pi}{6}$.

    Because velocity vectors are tangent to the graphs of the position vectors, we first find the velocity vector that will give us the direction vectors (and hence the direction numbers) for our line:

    $$\mathbf{v}(t) = \mathbf{r}'(t) = \langle -2\sin t, 2\cos t, 1 \rangle \quad \Rightarrow At \;\; t = \frac{\pi}{6} \quad \Rightarrow$$

    $$\mathbf{v}\left(\frac{\pi}{6}\right) = \left\langle -2\sin\frac{\pi}{6}, 2\cos\frac{\pi}{6}, 1 \right\rangle = \left\langle -2\left(\frac{1}{2}\right), 2\left(\frac{\sqrt{3}}{2}\right), 1 \right\rangle = \langle -1, \sqrt{3}, 1 \rangle$$

    The components of this vector are the direction numbers for our line. That is, $a = -1$, $b = \sqrt{3}$, and $c = 1$.

    We need to find a point P that "lives" on our line. We do this by simply plugging in $t = \dfrac{\pi}{6}$ into our position vector:

    $$\mathbf{r}\left(\frac{\pi}{6}\right) = \left\langle 2\cos\frac{\pi}{6}, 2\sin\frac{\pi}{6}, \frac{\pi}{6} \right\rangle = \left\langle 2\left(\frac{\sqrt{3}}{2}\right), 2\left(\frac{1}{2}\right), \frac{\pi}{6} \right\rangle = \left\langle \sqrt{3}, 1, \frac{\pi}{6} \right\rangle$$

    Now, we use the parametric equations for our line that have the form:

    $$\begin{cases} x = x_0 + at \\ y = y_0 + bt \\ z = z_0 + ct \end{cases} \Rightarrow \begin{cases} x = \sqrt{3} - t \\ y = 1 + \sqrt{3}t \\ z = \dfrac{\pi}{6} + t \end{cases}$$

2. **Find the tangential and normal components of acceleration at** $t = 0$ **for the position function given by** $\mathbf{r}(t) = \langle e^t, e^{-t}, \sqrt{2}\, t \rangle$.

    The tangential component of acceleration is denoted a_T and the normal component of acceleration is denoted a_N. From the Theorem, the acceleration vector $\mathbf{a}(t)$ can be written as a linear combination of the unit tangent and unit normal vectors in the following way: $\mathbf{a}(t) = a_T \mathbf{T} + a_N \mathbf{N}$. The formulas for a_T and a_N are as follows:

    $$a_T = \mathbf{a} \cdot \mathbf{T} \quad and \quad a_N = \sqrt{\|\mathbf{a}\|^2 - a_T^2}$$

    We need to find all of the ingredients first, then we'll plug everything in:

    $$\mathbf{v}(t) = \mathbf{r}'(t) = \langle e^t, -e^{-t}, \sqrt{2} \rangle \;\; \Rightarrow \;\; \mathbf{T}(t) = \frac{\mathbf{v}(t)}{\|\mathbf{v}(t)\|} = \frac{\langle e^t, -e^{-t}, \sqrt{2} \rangle}{\sqrt{(e^t)^2 + (e^{-t})^2 + (\sqrt{2})^2}} = \frac{\langle e^t, -e^{-t}, \sqrt{2} \rangle}{\sqrt{e^{2t} + e^{-2t} + 2}}$$

    $$\mathbf{a}(t) = \mathbf{v}'(t) = \langle e^t, e^{-t}, 0 \rangle \quad and \quad \|\mathbf{a}(t)\| = \sqrt{(e^t)^2 + (e^{-t})^2 + (0)^2} = \sqrt{e^{2t} + e^{-2t}}$$

CALCULUS III

Quiz 5 KEY, CONTINUED

Next, we will evaluate all of these at time $t = 0$:

$$\mathbf{v}(0) = \left\langle e^0, -e^{-0}, \sqrt{2} \right\rangle = \left\langle 1, -1, \sqrt{2} \right\rangle \implies \mathbf{T}(0) = \frac{\left\langle e^0, -e^{-0}, \sqrt{2} \right\rangle}{\sqrt{e^{2(0)} + e^{-2(0)} + 2}} = \frac{\left\langle 1, -1, \sqrt{2} \right\rangle}{\sqrt{1 + 1 + 2}} = \left\langle \frac{1}{2}, -\frac{1}{2}, \frac{\sqrt{2}}{2} \right\rangle$$

$$\mathbf{a}(0) = \left\langle e^0, e^{-0}, 0 \right\rangle = \left\langle 1, 1, 0 \right\rangle \quad and \quad \|\mathbf{a}(0)\| = \sqrt{e^{2(0)} + e^{-2(0)}} = \sqrt{1 + 1} = \sqrt{2}$$

Finally, we find our tangential and normal components of acceleration:

$$a_T = \mathbf{a} \cdot \mathbf{T} = \left\langle 1, 1, 0 \right\rangle \cdot \left\langle \frac{1}{2}, -\frac{1}{2}, \frac{\sqrt{2}}{2} \right\rangle = (1)\left(\frac{1}{2}\right) + (1)\left(-\frac{1}{2}\right) + (0)\left(\frac{\sqrt{2}}{2}\right) = 0 \quad and \quad a_N = \sqrt{\|\mathbf{a}\|^2 - a_T^2} = \sqrt{\left(\sqrt{2}\right)^2 - (0)^2} = \sqrt{2}$$

3. **A projectile is fired from the ground with initial speed of 160 feet per second at an angle of $\pi/6$.**

   (a) **How far did the projectile travel?**

   We need to use the Theorem for Position of Projectile Motion that gives us the position function:

   $$\mathbf{r}(t) = \left\langle v_0 \cos\theta \cdot t, \quad h + v_0 \sin\theta \cdot t - \frac{1}{2}(32)t^2 \right\rangle = \left\langle 160\cos(30°)t, \quad 0 + 160\sin(30°)t - 16t^2 \right\rangle$$

   $$= \left\langle \frac{\sqrt{3}}{2}(160)t, \quad \frac{1}{2}(160)t - 16t^2 \right\rangle = \left\langle 80\sqrt{3}t, \quad 80t - 16t^2 \right\rangle$$

   When the projectile hits the ground, the vertical component of position will be equal to zero. Therefore, we set the vertical component of position equal to zero and solve for time t. Then, to find the range (i.e., how far the projectile traveled) we will plug the value of time t we found into the horizontal component of position.

   $$80t - 16t^2 = 0 \implies 16t(5 - t) = 0 \implies t = 0 \ or \ 5 - t = 0 \implies t = 5\sec$$

   Plugging $t = 5$ sec into the horizontal component of position, we get:

   $$80\sqrt{3}(5) = 400\sqrt{3} \ feet \approx 693 \ feet$$

   (b) **Find the tangential and normal components of acceleration when $t = 1$ second.**

   The tangential component of acceleration is denoted a_T and the normal component of acceleration is denoted a_N. From the Theorem, the acceleration vector, $\mathbf{a}(t)$, can be written as a linear combination of the unit tangent and unit normal vectors in the following way: $\mathbf{a}(t) = a_T\mathbf{T} + a_N\mathbf{N}$. The formulas for and are as follows:

   $$a_T = \mathbf{a} \cdot \mathbf{T} \quad and \quad a_N = \sqrt{\|\mathbf{a}\|^2 - a_T^2}$$

   We need to find all of the ingredients first, then we'll plug everything in:

   $$\mathbf{v}(t) = \mathbf{r}'(t) = \left\langle 80\sqrt{3}, 80 - 32t \right\rangle \implies \mathbf{T}(t) = \frac{\mathbf{v}(t)}{\|\mathbf{v}(t)\|} = \frac{\left\langle 80\sqrt{3}, 80 - 32t \right\rangle}{\sqrt{\left(80\sqrt{3}\right)^2 + \left(80 - 32t\right)^2}}$$

   $$\mathbf{a}(t) = \mathbf{v}'(t) = \left\langle 0, -32 \right\rangle \quad and \quad \|\mathbf{a}(t)\| = \sqrt{(0)^2 + (32)^2} = 32$$

ANSWER 3, CONTINUED...

...*ANSWER 3, CONTINUED*

CALCULUS III   Quiz 5 KEY, CONTINUED

Next, we will evaluate all of these at time $t = 1$:

$$\mathbf{v}(1) = \left\langle 80\sqrt{3}, 80 - 32(1) \right\rangle = \left\langle 80\sqrt{3}, 48 \right\rangle \quad \Rightarrow$$

$$\mathbf{T}(1) = \frac{\left\langle 80\sqrt{3}, 48 \right\rangle}{\sqrt{(80\sqrt{3})^2 + (48)^2}} = \left\langle \frac{80\sqrt{3}}{\sqrt{21504}}, \frac{48}{\sqrt{21504}} \right\rangle = \left\langle \frac{80\sqrt{3}}{32\sqrt{21}}, \frac{48}{32\sqrt{21}} \right\rangle = \left\langle \frac{5}{2}\sqrt{\frac{3}{21}}, \frac{3}{2\sqrt{21}} \right\rangle$$

$$\mathbf{a}(1) = \left\langle 0, -32 \right\rangle \quad and \quad \|\mathbf{a}(1)\| = \sqrt{(0)^2 + (32)^2} = 32$$

Finally, we find our tangential and normal components of acceleration:

$$a_T = \mathbf{a} \cdot \mathbf{T} = \left\langle 0, -32 \right\rangle \cdot \left\langle \frac{5}{2}\sqrt{\frac{1}{7}}, \frac{3}{2\sqrt{21}} \right\rangle = (0)\left(\frac{5}{2}\sqrt{\frac{1}{7}} \right) + (-32)\left(\frac{3}{2\sqrt{21}} \right) = -\frac{48}{\sqrt{21}}$$

$$and \quad a_N = \sqrt{\|\mathbf{a}\|^2 - a_T^2} = \sqrt{(32)^2 - \left(\frac{48}{\sqrt{21}} \right)^2} = \sqrt{1024 - \frac{2304}{21}} = \sqrt{\frac{19200}{21}} = \frac{80}{\sqrt{7}}$$

4. Let $\mathbf{r}(t) = \left\langle t^2, t^3 \right\rangle$

   (a)   **Find T(2).**

   First, we will find the unit tangent vector $\mathbf{T}(t)$. Then, we will evaluate at $t = 2$:

$$\mathbf{T}(t) = \frac{\mathbf{r}'(t)}{\|\mathbf{r}'(t)\|} = \frac{\left\langle 2t, 3t^2 \right\rangle}{\sqrt{(2t)^2 + (3t^2)^2}} = \frac{\left\langle 2t, 3t^2 \right\rangle}{\sqrt{4t^2 + 9t^4}} = \left\langle \frac{2t}{\sqrt{t^2(4 + 9t^2)}}, \frac{3t^2}{\sqrt{t^2(4 + 9t^2)}} \right\rangle = \left\langle \frac{2t}{t\sqrt{4 + 9t^2}}, \frac{3t^2}{t\sqrt{4 + 9t^2}} \right\rangle$$

$$= \left\langle \frac{2}{\sqrt{4 + 9t^2}}, \frac{3t}{\sqrt{4 + 9t^2}} \right\rangle$$

Now, plug in $t = 2$:

$$\mathbf{T}(2) = \left\langle \frac{2}{\sqrt{4 + 9(2)^2}}, \frac{3(2)}{\sqrt{4 + 9(2)^2}} \right\rangle = \left\langle \frac{2}{\sqrt{40}}, \frac{6}{\sqrt{40}} \right\rangle = \left\langle \frac{2}{2\sqrt{10}}, \frac{6}{2\sqrt{10}} \right\rangle = \left\langle \frac{1}{\sqrt{10}}, \frac{3}{\sqrt{10}} \right\rangle \quad (answer)$$

Careful! Never make the mistake of saying that the square root of a sum of squares is equal to the sum of the square roots. In other words, $\sqrt{4 + 9t^2} \neq \sqrt{4} + \sqrt{9t^2}$. *No! No! No! No!*

If you are still not convinced, ask your self the following question: *Does* $\sqrt{4 + 9} = \sqrt{4} + \sqrt{9}$? That is, does $\sqrt{13} = 2 + 3$? Absolutely *not!*

√ *PART III*   TEST ANSWER KEY

CALCULUS **III**   **Quiz 5 KEY,** CONTINUED

(b)   **Find N(2). (*Hint:* Since this is a plane curve, the unit normal vector is easy to get once you already have the unit tangent vector. That is, if T(t) = <x, y > then N(t) = <–y, x>, since we must have that these two vectors are orthogonal to each other, or T · N = 0! So, use your answer from part (a).)**

To decide whether $\mathbf{N}(2) = \left\langle -\dfrac{3}{\sqrt{10}}, \dfrac{1}{\sqrt{10}} \right\rangle$ *OR* $\mathbf{N}(2) = \left\langle \dfrac{3}{\sqrt{10}}, -\dfrac{1}{\sqrt{10}} \right\rangle$, we choose the one

that is pointing in the same direction that the curve is turning. In other words, we need to do

part **(d)** below and sketch the graph of the vector-valued function **r**(*t*) before deciding the answer.

It turns out that $\mathbf{N}(2) = \left\langle -\dfrac{3}{\sqrt{10}}, \dfrac{1}{\sqrt{10}} \right\rangle$ is the one we desire. (See below.)

(c)   **Write the acceleration vector corresponding to *t* = 2 as a linear combination of T and N.**

$$\mathbf{a}(t) = \mathbf{r}''(t) = \langle 2, 6t \rangle \quad \Rightarrow \quad \mathbf{a}(2) = \langle 2, 6(2) \rangle = \langle 2, 12 \rangle$$

$$a_T = \mathbf{a} \cdot \mathbf{T} = \langle 2, 12 \rangle \cdot \left\langle \frac{1}{\sqrt{10}}, \frac{3}{\sqrt{10}} \right\rangle = (2)\left(\frac{1}{\sqrt{10}}\right) + (12)\left(\frac{3}{\sqrt{10}}\right) = \frac{38}{\sqrt{10}}$$

$$\text{and} \quad a_N = \mathbf{a} \cdot \mathbf{N} = \langle 2, 12 \rangle \cdot \left\langle \frac{-3}{\sqrt{10}}, \frac{1}{\sqrt{10}} \right\rangle = (2)\left(\frac{-3}{\sqrt{10}}\right) + (12)\left(\frac{1}{\sqrt{10}}\right) = \frac{6}{\sqrt{10}}$$

$$\textit{Therefore}: \quad \mathbf{a} = a_T \mathbf{T} + a_N \mathbf{N} \;=\; \left(\frac{38}{\sqrt{10}}\right)\mathbf{T} + \left(\frac{6}{\sqrt{10}}\right)\mathbf{N}$$

(b)   (d) no arrows   rve. At the point when *t* = 2, sketch the vectors T(2) and N(2). Be sure and
            ...rve using arrows.

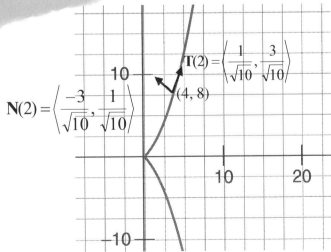

When *t* = 2, the position is at the point **r**(2)= <4, 8>. The unit vectors **T**(2) and **N**(2) we draw will "spring from" this point. It is best to first draw the vectors as algebraic vectors. In other words, first sketch the vectors where (0, 0) is the initial point for each. Then, translate the vectors so that they "spring" from the point (4, 8).

CALCULUS III   **Quiz 5 KEY, CONTINUED**

5. **Find the arc length of the curve C given by $\mathbf{r}(t) = \left\langle t, \dfrac{\sqrt{6}}{2}t^2, t^3 \right\rangle$ from $-1 \le t \le 1$.**
    Use the arclength formula:

$$s = \int_a^b \|\mathbf{r}'(t)\| \, dt = \int_{-1}^1 \left\| \left\langle 1, \sqrt{6}\,t, 3t^2 \right\rangle \right\| dt = \int_{-1}^1 \sqrt{(1)^2 + (\sqrt{6}\,t)^2 + (3t^2)^2} \, dt = \int_{-1}^1 \sqrt{1 + 6t^2 + 9t^4} \, dt$$

Believe it or not, the trinomial in the radicand factors and is a perfect square:

$$s = \int_{-1}^1 \sqrt{9t^4 + 3t^2 + 3t^2 + 1}\, dt = \int_{-1}^1 \sqrt{3t^2(3t^2+1) + 1(3t^2+1)} \, dt = \int_{-1}^1 \sqrt{(3t^2+1)^2}\, dt = \int_{-1}^1 (3t^2+1)\, dt$$

$$= \ \left[t^3 + t \right]_{-1}^1 = \left[(1)^3 + (1) \right] - \left[(-1)^3 - (-1) \right] = 2 - (-2) = 4$$

6. **Consider the curve C given by $\mathbf{r}(t) = \left\langle 2t, t^2, \ln t \right\rangle$.**
    (a) **Calculate the curvature, K, of the curve. Use $K = \dfrac{\|\mathbf{T}'(t)\|}{\|\mathbf{r}'(t)\|}$.**
    First, we find the unit tangent vector, $\mathbf{T}(t)$:

$$\mathbf{T}(t) = \frac{\mathbf{r}'(t)}{\|\mathbf{r}'(t)\|} = \frac{\left\langle 2, 2t, \dfrac{1}{t} \right\rangle}{\sqrt{(2)^2 + (2t)^2 + \left(\dfrac{1}{t} \right)^2}} = \frac{\left\langle 2, 2t, \dfrac{1}{t} \right\rangle}{\sqrt{4 + 4t^2 + \dfrac{1}{t^2}}} = \frac{\left\langle 2, 2t, \dfrac{1}{t} \right\rangle}{\sqrt{\dfrac{4t^2 + 4t^4 + 1}{t^2}}} = \frac{\left\langle 2, 2t, \dfrac{1}{t} \right\rangle}{\dfrac{\sqrt{(2t^2+1)^2}}{\sqrt{t^2}}} = \frac{\left\langle 2, 2t, \dfrac{1}{t} \right\rangle}{\dfrac{2t^2+1}{t}}$$

$$= \left\langle 2, 2t, \frac{1}{t} \right\rangle \left(\frac{t}{2t^2+1} \right) = \left\langle \frac{2t}{2t^2+1}, \frac{2t^2}{2t^2+1}, \frac{1}{2t^2+1} \right\rangle$$

Next, take the derivative of this to find $\mathbf{T}'(t)$. We'll need to use the Quotient Rule on the first two components and the Chain Rule on the last component:

$$\mathbf{T}'(t) = \left\langle \frac{2(2t^2+1) - (4t)(2t)}{(2t^2+1)^2}, \frac{(4t)(2t^2+1) - (4t)(2t^2)}{(2t^2+1)^2}, (-1)(2t^2+1)^{-2}(4t) \right\rangle$$

$$= \ \left\langle \frac{4t^2 + 2 - 8t^2}{(2t^2+1)^2}, \frac{8t^3 + 4t - 8t^3}{(2t^2+1)^2}, \frac{-4t}{(2t^2+1)^2} \right\rangle = \left\langle \frac{-4t^2+2}{(2t^2+1)^2}, \frac{4t}{(2t^2+1)^2}, \frac{-4t}{(2t^2+1)^2} \right\rangle$$

Next, take the magnitude of this last result to get $\|\mathbf{T}'(t)\|$ that will give us the numerator of the curvature K:

$$\|\mathbf{T}'(t)\| = \sqrt{ \left(\frac{-4t^2+2}{(2t^2+1)^2} \right)^2 + \left(\frac{-4t}{(2t^2+1)^2} \right)^2 + \left(\frac{4t}{(2t^2+1)^2} \right)^2 } = \sqrt{ \frac{(16t^4 - 16t^2 + 4) + 16t^2 + 16t^2}{(2t^2+1)^4} }$$

$$= \ \frac{\sqrt{4(4t^4 + 4t^2 + 1)}}{(2t^2+1)^2} = \frac{2\sqrt{(2t^2+1)^2}}{(2t^2+1)^2} = \frac{2(2t^2+1)}{(2t^2+1)^2} = \frac{2}{2t^2+1}$$

CALCULUS III   **Quiz 5 KEY, CONTINUED**

Now recall from above that:

$\|r'(t)\| = \dfrac{2t^2+1}{t}$ and find the curvature K:

$K = \dfrac{\|\mathbf{T}'(t)\|}{\|\mathbf{r}'(t)\|} = \dfrac{\left(\dfrac{2}{2t^2+1}\right)}{\left(\dfrac{2t^2+1}{t}\right)} = \dfrac{2}{2t^2+1}\cdot\left(\dfrac{t}{2t^2+1}\right) = \dfrac{2t}{(2t^2+1)^2}$

(b) **Find the arc length of the curve from $1 \le t \le 2$.**

We already obtained $\|r'(t)\| = \dfrac{2t^2+1}{t}$. We use this and plug it into the arclength formula:

$s = \int_a^b \|\mathbf{r}'(t)\|\,dt = \int_1^2\left(\dfrac{2t^2+1}{t}\right)dt = \int_1^2\left(\dfrac{2t^2}{t}+\dfrac{1}{t}\right)dt = \int_1^2\left(2t+\dfrac{1}{t}\right)dt = \left[t^2+\ln|t|\right]_1^2 = \left[2^2+\ln 2\right]-\left[1^2+\ln 1\right]$

$= 3 + \ln 2 \approx 3.69$

7. **Show that the helix given by $\mathbf{r}(t) = \langle \cos t, \sin t, t\rangle$ has constant curvature.**

We will show that $K = \dfrac{\|\mathbf{T}'(t)\|}{\|\mathbf{r}'(t)\|}$ = a constant.

$\mathbf{T}(t) = \dfrac{\mathbf{r}'(t)}{\|\mathbf{r}'(t)\|} = \dfrac{\langle-\sin t, \cos t, 1\rangle}{\sqrt{(\sin t)^2+(\cos t)^2+(1)^2}} = \dfrac{\langle-\sin t, \cos t, 1\rangle}{\sqrt{1+1}} = \dfrac{\langle-\sin t, \cos t, 1\rangle}{\sqrt{2}} = \left\langle\dfrac{-\sin t}{\sqrt{2}}, \dfrac{\cos t}{\sqrt{2}}, \dfrac{1}{\sqrt{2}}\right\rangle$

$\mathbf{T}'(t) = \left\langle\dfrac{-\cos t}{\sqrt{2}}, \dfrac{-\sin t}{\sqrt{2}}, 0\right\rangle \Rightarrow \|\mathbf{T}'(t)\| = \sqrt{\left(\dfrac{-\cos t}{\sqrt{2}}\right)^2+\left(\dfrac{-\sin t}{\sqrt{2}}\right)^2} = \sqrt{\dfrac{\cos^2 t+\sin^2 t}{2}} = \dfrac{1}{\sqrt{2}}$

$K = \dfrac{\|\mathbf{T}'(t)\|}{\|\mathbf{r}'(t)\|} = \dfrac{1/\sqrt{2}}{\sqrt{2}} = \dfrac{1}{\sqrt{2}}\cdot\dfrac{1}{\sqrt{2}} = \dfrac{1}{2}$

8. **Find the length of the curve $\mathbf{r}(t) = \langle e^t\cos t, e^t\sin t, e^t\rangle$ over the interval [0, 1].**

First find the derivative, or velocity vector $\mathbf{r}'(t)$. We will need to use the Product Rule to differentiate the first two components $\mathbf{r}'(t) = \langle e^t\cos t - e^t\sin t, e^t\sin t + e^t\cos t, e^t\rangle$. Then, use the arclength formula:

$s = \int_a^b \|\mathbf{r}'(t)\|\,dt = \int_0^1\|\langle e^t\cos t - e^t\sin t, e^t\sin t + e^t\cos t, e^t\rangle\|\,dt$

$= \int_0^1\sqrt{[e^t(\cos t - \sin t)]^2 + [e^t(\sin t + \cos t)]^2 + (e^t)^2}\,dt$

$= \int_0^1\sqrt{e^{2t}(\cos^2 t - 2\sin t\cos t + \sin^2 t) + e^{2t}(\sin^2 t + 2\sin t\cos t + \cos^2 t) + e^{2t}}\,dt$

$= \int_0^1\sqrt{e^{2t}(1 - 2\sin t\cos t + 1 + 2\sin t\cos t + 1)}\,dt$

$= \int_0^1 e^t\sqrt{3}\,dt = \sqrt{3}\left[e^t\right]_0^1 = \sqrt{3}\left(e^1-e^0\right) = \sqrt{3}(e-1) \approx 2.98$

9.   Consider the curve given by $r(t) = \langle t^2, 1 \rangle$.

(a)   **Sketch the curve. (Use the parametric mode on your calculator.) Be sure and show orientation using arrows.**

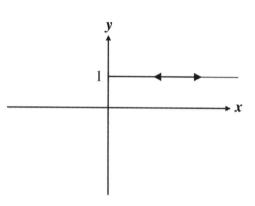

Note: The arrows point in both directions because when t is going from negative to positive, the curve heads towards the y-axis. When $t = 0$, the curve is at the point $(0, 1)$ on the y-axis. As t increases positively, the curve heads away from the y-axis.

(b)   **Calculate the curvature. Explain the result.**

$$T(t) = \frac{r'(t)}{\|r'(t)\|} = \frac{\langle 2t, 0 \rangle}{\sqrt{(2t)^2 + (0)^2}} = \frac{\langle 2t, 0 \rangle}{\sqrt{4t^2}} = \left\langle \frac{2t}{2t}, \frac{0}{2t} \right\rangle = \langle 1, 0 \rangle$$

$$T'(t) = \langle 0, 0 \rangle$$

$$\|T'(t)\| = \sqrt{0^2 + 0^2}$$

$$K = \frac{\|T'(t)\|}{\|r'(t)\|} = \frac{0}{2t} = 0$$

The curvature equals zero because the graph of the vector-valued function is a straight line that has no curve to it!

10.   Consider the curve given by $r(t) = \langle -1, -6t, 8t \rangle$.

(a)   **Find the arc length function, s, as a function of t.**

First, we find $r'(t)$ $r'(t) = \langle 0, -6, 8 \rangle$.

The arc length function has the form:

$$s = \int_0^t \|r'(x)\| dx = \int_0^t \sqrt{(1)^2 + (6)^2 + (8)^2} \, dx = \int_0^t \sqrt{100} \, dt = \int_0^t 10 \, dx = [10x]_0^t = 10t$$

(b)   **Express r as a function of the parameter s. In other words, find r(s). (Hint: Use your answer to part (a) and solve the equation for t. Then, substitute that version of t into r.)**

$$s = 10t \implies t = \frac{s}{10} \implies r(s) = \left\langle -1, -6 \left(\frac{s}{10} \right), 8 \left(\frac{s}{10} \right) \right\rangle = \left\langle -1, -\frac{3}{5}s, \frac{4}{5}s \right\rangle$$

(c)   **Show that $\|r'(s)\| = 1$, as it should! (Hint: First, use your answer from part (b) to obtain $r'(s)$.)**

$$r'(s) = \left\langle 0, -\frac{3}{5}, \frac{4}{5} \right\rangle \implies \|r'(s)\| = \sqrt{0^2 + \left(\frac{3}{5} \right)^2 + \left(\frac{4}{5} \right)^2} = \sqrt{\frac{9}{25} + \frac{16}{25}} = \sqrt{\frac{25}{25}} = 1$$

CALCULUS III   **Quiz 5 KEY, CONTINUED**

11. **Consider the curve given by $\mathbf{r}(s) = \langle 3+s, 1 \rangle$. Find the curvature using the formula $K = \|\mathbf{T}'(s)\|$ and explain the result.**

$$\mathbf{T}(s) = \frac{\mathbf{r}'(s)}{\|\mathbf{r}'(s)\|} = \frac{\langle 1, 0 \rangle}{\sqrt{(1)^2 + (0)^2}} = \frac{\langle 1, 0 \rangle}{1} = \langle 1, 0 \rangle$$

$$\mathbf{T}'(s) = \langle 0, 0 \rangle \Rightarrow \|\mathbf{T}'(s)\| = \sqrt{0^2 + 0^2}$$

$$Therefore: \ K = \|\mathbf{T}'(s)\| = 0$$

The curvature equals zero because the graph of $\mathbf{r}(s)$ is a straight line that has no curve at all.

12. **Find the curvature of the plane curve given by $y = \dfrac{1}{x}$ at the point $(-1, -1)$ using the formula:**

$$K = \frac{|y''|}{\left[1 + (y')^2\right]^{3/2}}$$

Rewrite the function y using negative exponents, then differentiate to find its first and second derivatives:

$$y = x^{-1} \Rightarrow y' = (-1)x^{-2} = -\frac{1}{x^2} \Rightarrow y'' = (-1)(-2)x^{-3} = \frac{2}{x^3}$$

Next, evaluate the derivatives when $x = -1$:

$$y'(-1) = -\frac{1}{(-1)^2} = 1 \Rightarrow y''(-1) = \frac{2}{(-1)^3} = -2$$

Plug these into the curvature formula:

$$K = \frac{|y''|}{\left[1 + (y')^2\right]^{3/2}} = \frac{|-2|}{\left[1 + (1)^2\right]^{3/2}} = \frac{2}{2^{3/2}} = 2^{-1/2} = \frac{1}{\sqrt{2}}$$

13. **Describe the level surfaces of the function $f(x, y, z) = x^2 + y^2 + z^2 - 2y + 4z + 5$ (That is, are they parabolas? circles? spheres? cones? hyperbolas?)**

To find the level surfaces, we set the function equal to a constant k, and examine the resulting equation. Complete the square to get the equation for a sphere:

$$f(x, y, z) = k$$
$$x^2 + y^2 + z^2 - 2y + 4z + 5 = k$$
$$x^2 + (y^2 - 2y + M_1) + (z^2 + 4z + M_2) = k - 5 + M_1 + M_2$$

Recall that $M_1 = \left(\dfrac{b}{2}\right)^2 = \left(\dfrac{-2}{2}\right)^2 = 1$ and $M_2 = \left(\dfrac{4}{2}\right)^2 = (2)^2 = 4$.

$$x^2 + (y^2 - 2y + 1) + (z^2 + 4z + 4) = k - 5 + 1 + 4$$
$$x^2 + (y-1)^2 + (z+2)^2 = k$$

The level surfaces are spheres centered at the point $(0, 1, -2)$ with radius of $\sqrt{k}$.

Calculus III   Quiz 5 KEY, continued

14. *Multiple choice:* **Describe the level surfaces of the function** $f(x,y,z) = 4x - 3y$. *(Please circle your choice.)*

  (a) **Planes parallel to the *xy*-plane**   (b) **Lines in the *xy*-plane**

  (c) **Planes parallel to the *z*-axis**   (d) **Lines parallel to the *z*-axis**

  (e) **None of these**

  Set the function equal to a constant and examine the resulting equation: $f(x,y,z) = k \implies 4x - 3y = k$. This is a linear equation—meaning that it is an equation for a plane that is parallel to the *z*-axis, since the equation is missing the *z* variable.

15. *Multiple choice:* **Find the domain of the function** $f(x,y) = \dfrac{2}{x - y^2}$. *(Please circle your choice.)*

  (a) **All (*x*, *y*) such that** $x = y^2$ (b) **All (*x*, *y*) such that** $x \neq y$.

  (c) **All (*x*, *y*) such that** $y \neq x^2$ (d) **All (*x*, *y*) such that** $x \neq y^2$

  (e) **None of these**

  We want to avoid having denominators be equal to zero!

16. **Find the domain of the function** $f(x,y) = \ln(x^2 + y^2 - 4)$.

  We want the argument of the logarithm to be positive in order for the function to be defined:

  $$x^2 + y^2 - 4 > 0 \quad OR \quad x^2 + y^2 > 4 \implies \left\{ (x,y) \mid x^2 + y^2 > 4 \right\}$$

17. **Find** $\dfrac{f(x + \Delta x, y) - f(x,y)}{\Delta x}$ **for** $f(x,y) = -x^2 + y^2 - 2x$.

  Note that the first term in the numerator will be $f(x + \Delta x, y) = -(x + \Delta x)^2 + y^2 - 2(x + \Delta x)$:

  $$\frac{f(x + \Delta x, y) - f(x,y)}{\Delta x} = \frac{\left[-(x+\Delta x)^2 + y^2 - 2(x + \Delta x) \right] - \left[-x^2 + y^2 - 2x \right]}{\Delta x}$$

  $$= \frac{\left[-(x+\Delta x)(x+\Delta x) + y^2 - 2x - 2\Delta x \right] + x^2 - y^2 + 2x}{\Delta x}$$

  $$= \frac{-\left(x^2 + 2x\Delta x + (\Delta x)^2 \right) - 2\Delta x + x^2}{\Delta x} = \frac{-x^2 - 2x\Delta x - (\Delta x)^2 - 2\Delta x + x^2}{\Delta x}$$

  $$= \frac{-2x\Delta x - (\Delta x)^2 - 2\Delta x}{\Delta x} = \frac{\Delta x(-2x - \Delta x - 2)}{\Delta x} = -2x - \Delta x - 2$$

18. *Multiple choice:* **Describe the level curves of** $f(x,y) = \sqrt{1 + x^2 - y}$.

  (a) **The level curves are circle.**

  (b) **The level curves are straight lines passing through the origin.**

  (c) **The level curves are parabolas with vertices on the *x*-axis.**

  (d) **The level curves are parabolas with vertices on the *y*-axis.**

  $f(x,y) = k$

  $\sqrt{1 + x^2 - y} = k$

  Set the function equal to a constant k and examine the resulting equation:

  Since the *x* is squared but the *y*-variable isn't, we have a family of vertical parabolas for the level curves!

  $\left(\sqrt{1 + x^2 - y} \right)^2 = (k)^2$

  $1 + x^2 - y = k^2$

CALCULUS III **Quiz 5 KEY,** CONTINUED

---

19. **Let** $f(x,y) = x^2 e^{xy^2}$.

(a) **Find the domain *and* range of f.**

The domain has no restrictions since there are no variables contained inside an even radical or the argument of a logarithm or the denominator of a quotient.

Thus, the domain is $\{(x,y)\mid x \in \Re, \, y \in \Re\}$

The range (output) will always be a positive value because the factor x^2 will always be non-negative, and the factor of e^{xy^2} will always be positive.

Therefore, the range is $f(x,y) = z \geq 0$

(b) **Find** $f(1,0) = (1)^2 e^{(1)(0)^2} = (1)(1) = 1$.

(c) **Find** $f(-2,2) = (-2)^2 e^{(-2)(2)^2} = (4)e^{-8} \approx 0.001$.

CALCULUS III ✠ **QUIZ 6 KEY** ✠

---

1. **Evaluate the limit** $\displaystyle \lim_{(x,y)\to(0,e)} e^x \ln y$.

$$\lim_{(x,y)\to(0,e)} e^x \ln y = e^0 \ln(e) = (1)(1) = 1$$

2. **Evaluate the limit** $\displaystyle \lim_{(x,y)\to(1,1)} \frac{x^2 - y^2}{x - y}$ *(Hint: Factor the numerator, then use direct substitution).*

$$\lim_{(x,y)\to(1,1)} \frac{x^2 - y^2}{x - y} = \lim_{(x,y)\to(1,1)} \frac{(x+y)(x-y)}{x-y} = \lim_{(x,y)\to(1,1)} (x+y) = 1 + 1 = 2$$

3. **Where is the function** $f(x,y) = \dfrac{x}{y\sqrt{4 - x^2 - y^2}}$ **continuous?**

Simply determine the domain of the function.

There are two things to worry about for this function:

1) We don't want the denominator to be equal to zero

2) We need the radicand to be greater than or equal to zero. In this case, we need the radicand to be strictly greater than zero so that the denominator won't be equal to zero.

$$4 - x^2 - y^2 > 0 \quad \Rightarrow \quad x^2 + y^2 < 4 \ \textit{plus} \text{ we want } y \neq 0.$$

Therefore, the domain (that is the same as where the function will be continuous) is:

$$\{(x,y)\mid x^2 + y^2 < 4, \, y \neq 0\}$$

4. **Prove that the limit does not exist by considering the two paths given below approaching the point (0, 0) for:**

$$\lim_{(x,y)\to(0,0)} \frac{xy^2}{x^2 + y^4}$$

ANSWER 4, CONTINUED...

...ANSWER 4, CONTINUED

CALCULUS III   Quiz 6 KEY, CONTINUED

Use the path where $x = 0$:

$$\lim_{(0,y)\to(0,0)} \frac{(0)y^2}{(0)^2 + y^4} = \lim_{(0,y)\to(0,0)} \frac{0}{y^4} = 0$$

Use the path where $x = y^2$

$$\lim_{(y^2, y)\to(0,0)} \frac{(y^2)y^2}{(y^2)^2 + y^4} = \lim_{(y^2, y)\to(0,0)} \frac{y^4}{y^4 + y^4} = \lim_{(y^2, y)\to(0,0)} \frac{y^4}{2y^4} = \lim_{(y^2, y)\to(0,0)} \frac{1}{2} = \frac{1}{2}$$

Since the two limits do not agree, that is – because $0 \neq \dfrac{1}{2}$, we can conclude that the limit does not exist!

5.   **Find** $f_{xy}(x,y)$ **for** $f(x,y) = \dfrac{4x^2}{y} + \dfrac{y^2}{2x}$.

First, we find the first partial derivative: $f_x(x,y)$.

It will be easier to do this if we rewrite the function using exponents so we can use the Power Rule when appropriate to differentiate:

$$f(x,y) = 4x^2 y^{-1} + 2x^{-1}y^2$$

Now, we find the first partial with respect to *x*. Be sure and treat the variable *y* as a constant when you do this:
$$f_x(x,y) = (4)(2x)y^{-1} + \left(\frac{1}{2}\right)(-1)x^{-2}y^2 = 8xy^{-1} - \left(\frac{1}{2}\right)x^{-2}y^2$$

Next, we take the second derivative. In other words, we now will find the mixed second partial derivative $f_{xy}(x,y)$. We do this by taking the derivative of $f_x(x,y)$ with respect to *y* (i.e., we will hold x constant).

$$f_{xy}(x,y) = (8x)(-1)y^{-2} - \left(\frac{1}{2}\right)x^{-2}(2y) = \frac{-8x}{y^2} - \frac{y}{x^2} \quad (answer)$$

6.   **Find the total differential of** $f(x,y) = 2x^2 y^3$

We know that $z = f(x,y) = 2x^2 y^3$, and the formula for the total differential dz is given by:

$$dz = \frac{\partial z}{\partial x}dx + \frac{\partial z}{\partial y}dy \quad \text{where} \quad \frac{\partial z}{\partial x} = 4xy^3 \quad and \quad \frac{\partial z}{\partial y} = 6x^2 y^2$$

So, now we have:

$$dz = 4xy^3 dx + 6x^2 y^2 dy \quad (answer)$$

7.   **Given** $z = \ln(x^2 + y^2)$ ...

(a)   **Use the differential dz to *approximate* the change in z as (x, y) moves from $(0, 1)$ to the point $(0.01, 0.98)$.**

Again, we use the formula for the total differential dz is given by:

$$dz = \frac{\partial z}{\partial x}dx + \frac{\partial z}{\partial y}dy \quad \text{where} \quad \frac{\partial z}{\partial x} = \frac{2x}{x^2 + y^2} \quad and \quad \frac{\partial z}{\partial y} = \frac{2y}{x^2 + y^2} .$$

Also, we have that $dx = \Delta x = 0.01 - 0 = 0.01$ and $dy = \Delta y = 0.98 - 1 = -0.02$.

CALCULUS III   **Quiz 6 KEY,** CONTINUED

We use the coordinates of the starting point for *x* and *y*. In other words, $x = 0$, and $y = 1$ in the total differential formula:

$$dz = \frac{\partial z}{\partial x}dx + \frac{\partial z}{\partial y}dy = \frac{2x}{x^2 + y^2}dx + \frac{2y}{x^2 + y^2}dy \quad \Rightarrow \quad dz = \frac{2(0)}{0^2 + 1^2}(0.01) + \frac{2(1)}{0^2 + 1^2}(-0.02) = -0.04$$

(b)   **Find the *actual* change in *z* as (*x*, *y*) moves from (0, 1) to the point (0.01, 0.98).**

Actual change in *z* is given by:

$$\Delta z = f(x + \Delta x, y + \Delta y) - f(x, y)$$
$$f(0 + 0.01, 1 + -0.02) - f(0,1)$$
$$f(0.01, 0.98) - f(0,1)$$
$$\ln(0.01^2 + 0.98^2) - \ln(0^2 + 1^2)$$
$$\ln(0.96) - \ln 1 = \ln(0.96) \approx -0.04$$

Note that the actual change and the approximation (a.k.a., the total differential) are exactly the same!

CALCULUS III   ✠ **PRACTICE MIDTERM 2 KEY** ✠

1.   **Complete the chart and convert the coordinates from one system to another among the rectangular, cylindrical, and spherical coordinate systems.**

RECTANGULAR	CYLINDRICAL	SPHERICAL
$\left(\dfrac{\sqrt{3}}{4}, \dfrac{3}{4}, \dfrac{3\sqrt{3}}{2}\right)$	$\left(\dfrac{\sqrt{3}}{2}, \dfrac{\pi}{3}, \dfrac{3\sqrt{3}}{2}\right)$	$\left(\sqrt{\dfrac{15}{2}}, \dfrac{\pi}{3}, 18.4°\right)$
$\left(-\dfrac{81\sqrt{3}}{2}, -\dfrac{81}{2}, 27\sqrt{3}\right)$	$\left(81, -\dfrac{5\pi}{6}, 27\sqrt{3}\right)$	$\left(54\sqrt{3}, -\dfrac{5\pi}{6}, \dfrac{\pi}{3}\right)$
$\left(0, -6\sqrt{3}, -6\right)$	$\left(6\sqrt{3}, -\dfrac{\pi}{2}, -6\right)$	$\left(12, -\dfrac{\pi}{2}, \dfrac{2\pi}{3}\right)$

ROW #1:

$$r = \sqrt{x^2 + y^2} = \sqrt{\left(\frac{\sqrt{3}}{4}\right)^2 + \left(\frac{3}{4}\right)^2} = \sqrt{\frac{3}{16} + \frac{9}{16}} = \sqrt{\frac{12}{16}}$$
$$= \frac{2\sqrt{3}}{4} = \frac{\sqrt{3}}{2}$$

$$\theta = \arctan\frac{y}{x} = \arctan\frac{3/4}{\sqrt{3}/4} = \arctan\left(\frac{3}{4}\right)\left(\frac{4}{\sqrt{3}}\right) = 60° = \frac{\pi}{3}$$

$$z = z = \frac{3\sqrt{3}}{2}$$

$$\rho = \sqrt{x^2 + y^2 + z^2} = \sqrt{\left(\frac{\sqrt{3}}{4}\right)^2 + \left(\frac{3}{4}\right)^2 + \left(\frac{3\sqrt{3}}{2}\right)^2}$$
$$= \sqrt{\frac{3}{16} + \frac{9}{16} + \frac{27}{4}} = \sqrt{\frac{15}{2}}$$

$$\theta = \arctan\frac{y}{x} = \arctan\frac{3/4}{\sqrt{3}/4} = 60° = \frac{\pi}{3}$$

$$\phi = \arccos\left(\frac{z}{\rho}\right) = \arccos\left(\frac{3\sqrt{3}/2}{\sqrt{15/2}}\right) \approx 18.4°$$

CALCULUS III   **Practice Midterm 2 KEY, CONTINUED**

ROW #2:

$$\begin{cases} x = r\cos\theta = 81\cos(-5\pi/6) = (81)\left(-\sqrt{3}/2\right) = -81\sqrt{3}/2 \\[2mm] y = r\sin\theta = 81\sin(-5\pi/6) = (81)(-1/2) = -81/2, \quad z = z = 27\sqrt{3} \\[2mm] \rho = \sqrt{r^2 + z^2} = \sqrt{81^2 + \left(27\sqrt{3}\right)^2} = \sqrt{6561 + 2187} \\[1mm] \quad = \sqrt{8748} = 54\sqrt{3} \\[2mm] \theta = \theta = -5\pi/6, \quad \phi = \arccos(z/\rho) = \arccos\left(27\sqrt{3}/54\sqrt{3}\right) = \dfrac{\pi}{3} \end{cases}$$

ROW #3:

$$\begin{cases} r = \rho\sin\phi = 12\sin(2\pi/3) = 12\left(\dfrac{\sqrt{3}}{2}\right) = 6\sqrt{3} \\[3mm] \theta = \theta = -\dfrac{\pi}{2}, \quad z = \rho\cos\phi = 12\cos\left(\dfrac{2\pi}{3}\right) = 12\left(-\dfrac{1}{2}\right) = -6 \\[4mm] x = \rho\sin\phi\cos\theta = 12\sin(2\pi/3)\cos(-\pi/2) = 0 \\[1mm] y = \rho\sin\phi\sin\theta = 12\sin(2\pi/3)\sin(-\pi/2) = (12)(\sqrt{3}/2)(-1) = -6\sqrt{3} \\[1mm] z = \rho\cos\phi = 12\cos(2\pi/3) = 12\left(-\dfrac{1}{2}\right) = -6 \end{cases}$$

2.   **Find an equation in rectangular coordinates from the given coordinate system.**

    (a)   **Spherical**

        This one's tricky! Use the cylindrical conversion formula (I know, I know) that tells us that $x = r\cos\theta$, then solve for $\cos\theta$: $\cos\theta = \dfrac{x}{r} = \dfrac{x}{\sqrt{x^2 + y^2}}$

        Then, use the spherical coordinate conversion formula that tells us that…

$$\rho = \sqrt{x^2 + y^2 + z^2}$$

        Put it all together by substituting into the original equation to get…

$$\sqrt{x^2 + y^2 + z^2} = 2\frac{x}{\sqrt{x^2 + y^2}} \quad (answer)$$

    (b)   **Spherical:** $\theta = \dfrac{\pi}{4}$

        Take tangent of both sides of the equation: $\tan\theta = \tan\dfrac{\pi}{4} = 1$.

        Now, use the fact that $\tan\theta = \dfrac{y}{x}$ (substitute) $\dfrac{y}{x} = 1 \implies y = x$

CALCULUS III   **Practice Midterm 2 KEY, CONTINUED**

(c)   **Cylindrical:** $r = 4\sin\theta$.

Start by multiplying both sides by r (we can assume r does not equal zero!):

$$r^2 = 4r\sin\theta \;\Rightarrow\; x^2 + y^2 = 4y \;\Rightarrow\; x^2 + y^2 - 4y = 0 \;\Rightarrow x^2 + y^2 - 4y + 4 = 4 \;\Rightarrow\; x^2 + (y-2)^2 = 4$$

This is an equation of a circle centered at the point (0, 2) with a radius of 2.

3.   **Convert the rectangular equation** $x^2 - y^2 = 2z$ **to:**

(a)   **Cylindrical coordinates:** $(r\cos\theta)^2 - (r\sin\theta)^2 = 2z \;\Rightarrow\; r^2(\cos^2\theta - \sin^2\theta) = 2z$

This answer is OK, but you could go one step further and use a Trig. ID and write $r^2\cos 2\theta = 2z$.

(b)   **Spherical coordinates:** $(\rho\sin\phi\cos\theta)^2 - (\rho\sin\phi\sin\theta)^2 = 2\rho\cos\phi$

$$\rho^2\sin^2\phi\cos^2\theta\,\rho^2\sin^2\phi\sin^2\theta = 2\rho\cos\phi$$

(factor out a common factor from the first two terms on the left side):

$$\rho^2\sin^2\phi(\cos^2\theta - \sin^2\theta) = 2\rho\cos\phi \;\Rightarrow\; \rho^2\sin^2\phi\cos(2\theta) = 2\rho\cos\phi \text{ (divide both sides by } \rho \text{ now)}$$

$$\rho\sin^2\phi\cos(2\theta) = 2\cos\phi \;\Rightarrow\; \rho\cos(2\theta) = 2\frac{\cos\phi}{\sin^2\phi} = 2\frac{1}{\sin\phi}\cdot\frac{\cos\phi}{\sin\phi} \;\Rightarrow\; \rho = \frac{2\csc\phi\cot\phi}{\cos(2\theta)}$$

4.   **Find the domain of the vector-valued functions. Express your answer in interval notation, please.**

(a) $\mathbf{r}(t) = e^{-t}\mathbf{i} + \dfrac{1}{t}\mathbf{j} - \sqrt{1-t}\,\mathbf{k}$

The domain of the first component has no restrictions. In other words, $t \in (-\infty, \infty)$.

The second component has a variable in the denominator, so we need to be sure our denominator never equals zero by having $t \neq 0$.

The last component has a variable under an even radical, so we need to insure that the radicand is never negative by writing the inequality $1 - t \geq 0 \;\Rightarrow\; t \leq 1$.

Finally, we take the intersection of all three sets to get the final answer that is $t \in (-\infty, 0) \cup (0, 1]$

(b) $\mathbf{r}(t) = |t|\mathbf{i} + \ln t\mathbf{j} + \dfrac{1}{3t^2 - 13t - 10}\mathbf{k}$

The domain of the first component has no restrictions. In other words, $t \in (-\infty, \infty)$.

The second component has a variable in the argument of the log. Arguments of logs must always be positive. So, we write the inequality $t > 0$.

The last component needs attention since there are variables in the denominator. We do not want the denominator equal to zero. We find all the values of t that would make the denominator equal to zero by solving the quadratic equation $3t^2 - 13t - 10 = 0$. We could use the quadratic formula of course, but this trinomial factors. So, I will factor and apply the ZPP to solve instead:

$$3t^2 - 15t + 2t - 10 = 0 \;\Rightarrow\; 3t(t-5) + 2(t-5) = 0 \;\Rightarrow (3t+2)(t-5) = 0 \;\Rightarrow 3t + 2 = 0 \quad and \quad t - 5 = 0$$

$$t = -\frac{2}{3} \quad and \quad t = 5$$

ANSWER 4, CONTINUED...

...*ANSWER 4, CONTINUED*

CALCULUS III   Practice Midterm 2 KEY, CONTINUED

These are the values we will exclude. That is, the domain for the last component is:

$$t \in \left(-\infty, -\frac{2}{3}\right) \cup \left(-\frac{2}{3}, 5\right) \cup (5, \infty)$$

We now take the intersection of all three domains to get the final answer: $t \in (0, 5) \cup (5, \infty)$.

5.   **Sketch the curve represented by the vector-valued function $\mathbf{r}(t) = (2 - t)\mathbf{i} + t^2\mathbf{j}$ and indicate the orientation of the curve. Then, write the corresponding rectangular equation. (Hint: eliminate the parameter.)**

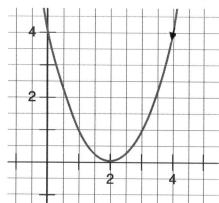

Since $x = 2 - t$, solve for t to get $t = 2 - x$. Substitute this into the y-component to get $y = t^2 = (2 - x)^2$

So, we have that $y = (2 - x)^2$ *(answer)*

6.   **Find the limits (if they exist). If they do not exist, say so.**

(a) $\displaystyle\lim_{t \to 1}\left(\ln t\mathbf{i} + \frac{t^2 + 7t - 8}{t - 1}\mathbf{j} + \cos(\pi t)\mathbf{k} \right)$

Always try direct substitution first. If that doesn't work, then try other methods such as factoring and reducing, and/or L'Hôpital's Rule.

Let us take the limit of the first component:

$$\lim_{t \to 1} \ln t = \ln(1) = 0$$

Now, take the limit of the second component:

$$\lim_{t \to 1} \frac{t^2 + 7t - 8}{t - 1} = \frac{1^2 + 7(1) - 8}{1 - 1} = \frac{0}{0}$$

This is an indeterminate form that we can apply L'Hôpital's Rule to:

$$\lim_{t \to 1} \frac{t^2 + 7t - 8}{t - 1} = \lim_{t \to 1} \frac{2t + 7}{1} = \frac{2(1) + 7}{1} = 9$$

Now, take the limit of the last component:

$$\lim_{t \to 1} \cos(\pi t) = \cos(\pi \cdot 1) = -1$$

The final answer must be written as a vector

$$0\mathbf{i} + 9\mathbf{j} - \mathbf{k} \quad OR \quad \langle 0, 9, -1 \rangle$$

CALCULUS III   **Practice Midterm 2 KEY, CONTINUED**

(b) $\lim\limits_{t\to 0}\left(\dfrac{\sin t}{t}\mathbf{i}+e^t\mathbf{j}+\dfrac{t+3}{t-2}\mathbf{k}\right)$

Let us take the limit of the first component: $\lim\limits_{t\to 0}\dfrac{\sin t}{t}=\dfrac{\sin(0)}{0}=\dfrac{0}{0}$

This is an indeterminate form that we can apply L'Hôpital's Rule to:

$$\lim_{t\to 0}\frac{\sin t}{t}=\lim_{t\to 0}\frac{\cos t}{1}=\frac{\cos 0}{1}=\frac{1}{1}=1$$

Now take the limit of the second component:

$$\lim_{t\to 0}e^t=e^0=1$$

Now take the limit of the last component:

$$\lim_{t\to 0}\frac{t+3}{t-2}=\frac{0+3}{0-2}=-\frac{3}{2}$$

The final answer must be written as a vector:

$$\mathbf{i}+\mathbf{j}-\frac{3}{2}\mathbf{k}\quad OR\quad \left\langle 1,1,-\frac{3}{2}\right\rangle$$

(c) $\lim\limits_{t\to\infty}\left(\dfrac{t}{t^2-2}\mathbf{i}+e^{-t}\mathbf{j}+\dfrac{t+3}{t-2}\mathbf{k}\right)$

Let us take the limit of the first component: $\lim\limits_{t\to\infty}\dfrac{t}{t^2-2}=\dfrac{\infty}{\infty^2-2}=\dfrac{\infty}{\infty}.$

This is an indeterminate form that we can apply L'Hôpital's Rule to:

$$\lim_{t\to\infty}\frac{t}{t^2-2}=\lim_{t\to\infty}\frac{1}{2t}=\frac{1}{2\infty}=\frac{1}{\infty}=0$$

Now, take the limit of the second component:

$$\lim_{t\to\infty}e^{-t}=e^{-\infty}=\frac{1}{e\infty}=\frac{1}{\infty}=0$$

Now take the limit of the last component:

$$\lim_{t\to\infty}\frac{t+3}{t-2}=\frac{\infty+3}{\infty-2}=\frac{\infty}{\infty}$$

This is an indeterminate form that we can apply L'Hôpital's Rule to:

$$\lim_{t\to\infty}\frac{t+3}{t-2}=\lim_{t\to\infty}\frac{1}{1}=\frac{1}{1}=1$$

The final answer must be written as a vector:

$$0\mathbf{i}+0\mathbf{j}+\mathbf{k}=\mathbf{k}\quad OR\quad \left\langle 0,0,1\right\rangle$$

Calculus III   **Practice Midterm 2 KEY, continued**

(d) $\lim_{t \to 1}\left(\dfrac{t^2+1}{t-1}\mathbf{i} + \sin\dfrac{\pi t}{2}\mathbf{j} + t^2\mathbf{k} \right)$

Let us take the limit of the first component:

$$\lim_{t \to 1}\frac{t^2+1}{t-1} = \frac{1^2+1}{1-1} = \frac{2}{0} = \infty \quad \text{or}: \ D.N.E.$$

(When we use the limit d.n.e. for even one component, we can say that the final answer **does not exist!**)

7. Find $\mathbf{r}'(t)$

(a) $\mathbf{r}(t) = e^{-t}\mathbf{i} + \dfrac{1}{t}\mathbf{j} - \sqrt{1-t}\,\mathbf{k}$

I will rewrite the function first using exponents for the last two components to make it easier to see using the General Power Rule when we differentiate:

$$\mathbf{r}(t) = \left\langle e^{-t},\, t^{-1},\, -(1-t)^{1/2} \right\rangle$$

Now we will take the derivative:

$$\mathbf{r}'(t) = \left\langle -e^{-t},\, -t^{-2},\, -\frac{1}{2}(1-t)^{-1/2}(-1) \right\rangle = \left\langle -e^{-t},\, -\frac{1}{t^2},\, \frac{1}{2\sqrt{1-t}} \right\rangle \ (answer)$$

(b) $\mathbf{r}(t) = \sin^3 t\,\mathbf{i} + \ln\sqrt[3]{\dfrac{5}{t^2}}\,\mathbf{j} - \arctan t\ \mathbf{k}$

I will rewrite the first component to make it clear we need the Chain Rule to take the derivative and simplify the second component using properties of logs in order to make it easier to differentiate:

$$\mathbf{r}(t) = \left\langle (\sin t)^3,\, \ln\left(\frac{5}{t^2}\right)^{1/3},\, -\arctan t \right\rangle = \left\langle (\sin t)^3,\, \frac{1}{3}\ln\left(\frac{5}{t^2}\right),\, -\arctan t \right\rangle = \left\langle (\sin t)^3,\, \frac{1}{3}\ln 5 - \frac{2}{3}\ln t,\, -\arctan t \right\rangle$$

Now we will take the derivative:

$$\mathbf{r}'(t) = \left\langle 3(\sin t)^2(\cos t),\, 0 - \frac{2}{3}\left(\frac{1}{t}\right),\, -\frac{1}{t^2+1} \right\rangle = \left\langle 3\sin^2 t \cos t,\, -\frac{2}{3t},\, -\frac{1}{t^2+1} \right\rangle$$

8. **Find the open intervals on which the curve given by the vector-valued function is smooth:**

(a) $\mathbf{r}(t) = (t + \sin t)\mathbf{i} + (1 - \cos t)\mathbf{j}$

The curve is not smooth when $\mathbf{r}'(t)$ is equal to the zero vector. So, first find $\mathbf{r}'(t)$:

$$\mathbf{r}'(t) = \left\langle 1 + \cos t,\, \sin t \right\rangle$$

This will be equal to the zero vector when both components are simultaneously equal to zero:

$$1 + \cos t = 0 \ \Rightarrow \ \cos t = -1 \ \Rightarrow \ t = \arccos(-1) = (2n-1)\pi$$

(i.e., all *odd* multiples of π, where n is an integer.)

CALCULUS III   **Practice Midterm 2 KEY, CONTINUED**

The second component will be equal to zero when:

$\sin t = 0 \quad \Rightarrow \quad t = \arcsin(0) = n\pi$ (all multiples of p, where n is an integer)

The only values that makes both components equal to zero at the same time is $t = (2n-1)\pi$. (odd multiples of p, where n is an integer)

Therefore, the curve is smooth when $t \neq (2n-1)\pi$.

Using interval notation, we say the curve is smooth on the intervals:

$$t \in \left((2n-1)\pi, (2n+1)\pi\right) \text{ (answer)}$$

(b) $\mathbf{r}(t) = \sqrt{t}\,\mathbf{i} + (t^2-1)\mathbf{j} + \dfrac{1}{4}t\mathbf{k}$

The curve is not smooth when $\mathbf{r}'(t)$ is equal to the zero vector. So, first find $\mathbf{r}'(t)$:

$$\mathbf{r}'(t) = \left\langle \frac{1}{2}t^{-1/2},\, 2t,\, \frac{1}{4} \right\rangle$$

Since this vector can *never* be equal to the zero vector (because the last component is a constant not equal to zero), the curve will *always* be smooth on the function's domain! In other words, the curve is smooth on all intervals so long as t is greater than or equal to zero, so $t \in [0, \infty)$ *(answer)*.

9.   **Given the functions** $\mathbf{r}(t) = (t^2-1)\mathbf{i} + (t^3-3t^2)\mathbf{j} + 5t\mathbf{k}$ **and** $\mathbf{u}(t) = \sqrt{2}\,t\mathbf{i} + e^t\mathbf{j} + e^{-t}\mathbf{k}$ ...

Find $\dfrac{d}{dt}\left(\mathbf{r}(t) \cdot \mathbf{u}(t)\right)$

We will use the Product Rule for derivatives, so that $\dfrac{d}{dt}\left(\mathbf{r}(t) \cdot \mathbf{u}(t)\right) = \mathbf{r}'(t) \cdot \mathbf{u}(t) + \mathbf{u}'(t) \cdot \mathbf{r}(t)$.

(And yes, those are indeed dot products contained within both terms.)

We have $\mathbf{r}'(t) = \left\langle 2t,\, 3t^2-6t,\, 5 \right\rangle$ *and* $\mathbf{u}'(t) = \left\langle \sqrt{2},\, e^t,\, -e^{-t} \right\rangle$.

So then, we can say that:

$$\frac{d}{dt}\left(\mathbf{r}(t) \cdot \mathbf{u}(t)\right) = \left\langle 2t,\, 3t^2-6t,\, 5 \right\rangle \cdot \left\langle \sqrt{2}t,\, e^t,\, e^{-t} \right\rangle + \left\langle \sqrt{2},\, e^t,\, -e^{-t} \right\rangle \cdot \left\langle t^2-1,\, t^3-3t^2,\, 5t \right\rangle$$

$$= (2t)(\sqrt{2}t) + \left(3t^2-6t\right)(e^t) + 5e^{-t} + \sqrt{2}\left(t^2-1\right) + e^t\left(t^3-3t^2\right) - 5te^{-t}$$

$$= 3\sqrt{2}\,t^2 - \sqrt{2} + e^t(t^3-6t) + (5-5t)e^{-t}$$

10.   **Find** $\mathbf{r}(t)$ **if** $\mathbf{r}'(t) = \dfrac{1}{1+t^2}\mathbf{i} + e^{-t}\mathbf{k}$ **and** $\mathbf{r}(0) = \mathbf{i} - \mathbf{j}$

We will need to anti-differentiate:

$$\mathbf{r}(t) = \int \mathbf{r}'(t)dt = \int \left\langle \frac{1}{1+t^2},\, 0,\, e^{-t} \right\rangle dt = \left\langle \int \frac{1}{1+t^2}dt,\, \int 0\,dt,\, \int e^{-t}dt \right\rangle = \left\langle \arctan t + C_1,\, C_2,\, -e^{-t} + C_3 \right\rangle$$

To find the constants of integration, we use the initial condition that tells us $\mathbf{r}(0) = \mathbf{i} - \mathbf{j} = \left\langle 1, -1, 0 \right\rangle$.

$\mathbf{r}(0) = \left\langle \arctan 0 + C_1,\, C_2,\, -e^{-0} + C_3 \right\rangle = \left\langle 1, -1, \ 0 \right\rangle$

$\arctan 0 + C_1 = 1 \ \Rightarrow C_1 = 1$

$C_2 = -1$

$-e^{-0} + C_3 = 0 \ \Rightarrow \ -1 + C_3 = 0 \ \Rightarrow C_3 = 1$

So, our *final answer is:*

$$\mathbf{r}(t) = \left\langle \arctan t + 1,\, -1,\, -e^{-t} + 1 \right\rangle$$

CALCULUS III   **Practice Midterm 2 KEY,** CONTINUED

11.   **Find r (t) given the following conditions. (*Hint:* you will need to anti-differentiate twice.)**

$$\mathbf{r}''(t) = -4\cos t\mathbf{i} - 3\sin t\mathbf{k}, \quad \mathbf{r}'(0) = 2\mathbf{k}, \quad \mathbf{r}(0) = 4\mathbf{j}$$

$$\mathbf{r}'(t) = \int \mathbf{r}''(t)dt = \int \left(-4\cos t\mathbf{i} - 3\sin t\mathbf{k}\right)dt \;=\; \left\langle \int -4\cos t\,dt, \int 0\,dt, \int -3\sin t\,dt \right\rangle$$

$$= \left\langle -4\sin t + C_1, C_2, 3\cos t + C_3 \right\rangle$$

Use the initial condition to get the constants of integration:

$$\mathbf{r}'(0) = \left\langle -4\sin 0 + C_1, C_2, 3\cos 0 + C_3 \right\rangle = \left\langle 0, 0, 2 \right\rangle$$
$$-4\sin 0 + C_1 = 0 \;\Rightarrow\; C_1 = 0$$
$$C_2 = 0$$
$$3\cos 0 + C_3 = 2 \;\Rightarrow\; C_3 = -1$$

Therefore, we have that $\mathbf{r}'(t) = \left\langle -4\sin t, 0, 3\cos t + -1 \right\rangle$.

Now we can find $\mathbf{r}(t)$ by integrating:

$$\mathbf{r}(t) = \int \mathbf{r}'(t)dt = \int \left\langle -4\sin t, 0, 3\cos t - 1 \right\rangle dt \;=\; \left\langle \int -4\sin t\,dt, \int 0\,dt, \int (3\cos t - 1)dt \right\rangle$$

$$= \left\langle 4\cos t + C_{11}, C_{22}, 3\sin t - t + C_{33} \right\rangle$$

Use the initial condition to get the constants of integration:

$$\mathbf{r}(0) = \left\langle 4\cos 0 + C_{11}, C_{22}, 3\sin 0 - 0 + C_{33} \right\rangle = \left\langle 0, 4, 0 \right\rangle$$
$$4\cos 0 + C_{11} = 0 \;\Rightarrow\; C_{11} = -4 \quad and \quad C_{22} = 4 \quad and \quad 3\sin 0 - 0 + C_{33} = 0 \;\Rightarrow C_{33} = 0$$
Therefore : $\mathbf{r}(t) = \left\langle 4\cos t - 4, 4, 3\sin t - t \right\rangle$

12.   **Evaluate the definite integrals:**

(a) $\displaystyle\int_0^1 \left(\frac{t}{\sqrt{t^2 + 1}}\mathbf{i} + t\mathbf{j} - \mathbf{k} \right) dt$

Use the General Power Rule to integrate the first component, where $u = t^2 + 1$, and $du = 2t\,dt$.
We need a factor of 2 in the integrand. We'll put one in (and divide one out):

$$\frac{1}{2}\int_0^1 \frac{2t}{\sqrt{t^2 + 1}}dt \;=\; \frac{1}{2}\int \frac{du}{\sqrt{u}} \;=\; \frac{1}{2}\int u^{-1/2}du \;=\; \frac{1}{2}(2)u^{1/2} = \left[\sqrt{t^2 + 1}\right]_0^1 = \sqrt{1^2 + 1} - \sqrt{0^2 + 1} = \sqrt{2} - 1$$

The last two components are straightforward:

$$\int_0^1 t\,dt \;=\; \left[\frac{t^2}{2}\right]_0^1 \;=\; \frac{1^2}{2} - \frac{0^2}{2} \;=\; \frac{1}{2}$$

$$\int_0^1 (-1)dt \;=\; \left[-t\right]_0^1 \;=\; -1 \qquad \Rightarrow \text{ We write the final answer as a vector: } \left\langle \sqrt{2} - 1, \frac{1}{2}, -1 \right\rangle$$

CALCULUS III   Practice Midterm 2 KEY, CONTINUED

(b) $\int_{-1}^{1} \left(t\mathbf{i} + t^3\mathbf{j} + \sqrt[3]{t}\,\mathbf{k} \right) dt$

$= \left\langle \left[\frac{t^2}{2}\right]_{-1}^{1}, \left[\frac{t^4}{4}\right]_{-1}^{1}, \left[\frac{3}{4}t^{4/3}\right]_{-1}^{1} \right\rangle = \left\langle \frac{1^2}{2} - \frac{(-1)^2}{2}, \frac{1^4}{4} - \frac{(-1)^4}{4}, \frac{3}{4}\left[1^{4/3} - (-1)^{4/3}\right] \right\rangle$

$= \langle 0, 0, 0 \rangle = \mathbf{0}$ (answer)

(c) $\int_{0}^{\pi/2} \left[(2\cos t)\mathbf{i} + (2\sin t)\mathbf{j} + \mathbf{k} \right] dt$

$= \left\langle \int_{0}^{\pi/2} 2\cos t\,dt, \int_{0}^{\pi/2} 2\sin t\,dt, \int_{0}^{\pi/2} dt \right\rangle = \left\langle [2\sin t]_{0}^{\pi/2}, [-2\cos t]_{0}^{\pi/2}, [t]_{0}^{\pi/2} \right\rangle$

$= \left\langle 2\sin\frac{\pi}{2} - 2\sin 0, -2\cos\frac{\pi}{2} - (-2\cos 0), \frac{\pi}{2} \right\rangle = \left\langle 2, 2, \frac{\pi}{2} \right\rangle$

13. **Find a set of parametric equations for the tangent line to the graph of the vector-valued function** $\mathbf{r}(t) = \left\langle \ln(t-3), t^2, \frac{1}{2}t \right\rangle$ **at the point (0, 16, 2).**

First, we need to determine the value of time t at which the function is at the position of the point given. We equate the components of the function with the coordinates of the point to get three equations. What value of time t makes all three equations true simultaneously?

$$\ln(t-3) = 0, \quad t^2 = 16, \quad \frac{1}{2}t = 2 \quad \Rightarrow \quad t = 4$$

Now, we need to get the direction vector for the line. This will be the tangent vector to the graph, a.k.a., the velocity vector $\mathbf{r}'(t) = \left\langle \frac{1}{t-3}, 2t, \frac{1}{2} \right\rangle$.

Next, evaluate this at $t = 4$ to get the direction numbers for the line:

$$\mathbf{r}'(4) = \left\langle \frac{1}{4-3}, 2(4), \frac{1}{2} \right\rangle = \left\langle 1, 8, \frac{1}{2} \right\rangle \quad \Rightarrow a = 1, b = 8, c = \frac{1}{2}$$

Now, we plug in the direction numbers and the coordinates of the point into the parametric equations for a line:

$$\begin{cases} x = x_0 + at \\ y = y_0 + bt \\ z = z_0 + ct \end{cases} \Rightarrow \begin{cases} x = 0 + t \\ y = 16 + 8t \\ z = 2 + \frac{1}{2}t \end{cases}$$

Remember that parametric equations for a line are not unique; your answer may vary depending on if you used a multiple of the direction vector (i.e., a vector parallel to the vector I used) and/or a different point.

CALCULUS III   Practice Midterm 2 KEY, CONTINUED

14. Given the vector-valued function $\mathbf{r}(t) = \left\langle t - 1, t, \dfrac{1}{t} \right\rangle$.

(a) Find the velocity, speed, and acceleration of the function at time t.

$$\textit{velocity} = \mathbf{r}'(t) = \left\langle 1, 1, -\dfrac{1}{t^2} \right\rangle$$

$$\textit{speed} = \|\mathbf{r}'(t)\| = \sqrt{1^2 + 1^2 + \left(-\dfrac{1}{t^2}\right)^2} = \sqrt{2 + \dfrac{1}{t^4}} = \sqrt{\dfrac{2t^4 + 1}{t^4}} = \dfrac{\sqrt{2t^4 + 1}}{\sqrt{t^4}} = \dfrac{\sqrt{2t^4 + 1}}{t^2}$$

$$\textit{acceleration} = \mathbf{r}''(t) = \left\langle 0, 0, \dfrac{2}{t^3} \right\rangle$$

(b) Find the unit tangent vector, $\mathbf{T}(t)$, the unit normal vector, $\mathbf{N}(t)$, as well as the tangential and normal components of acceleration, a_T and a_N, at time t.

The unit tangent vector is:

$$\mathbf{T}(t) = \dfrac{\mathbf{r}'(t)}{\|\mathbf{r}'(t)\|} = \dfrac{\left\langle 1, 1, -\dfrac{1}{t^2} \right\rangle}{\dfrac{\sqrt{2t^4 + 1}}{t^2}} = \dfrac{t^2}{\sqrt{2t^4 + 1}} \left\langle 1, 1, -\dfrac{1}{t^2} \right\rangle = \left\langle \dfrac{t^2}{\sqrt{2t^4 + 1}}, \dfrac{t^2}{\sqrt{2t^4 + 1}}, -\dfrac{1}{\sqrt{2t^4 + 1}} \right\rangle$$

The unit normal vector is $\mathbf{N}(t) = \dfrac{\mathbf{T}'(t)}{\|\mathbf{T}'(t)\|}$.

Let's find the numerator of this first. In other words, let's find $\mathbf{T}'(t)$ now. We will use the Quotient Rule on the first two components and the Chain Rule on the last component (after rewriting it using negative rational exponents *first*):

$$\mathbf{T}'(t) = \left\langle \dfrac{2t\sqrt{2t^4 + 1} - \dfrac{1}{2}(2t^4 + 1)^{-1/2}(8t^3)(t^2)}{\left(\sqrt{2t^4 + 1}\right)^2}, \dfrac{2t\sqrt{2t^4 + 1} - \dfrac{1}{2}(2t^4 + 1)^{-1/2}(8t^3)(t^2)}{\left(\sqrt{2t^4 + 1}\right)^2}, \dfrac{1}{2}(2t^4 + 1)^{-3/2}(8t^3) \right\rangle$$

$$= \left\langle \dfrac{2t\sqrt{2t^4 + 1} - \dfrac{4t^5}{\sqrt{2t^4 + 1}}}{2t^4 + 1}, \dfrac{2t\sqrt{2t^4 + 1} - \dfrac{4t^5}{\sqrt{2t^4 + 1}}}{2t^4 + 1}, \dfrac{4t^3}{(2t^4 + 1)^{3/2}} \right\rangle$$

We need to simplify the complex fractions in the first two components. We do this by multiplying both numerator and denominator by $\sqrt{2t^4 + 1}$.

CALCULUS **III**   **Practice Midterm 2 KEY,** CONTINUED

$$\mathbf{T}'(t) = \left\langle \left(\frac{2t\sqrt{2t^4+1} - \dfrac{4t^5}{\sqrt{2t^4+1}}}{2t^4+1} \right) \left(\frac{\sqrt{2t^4+1}}{\sqrt{2t^4+1}} \right), \left(\frac{2t\sqrt{2t^4+1} - \dfrac{4t^5}{\sqrt{2t^4+1}}}{2t^4+1} \right) \left(\frac{\sqrt{2t^4+1}}{\sqrt{2t^4+1}} \right), \frac{4t^3}{(2t^4+1)^{3/2}} \right\rangle$$

$$= \left\langle \frac{2t(2t^4+1)-4t^5}{(2t^4+1)^{3/2}}, \frac{2t(2t^4+1)-4t^5}{(2t^4+1)^{3/2}}, \frac{4t^3}{(2t^4+1)^{3/2}} \right\rangle$$

$$= \left\langle \frac{4t^5+2t-4t^5}{(2t^4+1)^{3/2}}, \frac{4t^5+2t-4t^5}{(2t^4+1)^{3/2}}, \frac{4t^3}{(2t^4+1)^{3/2}} \right\rangle = \left\langle \frac{2t}{(2t^4+1)^{3/2}}, \frac{2t}{(2t^4+1)^{3/2}}, \frac{4t^3}{(2t^4+1)^{3/2}} \right\rangle$$

Next, we find the denominator of the unit normal vector by taking the magnitude of this result:

$$\|\mathbf{T}'(t)\| = \sqrt{\left(\frac{2t}{(2t^4+1)^{3/2}} \right)^2 + \left(\frac{2t}{(2t^4+1)^{3/2}} \right)^2 + \left(\frac{4t^3}{(2t^4+1)^{3/2}} \right)^2}$$

$$= \sqrt{\frac{4t^2+4t^2+16t^6}{(2t^4+1)^3}} = \sqrt{\frac{8t^2+16t^6}{(2t^4+1)^3}} = \sqrt{\frac{8t^2(1+2t^4)}{(2t^4+1)^3}} = \sqrt{\frac{(4)(2)t^2}{(2t^4+1)^2}}$$

$$= \frac{2t\sqrt{2}}{2t^4+1}$$

Finally, we form the unit normal vector:

$$\mathbf{N}(t) = \frac{\mathbf{T}'(t)}{\|\mathbf{T}'(t)\|} = \frac{\left\langle \dfrac{2t}{(2t^4+1)^{3/2}}, \dfrac{2t}{(2t^4+1)^{3/2}}, \dfrac{4t^3}{(2t^4+1)^{3/2}} \right\rangle}{\dfrac{2t\sqrt{2}}{2t^4+1}} = \left(\frac{2t^4+1}{2t\sqrt{2}} \right) \left\langle \frac{2t}{(2t^4+1)^{3/2}}, \frac{2t}{(2t^4+1)^{3/2}}, \frac{4t^3}{(2t^4+1)^{3/2}} \right\rangle$$

$$= \left\langle \frac{1}{\sqrt{2}(2t^4+1)^{1/2}}, \frac{1}{\sqrt{2}(2t^4+1)^{1/2}}, \frac{2t^2}{\sqrt{2}(2t^4+1)^{1/2}} \right\rangle = \left\langle \frac{1}{(4t^4+2)^{1/2}}, \frac{1}{(4t^4+2)^{1/2}}, \frac{2t^2}{(4t^4+2)^{1/2}} \right\rangle$$

To get the tangential and normal components of acceleration, a_T and a_N, we use formulas:

$$a_T = \mathbf{a} \cdot \mathbf{T} \quad \text{and} \quad a_N = \mathbf{a} \cdot \mathbf{N}$$

$$a_T = \mathbf{a} \cdot \mathbf{T} = \left\langle 0, 0, \frac{2}{t^3} \right\rangle \cdot \left\langle \frac{t^2}{\sqrt{2t^4+1}}, \frac{t^2}{\sqrt{2t^4+1}}, \frac{-1}{\sqrt{2t^4+1}} \right\rangle = \left(\frac{2}{t^3} \right) \left(\frac{-1}{\sqrt{2t^4+1}} \right) = \frac{-2}{t^3\sqrt{2t^4+1}}$$

$$a_N = \mathbf{a} \cdot \mathbf{N} = \left\langle 0, 0, \frac{2}{t^3} \right\rangle \cdot \left\langle \frac{1}{\sqrt{4t^4+2}}, \frac{1}{\sqrt{4t^4+2}}, \frac{2t^2}{\sqrt{4t^4+2}} \right\rangle = \left(\frac{2}{t^3} \right) \left(\frac{2t^2}{\sqrt{4t^4+2}} \right) = \frac{4}{t\sqrt{4t^4+2}}$$

ANSWER 14, CONTINUED...

...ANSWER 14, CONTINUED

CALCULUS III   Practice Midterm 2 KEY, CONTINUED

(c)   Find the curvature of the function at time *t*.

We use the formula for curvature:

$$K = \frac{\|\mathbf{T}'(t)\|}{\|\mathbf{r}'(t)\|} = \frac{\dfrac{2t\sqrt{2}}{2t^4+1}}{\dfrac{\sqrt{2t^4+1}}{t^2}} = \left(\frac{2t\sqrt{2}}{2t^4+1}\right)\left(\frac{t^2}{\sqrt{2t^4+1}}\right) = \frac{2\sqrt{2}t^3}{\left(2t^4+1\right)^{3/2}}$$

15.   **Find the maximum height and the range of a projectile fired from ground-level at an angle of elevation of 30°, if the initial velocity is 76 feet per second. (Use** $g = 32\,f/s^2$ **.)**

Initial height $h = 0$ (since it was fired from the ground). Use the Theorem to get the vector-valued function describing position:

$$\mathbf{r}(t) = \left\langle v_0 \cos\theta \cdot t,\ h + v_0 \sin\theta \cdot t - \frac{1}{2}(32)t^2 \right\rangle = \left\langle 76\cos(30°)t,\ 76\sin(30°)t - 16t^2 \right\rangle$$

$$= \left\langle 76\left(\frac{\sqrt{3}}{2}\right)t,\ 76\left(\frac{1}{2}\right)t - 16t^2 \right\rangle = \left\langle 38\sqrt{3}t,\ 38t - 16t^2 \right\rangle$$

The projectile reaches maximum height when the vertical component of velocity equals zero. Set it equal to zero and solve for time, *t*. Of course, we need to differentiate the position function first to get the velocity vector:

$$\mathbf{v}(t) = \mathbf{r}'(t) = \left\langle 38\sqrt{3},\ 38 - 32t \right\rangle \ \Rightarrow 38 - 32t = 0 \ \Rightarrow t = \frac{38}{32} \approx 1.188\,\text{sec}$$

The maximum height can be found by substituting this value of time into the vertical component of position:

$$38(1.188) - 16(1.188)^2 \approx 22.56\,ft$$

To find the range (or how far the projectile went), we first need to find when the height is equal to zero, i.e., when the vertical component of position equals zero. So, set it equal to zero and solve for time *t*:

$$38t - 16t^2 = 0$$

This is a quadratic equation; we will factor and apply the ZPP to solve:

$$2t(19 - 8t) = 0 \ \Rightarrow \ 2t = 0 \ or \ 19 - 8t = 0 \ \Rightarrow t = \frac{19}{8} \approx 2.38\,\text{sec}$$

(Height did in fact equal zero when time $t = 0$, but this is not the time we are interested in.)

The range of the projectile can be found by substituting the time $t = 2.38$ seconds into the horizontal component of position:

$$38\sqrt{3}(2.38) \approx 156.3\,ft$$

CALCULUS **III**   **Practice Midterm 2 KEY,** CONTINUED

16. **Find the arclength of the space curve given by** $\mathbf{r}(t) = \langle 10\cos^3 t, 10\sin^3 t \rangle$ **from** $0 \le t \le 2\pi$.

We will need to find the derivative of the function in order to use the formula for arclength:

$$s = \int_a^b \|\mathbf{r}'(t)\| dt$$

There is a slight problem, as this formula only works on curves that are smooth (i.e., have no cusps or sharp corners). Use your graphing calculator in parametric mode and look at the graph. Use a range for time t of $0 \le t \le 2\pi$.

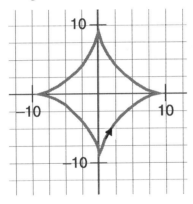

Note the four sharp corners (cusps). We will figure out the range of time t that traces out a piece of the curve and then use symmetry of the curve. We will multiply our result for arclength by 4 at the end. It turns out the range of time t to trace out a smooth piece of the curve with no cusps is $0 \le t \le \dfrac{\pi}{2}$.

To take the derivative, we will need to use the Chain Rule for both components. To see this better, I will first rewrite the original function's components:

$$\mathbf{r}(t) = \langle 10(\cos t)^3, 10(\sin t)^3 \rangle$$

Now, I find the derivative:

$$\mathbf{r}'(t) = \langle 30(\cos t)^2(-\sin t), 30(\sin t)^2(\cos t) \rangle = \langle -30\cos^2 t \sin t, 30\sin^2 t \cos t \rangle$$

Next, we can find arclength:

$$s = \int_0^{\pi/2} \|\langle -30\cos^2 t \sin t, 30\sin^2 t \cos t \rangle\| dt = \int_0^{\pi/2} \sqrt{\left(-30\cos^2 t \sin t\right)^2 + \left(30\sin^2 t \cos t\right)^2} \, dt$$

$$= \int_0^{\pi/2} \sqrt{30^2(\cos^4 t \sin^2 t + \sin^4 t \cos^2 t)} dt \quad = \quad \int_0^{\pi/2} 30\sqrt{\cos^2 t \sin^2 t(\cos^2 t + \sin^2 t)} dt$$

$$= \int_0^{\pi/2} 30\sqrt{\cos^2 t \sin^2 t(1)} dt \quad = \quad \int_0^{\pi/2} 30|\cos t \sin t| dt$$

The last step we did uses the Principal Square Root Property that says for any variable a: $\sqrt{a^2} = |a|$. To integrate, we will assume the expression is positive and use the Power Rule where: $u = \sin t, du = \cos t$. Then, we have that:

$$30\int_0^{\pi/2} u \, du \quad 30\left(\frac{u^2}{2}\right) = 15\left[\sin^2 t\right]_0^{\pi/2} = 15\left(\sin^2\left(\frac{\pi}{2}\right) - \sin^2(0)\right) = 15$$

Use symmetry of the curve to and multiply this result to get the total arclength:

$$(4)(15) = 60 \; (\textit{answer})$$

17. **Given the function** $f(x,y) = e^{\frac{x}{y}}$ **...**

   (a) **Find the domain.**

   Because the numerator of the exponent has a denominator with a variable in it, we need to restrict the domain. In other words, we cannot have y equal to zero. Therefore the domain written using set notation is $\{(x,y)\mid y \neq 0\}$.

   (b) **Find the range.**

   Because exponential functions have outputs that are always positive, the range of this function will be $z \in (0, \infty)$.

18. **Discuss the continuity of the function** $f(x,y,z) = \dfrac{1}{\sqrt{x^2 + y^2 + z^2}}$.

   Finding where a function is continuous is essentially the same as determining its domain. Since we have a radicand, we need to restrict the domain in order to insure that the radicand is non-negative. Further, since the radical is in the denominator, we need to have the radicand be positive *only*. Thus, the domain must satisfy the inequality $x^2 + y^2 + z^2 > 0$.

   Writing the final answer using set notation, we have the function continuous everywhere where:

$$\{(x, y, z)\mid x^2 + y^2 + z^2 > 0\}$$

19. **Find the limits (if they exist):**

   (a) $\displaystyle\lim_{(x,y)\to(-2,1)} \dfrac{x^2 - 4y^2}{x + 2y}$

   After applying direct substitution, we get the indeterminate form $\dfrac{0}{0}$.

   Unfortunately, we cannot use L'Hôpital's Rule on limits involving more than one variable. So, instead, we notice that the quotient can be factored and reduced as the numerator is a difference of squares. Afterwards, we can apply direct substitution again and get our result:

$$\lim_{(x,y)\to(-2,1)} \dfrac{x^2 - 4y^2}{x + 2y} = \lim_{(x,y)\to(-2,1)} \dfrac{(x+2y)(x-2y)}{x + 2y} = \lim_{(x,y)\to(-2,1)} (x-2y) = -2 - 2(1) = -4$$

   (b) $\displaystyle\lim_{(x,y)\to(0,0)} \dfrac{y + xe^{-y^2}}{1 + x^2}$

   We can simply use direct substitution for this problem: $\displaystyle\lim_{(x,y)\to(0,0)} \dfrac{y + xe^{-y^2}}{1 + x^2} = \dfrac{0 + (0)e^0}{1 + 0^2} = \dfrac{0}{1} = 0$

   (c) $\displaystyle\lim_{(x,y)\to(0,0)} \dfrac{\sin(x - y)}{x - y}$ (**Hint:** Use a theorem we learned in Calculus I that $\displaystyle\lim_{\theta \to 0} \dfrac{\sin\theta}{\theta} = 1$.

   If we don't take the hint, then after applying direct substitution, we will obtain the indeterminate form $\dfrac{0}{0}$ again, where we cannot apply use L'Hôpital's Rule. So, we take the hint and notice the argument of the sine function that is $x - y$ is the same as the denominator. Thus, the final answer will be 1.

CALCULUS III   **Practice Midterm 2 KEY,** CONTINUED

20. Given the function $f(x, y) = x^2 + y^2 \ldots$

(a) **Sketch the graph of the function.** *(Hint:* **Quadric surface!)**

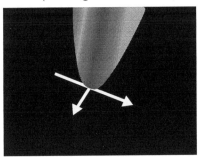

This is a paraboloid.

(b) **Describe the level curves for the function. Then sketch the level curves for the values of $c = 0, 2, 4$.**

First, we set the function equal to a constant and examine the resulting equation:

$$f(x, y) = c \implies x^2 + y^2 = c$$

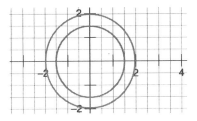

This is an equation for a circle of radius $\sqrt{c}$.

If $c = 0$, the graph is a point. For $c = 2$, the graph is a circle of radius $\sqrt{2}$.

For $c = 4$, the graph is a circle of radius 2.

21. **Sketch the level surface for the function of three variables $f(x, y, z) = 2y + 3z$ for the value of $c = 6$.**

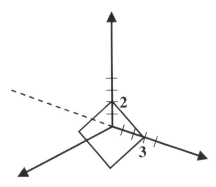

First, we set the function equal to a constant and examine the resulting equation:

$$f(x, y, z) = c \implies 2y + 3z = c$$

These are planes that are parallel to the x-axis (since the missing variable is x).

When $c = 6$, the y-intercept is 3, and the z-intercept is $z = 2$.

CALCULUS III   ✖ QUIZ 7 KEY ✖

1. **Consider the surface given by** $f(x,y) = \dfrac{1}{7}\left(4x^2 - 2y^2\right)$.

    (a) **Find an equation of the tangent plane (in general form) to the surface at the point (–2, –1, 2).**

    First we need to rewrite the equation in the form $F(x,y,z) = 0$.

    We first recognize that $f(x,y) = z$, and we do that substitution *first*. Then we can write:

    $$F(x,y,z) = \frac{1}{7}\left(4x^2 - 2y^2\right) - z = 0$$

    Next, we find the first partial of the function of three variables $F(x,y,z)$.

    $$F_x(x,y,z) = \frac{1}{7}(8x), \quad F_y(x,y,z) = \frac{1}{7}(-4y), \quad F_z(x,y,z) = -1$$

    We now evaluate each of these three partial derivatives at the point (–2, –1, 2). This will give us a vector that is normal to the level surface of $F(x,y,z)$. We will use the components of this normal vector to help us get the equation of our tangent plane:

    $$F_x(-2,-1,2) = \frac{8}{7}(-2) = \frac{-16}{7}, \quad F_y(-2,-1,2) = \frac{-4}{7}(-1) = \frac{4}{7}, \quad F_z(-2,-1,2) = -1$$

    So, a vector normal to the surface at the given point is:

    $$\mathbf{n} = \langle a,b,c \rangle = \left\langle \frac{-16}{7}, \frac{4}{7}, -1 \right\rangle$$

    Finally we recall that the standard equation of a plane is:

    $$a(x - x_0) + b(y - y_0) + c(z - z_0) = 0 \quad \Rightarrow \quad \frac{-16}{7}(x - (-2)) + \frac{4}{7}(y - (-1)) + (-1)(z - 2) = 0$$

    $$-16(x + 2) + 4(y + 1) - 7(z - 2) = 0$$
    $$-16x - 32 + 4y + 4 - 7z + 14 = 0$$
    $$16x - 4y + 7z + 14 = 0$$

    (b) **Find a set of parametric equations for the normal line to the surface at the point (–2, –1, 2).**

    A direction vector for the line is exactly the vector we found above that is normal to the surface, or $\mathbf{n} = \langle a,b,c \rangle = \left\langle \dfrac{-16}{7}, \dfrac{4}{7}, -1 \right\rangle$.

    The parametric equations for a line may now be used:

    $$\begin{cases} x = x_0 + at \\ y = y_0 + bt \\ z = z_0 + ct \end{cases} \quad \Rightarrow \quad \begin{cases} x = -2 - \dfrac{16}{7}t \\ y = -1 + \dfrac{4}{7}t \\ z = 2 - t \end{cases}$$

√ *PART III*   *Test Answer Key*

Calculus III   **QUIZ 7 KEY,** continued

(c)   **Find the angle of inclination of the tangent plane to the surface at the point (–2, –1, 2)**

We use the formula that says:

$$\cos\theta = \frac{|\nabla \mathbf{F} \cdot \mathbf{k}|}{\|\nabla \mathbf{F}\|} = \frac{\left|\left\langle \dfrac{-16}{7}, \dfrac{4}{7}, -1 \right\rangle \cdot \langle 0,0,1 \rangle\right|}{\sqrt{\left(\dfrac{-16}{7}\right)^2 + \left(\dfrac{4}{7}\right)^2 + (-1)^2}} = \frac{|-1|}{\sqrt{\dfrac{321}{49}}} = \frac{\sqrt{49}}{\sqrt{321}} = \frac{7}{\sqrt{321}}$$

$$\theta = \arccos\frac{7}{\sqrt{321}} \approx 67°$$

CALCULUS III   ✖ **QUIZ 8 KEY** ✖

1. **Given the equations of two surfaces** $z = x^2 + y^2$ **and** $z = 4 - y$ **and a point of intersection of the two surfaces (2, –1, 5), do the following:**

    (a) **Find a set of parametric equations of the tangent line to the curve of intersection of the surfaces at the given point.**

    We will use the hint. We will first write each equation in the form: $F(x, y, z) = 0$, where:

    $$F(x, y, z) = x^2 + y^2 - z \quad and \quad G(x, y, z) = 4 - y - z$$

    The gradients for these two functions, $\nabla \mathbf{F}$ *and* $\nabla \mathbf{G}$ will give vectors that are normal to each of the level surfaces:

    $$\nabla \mathbf{F}(x, y, z) = \langle F_x, F_y, F_z \rangle = \langle 2x, 2y, -1 \rangle \quad and \quad \nabla \mathbf{G}(x, y, z) = \langle G_x, G_y, G_z \rangle = \langle 0, -1, -1 \rangle$$

    Next, we evaluate these gradients at the given point of (2, –1, 5).

    $$\nabla \mathbf{F}(2, -1, 5) = \langle 2(2), 2(-1), -1 \rangle = \langle 4, -2, -1 \rangle \quad and \quad \nabla \mathbf{G}(2, -1, 5) = \langle 0, -1, -1 \rangle$$

    Now, we take the cross-product of these two vectors. This will give us a vector orthogonal to both of the surfaces:

    $$\nabla \mathbf{F} \times \nabla \mathbf{G} = \begin{vmatrix} \mathbf{i} & \mathbf{j} & \mathbf{k} \\ 4 & -2 & -1 \\ 0 & -1 & -1 \end{vmatrix} = \begin{vmatrix} -2 & -1 \\ -1 & -1 \end{vmatrix} \mathbf{i} - \begin{vmatrix} 4 & -1 \\ 0 & -1 \end{vmatrix} \mathbf{j} + \begin{vmatrix} 4 & -2 \\ 0 & -1 \end{vmatrix} \mathbf{k}$$

    $$= [(-2)(-1) - (-1)(-1)]\mathbf{i} - [(4)(-1) - (-1)(0)]\mathbf{j} + [(4)(-1) - (-2)(0)]\mathbf{k}$$

    $$= 1\mathbf{i} - (-4)\mathbf{j} + -4\mathbf{k} \quad OR: \quad \langle 1, 4, -4 \rangle$$

    Finally, we use the parametric equations for a line where the cross-product we just found gives the direction numbers for the line:

    $$\begin{cases} x = x_0 + at \\ y = y_0 + bt \\ z = z_0 + ct \end{cases} \Rightarrow \begin{cases} x = 2 + 1t \\ y = -1 + 4t \\ z = 5 - 4t \end{cases}$$

    (b) **Find the cosine of the angle between the two gradient vectors at the given point.**

    We use the formula that says:

    $$\cos\theta = \frac{|\nabla \mathbf{F} \cdot \nabla \mathbf{G}|}{\|\nabla \mathbf{F}\|\|\nabla \mathbf{G}\|} = \frac{|\langle 4, -2, -1 \rangle \cdot \langle 0, -1, -1 \rangle|}{\sqrt{(4)^2 + (-2)^2 + (-1)^2} \sqrt{0^2 + 1^2 + 1^2}} = \frac{|3|}{\sqrt{21}\sqrt{2}} = \frac{3}{\sqrt{42}}$$

    $$\theta = \arccos\frac{3}{\sqrt{42}} \approx 62.4°$$

    (c) **State whether or not the surfaces are orthogonal at the given point.**

    The surfaces are not orthogonal since the angle between them is not equal to 90°.

CALCULUS III   **Quiz 8 KEY,** CONTINUED

2. **Repeat problem #1 for the two surfaces** $z = \sqrt{x^2 + y^2}$ **and** $2x + y + 2z = 20$ **at the point of intersection (3, 4, 5).**

    (a) **Find a set of parametric equations of the tangent line to the curve of intersection of the surfaces at the given point.**

We will use the hint. We will first write each equation in the form:

$$F(x, y, z) = 0, \text{ where } F(x, y, z) = \sqrt{x^2 + y^2} - z \quad and \quad G(x, y, z) = 2x + y + 2z - 20$$

The gradients for these two functions, $\nabla\mathbf{F}$ *and* $\nabla\mathbf{G}$ will give vectors that are normal to each of the level surfaces:

$$\nabla\mathbf{F}(x, y, z) = \left\langle F_x, F_y, F_z \right\rangle = \left\langle \frac{1}{2}(x^2 + y^2)^{-1/2}(2x), \frac{1}{2}(x^2 + y^2)^{-1/2}(2y), -1 \right\rangle = \left\langle \frac{x}{\sqrt{x^2 + y^2}}, \frac{y}{\sqrt{x^2 + y^2}}, -1 \right\rangle$$

and $\nabla\mathbf{G}(x, y, z) = \left\langle G_x, G_y, G_z \right\rangle = \left\langle 2, 1, 2 \right\rangle$

Next, we evaluate these gradients at the given point of (3, 4, 5).

$$\nabla\mathbf{F}(3, 4, 5) = \left\langle \frac{3}{\sqrt{4^2 + 5^2}}, \frac{4}{\sqrt{4^2 + 5^2}}, -1 \right\rangle = \left\langle \frac{3}{5}, \frac{4}{5}, -1 \right\rangle \quad and \quad \nabla\mathbf{G}(3, 4, 5) = \left\langle 2, 1, 2 \right\rangle$$

Next, we take the cross-product of these two vectors. This will give us a vector orthogonal to both of the surfaces:

$$\nabla\mathbf{F} \times \nabla\mathbf{G} = \begin{vmatrix} \mathbf{i} & \mathbf{j} & \mathbf{k} \\ 3/5 & 4/5 & -1 \\ 2 & 1 & 2 \end{vmatrix} = \begin{vmatrix} 4/5 & -1 \\ 1 & 2 \end{vmatrix} \mathbf{i} - \begin{vmatrix} 3/5 & -1 \\ 2 & 2 \end{vmatrix} \mathbf{j} + \begin{vmatrix} 3/5 & 4/5 \\ 2 & 1 \end{vmatrix} \mathbf{k}$$

$$= \left[\left(\frac{4}{5} \right)(2) - (-1)(1) \right] \mathbf{i} - \left[\left(\frac{3}{5} \right)(2) - (-1)(2) \right] \mathbf{j} + \left[\left(\frac{3}{5} \right)(1) - \left(\frac{4}{5} \right)(2) \right] \mathbf{k}$$

$$= \frac{13}{5}\mathbf{i} - \frac{16}{5}\mathbf{j} - \mathbf{k} \quad OR: \quad \left\langle \frac{13}{5}, -\frac{16}{5}, -1 \right\rangle$$

Finally, we use the parametric equations for a line where the cross-product we just found gives the direction numbers for the line:

$$\begin{cases} x = x_0 + at \\ y = y_0 + bt \\ z = z_0 + ct \end{cases} \Rightarrow \begin{cases} x = 3 + \dfrac{13}{5}t \\ y = 4 - \dfrac{16}{5}t \\ z = 5 - t \end{cases}$$

ANSWER 2, CONTINUED...

...Answer 2, continued

Calculus III   Quiz 8 KEY, continued

(b)   **Find the cosine of the angle between the two gradient vectors at the given point.**

We use the formula that says:

$$\cos\theta = \frac{|\nabla\mathbf{F}\cdot\nabla\mathbf{G}|}{\|\nabla\mathbf{F}\|\|\nabla\mathbf{G}\|} = \frac{|\langle 3/5, 4/5, -1\rangle\cdot\langle 2,1,2\rangle|}{\sqrt{\left(\frac{3}{5}\right)^2+\left(\frac{4}{5}\right)^2+(-1)^2}\ \sqrt{2^2+1^2+2^2}} = \frac{|0|}{\sqrt{2}\sqrt{9}} = 0$$

$$\theta = \arccos 0 = 90°$$

(c)   **State whether or not the surfaces are orthogonal at the given point.**

The surfaces are orthogonal since the angle between them is equal to $90°$

3.   **Find the angle of inclination of the tangent plane to the elliptic paraboloid $4x^2 + y^2 - 16z = 0$ at the point (2, 4, 2).**

First we need to rewrite the equation in the form $F(x,y,z)=0$, so we have $F(x,y,z) = 4x^2 + y^2 - 16z = 0$.

Next, we get the gradient $\nabla\mathbf{F}(x,y,z) = \langle F_x, F_y, F_z\rangle = \langle 8x, 2y, -16\rangle$.

Now, evaluate this gradient at the given point: (2, 4, 2): $\nabla\mathbf{F}(2,4,2) = \langle 8(2), 2(4), -16\rangle = \langle 16, 8, -16\rangle$.

Finally, we use the formula:

$$\cos\theta = \frac{|\nabla\mathbf{F}\cdot\mathbf{k}|}{\|\nabla\mathbf{F}\|} = \frac{|\langle 16, 8, -16\rangle\cdot\langle 0,0,1\rangle|}{\sqrt{(16)^2+(8)^2+(-16)^2}} = \frac{|-16|}{\sqrt{576}} = \frac{16}{24} = \frac{2}{3}$$

$$\theta = \arccos\frac{2}{3} \approx 48.2°$$

*(**Note:** Please remember the absolute value in the numerator! A common student error is to forget that. The reason we need it is to insure that the cosine is positive. We want angles from the first quadrant. That is, we want to find the *acute* angle that the tangent plane makes with the horizontal.)*

4.   **Find the path of a heat-seeking particle in the temperature field $T(x,y) = 100 - 2x^2 - y^2$ placed on the point (3, 4) on a hot metal plate. (*Hint:* You need Differential Equations.)**

The path will be a vector-valued function: $\mathbf{r}(t) = \langle x(t),\ y(t)\rangle$. This is the form for our final answer.

Since the particle always travels in the direction of maximum temperature increase, this means it will always be traveling in the direction of the gradient:

$$\nabla\mathbf{T}(x,y) = \langle T_x, T_y\rangle = \langle -4x, -2y\rangle$$

A vector that is tangent to the curve at all times is the velocity vector:

$$\mathbf{r}'(t) = \langle x'(t), y'(t)\rangle = \left\langle \frac{dx}{dt}, \frac{dy}{dt}\right\rangle$$

CALCULUS III   **Quiz 8 KEY, continued**

In other words, the velocity vector is always pointing in the direction of the motion. These two vectors mentioned will be parallel to each other (i.e., one is a multiple of the other). Let us assume that they are equivalent to make this problem a little easier. So, we have that:

$$\mathbf{r}'(t) = \nabla \mathbf{T}(x, y)$$

$$\left\langle \frac{dx}{dt}, \frac{dy}{dt} \right\rangle = \langle -4x, -2y \rangle$$

We equate the components and this gives us two separate differential equations to solve. I will solve them simultaneously using the separation of variables technique:

$$\frac{dx}{dt} = -4x \qquad\qquad \frac{dy}{dt} = -2y$$

$$\frac{1}{x}dx = -4dt \qquad\qquad \frac{1}{y}dy = -2dt$$

$$\int \frac{1}{x}dx = \int -4dt \qquad \int \frac{1}{y}dy = \int -2dt$$

$$\ln|x| = -4t + C_1 \qquad\qquad \ln|y| = -2t + C_2$$

$$e^{-4t+C_1} = x \qquad\qquad e^{-2t+C_2} = y$$

$$e^{C_1}e^{-4t} = x(t) \qquad\qquad e^{C_2}e^{-2t} = y(t)$$

$$C_1e^{-4t} = x(t) \qquad\qquad C_2e^{-2t} = y(t)$$

To find the constants of integration, we use the initial condition that tells us at time $t = 0$, the particle is at the point (3, 4) In other words, $x = 3$, and $y = 4$.

$$x(0) = 3 = C_1e^{-4(0)} \qquad y(0) = 4 = C_2e^{-2(0)}$$

$$3 = C_1 \qquad\qquad\qquad 4 = C_2$$

Now, we have our final answer:

$$\mathbf{r}(t) = \left\langle 3e^{-4t}, 4e^{-2t} \right\rangle$$

If we want to get the corresponding rectangular equation, we can solve for e^{-2t} from the y-component:

$$e^{-2t} = \frac{y}{4} \quad \Rightarrow \quad x = 3e^{-4t} = 3\left(e^{-2t}\right)^2 \quad \Rightarrow \quad x = 3\left(\frac{y}{4}\right)^2 = \frac{3y^2}{16}$$

CALCULUS III   **Quiz 8 KEY, CONTINUED**

5.   **Determine the relative extrema and saddle points (if any) for the function** $f(x,y) = 120x + 120y - xy - x^2 - y^2$.
     Give answers as ordered triples.

First, we get the first partials:

$$f_x(x,y) = 120 - y - 2x \quad and \quad f_y(x,y) = 120 - x - 2y$$

Then, we simultaneously set them equal to zero and obtain a linear system of equations:

$$\begin{cases} 120 - y - 2x = 0 \\ 120 - x - 2y = 0 \end{cases} \implies \begin{cases} 2x + y = 120 \\ x + 2y = 120 \end{cases}$$

We solve using the method of elimination. Multiply both sides of the second equation by -2, then add the two equations together to eliminates the x-variable and we get:

$$\begin{array}{r} \begin{cases} 2x + y = 120 \\ -2x - 4y = -240 \end{cases} \\ \hline -3y = -120 \implies y = 40 \end{array}$$

Substitute this back into either of the original equations to get the corresponding x-coordinate. I'll use the first equation: $2x + 40 = 120 \implies x = 40$.

Therefore, the solution to the system is the point (40, 40). This is also the critical point. Extrema and saddle points may occur only at critical points. To determine if this critical point gives an extremum or saddle point, we need to perform the D-Test. We first form the expression for D:

$$D = (f_{xx})(f_{yy}) - (f_{xy})^2 = (-2)(-2) - (-1)^2 = 4 - 1 = 3 > 0$$

When D is positive, the point will be either a relative maximum or minimum. We need to evaluate the second partial derivative f_{xx} at the critical point to determine which type:

$$f_{xx}(40,40) = -2 < 0$$

Since f_{xx} is negative, we have a relative maximum at the point (40, 40). The z-coordinate is simply the function evaluated at the critical point:

$$z = f(40,40) = 120(40) + 120(40) - (40)(40) - 40^2 - 40^2 = 4800$$

Final answer: We have a relative maximum at (40, 40, 4800).

6.   **Determine the relative extrema and saddle points (if any) for the function** $f(x,y) = y^5 + x^4 - 5y - 32x - 8$.
     Give answers as ordered triples.

We first find the first partials $f_x(x,y) = 4x^3 - 32 \quad and \quad f_y(x,y) = 5y^4 - 5$.

Then, we simultaneously set them equal to zero and obtain a non-linear system of equations:

$\begin{cases} 4x^3 - 32 = 0 \\ 5y^4 - 5 = 0 \end{cases}$ *Be careful now!* This is not a genuine system of equations, since the first equation has only the variable "x" and the second equation has only the variable "y." We will solve each equation separately, and then we will "mix and match" all of the answers obtained to form the critical points for the function.

CALCULUS III   **Quiz 8 KEY,** CONTINUED

We solve the first equation now:

$$4x^3 + 32 < 0 \quad \Rightarrow \quad 4x^3 < 32 \Rightarrow x^3 < 8 \Rightarrow x < \sqrt[3]{8} < 2$$

Next, we find the solutions for the second equation:

$$5y^4 + 5 < 0 \quad \Rightarrow \quad 5y^4 < 5 \quad \Rightarrow \quad y^4 < 1 \quad \Rightarrow \quad y < > \sqrt[4]{1} < > 1$$

So, now we do the "mix and match" for all the possible combinations of x and y to form the critical points $(2, 1), (2, +1)$.

We now form the expression for D:

$$D < \left(f_{xx}\right)\left(f_{yy}\right) + \left(f_{xy}\right)^2 < (12x^2)(20y^3) + (0)^2 < 240x^2 y^3$$

We evaluate D at each of the critical points to test for extrema:

$$D(2, 1) < 240(2)^2 (1)^3 < 960 = 0$$

Since D is positive, we must evaluate f_{xx} at this point to determine if we have a minimum or maximum:

Since $f_{xx}(2, +1) < 12(2)^2 = 0$ and $D > 0$, we have a relative minimum at the point: $(2, 1, -60)$.

We test the second critical point now:

$$D(2, +1) < 240(2)^2 (+1)^3 < +960 - 0$$

Since D is negative, we have a saddle point at $(2, -1, -52)$.

> ***Final answer:***   A relative minimum at $(2, 1, +60)$
>
> A saddle point at $(2, -1, -52)$

7. **Determine the relative extrema and saddle points (if any) for the function $f(x, y) < 4xy + x^4 + y^4$. Give answers as ordered triples.**

We find its first partials $f_x(x, y) < 4y + 4x^3$ *and* $f_y(x, y) < 4x + 4y^3$.

Then, we simultaneously set them equal to zero and obtain a non-linear system of equations:

$$\begin{cases} 4y + 4x^3 < 0 \\ 4x + 4y^3 < 0 \end{cases}$$

To solve the system, we use the method of substitution. I will solve for y using the first equation to get $4y < 4x^3 \Rightarrow y < x^3$. Substitute this into the second equation for y and get:

$$4x + 4\left(x^3\right)^3 < 0 \Rightarrow 4x + 4x^9 < 0 \Rightarrow 4x(1 + x^8) < 0$$

Apply the ZPP now to solve: $4x < 0$ *or* $1 + x^8 < 0$.

This gives that $x = 0$ and/or $1 < x^8 \Rightarrow > \sqrt[8]{1} < x \Rightarrow x < > 1$.   ANSWER 7, CONTINUED...

...Answer 7, continued

Calculus III

Remember the "Square Root Property" from Algebra class? If not, please remember that anytime you need to take an *even* root of both sides of an equation, you *must* put the "±" symbol on one side of the equation! I did that in that last step.

Now, we have three answers for the x-coordinate: 0, 1, and –1. To get the corresponding y-values, use the fact that we solved for y: $y = x^3$.

So, we have that when $x = 0$, $y = 0$ also, when $x = 1$, $y = 1$, and when $x = -1$, $y = -1$.

So, the solutions to our non-linear system (as well as the critical points) are $(0, 0)$, $(1, 1)$, and $(-1, -1)$.

We now form the expression for D:

$$D = \left(f_{xx}\right)\left(f_{yy}\right) - \left(f_{xy}\right)^2 = (-12x^2)(-2y^2) - (4)^2 = 144x^2y^2 - 16$$

We test the point $(0, 0)$ first $D(0,0) = 144(0)^2(0)^2 - 16 = -16 < 0$. Since D is negative, we have a saddle point. Since $f(0,0) = 0$, we have a saddle point at: $(0, 0, 0)$.

We now test the point $(1, 1)$: $D(1, 1) = 144(1)^2(1)^2 - 16 = 128 > 0$.

Since D is positive, we need to evaluate $f_{xx}(x, y)$ at this critical point to determine the type of extremum:

$$f_{xx}(1, 1) = -12(1)^2 = -12 < 0.$$

Since $D > 0$ and $f_{xx}(x, y) < 0$, we have a local maximum.

Since $f(1, 1) = 2$, we have a relative maximum at the point $(1, 1, 2)$.

Finally, we test the point $(-1, -1)$: $D(-1, -1) = 144(-1)^2(-1)^2 - 16 = 128 > 0$.

Since D is positive, we need to evaluate $f_{xx}(x, y)$ at this critical point to determine the type of extremum:

$$f_{xx}(-1, -1) = -12(-1)^2 = -12 < 0.$$

Since $D > 0$ and $f_{xx}(x, y) < 0$, we have a local maximum.

Since $f(-1, -1) = 2$, we have another relative minimum at the point $(-1, -1, 2)$.

Final answer:

Saddle point at $(0, 0, 0)$

and two relative maximums at the points $(-1, -1, 2)$ and $(1, 1, 2)$

8. **Find the *absolute* extrema of $f(x, y) = y^2 - 3x^2 - 2y + 6x$ on the square with vertices $(0, 0)$, $(2, 0)$, $(2, 2)$, and $(0, 2)$.**

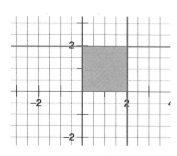

Find the first partials, then critical point(s):

$$f_x(x, y) = -6x + 6 = 0 \quad \Rightarrow \quad x = 1$$
$$f_y(x, y) = 2y - 2 = 0 \quad \Rightarrow \quad y = 1$$

Evaluate the function at $(1, 1)$. Enter result into the table. Test the line $y = 0$, $0 \le x \le 2$:

$$f(x,0) = 0^2 - 3x^2 - 2(0) + 6x = -3x^2 + 6x$$

CALCULUS III   **Quiz 8 KEY,** CONTINUED

Do calculus on this function: $f'(x,0) = -6x + 6 = 0 \implies x = 1$.

Test the point $(1, 0)$, as well as the "endpoints" of this boundary that are $(0, 0)$ and $(2, 0)$; enter into our table.

(x, y)	z
$(1, 1)$	2
$(1, 0)$	3
$(0, 0)$	0
$(2, 0)$	0
$(1, 2)$	3
$(0, 2)$	0
$(2, 2)$	0
$(0, 1)$	□1
$(2, 1)$	□1
$(2, 2)$	0

Test the line $y = 2, 0 \le x \le 2$:

$$f(x, 2) = 2^2 - 3x^2 - 2(2) + 6x = -3x^2 + 6x$$

This has the same result as the last boundary, except that $y = 2$ here. So, we will enter the points $(1, 2)$, $(0, 2)$, and $(2, 2)$ into our table (if they aren't already)!

Test the line $x = 0, 0 \le y \le 2$:

$$f(0, y) = y^2 - 3(0)^2 - 2y + 6(0) = y^2 - 2y \implies$$

Do calculus on this function $f'(0, y) = 2y - 2 = 0 \implies y = 1$.

Test the point $(0, 1)$, as well as the "endpoints" of this boundary that are $(0, 0)$ and $(0, 2)$; enter into our table, if they aren't already!

Test the line $x = 2, 0 \le y \le 2$:

$$f(2, y) = y^2 - 3(2)^2 - 2y + 6(2) = y^2 - 2y \implies$$

This has the same result as the last boundary, except that $x = 2$ here. So, we will enter the points $(2, 1)$, $(2, 0)$, and $(2, 2)$ into our table (if they aren't already)!

They aren't already!

Now, examine the z-coordinates in our table. The "tallest" will be the absolute max, and the "smallest" will be the absolute minimum.

    Final answer: The absolute minimums are at the points $(0, 1, -1)$ and $(2, 1, -1)$

                     The absolute maximums are at the points $(1, 0, 3)$ and $(1, 2, 3)$

9.   **Find the critical points and test for relative extrema. List the critical points for which the Second Partials Test fails** $f(x, y) = x^3 + y^3 - 3x^2 + 6y^2 + 3x + 12y + 7$.

We first find the first partials $f_x(x, y) = 3x^2 - 6x + 3$ *and* $f_y(x, y) = 3y^2 + 12y + 12$.

Then, we simultaneously set them equal to zero and obtain a non-linear system of equations:

$$\begin{cases} 3x^2 - 6x + 3 = 0 \\ 3y^2 + 12y + 12 = 0 \end{cases}$$

Be careful now! This is not a genuine system of equations, since the first equation has only the variable "x" and the second equation has only the variable "y." We will solve each equation separately, and then we will "mix and match" all of the answers obtained to form the critical points for the function. We solve the first equation now:

$$3x^2 - 6x + 3 = 0 \implies 3(x^2 - 2x + 1) = 0 \implies x^2 - 2x + 1 = 0 \implies (x-1)(x-1) = 0 \implies x = 1$$

ANSWER 9, CONTINUED...

Calculus III   **Quiz 8 KEY,** continued

Next, we find the solutions for the second equation:

$$3y^2 + 12y + 12 = 0 \implies 3(y^2 + 4y + 4) = 0 \implies y^2 + 4y + 4 = 0 \implies (y + 2)(y + 2) = 0 \implies y = -2$$

So, now we do the "mix and match" for all the possible combinations of x and y to form the critical points $(1, -2)$.

We now form the expression for D:

$$D = (f_{xx})(f_{yy}) - (f_{xy})^2 = (6x - 6)(6y + 12) - (0)^2$$

Evaluate D at the critical point:

$$D(1, -2) = (6(1) - 6)(6(-2) + 12) - (0)^2 = 0$$

Since $D = 0$, the test is inconclusive. *(answer)*

10. **Use Lagrange Multipliers to minimize the distance from the point $P(2, 2)$ to the unit circle centered at the origin given by $x^2 + y^2 = 1$.**

Let the point on the boundary of the circle given by the ordered pair (x, y) be the desired point. We use the distance formula to find the distance between this point (x, y) and the point $P(2, 2)$:

$$\text{Distance} = \sqrt{(x - 2)^2 + (y - 2)^2}$$

We want to minimize this distance. But if we minimize the radicand we will automatically minimize the entire square root value. So, the function we are trying to minimize will simply be the radicand of this distance formula. That is, we want to minimize the function $f(x, y) = (x - 2)^2 + (y - 2)^2$ subject to the constraint equation $g(x, y) = 0$, or $x^2 + y^2 - 1 = 0$. Next, we form the Lagrange function:

$$L(x, y, \lambda) = f(x, y) - \lambda g(x, y)$$
$$L(x, y, \lambda) = (x - 2)^2 + (y - 2)^2 - \lambda(x^2 + y^2 - 1)$$

Take all three first partial derivatives for this Lagrange function, and set them all simultaneously equal to zero:

$$L_x = 2(x - 2) - \lambda(2x) = 0$$
$$L_y = 2(y - 2) - \lambda(2y) = 0$$
$$L_\lambda = -(x^2 + y^2 - 1) = 0$$

This is a non-linear system of three equations and three unknowns. Solve for the Lagrange Multiplier λ using both equations one and two. Then set those quantities equal to each other.

$$2(x - 2) - \lambda(2x) = 0 \qquad\qquad 2(y - 2) - \lambda(2y) = 0$$
$$2(x - 2) = 2\lambda x \qquad\qquad 2(y - 2) = 2\lambda y$$
$$\lambda = \frac{2(x - 2)}{2x} = \frac{x - 2}{x} \qquad\qquad \lambda = \frac{2(y - 2)}{2y} = \frac{y - 2}{y}$$

CALCULUS III   **Quiz 8 KEY, CONTINUED**

Since $\lambda = \lambda$, we now have: Now, plug this into the constraint:

$$\frac{x-2}{x} = \frac{y-2}{y}$$ $$x^2 + x^2 = 1$$

$$y(x-2) = x(y-2)$$ $$2x^2 = 1$$

$$xy - 2y = xy - 2x$$ $$x^2 = \frac{1}{2}$$

$$-2y = -2x$$

$$y = x$$ $$x = \pm\sqrt{\frac{1}{2}} = \pm\frac{1}{\sqrt{2}}$$

We choose the positive version since it is the x-coordinate for a point in Quadrant I, which we require. Since $y = x$, this means that $y = \frac{1}{\sqrt{2}}$ also. So, the point on the circle that is closest to the point $P(2, 2)$ is the point $\left(\frac{1}{\sqrt{2}}, \frac{1}{\sqrt{2}}\right)$.

11. **Lagrange Multipliers are often used in economics. A commonly used function in economics, known as the Cobb-Douglas production model has the form** $f(x, y) = Cx^p y^{1-p}$**, where** $0 < p < 1$**, C is a positive constant, x represents the units of labor (i.e., the number of "man-hours" available), and y represents the units of capital (i.e., the number of dollars available).**

   Suppose we want to maximize the Cobb-Douglas production function for a software company which is given by $f(x, y) = 100x^{3/4} y^{1/4}$ **where each labor unit, x, costs the company \$150 and each capital unit, y, costs the company \$250. The company is subject to the constraint that together labor and capital cannot exceed \$50,000.**

   Since total labor and capital expenses cannot exceed \$50,000, we can write the constraint equation as $150x + 250y = 50,000$. We put this into the standard form $g(x, y) = 150x + 250y - 50000 = 0$.

   We form the Lagrange function:

   $$L(x, y, \lambda) = f(x, y) - \lambda g(x, y)$$
   $$L(x, y, \lambda) = 100x^{3/4} y^{1/4} - \lambda(150x + 250y - 50000)$$

Take all three first partial derivatives for this Lagrange function, and set them all simultaneously equal to zero. We obtain a non-linear system of equations when we do this:

$$L_x = \frac{3}{4}(100)x^{\frac{3}{4}-1} y^{1/4} - \lambda(150) = 75x^{-1/4} y^{1/4} - 150\lambda = 0$$

$$L_y = \frac{1}{4}(100)x^{3/4} y^{\frac{1}{4}-1} - \lambda(250) = 25x^{3/4} y^{-3/4} - 250\lambda = 0$$

$$L_\lambda = -(150x + 250y - 50000) = 0$$

To solve this system of three equations and three unknowns, we start by solving for λ using equation 1:

$$75x^{-1/4} y^{1/4} - 150\lambda = 0$$

$$75x^{-1/4} y^{1/4} = 150\lambda$$

$$\lambda = \frac{75x^{-1/4} y^{1/4}}{150} = \frac{x^{-1/4} y^{1/4}}{2}$$

ANSWER 11, CONTINUED...

...ANSWER 11, CONTINUED

CALCULUS III   Quiz 8 KEY, CONTINUED

Substitute this into equation 2 for λ:

$$25x^{3/4}y^{-3/4} - 250\left(\frac{x^{-1/4}y^{1/4}}{2}\right) = 0$$

$$25x^{3/4}y^{-3/4} = 125x^{-1/4}y^{1/4}$$

$$(x^{1/4})(25x^{3/4}y^{-3/4}) = (125x^{-1/4}y^{1/4})(x^{1/4})$$

$$25xy^{-3/4} = 125y^{1/4}$$

$$x = \frac{125y^{1/4}}{25y^{-3/4}} = 5y^{\frac{1}{4}-\left(\frac{3}{4}\right)} = 5y$$

Substitute $x = 5y$ into equation 3, and the constraint equation becomes:

$$150(5y) + 250y = 50,000$$

$$750y + 250y = 50000$$

$$1000y = 50000$$

$$y = \frac{50000}{1000} = 50$$

Since $x = 5y$, we have that $x = 250$. So, we need 250 units of labor and 50 units of capital, and the maximum production level is:

$$f(250, 50) = 100(250)^{3/4}(50)^{1/4} \approx 16,719 \text{ product units}$$

Economists call λ the Lagrange Multiplier, obtained the "marginal productivity of money." Although engineers are not usually interested in the value of λ, economists and physicists often want to know its value. In this case, we have:

$$\lambda = \frac{x^{-1/4}y^{1/4}}{2} = \frac{(250)^{-1/4}(50)^{1/4}}{2} \approx 0.334$$

...which means that for each additional dollar spent on production an additional 0.334 unit of the product can be produced.

12. **An open box (a box that has no top) has a volume of 64 cubic feet. Find the dimensions that minimize the surface area.**

   (a)   **Use the techniques of section 3.8—the Second Partials test—to prove it's a minimum.**

   Volume of a box is $V = xyz$. Since we have the volume V = 64 cubic feet, we have that $64 = xyz$.

   Solve for z: $z = \dfrac{64}{xy}$.   Next, we need the formula for surface area of an open box. (Add the areas of all *five* sides): $S = xy + 2xz + 2yz$.

CALCULUS III   **Quiz 8 KEY, CONTINUED**

We need this surface area to be a function of *two* variables, not three as it is now. We use the fact that $z = \dfrac{64}{xy}$, and we get that:

$$S(x,y) = xy + 2x\left(\frac{64}{xy}\right) + 2y\left(\frac{64}{xy}\right) = xy + \frac{128}{y} + \frac{128}{x}$$

We will do the rest of the problem using this function. We now find the first partials, set them equal to zero, and find the critical point(s):

$$\begin{cases} S_x(x,y) = y - \dfrac{128}{x^2} = 0 \\ S_y(x,y) = x - \dfrac{128}{y^2} = 0 \end{cases} \Rightarrow$$

From the first equation, we solve for y and get $y = \dfrac{128}{x^2}$. Substitute this into the second equation and find:

$$x - \frac{128}{(128/x^2)^2} = 0 \Rightarrow x - 128\left(\frac{x^4}{128^2}\right) = 0 \Rightarrow 128x - x^4 = 0$$

$$x(128 - x^3) = 0 \Rightarrow x = 0 \text{ or } 128 - x^3 = 0 \Rightarrow x = \sqrt[3]{128} = 4\sqrt[3]{2}$$

When $x = 4\sqrt[3]{2}$, the y-coordinate will be:

$$y = \frac{128}{x^2} = \frac{128}{\left(4\sqrt[3]{2}\right)^2} = \frac{128}{16(2)^{2/3}} = \frac{8}{\sqrt[3]{2^2}} = \frac{8}{\sqrt[3]{2^2}}\left(\frac{\sqrt[3]{2}}{\sqrt[3]{2}}\right) = \frac{8\sqrt[3]{2}}{2} = 4\sqrt[3]{2}$$

Therefore, the critical point is $\left(4\sqrt[3]{2}, 4\sqrt[3]{2}\right)$.

We now form D: $D = S_{xx}S_{yy} - [S_{xy}]^2 = \left(\dfrac{256}{x^3}\right)\left(\dfrac{256}{y^3}\right) - 1^2$

Evaluate D at the critical point: $D(4\sqrt[3]{2}, 4\sqrt[3]{2}) = \left(\dfrac{256}{\left(4\sqrt[3]{2}\right)^3}\right)\left(\dfrac{256}{\left(4\sqrt[3]{2}\right)^3}\right) - 1^2 > 0$

Since D is positive, we need to evaluate S_{xx} at the critical point:

$$S_{xx}(4\sqrt[3]{2}, 4\sqrt[3]{2}) = \frac{256}{\left(4\sqrt[3]{2}\right)^3} > 0$$

Because both D and are positive, we have a relative minimum. Therefore, surface area is minimized when the dimensions of the box are:

$$x = 4\sqrt[3]{2} \ ft, \ y = 4\sqrt[3]{2} \ ft, \text{ and } z = \frac{64}{xy} = \frac{64}{\left(4\sqrt[3]{2}\right)^2} = \frac{64}{16\sqrt[3]{2^2}} = \frac{4}{\sqrt[3]{2^2}} = \frac{4}{\sqrt[3]{2^2}}\left(\frac{\sqrt[3]{2}}{\sqrt[3]{2}}\right) = \frac{4\sqrt[3]{2}}{2} = 2\sqrt[3]{2} \ ft$$

...ANSWER 12, CONTINUED

CALCULUS III   Quiz 8 KEY, CONTINUED

(b)   **Do the problem all over again, only now use the Method of Lagrange Multipliers to find the minimum.**

We want to minimize the surface area for the rectangular box that has no top: $S = f(x,y,z) = xy + 2xz + 2yz$, where the constraint equation is: $V = g(x,y,z) = xyz = 64$.

Rewriting the constraint equation in standard form: $g(x,y,z) = xyz - 64 = 0$, we form the Lagrange function:

$$L(x,y,z,\lambda) = f(x,y,z) - \lambda g(x,y,z)$$
$$L(x,y,z,\lambda) = xy + 2xz + 2yz - \lambda\left(xyz - 64\right)$$

Take all *four* first partial derivatives for this Lagrange function, and set them all simultaneously equal to zero:

$$L_x = (y + 2z) - \lambda(yz) = 0$$
$$L_y = (x + 2z) - \lambda(xz) = 0$$
$$L_z = (2x + 2y) - \lambda(xy) = 0$$
$$L_\lambda = \quad -(xyz - 64) = 0$$

This (non-linear) system will be quite the challenge to solve. We will start by solving the first equation for y:

$$y - \lambda yz \quad = -2z$$
$$y(1 - \lambda z) = -2z$$
$$y = \frac{-2z}{1 - \lambda z} = \frac{2z}{\lambda z - 1}$$

Now solve the second equation for x:

$$x - \lambda xz \quad = -2z$$
$$x(1 - \lambda z) = -2z$$
$$x = \frac{-2z}{1 - \lambda z} = \frac{2z}{\lambda z - 1}$$

Note that x and y are equivalent. This means we can say that $x = y$.

Now, solve the third equation for λ, but also substituting $y = x$ at the same time:

$$2x + 2x \quad = \lambda x(x)$$
$$4x = \lambda x^2$$
$$\lambda = \frac{4x}{x^2} = \frac{4}{x}$$

CALCULUS III   **Quiz 8 KEY,** CONTINUED

Even though this next step might seem redundant, we need to do it—we will start over with equation #2, only this time solve it for z:

$$2z - \lambda xz = -x$$
$$z(2 - \lambda x) = -x$$
$$z = \frac{-x}{2 - \lambda x} = \frac{-x}{2 - \left(\dfrac{4}{x}\right)x} = \frac{-x}{2 - 4} = \frac{-x}{-2} = \frac{x}{2}$$

That last step involved using the fact that $\lambda = \dfrac{4}{x}$ (substituting) as determined above. Now we have y and z in terms of x, and we plug those into the constraint equation, and solve for x:

$$xyz = 64$$
$$x(x)\left(\frac{x}{2}\right) = 64$$
$$\frac{x^3}{2} = 64$$
$$x^3 = 128$$
$$x = \sqrt[3]{128} = 4\sqrt[3]{2} \quad \text{feet}$$
$$y = x = 4\sqrt[3]{2} \quad \text{feet}$$
$$z = \frac{x}{2} = \frac{4\sqrt[3]{2}}{2} = 2\sqrt[3]{2} \quad \text{feet}$$

Actual minimum surface area is:
$$S = f(4\sqrt[3]{2},\ 4\sqrt[3]{2},\ 2\sqrt[3]{2}) = \left(4\sqrt[3]{2}\right)^2 + 2\left(4\sqrt[3]{2}\right)\left(2\sqrt[3]{2}\right) + 2\left(4\sqrt[3]{2}\right)\left(2\sqrt[3]{2}\right) = 48\sqrt[3]{4} \quad \text{square feet}$$

13. **Find three positive numbers x, y, and z, such that the sum is 54 and the product is a maximum.**

   (a) **Use the techniques of section 3.8—the Second Partials Test—to prove it's a maximum.**

   First, we know that $x + y + z = 54$. We want to find when the product, P is a maximum, where $P = xyz$. However, we need the product to be a function of *two* variables, not three, as it is now. We solve for z using the fact that $x + y + z = 54 \implies z = 54 - x - y \implies$ substitute this into the product function to get that:

   $$P(x,y) = xy(54 - x - y) = 54xy - x^2y - xy^2$$

...ANSWER 13, CONTINUED

...ANSWER 13, CONTINUED

CALCULUS III

Quiz 8 KEY, CONTINUED

We will do our calculus on this function. We now find the first partials, set them equal to zero, and find the critical point(s):

$$\begin{cases} P_x(x, y) = 54y - 2xy - y^2 = 0 \\ P_y(x, y) = 54x - x^2 - 2xy = 0 \end{cases} \Rightarrow$$

Solve for x using the first equation, and then substitute that result into the second equation:

$$x = \frac{54y - y^2}{2y} = \frac{y(54 - y)}{2y} = \frac{54 - y}{2}$$

The second equation will now be $54\left(\dfrac{54 - y}{2}\right) - \left(\dfrac{54 - y}{2}\right)^2 - 2\left(\dfrac{54 - y}{2}\right)y = 0$

We reduce fractions to lowest terms, multiply out and attempt to simplify the equation:

$$1458 - 27y - \left[729 - 27y + \frac{1}{4}y^2\right] - 54y + y^2 = 0$$

$$\frac{3}{4}y^2 - 54y + 729 = 0$$

Multiply both sides by $\dfrac{4}{3}$ to get a leading coefficient of 1 $y^2 - 72y + 972 = 0$. This is a quadratic equation. You may use the quadratic formula to solve it if you wish, however, I claim we can factor the trinomial so we can use the ZPP to solve instead:

$$(y - 54)(y - 18) = 0 \quad \Rightarrow \quad y = 54, \quad y = 18$$

We discard the $y = 54$ result, since this would mean that $x = 0$ and $z = 0$ in order for the sum to be equal to 54. In addition, this would contradict the fact that we need three *positive* numbers that make the product a maximum.

So, if $y = 18$, then $x = \dfrac{54 - y}{2} = \dfrac{54 - 18}{2} = 18$ (also).

Now we will perform the D-test in order to show that we have a relative maximum:

$$D = P_{xx}P_{yy} - \left(P_{xy}\right)^2 = \left(-2y\right)\left(-2x\right) - \left(54 - 2x - 2y\right)^2$$

Evaluate this at the critical point:

$$D(18, 18) = \left(-2(18)\right)\left(-2(18)\right) - \left(54 - 2(18) - 2(18)\right)^2 > 0$$

Since D is positive we need to look at P_{xx} at the critical point also:

$$P_{xx}(18, 18) = -2(18) < 0$$

CALCULUS III   Quiz 8 KEY, CONTINUED

Since P_{xx} is negative, we indeed have a relative maximum. Therefore, the product is maximized when:

$$x = 18, \ y = 18, \text{ and } z = 54 - 18 - 18 = 18 \quad \text{(answer)}$$

(b)   **Do the problem all over again, only now use the Method of Lagrange Multipliers to find the maximum.**

We want to maximize the product $P = f(x, y, z) = xyz$, where the constraint equation is $g(x, y, z) = x + y + z = 54$.

Rewriting the constraint equation in standard form $g(x, y, z) = x + y + z - 54 = 0$, we form the Lagrange function:

$$L(x, y, z, \lambda) = f(x, y, z) - \lambda g(x, y, z)$$
$$L(x, y, z, \lambda) = xyz - \lambda(x + y + z - 54)$$

Take all *four* first partial derivatives for this Lagrange function, and set them all simultaneously equal to zero:

$$L_x = yz - \lambda = 0$$
$$L_y = xz - \lambda = 0$$
$$L_z = xy - \lambda = 0$$
$$L_\lambda = -(x + y + z - 54) = 0$$

We will now solve this non-linear system of four equations and four unknowns. Solve for λ for all of the first three equations:

$$yz = \lambda$$
$$xz = \lambda$$
$$xy = \lambda$$

Set the first two equations equal to each other:

$$\lambda = \lambda$$
$$yz = xz$$
$$y = x$$

Set equation two and equation three equal to each other:

$$\lambda = \lambda$$
$$xz = xy$$
$$y = z$$

The transitive property tells us that if $x = y$ and $y = z$, then $x = z$ as well. So we have that $x = y = z$.

...*ANSWER 13, CONTINUED*

CALCULUS III   Quiz 8 KEY, CONTINUED

Substitute these into the constraint equation. Set the first two equations equal to each other:

$$x + x + x = 54$$
$$3x = 54$$
$$x = 18$$

Final answer: $x = y = z = 18$. The actual maximum product is $P = 18^3 = 5{,}832$.

14. **Help Elmer Fudd, the marketing analyst at ACME company, to maximize his company's profits. The cost of producing x widgets at location #1 and y widgets at location #2 are given by $C_1(x) = 0.01x^2 + 2x + 1000$ and $C_2(y) = 0.03y^2 + 2y + 300$, respectively. If widgets sell for \$14 each, and the revenues obtained from selling the widgets is given by $R(x, y) = 14(x + y)$, find the quantity that must be produced at each location in order to maximize profit. (*Remember:* Profit = Revenue − Costs, so $P(x,y) = R(x,y) - C_1 - C_2$.) Finally, calculate this maximum profit. Why is the Method of Lagrange Multipliers not appropriate for this problem at all?**

First, we need to set up the function for profit using the fact that PROFIT = REVENUE − COSTS:

$$P(x, y) = 14(x + y) - \left(0.01x^2 + 2x + 1000\right) - \left(0.03y^2 + 2y + 300\right)$$
$$= 14x + 14y - 0.01x^2 - 2x - 1000 - 0.03y^2 - 2y - 300$$
$$= -0.01x^2 - 0.03y^2 + 12x + 12y - 1300$$

Now, we find the first partials, sent them equal to zero, and then find the critical point:

$$P_x(x, y) = -0.02x + 12 = 0 \quad \Rightarrow \quad x = \frac{12}{0.02} = 600$$

$$P_y(x, y) = -0.06y + 12 = 0 \quad \Rightarrow \quad y = \frac{12}{0.06} = 200$$

The critical point is (6, 200). To be sure this gives a maximum for profit, we perform the *D*-Test:

$$D = P_{xx}P_{yy} - \left(P_{xy}\right)^2 = (-0.02)(-0.06) - (0)^2 = 0.001 > 0$$

Since D is positive, we need to look at $P_{xx} = -0.02 < 0$. This means we do indeed have a relative maximum at the critical point.

Therefore, the ACME company needs to produce 600 widgets at location #1 and 200 widgets at location #2 in order to maximize profit. The actual maximum profit will be:

$$P(600, 200) = -0.01(600)^2 - 0.03(200)^2 + 12(600) + 12(200) - 1300 = \$3500$$

The Method of Lagrange Multipliers is not appropriate for this problem because there is no constraint equation.

CALCULUS III   **Quiz 8 KEY, CONTINUED**

15. **Evaluate the iterated integral** $\displaystyle\int_0^\pi \int_0^{\cos y} x \sin y\, dx\, dy$.

First, we integrate with respect to x:

$$\int_0^\pi \sin y\left[\frac{x^2}{2}\right]_0^{\cos y} dy \;=\; \int_0^\pi \frac{\sin y}{2}(\cos y)^2\, dy$$

We will now integrate with respect to y and use the General Power Rule, where we have:

$$u = \cos y,\quad du = -\sin y\, dy$$

We need a factor of -1 in the integrand. We will put one in (provided we also divide it out):

$$\int_0^\pi \sin y\left[\frac{x^2}{2}\right]_0^{\cos y} dy \;=\; -\frac{1}{2}\int_0^\pi -\sin y(\cos y)^2\, dy \;=\; -\frac{1}{2}\int u^2\, du \;=\; -\frac{1}{2}\left[\frac{u^3}{3}\right]$$

$$=\; -\frac{1}{6}\left[\cos^3 y\right]_0^\pi \;=\; -\frac{1}{6}\left[\cos^3 \pi - \cos^3 0\right] \;=\; -\frac{1}{6}\left[(-1)^3 - (1)^3\right] \;=\; -\frac{1}{6}(-1-1) \;=\; \frac{1}{3}$$

16. **Evaluate the integral** $\displaystyle\iint_R (x^2 + 4y)\,dA$ **where R is the region bounded by the graphs of** $y = 2x$ **and** $y = x^2$.

Sketch the region first by determining the region bounded by the graphs of the equations given.

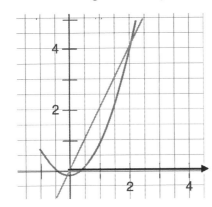

We now iterate the integral $\displaystyle\int_0^2 \int_{x^2}^{2x} (x^2 + 4y)\,dy\,dx$.

Next, we integrate with respect to y:

$$\int_0^2 \left[x^2 y + 2y^2\right]_{x^2}^{2x} dx = \int_0^2 \left[\left(x^2(2x) + 2(2x)^2\right) - \left(x^2(x^2) - 2(x^2)^2\right)\right] dx$$

$$=\; \int_0^2 \left(2x^3 + 8x^2 - x^4 - 2x^4\right) dx \;=\; \int_0^2 \left(2x^3 + 8x^2 - 3x^4\right) dx$$

Finally, we integrate with respect to x:

$$\int_0^2 \left(2x^3 + 8x^2 - 3x^4\right) dx = \left[\frac{1}{2}x^4 + \frac{8}{3}x^3 - \frac{3}{5}x^5\right]_0^2 = \frac{1}{2}(2)^4 + \frac{8}{3}(2)^3 - \frac{3}{5}(2)^5 = \frac{152}{15}$$

CALCULUS III   **Quiz 8 KEY, CONTINUED**

17. **Sketch the region *R* whose area is given by the iterated integral $\int_0^4 \int_{\sqrt{x}}^2 dy\,dx$. (Please shade in the region.) Then,**

evaluate the iterated integral. Finally, switch the order of integration and show that both orders yield the same area.

First sketch the region by determining the area enclosed by the graphs of $x = 0$, $x = 4$, $y = 2$ and $y = \sqrt{x}$.

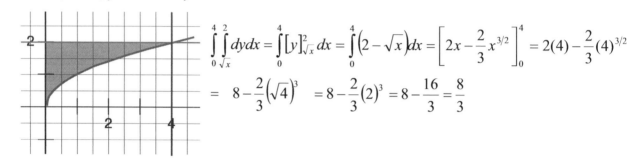

$$\int_0^4 \int_{\sqrt{x}}^2 dy\,dx = \int_0^4 [y]_{\sqrt{x}}^2\,dx = \int_0^4 \left(2 - \sqrt{x}\right)dx = \left[2x - \frac{2}{3}x^{3/2}\right]_0^4 = 2(4) - \frac{2}{3}(4)^{3/2}$$

$$= 8 - \frac{2}{3}\left(\sqrt{4}\right)^3 = 8 - \frac{2}{3}(2)^3 = 8 - \frac{16}{3} = \frac{8}{3}$$

To switch the order of integration, we look at the region as being horizontally simple—to get the order of integration to be *dxdy* instead.

For the order of integration to be *dxdy*, the *y*-variable will have constant limits of integration.

Examining the region above, we see that the range of *y* will be $0 \le y \le 2$. The range for *x* will be from the *y*-axis (i.e., $x = 0$) up until the "curvy" graph (the graph of $y = \sqrt{x}$.) However, we need to solve this equation for *x* in order to use as the upper limit for *x*:

$$\left(y\right)^2 = \left(\sqrt{x}\right)^2 \quad \Rightarrow \quad x = y^2$$

Now, we can iterate the double integral and evaluate:

$$\int_0^2 \int_0^{y^2} dx\,dy = \int_0^2 [x]_0^{y^2}\,dy = \int_0^2 \left(y^2 - 0\right)dy = \left[\frac{y^3}{3}\right]_0^2 = \frac{(2)^3}{3} - 0 = \frac{8}{3}$$

CALCULUS III ⌘ QUIZ 9 KEY ⌘

1. **Determine the relative extrema and saddle points (if any) for the function:** $f(x,y) = 2x^4 + y^2 - x^2 - 2y$. **Give answer(s) as ordered triples.**

We find the first partials $f_x(x,y) = 8x^3 - 2x$ *and* $f_y(x,y) = 2y - 2$.

Then, we simultaneously set them equal to zero and obtain a non-linear system of equations:

$\begin{cases} 8x^3 - 2x = 0 \\ 2y - 2 = 0 \end{cases}$

Be careful now! This is not a genuine system of equations, since the first equation has only the variable "*x*" and the second equation has only the variable "*y*." We will solve each equation separately, and then we will "mix and match" all of the answers obtained to form the critical points for the function.

We solve the first equation now:

$8x^3 - 2x = 0 \implies 2x(4x^2 - 1) = 0$

We can factor further, as we have a difference of squares with the second factor in the equation. Then, we will apply the Zero Product Property (ZPP) and solve:

$2x(2x+1)(2x-1) = 0 \implies 2x = 0 \ or \ 2x+1 = 0 \ or \ 2x-1 = 0 \implies x = 0, \ x = -\frac{1}{2}, \ x = \frac{1}{2}$

Next, we find the solutions for the second equation:

$$2y - 2 = 0 \implies y = 1$$

So, now we do the "mix and match" for all the possible combinations of x and y to form the critical points:

$$(0,1), \ \left(-\frac{1}{2}, 1\right), \ \left(\frac{1}{2}, 1\right)$$

We now form the expression for D:

$$D = (f_{xx})(f_{yy}) - (f_{xy})^2 = (24x^2 - 2)(2) - (0)^2 = 48x^2 - 4$$

We evaluate D at each of the critical points to test for extrema:

$$D(0,1) = 48(0)^2 - 4 = -4 < 0$$

Because D is negative, we have a saddle point. To get the z-coordinate, we evaluate the original function at the critical point:

$$f(0,1) = 2(0)^4 + (1)^2 - (0)^2 - 2(1) = -1$$

Therefore, the point (0, 1, −1) is a saddle point.

$$D\left(-\frac{1}{2}, 1\right) = 48\left(-\frac{1}{2}\right)^2 - 4 = 4 > 0$$

Since D is positive, we must evaluate f_{xx} at this point to determine if we have a minimum or maximum. Since $f_{xx}\left(-\frac{1}{2}, 1\right) = 24\left(-\frac{1}{2}\right)^2 - 2 = 4 > 0$ and $D > 0$, we have a relative minimum.

To get the z-coordinate, we evaluate the original function at the critical point:

$$f\left(-\frac{1}{2}, 1\right) = 2\left(-\frac{1}{2}\right)^4 + (1)^2 - \left(-\frac{1}{2}\right)^2 - 2(1) = -\frac{9}{8}$$

ANSWER 1, CONTINUED...

...*ANSWER 1, CONTINUED*

CALCULUS III   Quiz 9 KEY, CONTINUED

Therefore, we have a relative minimum at the point:

$$\left(-\frac{1}{2}, 1, -\frac{9}{8}\right)$$

We test the third critical point now:

$$D\left(\frac{1}{2}, 1\right) = 48\left(\frac{1}{2}\right)^2 - 4 = 4 > 0$$

Since D is positive, we must evaluate f_{xx} at this point to determine if we have a minimum or maximum.

Since $f_{xx}\left(\frac{1}{2}, 1\right) = 24\left(\frac{1}{2}\right)^2 - 2 = 4 > 0$ and $D > 0$, we have a relative minimum.

To get the z-coordinate, we evaluate the original function at the critical point:

$$f\left(\frac{1}{2}, 1\right) = 2\left(\frac{1}{2}\right)^4 + (1)^2 - \left(\frac{1}{2}\right)^2 - 2(1) = -\frac{9}{8}$$

Therefore, we have another relative minimum at the point $\left(\frac{1}{2}, 1, -\frac{9}{8}\right)$.

Final answer: Two relative minimums at $\left(\frac{1}{2}, 1, -\frac{9}{8}\right)$ and $\left(-\frac{1}{2}, 1, -\frac{9}{8}\right)$,

and a saddle point at $(0, 1, -1)$

2. **Find the *absolute* extrema for** $f(x, y) = x^2 + y^2$ **on the disc** $x^2 + y^2 \leq 1$.

First, we find the critical points and evaluate the function at the critical points to get the corresponding z-values. We put our results into a table.

Find the first partials, then critical point(s):

$$f_x(x, y) = 2x = 0 \quad \Rightarrow \quad x = 0 \quad f_y(x, y) = 2y = 0 \quad \Rightarrow \quad y = 0$$

We have a single critical point at $(0, 0)$. Evaluate the function at $(0, 0)$. Enter result into the table.

(x, y)	z
$(0, 0)$	0
$(1, 0)$	1
$(0, 1)$	1
$(-1, 0)$	1
$(0, -1)$	1

Now, we test the border of our region that is the perimeter of a circle. Solve for y using the equation:

$$x^2 + y^2 = 1 \quad \Rightarrow \quad y = \pm\sqrt{1 - x^2}$$

Test the function along this border where we also have that $-1 \leq x \leq 1$:

$$f(x, \pm\sqrt{1 - x^2}) = x^2 + \left(\pm\sqrt{1 - x^2}\right)^2 = x^2 + (1 - x^2) = 1$$

CALCULUS **III**   **Quiz 9 KEY,** CONTINUED

This means that the function is equal to 1 along the entire border. We can put some points into our table that "live" on the circle of radius 1, and all corresponding z-values equal 1.

The absolute maximums occur along the entire border of a circle of radius 1 and have a value of $z = 1$, and the critical point of $(0,0,0)$ gives our absolute minimum. *(answer)*

3.   **A rectangular box has its base resting on the xy-plane and a vertex at the origin. Find the dimensions of the box that would yield the maximum volume if another vertex of the box was in the plane $2x + y + z = 6$. Then, calculate this maximum volume.**

(a)   **Use the techniques of section 3.8—the Second Partials Test—to prove it's a maximum.**

We need to come up with the volume function we need to maximize. We know the volume of a box is $V = xyz$, where x, y and z are the width, length, and height of a box. But, we need this function to be a function of *two* variables, and not three. We use the fact that one of the vertices of the box must lie on the plane $2x + y + z = 6$.

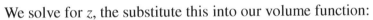

We solve for z, the substitute this into our volume function:

$$z = 6 - y - 2x \quad \Rightarrow \quad V(x, y) = xy(6 - y - 2x) = 6xy - xy^2 - 2x^2 y$$

Next, we take the first partials, set them equal to zero, and solve the system in order to get the critical point(s):

$$V_x(x, y) = 6y - 4xy - y^2 = 0 \quad \Rightarrow \quad y(6 - 4x - y) = 0$$
$$V_y(x, y) = 6x - 2x^2 - 2xy = 0 \quad \Rightarrow \quad x(6 - 2x - 2y) = 0$$

Notice that the point $(0, 0)$ will be a critical point (although an uninteresting one).

We will solve the linear system of equations to get another more interesting critical point:

$$\begin{cases} 6 - 4x - y = 0 \\ 6 - 2x - 2y = 0 \end{cases} \Rightarrow \begin{cases} 4x + y = 6 \\ 2x + 2y = 6 \end{cases} \Rightarrow \begin{array}{c} \begin{cases} 4x + y = 6 \\ -4x - 4y = -12 \end{cases} \\ \hline -3y = -6 \end{array} \Rightarrow y = 2 \Rightarrow x = 1$$

The point $(1, 2)$ is a critical point that we will perform the D-Test on now:

$$D = (V_{xx})(V_{yy}) - (V_{xy})^2 = (-4y)(-2x) - (6 - 4x - 2y)^2 = 8xy - (6 - 4x - 2y)^2$$

Evaluate D at the critical point:

$$D(1, 2) = 8(1)(2) - (6 - 4(1) - 2(2))^2 = 16 - (-2)^2 = 12 > 0$$

Since D is positive, we need to evaluate the second partial V_{xx} at the critical point also: $V_{xx}(1,2) = -12(2) = -24 < 0$

Since V_{xx} is negative, we do indeed have a local maximum for the volume when $x = 2$, $y = 1$, and $z = 6 - y - 2x = 6 - 2 - 2(1) = 2$. The actual maximum volume will be: $V = (2)(1)(2) = 4$ cu units.   *ANSWER 3, CONTINUED...*

CALCULUS III   Quiz 9 KEY, CONTINUED

(b) **Do the problem all over again, only now use the Method of Lagrange Multipliers to find the maximum.**

We want to maximize the volume $V = f(x,y,z) = xyz$, where the constraint equation is $g(x,y,z) = 2x + y + z = 6$.

Rewriting the constraint equation in standard form $g(x,y,z) = 2x + y + z - 6 = 0$, we form the Lagrange function:

$$L(x,y,z,\lambda) = f(x,y,z) - \lambda g(x,y,z)$$
$$L(x,y,z,\lambda) = xyz - \lambda(2x + y + z - 6)$$

Take all *four* first partial derivatives for this Lagrange function, and set them all simultaneously equal to zero:

$$L_x = yz - 2\lambda = 0$$
$$L_y = xz - \lambda = 0$$
$$L_z = xy - \lambda = 0$$
$$L_\lambda = -(2x + y + z - 6) = 0$$

We need to solve this non-linear system of four equations and four unknowns. Solve for λ for the first three equations:

$$\frac{yz}{2} = \lambda$$
$$xz = \lambda$$
$$xy = \lambda$$

Set the first two equations equal to each other:

$$\lambda = \lambda$$
$$\frac{yz}{2} = xz$$
$$y = 2x$$

Set equation two and equation three equal to each other:

$$\lambda = \lambda$$
$$xz = xy$$
$$y = z$$

The transitive property tells us that if $y = 2x$ and $y = z$, then $z = 2x$ as well. Substitute these into the constraint equation:

$$2x + 2x + 2x = 6$$
$$6x = 6$$
$$x = 1$$

Then, since $y = z = 2x$, $y = z = 2$. Actual maximum volume will be $V = (1)(2)(2) = 4$ cubic units.

CALCULUS III   **Quiz 9 KEY,** CONTINUED

4.   *Minimizing the cost of a container.* **A rectangular box with no top is to be constructed having a volume of 12 cubic feet. The cost per square foot of the material to be used is \$4 for the bottom, \$3 for two of the opposite sides, and \$2 for the remaining pair of sides. Use the Method of Lagrange Multipliers to find the dimensions of the box that will minimize the cost.**

We want to minimize the cost of the material covering the surface area for the rectangular box that has no top: $Cost = f(x,y,z) = \$4(xy) + (\$3)(2xz) + (\$2)(2yz) = 4xy + 6xz + 4yz$, where the constraint equation is $V = g(x,y,z) = xyz = 12$.

Rewriting the constraint equation in standard form $g(x,y,z) = xyz - 12 = 0$, we form the Lagrange function:

$$L(x,y,z,\lambda) = f(x,y,z) - \lambda g(x,y,z)$$
$$L(x,y,z,\lambda) = 4xy + 6xz + 4yz - \lambda(xyz - 12)$$

Take all *four* first partial derivatives for this Lagrange function, and set them all simultaneously equal to zero:

$$L_x = (4y + 6z) - \lambda(yz) = 0$$
$$L_y = (4x + 4z) - \lambda(xz) = 0$$
$$L_z = (6x + 4y) - \lambda(xy) = 0$$
$$L_\lambda = \quad -(xyz - 12) = 0$$

This non-linear system of four equations and four unknowns will be quite the challenge to solve. We will start by solving the first equation for y:

$$4y - \lambda yz = -6z$$
$$y(4 - \lambda z) = -6z$$
$$y = \frac{-6z}{4 - \lambda z} = \frac{6z}{\lambda z - 4}$$

Now solve the second equation for x:

$$4x - \lambda xz = -4z$$
$$x(4 - \lambda z) = -4z$$
$$x = \frac{-4z}{4 - \lambda z} = \frac{4z}{\lambda z - 4}$$

Note that x and y are related to each other; y is slightly larger than x. In fact, $y = \frac{3}{2}x$. Now, solve the third equation for λ, but also substituting $y = \frac{3}{2}x$ at the same time:

$$6x + 4\left(\frac{3}{2}x\right) = \lambda x\left(\frac{3}{2}x\right)$$
$$6x + 6x = \frac{3}{2}\lambda x^2$$
$$12x = \frac{3}{2}\lambda x^2$$
$$\lambda = 12x\left(\frac{2}{3x^2}\right) = \frac{8}{x} \quad \text{ANSWER 4, CONTINUED...}$$

Calculus III   Quiz 9 KEY, continued

Even though this next step might seem redundant, we need to do it—we will start over with equation #2, only this time solve it for z:

$$4z - \lambda xz = -4x$$

$$z(4 - \lambda x) = -4x$$

$$z = \frac{-4x}{4 - \lambda x} = \frac{-4x}{4 - \left(\dfrac{8}{x}\right)x} = \frac{-4x}{4 - 8} = \frac{-4x}{-4} = \frac{4x}{4} = x$$

That last step involved using the fact that $\lambda = \dfrac{8}{x}$ (substituting) that we found above. Now we have y and z in terms of x, and we plug those into the constraint equation, and solve for x:

$$xyz = 12$$

$$x\left(\frac{3x}{2}\right)(x) = 12$$

$$\frac{3x^3}{2} = 12$$

$$x^3 = 12\left(\frac{2}{3}\right) = 8$$

$$x = \sqrt[3]{8} = 2 \text{ feet}$$

$$y = \frac{3}{2}x = \frac{3}{2}(2) = 3 \text{ feet}$$

$$z = x = 2 \text{ feet}$$

Actual minimum cost for the materials to cover the surface of the box is:

$$Cost = f(2,\ 3,\ 2) = (4)(2)(3) + 6(2)(2) + 4(3)(2) = \$72$$

5. **Help Elmer Fudd, the marketing analyst at ACME company, to maximize his company's revenue. The company produces two competitive products, the Widget-2000 and the Widget-EX, the prices (in dollars) of which are p_1 and p_2, respectively. The total revenue obtained from selling both products is given by $R(p_1, p_2) = 300p_1 + 900p_2 + 1.8p_1p_2 - 1.5p_1^2 - p_2^2$.**

    (a)   Determine the maximum revenue.

    In order to maximize the revenue, we need to find the first partials, set equal to zero, solve the system and find the critical point:

$$R_{p_1}(p_1, p_2) = 300 + 1.8p_2 - 3p_1 = 0 \quad \Rightarrow \quad 3p_1 - 1.8p_2 = 300$$

$$R_{p_2}(p_1, p_2) = 900 + 1.8p_1 - 2p_2 = 0 \quad \Rightarrow \quad -1.8p_1 + 2p_2 = 900$$

CALCULUS III   **Quiz 9 KEY**, CONTINUED

We now have a linear system of equations. I will use the method of elimination to solve. I will multiply the first equation by 1.8 and the second equation by 3, then add the two equations together to eliminate the first variable, p_1:

$$\begin{cases}(1.8)(3p_1 - 1.8p_2) = (300)(1.8)\\(3)(-1.8p_1 + 2p_2) = (900)(3)\end{cases} \Rightarrow \;\; \begin{array}{l}\begin{cases}5.4p_1 - 3.24p_2 = 540\\+\begin{cases}-5.4p_1 + 6p_2 = 2700\end{cases}\end{cases}\\\hline\quad\;\; 2.76p_2 = 3240 \end{array} \Rightarrow \;\; p_2 = \frac{3240}{2.76} \approx 1173.91$$

Now we can find the value of the other variable by substituting this back into either of the two original equations and find that $p_1 = 804.35$. Therefore, the critical point is ($804.35, $1173.91).

To make sure that this indeed maximizes the revenue, we perform the *D*-test:

$$D = R_{p_1 p_1} R_{p_2 p_2} - \left(R_{p_1 p_2}\right)^2 = (-3)(-2) - (1.8)^2 > 0$$

Since *D* is positive, we check the second partial $R_{p_1 p_1} = -3 < 0$, and we do have a maximum.

Therefore, Elmer Fudd needs to charge $804.35 for his first product, the Widget-2000, and $1173.91 for his second product, the Widget-EX. His actual maximum revenue will be:

$$R(804.35, 1173.91) = 300(804.35) + 900(1173.91) + 1.8(804.35)(1173.91) - 1.5(804.35)^2 - (1173.91)^2 \approx \$648{,}913$$

(b)   What prices should Mr. Fudd charge for each product? (i.e., what do the values of p_1 and p_2 need to be in order to maximize revenue?)

As stated above: Elmer Fudd needs to charge $804.35 for his first product, the Widget-2000, and $1173.91 for his second product, the Widget-EX.

6.   Evaluate the iterated integral $\displaystyle\int_0^{\sqrt{\pi}} \int_{\pi/6}^{y^2} 2y \cos x \; dx dy$.

We will integrate with respect to *x* first:

$$\int_0^{\sqrt{\pi}} 2y[\sin x]_{\pi/6}^{y^2}\, dy = \int_0^{\sqrt{\pi}} 2y\left(\sin y^2 - \sin\frac{\pi}{6}\right)dy = \int_0^{\sqrt{\pi}} 2y\left(\sin y^2 - \frac{1}{2}\right)dy = \int_0^{\sqrt{\pi}}\left(2y\sin y^2 - y\right)dy$$

We will use *u*-substitution to integrate the first term where:

$$u = y^2, du = 2y\, dy \quad and \quad \int \sin u\, du = -\cos u + C$$

So, we now have that:

$$\left[-\cos y^2 - \frac{y^2}{2}\right]_0^{\sqrt{\pi}} = \left(-\cos\left(\sqrt{\pi}\right)^2 - \frac{\left(\sqrt{\pi}\right)^2}{2}\right) - \left(-\cos 0^2 - \frac{0^2}{2}\right) = -\cos\pi - \frac{\pi}{2} + \cos 0 = -(-1) - \frac{\pi}{2} + 1 = 2 - \frac{\pi}{2}$$

(answer)

7. Evaluate the integral $\iint\limits_R xy\,dA$, where R is the region bounded by the graphs of $y = \sqrt{x},\ y = \dfrac{1}{2}x,\ x = 2,$ and $x = 4$.

First, we sketch the region so that we can determine if it is vertically or horizontally simple. (This helps us to determine what the order of integration should be.)

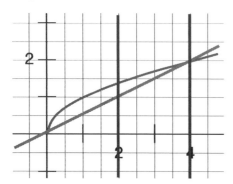

This is a vertically simple region, so the order of integration will be *dydx*. The ranges (limits) will be:

$$2 \le x \le 4 \quad \text{and} \quad \frac{1}{2}x \le y \le \sqrt{x}$$

Now we iterate the integral:

$$\int_2^4 \int_{x/2}^{\sqrt{x}} xy\,dy\,dx = \int_2^4 x\left[\frac{y^2}{2}\right]_{x/2}^{\sqrt{x}} dx = \int_2^4 x\left(\frac{\left(\sqrt{x}\right)^2}{2} - \frac{\left(x/2\right)^2}{2}\right)dx$$

Simplify, then distribute the "*x*" over the sum before integrating again:

$$\int_2^4 x\left(\frac{x}{2} - \frac{\left(x^2/4\right)}{2}\right)dx = \int_2^4 \left(\frac{x^2}{3} - \frac{x^3}{8}\right)dx = \left[\frac{x^3}{6} - \frac{x^4}{32}\right]_2^4 = \left(\frac{4^3}{6} - \frac{4^4}{32}\right) - \left(\frac{2^3}{6} - \frac{2^4}{32}\right) = \frac{11}{6}$$

8. Evaluate the integral $\int_0^1 \int_{2x}^2 e^{y^2}\,dy\,dx$ by reversing the order of integration. *(**Note: without** reversing the order; this integral is not integrable!!)*

Let's sketch the region of integration to help us visualize what is going on. This makes it easier to reverse the order of integration. We sketch the area bounded by the graphs of $y = 2x,\ y = 2,\ x = 1,$ and $x = 0$.

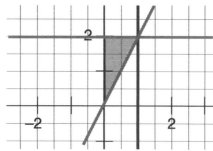

Currently the ranges (limits) are $0 \le x \le 1$ and $2x \le y \le 2$. The *y*-coordinate will have constant limits for the horizontally simple order of *dxdy*. After examining our region of integration, we will have that $0 \le y \le 2$. The value for *x* will start at a lower limit at the *y*-axis (i.e., $x = 0$) and the upper limit will stop at the graph of $y = 2x$. But, we need to solve this equation for *x* to get that upper limit $0 \le x \le \dfrac{y}{2}$.

Now we can iterate the integral and evaluate: $\int_0^2 \int_0^{y/2} e^{y^2}\,dx\,dy = \int_0^2 e^{y^2}\left[x\right]_0^{y/2} dy = \int_0^2 e^{y^2}\left(\frac{y}{2}\right)dy$

Next, we integrate using *u*-substitution and the exponential rule, where $u = y^2,\ du = 2y\,dy$. Thus, we need a factor of 2 in the integrand (provided we divide it out):

$$\left(\frac{1}{2}\right)\left(\frac{1}{2}\right)\int_0^2 e^{y^2}(2y)dy = \frac{1}{4}\int e^u\,du = \frac{1}{4}e^u = \frac{1}{4}\left[e^{y^2}\right]_0^2 = \frac{1}{4}\left(e^{2^2} - e^0\right) = \frac{1}{4}\left(e^4 - 1\right) \approx 13.4$$

CALCULUS III   **Quiz 9 KEY,** CONTINUED

9. **Evaluate the integral** $\int_0^1 \int_{\sqrt{y}}^1 \sin(x^3)\,dx\,dy$ **by reversing the order of integration.** *(Note: without reversing the order; this integral is not integrable!!)*

Let's sketch the region of integration to help us visualize what is going on. This makes it easier to reverse the order of integration. We sketch the area bounded by the graphs of:

$$y = 0, \quad y = 1, \quad x = 1, \quad \text{and} \quad x = \sqrt{y}$$

Can you tell that the region of integration will be the region under the graph of $x = \sqrt{y}$, but to the left of the vertical line $x = 1$?

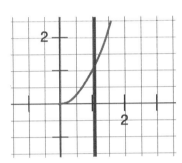

Currently the ranges (limits) are $0 \le y \le 1$ and $\sqrt{y} \le x \le 1$. The x-coordinate will have constant limits for the vertically simple order of $dx\,dy$. After examining our region of integration, we will have that $0 \le x \le 1$. The value for y will start at a lower limit at the x-axis (i.e., $y = 0$) and the upper limit will stop at the graph of $x = \sqrt{y}$.

But, we need to solve this equation for y to get that upper limit $0 \le y \le x^2$.

Now we can iterate the integral and evaluate:

$$\int_0^1 \int_{\sqrt{y}}^1 \sin(x^3)\,dx\,dy = \int_0^1 \int_0^{x^2} \sin(x^3)\,dy\,dx = \int_0^1 \sin(x^3)\left[y\right]_0^{x^2}\,dx = \int_0^1 \sin(x^3)x^2\,dx$$

Next, we integrate using u-substitution and the exponential rule, where $u = x^3$, $du = 3x^2\,dx$. Thus, we need a factor of 3 in the integrand (provided we divide it out):

$$= \frac{1}{3}\int_0^1 \sin(x^3)3x^2\,dx = \frac{1}{3}\int \sin u\,du = \frac{1}{3}\left[-\cos u\right] = \frac{-1}{3}\left[\cos(x^3)\right]_0^1 = \frac{-1}{3}\left[\cos(1^3) - \cos(0)\right] = \frac{-1}{3}(\cos 1 - 1) \approx 0.153$$

Note: Be sure you are in RADIANS mode when you evaluate cos(1) on your calculator!

10. **Multiple choice: Reverse the order of integration and rewrite the iterated integral** $\int_{-2}^0 \int_0^{\sqrt{4-x^2}} f(x,y)\,dy\,dx$. *(Please circle your choice.)* **Show your work!**

(a) $\int_0^2 \int_0^{\sqrt{4-y^2}} f(x,y)\,dx\,dy$ (b) $\int_2^0 \int_0^{\sqrt{4-y^2}} f(x,y)\,dx\,dy$

(c) $\int_2^0 \int_{\sqrt{4-y^2}}^0 f(x,y)\,dx\,dy$ (d) $\int_0^2 \int_{-\sqrt{4-y^2}}^0 f(x,y)\,dx\,dy$

(e)   **none of these**

ANSWER 9, CONTINUED...

...*ANSWER 9, CONTINUED*

CALCULUS III   Quiz 9 KEY, CONTINUED

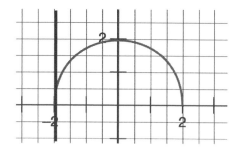

Let's sketch the region of integration to help us visualize what is going on. This makes it easier to reverse the order of integration. We sketch the area bounded by the graphs of $y = 0$, $y = \sqrt{4 - x^2}$, $x = -2$, and $x = 0$. Can you tell that the region of integration will be the region under the graph of the circle, but to the left of the y-axis?

To reverse the order, we must get the ranges for y to be constants $0 \le y \le 2$. Then we need to solve the equation $y = \sqrt{4 - x^2}$ for x, since this will be the lower limit for x

$$y^2 = 4 - x^2 \implies x = \pm\sqrt{4 - y^2}.$$

However, we take the negative of the radical, since we are on the left half of the circle (where the x-coordinates are negative).

So, the range for x will be $-\sqrt{4 - y^2} \le x \le 0$. Now we can iterate the integral. The answer is letter **(d)**.

11. **Evaluate the improper integral $\int\limits_{1}^{\infty}\int\limits_{x}^{2x} ye^{-x}\,dy\,dx$. (*Hint:* You will need to do repeated integration by parts!)**

We integrate with respect to y first (we don't need integration by parts yet):

$$\int\limits_{1}^{\infty} e^{-x}\left[\frac{y^2}{2}\right]_{x}^{2x} dx = \frac{1}{2}\int\limits_{1}^{\infty} e^{-x}\left[(2x)^2 - (x)^2\right]dx = \frac{1}{2}\int\limits_{1}^{\infty} e^{-x}(3x^2)dx = \frac{3}{2}\int\limits_{1}^{\infty} x^2 e^{-x}$$

Now, we need integration by parts $u = x^2$, $du = 2x\,dx$, $dv = e^{-x}\,dx$, $v = -e^{-x}$.

Recall that the integration by parts formula tells us that $\int u\,dv = uv - \int v\,du$. We do this now:

$$\frac{3}{2}\int\limits_{1}^{\infty} x^2 e^{-x} = \frac{3}{2}\left\{\left[x^2(-e^{-x})\right]_{1}^{\infty} - \int\limits_{1}^{\infty}(-e^{-x})(2x)dx\right\} = \frac{-3}{2}\left[x^2 e^{-x}\right]_{1}^{\infty} + \frac{3}{2}(2)\int\limits_{1}^{\infty} xe^{-x}dx$$

We need integration by parts again where $u = x$, $du = dx$, $dv = e^{-x}dx$, $v = -e^{-x}$:

$$= \frac{-3}{2}\left[\frac{x^2}{e^x}\right]_{1}^{\infty} + 3\left\{\left[-xe^{-x}\right]_{1}^{\infty} - \int\limits_{1}^{\infty} -e^{-x}dx\right\} = \frac{-3}{2}\left[\frac{x^2}{e^x}\right]_{1}^{\infty} - 3\left[\frac{x}{e^x}\right]_{1}^{\infty} - 3\left[e^{-x}\right]_{1}^{\infty}$$

Finally, we use our knowledge of limits (including L'Hôpital's Rule) where:

$$\left[\frac{x^2}{e^x}\right]_{1}^{\infty} = \lim_{x\to\infty}\frac{x^2}{e^x} - \frac{(1)^2}{e^{(1)}} = \frac{\infty}{\infty} - \frac{1}{e} \implies \lim_{x\to\infty}\frac{2x}{e^x} - \frac{1}{e} = \frac{\infty}{\infty} - \frac{1}{e} \implies \lim_{x\to\infty}\frac{2}{e^x} - \frac{1}{e} = \frac{2}{e^\infty} - \frac{1}{e} = \frac{2}{\infty} - \frac{1}{e} = 0 - \frac{1}{e} = -\frac{1}{e}$$

Do this similarly for the other limits, so that both of the other terms go to zero.

$$\textit{Final answer:} \quad \frac{-3}{2}\left(0 - \frac{1}{e}\right) - 3\left(0 - \frac{1}{e}\right) - 3\left(0 - \frac{1}{e}\right) = \frac{15}{2e}$$

CALCULUS III   **Quiz 9 KEY,** CONTINUED

---

12. **Evaluate the improper integral** $\int_0^3 \int_0^\infty \dfrac{x^2}{1+y^2}\,dy\,dx$. **(*Hint:* The Arctan Rule will be used here!)**

Using the Arctan Rule for integration immediately while integrating with respect to y we have:

$$\int_0^3 \int_0^\infty \frac{x^2}{1+y^2}\,dy\,dx = \int_0^3 x^2 \left[\arctan y\right]_0^\infty dx = \int_0^3 x^2 \left[\arctan(\infty) - \arctan(0)\right]dx$$

We use the facts that $\lim\limits_{x\to\infty} \arctan x = \arctan(\infty) = \dfrac{\pi}{2}$ *and* $\arctan(0) = 0$, so we can write:

$$\int_0^3 x^2 \left(\frac{\pi}{2}\right)dx = \left(\frac{\pi}{2}\right)\left[\frac{x^3}{3}\right]_0^3 = \left(\frac{\pi}{2}\right)\left(\frac{3^3}{3}\right) = \left(\frac{\pi}{2}\right)\left(\frac{27}{3}\right) = \frac{9\pi}{2}$$

13. **Use a double integral to find the volume of the solid bounded by the plane** $z = 4 - 2x - 4y$ **and the coordinate planes.**

We use the fact that volume is given by:

$$V = \iint\limits_R f(x,y)\,dA \text{, where } f(x,y) = z = 4 - 2x - 4y$$

To iterate the integral, we need to determine what is going on in the *xy*-plane. This is the same as determining what the base of the solid looks like. Since $z = 0$ in the *xy*-plane, we set $z = 0$ in the equation $z = 4 - 2x - 4y$ and get:

$$0 = 4 - 2x - 4y \quad \Rightarrow \quad y = \frac{4 - 2x}{4} = 1 - \frac{1}{2}x$$

Since we are also bound by the coordinate planes, we know the base of the solid is in the first octant and that the graphs of the lines $x = 0$ and $y = 0$ are also bounding the region. A graph of the region of integration is shown:

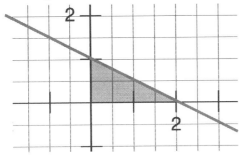

The ranges (limits) for x and y (we will use $dy\,dx$ as our order of integration) are:

$$0 \le x \le 2 \quad and \quad 0 \le y \le 1 - \frac{1}{2}x$$

We use these to iterate our integral and evaluate:

$$V = \int_0^2 \int_0^{1-x/2} (4 - 2x - 4y)\,dy\,dx = \int_0^2 \left[4y - 2xy - 2y^2\right]_0^{1-x/2} dx$$

$$= \int_0^2 \left[4\left(1 - \frac{x}{2}\right) - 2x\left(1 - \frac{x}{2}\right) - 2\left(1 - \frac{x}{2}\right)^2\right]_0^{1-x/2} dx = \int_0^2 \left(4 - 2x - 2x + x^2 - 2\left(1 - \frac{x}{2}\right)\left(1 - \frac{x}{2}\right)\right)dx$$

$$= \int_0^2 \left(4 - 2x - 2x + x^2 - 2\left(1 - x + \frac{x^2}{4}\right)\right)dx = \int_0^2 \left(-2x + 2 + \frac{x^2}{2}\right)dx = \left[-x^2 + 2x + \frac{x^3}{6}\right]_0^2$$

$$= -(2)^2 + 2(2) + \frac{(2)^3}{6} = \frac{4}{3}$$

---

14. **Evaluate:** $\displaystyle\int_0^{\pi/2}\int_0^{\pi/4}\cos x \sin^2 y \; dydx$. (*Hint:* **You will need to use a trig ID before integrating.**)

The trig ID we need to use before integrating with respect to y tells us that:

$$\sin^2 y = \frac{1-\cos(2y)}{2} = \frac{1}{2}(1-\cos(2y))$$

So now we rewrite:

$$= \frac{1}{2}\int_0^{\pi/2}\int_0^{\pi/4}\cos x(1-\cos(2y))dydx = \frac{1}{2}\int_0^{\pi/2}\cos x\left[y-\frac{1}{2}\sin(2y)\right]_0^{\pi/4}dx = \frac{1}{2}\int_0^{\pi/2}\cos x\left[\left(\frac{\pi}{4}\right)-\frac{1}{2}\sin\left(2\cdot\frac{\pi}{4}\right)\right]_0^{\pi/4}dx$$

$$= \frac{1}{2}\int_0^{\pi/2}\cos x\left[\left(\frac{\pi}{4}\right)-\frac{1}{2}(1)\right]_0^{\pi/4}dx = \left(\frac{\pi}{4}-\frac{1}{2}\right)\left(\frac{1}{2}\right)\int_0^{\pi/2}\cos x dx = \left(\frac{\pi}{8}-\frac{1}{4}\right)[\sin x]_0^{\pi/2} = \left(\frac{\pi}{8}-\frac{1}{4}\right)\left[\sin\frac{\pi}{2}-\sin 0\right]$$

$$= \frac{\pi}{8}-\frac{1}{4} \approx 0.143$$

15. **Evaluate** $\displaystyle\int_0^2\int_{y+5}^0 e^{2x+y} \; dxdy$.

We integrate with respect to x first and use u-substitution with the Exponential Rule, where $u = 2x+y, \; du = 2dx$.

Therefore, we need a factor of 2 in the integrand (provided we also divide it out):

$$\int_0^2\left(\frac{1}{2}\right)\int_{y+5}^0 e^{2x+y}2dxdy = \int_0^2\left(\frac{1}{2}\right)\left[e^{2x+y}\right]_{y+5}^0 dy = \frac{1}{2}\int_0^2\left[e^{2(0)+y}-e^{2(y+5)+y}\right]dy = \frac{1}{2}\int_0^2\left(e^y-e^{3y+10}\right)dy$$

We will now integrate with respect to y. We again need u-substitution with the Exponential Rule for the second term, where $u = 3y+10, \; du = 3dy$. Therefore, we need a factor of 3 in the integrand (provided we also divide it out). I am also going to split the integral of the sum into a sum of two integrals. The reason is because I have seen students make the mistake of dividing the entire sum out by the 3, where only the second term needs it! So, to avoid this mistake, I will "play it safe" for clarity:

$$= \frac{1}{2}\int_0^2\left(e^y-e^{3y+10}\right)dy = \frac{1}{2}\int_0^2 e^y dy - \frac{1}{2}\left(\frac{1}{3}\right)\int_0^2 e^{3y+10}(3)dy = \frac{1}{2}\left[e^y\right]_0^2 - \frac{1}{6}\left[e^{3y+10}\right]_0^2$$

$$= \frac{1}{2}e^2-\frac{1}{2}e^0-\frac{1}{6}\left(e^{3(2)+10}-e^{3(0)+10}\right) = \frac{1}{2}e^2-\frac{1}{2}-\frac{1}{6}e^{16}+\frac{1}{6}e^{10}$$

(answer)

CALCULUS III   **Quiz 9 KEY, CONTINUED**

---

16.   **Evaluate:** $\displaystyle\int_0^1\int_{-3}^0 \frac{1}{(x+2)(1-y)}\,dy\,dx$.

To integrate with respect to y, we use u-substitution with the Log Rule, where: $u = 1-y,\; du = (-1)dy$. This means we need a factor of -1 in the integrand, provided we divide it out:

$$= -\int_0^1\int_{-3}^0 \frac{-1}{(x+2)(1-y)}\,dy\,dx = -\int_0^1 \frac{1}{x+2}\Big[\ln|1-y|\Big]_{-3}^0 dx = -\int_0^1 \frac{1}{x+2}\Big[\ln|1-0| - \ln|1-(-3)|\Big]dx$$

$$= -\int_0^1 \frac{1}{x+2}\Big[\ln|1| - \ln|4|\Big]dx = \ln(4)\int_0^1 \frac{1}{x+2}\,dx = \ln(4)\Big[\ln|x+2|\Big]_0^1 = \ln(4)\Big[\ln|1+2| - \ln|0+2|\Big]$$

$$= \ln(4)\big[\ln 3 - \ln 2\big] = \ln(4)\ln\!\left(\frac{3}{2}\right) \approx 0.562$$

17.   **Evaluate** $\displaystyle\int_0^2\int_0^x \frac{1}{x+1}\,dy\,dx$. *(Hint: Do not* **reverse the order of integration—that will make it too hard. Instead, use long division to simplify the result of the inside integral (after integrating), remembering that the answer to a long division problem looks like:** $\text{Quotient} + \dfrac{\text{Remainder}}{\text{Divisor}}$ **. Another way you can do it instead (if you really despise long division) is to use Change of Variables.)**

We first integrate with respect to y:

$$\int_0^2 \frac{1}{x+1}\big[y\big]_0^x dx = \int_0^2 \frac{1}{x+1}\big[x\big]dx = \int_0^2 \frac{x}{x+1}\,dx$$

Now we have an improper rational function in the integrand, so we use polynomial long division:

$$
\begin{array}{r}
1 \\
x+1\overline{)\,x}\\
\underline{x+1}\\
-1
\end{array}
$$

We need to write this result as $1 + \dfrac{-1}{x+1}$, and then rewrite our integral:

$$\int_0^2\left(1 - \frac{1}{x+1}\right)dx = \Big[x - \ln|x+1|\Big]_0^2 = \big[2 - \ln|2+1|\big] - \big[0 - \ln|0+1|\big] = 2 - \ln 3$$
(answer)

To do the problem using Change of Variables instead:

Let $u = x+1$, then $x = u-1$ and $du = dx$. Substitute all of these "ingredients" into the integral to get that:

$$\int_0^2 \frac{x}{x+1}\,dx = \int \frac{u-1}{u}\,du = \int\left(\frac{u}{u} - \frac{1}{u}\right)du = \int\left(1 - \frac{1}{u}\right)du = u - \ln|u| = \Big[(x+1) - \ln|x+1|\Big]_0^2$$

$$= (2+1) - \ln|2+1| - \big[(0+1) - \ln|0+1|\big] = 3 - \ln(3) - 1 - \ln 1 = 2 - \ln 3$$

(answer)

CALCULUS III

18. **Reverse the order of integration, then evaluate** $\displaystyle\int_0^2 \int_{y^2}^4 \frac{y}{1+x^2}\,dx\,dy$.

Let's sketch the region of integration to help us visualize what is going on. This makes it easier to reverse the order of integration. We sketch the area bounded by the graphs of $y=0$, $y=2$, $x=y^2$, and $x=4$.

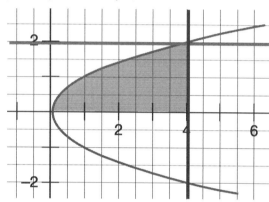

The region of integration is the region above the x-axis, but below the "curvy" graph of $x=y^2$. For the order of integration to be $dydx$, the range (limits) for x will be constants. From inspection of our region, we have that $0 \le x \le 4$. The limits for y will be bounded below by $y=0$ and above by the graph of $x=y^2$, which we need to solve for y with $y=\pm\sqrt{x}$. We choose the positive radical, since our region is in the first quadrant, where y is positive.

Thus, our limits for y will be $0 \le y \le \sqrt{x}$. Now we can iterate our integral.

$$\int_0^4 \int_0^{\sqrt{x}} \frac{y}{1+x^2}\,dy\,dx = \int_0^4 \left(\frac{1}{1+x^2}\right)\left[\frac{y^2}{2}\right]_0^{\sqrt{x}} dx = \int_0^4 \left(\frac{1}{1+x^2}\right)\left[\frac{(\sqrt{x})^2}{2}\right] dx = \frac{1}{2}\int_0^4 \frac{x}{1+x^2}\,dx$$

Using u-substitution with the Log Rule, where $u=1+x^2$, $du=2x\,dx$, we need to factor in a multiple of 2, provided we divide it out:

$$= \left(\frac{1}{2}\right)\left(\frac{1}{2}\right)\int_0^4 \frac{2x}{1+x^2}\,dx = \frac{1}{4}\left[\ln\left(1+x^2\right)\right]_0^4 = \frac{1}{4}\left[\ln(1+4^2) - \ln(1+0^2)\right] = \frac{1}{4}\ln 17$$

19. **Use a double integral to find the volume of the solid bounded by the three coordinate planes and** $z = 1 - x^2 - y^2$ **in the first octant (as shown).** *(Hint: Converting to polar coordinates makes this problem so much easier!)*

We use the fact that volume is given by $V = \displaystyle\iint_R f(x,y)\,dA$, where $f(x,y) = z = 1 - x^2 - y^2$. To iterate the integral, we need to determine what is going on in the xy-plane. This is the same as determining what the base of the solid looks like. Since $z = 0$ in the xy-plane, we set $z = 0$ in the equation $z = 1 - x^2 - y^2$ and get $0 = 1 - x^2 - y^2 \ \Rightarrow \ x^2 + y^2 = 1$.

Thus, the portion of a circle of radius 1 in the 1ST quadrant is the base of our solid:

CALCULUS III   **Quiz 9 KEY,** CONTINUED

With *dydx* as our choice of order of integration, the range for x is $0 \le x \le 1$. The range for y is

$0 \le y \le \sqrt{1-x^2}$. Then, volume will be $\int_0^1 \int_0^{\sqrt{1-x^2}} (1 - x^2 - y^2)dydx$. This integral would be difficult to

evaluate; it's much easier if we convert to polar coordinates.

The ranges for r and θ are $0 \le r \le 1$ *and* $0 \le \theta \le \pi/2$. Also, $1 - x^2 - y^2 = 1 - (x^2 + y^2) = 1 - r^2$.

(***Note:*** Always remember to replace "*dydx*" with "*rdrdθ*" when converting from rectangular to polar coordinates!)

So, after converting to polar coordinates, we have that:

$$\int_0^{\pi/2} \int_0^1 (1 - r^2)rdrd\theta = \int_0^{\pi/2} \int_0^1 (r - r^3)drd\theta = \int_0^{\pi/2} \left[\frac{r^2}{2} - \frac{r^4}{4} \right]_0^1 d\theta = \int_0^{\pi/2} \left[\frac{1^2}{2} - \frac{1^4}{4} \right] d\theta$$

$$= \frac{1}{4} \int_0^{\pi/2} d\theta = \frac{1}{4} [\theta]_0^{\pi/2} = \frac{1}{4}\left(\frac{\pi}{2} \right) = \frac{\pi}{8}$$

(***Note:*** Be sure and distribute the "*r*" from the "*rdrdq*" before you integrate!)

20.   **Evaluate the double integral by converting to polar coordinates** $\int_0^2 \int_0^{\sqrt{4-x^2}} x \, dydx$.

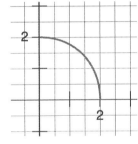

The region of integration is shown in the diagram. Using the conversion formula that tells us that $x = r\cos\theta$, the integral now has the form $\int_0^{\pi/2} \int_0^2 (r\cos\theta)rdrd\theta$.

Now we multiply the "*r*" and then evaluate:

$$\int_0^{\pi/2} \int_0^2 (r^2 \cos\theta)drd\theta = \int_0^{\pi/2} \cos\theta \left[\frac{r^3}{3} \right]_0^2 d\theta = \int_0^{\pi/2} \cos\theta \left[\frac{2^3}{3} \right] d\theta = \frac{8}{3} \int_0^{\pi/2} \cos\theta \, d\theta$$

$$= \frac{8}{3} [\sin\theta]_0^{\pi/2} = \frac{8}{3}\left(\sin\frac{\pi}{2} - \sin 0 \right) = \frac{8}{3}$$

21.   **Use a double integral in polar coordinates to find the volume of the solid bounded by the graphs of the equations** $z = 9 - x^2 - y^2$, $z = 0$, $x^2 + y^2 \ge 1$, $x^2 + y^2 \le 4$.

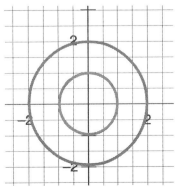

ANSWER 20, CONTINUED...

...Answer 20, continued

Calculus III   Quiz 9 KEY, continued

The region of integration is shown in the diagram. It is a "donut-shaped" region. We use the fact that volume is given by:

$$V = \iint_R f(x,y)\,dA, \text{ where } f(x,y) = z = 9 - x^2 - y^2$$

Using rectangular coordinates is very awkward here and it would be difficult to evaluate the integral. Because the region is a portion of a circle, it lends itself to polar coordinates quite nicely. We use polar coordinates to iterate the integral and find:

$$\int_0^{2\pi}\int_1^2 (9 - r^2)r\,dr\,d\theta \;=\; \int_0^{2\pi}\int_1^2 (9r - r^3)\,dr\,d\theta \;=\; \int_0^{2\pi}\left[\frac{9}{2}r^2 - \frac{r^4}{4}\right]_1^2 d\theta$$

$$= \int_0^{2\pi}\left[\left(\frac{9}{2}(2)^2 - \frac{(2)^4}{4}\right) - \left(\frac{9}{2}(1)^2 - \frac{(1)^4}{4}\right)\right]d\theta \;=\; \int_0^{2\pi}\left(18 - 4 - \frac{9}{2} + \frac{1}{4}\right)d\theta = \frac{39}{4}\int_0^{2\pi}d\theta = \frac{39}{4}\left[\theta\right]_0^{2\pi} = \frac{39\pi}{2}$$

22. **Find the mass of the triangular lamina with vertices (0, 0) , (0, 3), and (3, 0) given that the density at (x, y) is** $\rho(x,y) = 10y$.

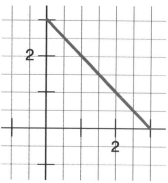

Mass is found using the formula: mass $= \iint_R \rho(x,y)\,dA$. The region of integration is shown. We iterate the integral and evaluate:

$$\int_0^3\int_0^{3-x}(10y)\,dy\,dx \;=\; \int_0^3\left[5y^2\right]_0^{3-x}dx \;=\; \int_0^3 5(3-x)^2\,dx$$

Use u-substitution and the Power Rule where $u = 3 - x$, $du = (-1)dx$. We need to put a factor of -1 in (and divide it out):

$$= -5\int_0^3 -(3-x)^2\,dx = -5\int u^2\,du = -5\left(\frac{u^3}{3}\right) = \frac{-5}{3}\left[(3-x)^3\right]_0^3$$

$$= \frac{-5}{3}\left[(3-3)^3 - (3-0)^3\right] = \frac{-5}{3}\left[0 - 3^3\right] = \frac{-5}{3}(-27) = 45$$

$$\sqrt{}$$ **PART III** TEST ANSWER KEY

CALCULUS III   ✠ **PRACTICE MIDTERM 3 KEY** ✠

1. **Find the limits (if they exist):**

   (a) $\displaystyle\lim_{(x,y)\to(4,5)}\frac{5x-4y}{15x^2-22xy+8y^2}$

   Using direct substitution, we get the indeterminate form $\dfrac{0}{0}$. But we cannot use L'Hopital's Rule since this is an expression with two variables. However, we can factor and reduce to lowest terms, and then do direct substitution again:

   $$=\lim_{(x,y)\to(4,5)}\frac{5x-4y}{15x^2-10xy-12xy+8y^2}=\lim_{(x,y)\to(4,5)}\frac{5x-4y}{5x(3x-2y)-4y(3x-2y)}=\lim_{(x,y)\to(4,5)}\frac{5x-4y}{(5x-4y)(3x-2y)}$$

   $$=\lim_{(x,y)\to(4,5)}\frac{1}{3x-2y}=\frac{1}{3(4)-2(5)}=\frac{1}{12-10}=\frac{1}{2}$$

   (b) $\displaystyle\lim_{(x,y)\to(\ln 8,\ln 8)}\frac{10e^y\sin(x^2-y^2)}{x^2-y^2}$

   Using direct substitution, we get the indeterminate form $\dfrac{0}{0}$. But we cannot use L'Hopital's Rule since this is an expression with two variables. However, we use the fact that $\displaystyle\lim_{\theta\to 0}\frac{\sin\theta}{\theta}=1$ where $\theta=x^2-y^2$ and direct substitution to get:

   $$\lim_{(x,y)\to(\ln 8,\ln 8)}\frac{10e^{\ln 8}\sin\theta}{\theta}=10(8)(1)=80$$

2. **Give the points at which the function is continuous in set notation:**

   $$f(x,y)=\ln(4-x-y)+\arccos\left(\frac{y}{x}\right)$$

   The function will be continuous at points in its domain. There are variables in the argument of the log, so we need to restrict the domain such that $4-x-y>0$. In addition, there is a denominator with a variable in it in the argument of the arccosine function. We require that $x\neq 0$. Finally, the domain of the arccosine function must be values on the interval $[-1,1]$.

   In other words, we need $-1\leq\dfrac{y}{x}\leq 1$.

   We write the final answer suing set notation and include all of these restrictions:

   $$\left\{(x,y)\,\middle|\,-1\leq\frac{y}{x}\leq 1,\ x\neq 0,\ 4-x-y>0\right\}$$

3. **Find the partial derivatives. Write final answers without negative exponents and simplify all complex fractions.**

   (a) **Find** f_y **for** $f(x,y)=e^{xy}(\cos x\sin y)\Rightarrow$ We need the Product Rule:

   $$f_y(x,y)=xe^{xy}(\cos x\sin y)+e^{xy}\cos x\cos y\ =\ e^{xy}\cos x(x\sin y+\cos y)$$

Calculus III   Practice Midterm 3 KEY, continued

(b)   Find f_x for $f(x, y) = \arctan \dfrac{y}{x}$

We use the Arctan Rule that says $\dfrac{d[\arctan u]}{dx} = \dfrac{u'}{1 + u^2}$ where:

$$u = \frac{y}{x} = yx^{-1} \quad and \quad u' = y(-1)x^{-2} = -\frac{y}{x^2}$$

So, we now have:

$$f(x, y) = \frac{-\dfrac{y}{x^2}}{1 + \left(\dfrac{y}{x}\right)^2} = \frac{-\dfrac{y}{x^2}}{1 + \dfrac{y^2}{x^2}} = \left(\frac{-\dfrac{y}{x^2}}{1 + \dfrac{y^2}{x^2}}\right)\left(\frac{x^2}{x^2}\right) = \frac{-y}{x^2 + y^2}$$

4.   **Show that the limit does not exist by considering two different paths approaching the point (0, 0).**

$$\lim_{(x,y)\to(0,0)} \frac{x - y}{x^2 + y^2}$$

(a)   **Path #1: along the line $y = x$**

Replace y with x:

$$\lim_{(x,x)\to(0,0)} \frac{x - (x)}{x^2 + (x)^2} = \lim_{(x,x)\to(0,0)} \frac{0}{2x^2} = 0$$

(b)   **Path #2: along the y-axis**

Replace x with 0:

$$\lim_{(0,y)\to(0,0)} \frac{(0) - y}{(0)^2 + y^2} = \lim_{(0,y)\to(0,0)} \frac{-y}{y^2} = \lim_{(0,y)\to(0,0)} \frac{-1}{y} = \frac{-1}{0} = \infty \ (d.n.e.)$$

Conclusion: Since the two limits do not agree, we can conclude that the limit does not exist.

5.   **Use implicit partial differentiation to find** $\dfrac{\partial z}{\partial x}$ **for** $x^2 y - 2yz - xz - z^2 = 0$.

First, we rewrite the equation as $F(x, y, z) = x^2 y - 2yz - xz - z^2 = 0$ and then we use the Theorem that says:

$$\frac{\partial z}{\partial x} = -\frac{F_x(x, y, z)}{F_z(x, y, z)}$$

So for this problem we will have that:

$$\frac{\partial z}{\partial x} = -\frac{F_x(x, y, z)}{F_z(x, y, z)} = -\frac{(2xy - z)}{-2y - x - 2z} \quad OR \quad \frac{z - 2xy}{-2y - x - 2z} \quad OR \quad \frac{2xy - z}{2y + x + 2z}$$

CALCULUS III   Practice Midterm 3 KEY, CONTINUED

6. Show that the function $z = \dfrac{y}{x^2 + y^2}$ satisfies LaPlace's equation $\dfrac{\partial^2 z}{\partial x^2} + \dfrac{\partial^2 z}{\partial y^2} = 0$.

Please note that the notation $\dfrac{\partial^2 z}{\partial x^2} = z_{xx}$ and $\dfrac{\partial^2 z}{\partial y^2} = z_{yy}$ is equivalent.

Therefore, we must find the second partials. Of course, this means we need to find the first partials as our first step. To do this, we will need the Quotient Rule:

$$z_x = \frac{(0)(x^2 + y^2) - (2x)(y)}{(x^2 + y^2)^2} = \frac{-2xy}{(x^2 + y^2)^2} \quad and \quad z_y = \frac{(1)(x^2 + y^2) - (2y)(y)}{(x^2 + y^2)^2} = \frac{x^2 - y^2}{(x^2 + y^2)^2}$$

We need the Quotient Rule again to find the second partials:

$$z_{xx} = \frac{(-2y)(x^2 + y^2)^2 - (2)(x^2 + y^2)(2x)(-2xy)}{\left[(x^2 + y^2)^2\right]^2} = \frac{-2y(x^2 + y^2)\left[(x^2 + y^2) - 4x^2\right]}{(x^2 + y^2)^4} = \frac{-2y(x^2 + y^2)\left[-3x^2 + y^2\right]}{(x^2 + y^2)^4}$$

$$z_{yy} = \frac{(-2y)(x^2 + y^2)^2 - (2)(x^2 + y^2)(2y)(x^2 - y^2)}{\left[(x^2 + y^2)^2\right]^2} = \frac{-2y(x^2 + y^2)\left[(x^2 + y^2) + 2(x^2 - y^2)\right]}{(x^2 + y^2)^4}$$

$$= \frac{-2y(x^2 + y^2)\left[3x^2 - y^2\right]}{(x^2 + y^2)^4}$$

We now add them together:

$$z_{xx} + z_{yy} = \frac{-2y(x^2 + y^2)\left[(x^2 + y^2) - 4x^2\right]}{(x^2 + y^2)^4} + \frac{-2y(x^2 + y^2)\left[3x^2 - y^2\right]}{(x^2 + y^2)^4}$$

$$= \frac{-2y(x^2 + y^2)\left[-3x^2 + y^2 + 3x^2 - y^2\right]}{(x^2 + y^2)^4} = 0$$

7. Find the mixed partials and verify that they are equal for the function $h(x,y) = x \sin y + y \cos x$.

Mixed partials are the second partials h_{xy} and h_{yx}. We must first find the first partial derivatives:

$$h_x(x, y) = \sin y - y \sin x \quad and \quad h_y(x, y) = x \cos y + \cos x$$

Now, we find the mixed (second) partials:

$$h_{xy} = \cos y - \sin x = h_{yx}$$

8. Given the function $f(x, y) = \dfrac{x}{x^2 + y^2}$ ...

(a) Use the total differential to approximate change in the function as (x, y) varies from the point $(1, 2)$ to the point $(0.98, 2.01)$.

The total differential is given by $dz = \dfrac{\partial z}{\partial x} dx + \dfrac{\partial z}{\partial y} dy$, where we have that

$dx = \Delta x = 0.98 - 1 = -0.02$ and $dy = \Delta y = 2.01 - 2 = 0.01$.

To find z_x we use the Quotient Rule. For z_y; we rewrite the function as $f(x, y) = x(x^2 + y^2)^{-1}$ and use the Chain Rule to differentiate:

$$z_x = f_x(x, y) = \frac{(1)(x^2 + y^2) - 2x(x)}{(x^2 + y^2)^2} = \frac{y^2 - x^2}{(x^2 + y^2)^2} \qquad z_y = f_y(x, y) = x(-1)(x^2 + y^2)^{-2}(2y) = \frac{-2xy}{(x^2 + y^2)^2}$$

Calculus III   Practice Midterm 3 KEY, continued

We use the initial point's coordinates for x and y when plugging into the total differential that now looks like:

$$dz = \frac{y^2 - x^2}{\left(x^2 + y^2\right)^2}\, dx + \frac{-2xy}{\left(x^2 + y^2\right)^2}\, dy \;=\; \frac{2^2 - 1^2}{\left(1^2 + 2^2\right)^2}(-0.02) + \frac{-2(1)(2)}{\left(1^2 + 2^2\right)^2}(0.01) \approx \boxminus 0.004$$

(b) **Calculate the actual change in the function as (x, y) varies from the point $(1, 2)$ to the point $(0.98, 2.01)$.**

The actual change for the function is given by $\Delta z = f(x + \Delta x, y + \Delta y) - f(x, y)$. So, we have for this problem:

$$\Delta z = f(0.98, 2.01) - f(1, 2) \;=\; \frac{2.01}{(0.98)^2 + (2.01)^2} - \frac{1}{(1)^2 + (2)^2} \;\approx\; -.004$$

Please note that the values for dz and Δz are the same. This means that the total differential, dz, makes for an excellent approximation for changes in z!

9. **The height and radius of a right circular cylinder are approximately 15 centimeters and 8 centimeters, respectively. The maximum error in each measurement is ±0.02 centimeters.**

(a) **Find the approximate volume of the cylinder.**

The formula for volume of a cylinder is $V(r, h) = \pi r^2 h$. So the volume of this particular cylinder is $V(8,15) = \pi(8)^2(15) \approx 3015.9\,$cu. cm.

(b) **Use the total differential to estimate the propagated error in the calculated volume of the cylinder.**

The total differential for this volume function is $dV = \dfrac{\partial V}{\partial h}\, dh + \dfrac{\partial V}{\partial r}\, dr$, where both dh and dr are ±0.02 cm.

After finding the first partials, we have $dV = \left(\pi r^2\right) dh + \left(2\pi r h\right) dr$.

Plugging all of the values in gives $dV = \left(\pi(8)^2\right)(\pm 0.02) + \left(2\pi(8)(15)\right)(\pm 0.02) \approx \pm 19$ cu. cm. *(answer)*

(c) **Use the total differential to estimate the relative error in the calculated volume of the cylinder and express this relative error in percent.**

Relative error $= \dfrac{propogated\ error}{volume\ measured} = \dfrac{\pm 19 cm^3}{3015.0 cm^3} = 0.00633 \;\Rightarrow\; 0.633\%$

10. **The length, width, and height of a rectangular chamber are changing at the rate of 3 feet per minute, 2 feet per minute, and ½-foot per minute, respectively. Find the rate at which the volume is changing at the instant the length is 10 feet, the width is 6 feet, and the height is 4 feet.**

We know that $\dfrac{dL}{dt} = 3\,ft/\min \quad \dfrac{dW}{dt} = 2\,ft/\min \quad \dfrac{dH}{dt} = \dfrac{1}{2}\,ft/\min$, at the moment when L = 10ft,

CALCULUS **III**   **Practice Midterm 3 KEY,** CONTINUED

W = 6ft, and H = 4 feet.

We know the volume of a box is given by $V(L,W,H) = LWH$.

We use the Chain Rule for a Function of Several Variables:

$$\frac{dV}{dt} = \frac{\partial V}{\partial L} \cdot \frac{dL}{dt} + \frac{\partial V}{\partial W} \cdot \frac{dW}{dt} + \frac{\partial V}{\partial H} \cdot \frac{dH}{dt} = (WH) \cdot \frac{dL}{dt} + (LH) \cdot \frac{dW}{dt} + (LW) \cdot \frac{dH}{dt}$$

Next, we plug in all of the values given:

$$\frac{dV}{dt} = (6)(4)(3) + (10)(4)(2) + (10)(6)\left(\frac{1}{2}\right) = 182 \; cu. \; ft./\min$$

11. **Use the Chain Rule to find** $\dfrac{dw}{dt}$ **when** $w = \ln(xy) + xy$ **and** $x = e^t, \; y = e^{-t}$. **You will not get any credit unless you use the Chain Rule.**

$$\frac{dw}{dt} = \frac{\partial w}{\partial x} \cdot \frac{dx}{dt} + \frac{\partial w}{\partial y} \cdot \frac{dy}{dt} = \left(\frac{1}{xy} \cdot (y) + y\right)e^t + \left(\frac{1}{xy} \cdot (x) + x\right)\left(-e^{-t}\right)$$

We reduce all quotients to lowest terms, and also replace $x = e^t$ *and* $y = e^{-t}$ to get that:

$$\frac{dw}{dt} = \left(\frac{1}{e^t} + e^{-t}\right)e^t + \left(\frac{1}{e^{-t}} + e^t\right)\left(-e^{-t}\right)$$

Finally, distribute over the sums and add:

$$\frac{dw}{dt} = (1 + 1) + (-1 - 1) = 0$$

12. **The surface of a mountain is described by the equation** $f(x,y) = 1500 - x^2 - 3y^2$. **If a skier is at the point (10, 20, 200), what direction should she move in order to descend at the greatest rate?**

The gradient gives the direction of steepest *ascent*. So, if we take the opposite of the gradient, we'll get the direction of steepest *descent*—the gradient is given by:

$$\nabla \mathbf{f}(x, y) = \langle f_x, f_y \rangle = \langle -2x, -6y \rangle$$

At the point given, we have that $x = 10$ and $y = 20$ (substitute):

$$\nabla \mathbf{f}(10,20) = \langle -2(10), -6(20) \rangle = \langle -20, -120 \rangle$$

Take the opposite of this to get the direction for steepest descent:

$$-\nabla \mathbf{f}(10,20) = -\langle -20, -120 \rangle = \langle 20, 120 \rangle$$

13. **Given the function** $f(x,y,z) = x^2 - 2y^2 + z^2 e^z$ ...
     (a)   **Find the gradient of** *f* **at the point (–1, 2, 1).**

The gradient is given by:

$$\nabla \mathbf{f}(x, y, z) = \langle f_x, f_y, f_z \rangle = \langle 2x, 4y, 2ze^z + z^2 e^z \rangle \quad \textit{(I used the Product Rule for the last component.)}$$

ANSWER *13,* CONTINUED...

...Answer 13, continued

Calculus III

At the point given, we have $\nabla f(-1,2,1) = \langle 2(-1), -4(2), 2(1)e^1 + (1)^2 e^1 \rangle = \langle -2, -8, 3e \rangle$.

(b) **Calculate the directional derivative in the direction of the vector** $\mathbf{v} = \langle -2, 7, 4 \rangle$ **at the point** (–1, 2, 1).

First, we need the direction vector written as a unit vector:

$$\mathbf{u} = \frac{\mathbf{v}}{\|\mathbf{v}\|} = \frac{\langle -2, 7, 4 \rangle}{\sqrt{2^2 + 7^2 + 4^2}} = \frac{\langle -2, 7, 4 \rangle}{\sqrt{69}} = \left\langle \frac{-2}{\sqrt{69}}, \frac{7}{\sqrt{69}}, \frac{4}{\sqrt{69}} \right\rangle$$

The directional derivative is the dot product of the normalized direction vector with the gradient (at the point given):

$$D_{\mathbf{u}}f(-1,2,1) = \nabla f(-1,2,1) \cdot \mathbf{u} = \langle -2, -8, 3e \rangle \cdot \left\langle \frac{-2}{\sqrt{69}}, \frac{7}{\sqrt{69}}, \frac{4}{\sqrt{69}} \right\rangle = (-2)\left(\frac{-2}{\sqrt{69}}\right) + (-8)\left(\frac{7}{\sqrt{69}}\right) + (3e)\left(\frac{4}{\sqrt{69}}\right)$$

$$= \frac{-52 + 12e}{\sqrt{69}} \approx -2.333$$

(c) **Find the maximum value of the directional derivative at the point** (–1, 2, 1).

The maximum value of a directional derivative at a given point is simply equal to the magnitude of the gradient at that point:

$$\textbf{\textit{Maximum Value}} = \|\nabla f(-1,2,1)\| = \sqrt{(2)^2 + (8)^2 + (3e)^2} = \sqrt{4 + 64 + 9e^2} \approx 11.6$$

14. **Given the function** $f(x,y) = \ln(x^2 + y^2 + 1) + e^{2xy}$ ...

(a) **Find** ∇f **at the point** (0, –2).

$$\nabla f(x,y) = \langle f_x, f_y \rangle = \left\langle \frac{2x}{x^2 + y^2 + 1} + 2ye^{2xy}, \frac{2y}{x^2 + y^2 + 1} + 2xe^{2xy} \right\rangle$$

$$\nabla f(0, -2) = \left\langle \frac{2(0)}{0^2 + (-2)^2 + 1} + 2(-2)e^{2(0)(-2)}, \frac{2(-2)}{0^2 + (-2)^2 + 1} + 2(0)e^{2(0)(-2)} \right\rangle = \left\langle 0 + -4(1), \frac{-4}{5} + 0 \right\rangle = \left\langle -4, -\frac{4}{5} \right\rangle$$

(b) **Calculate the directional derivative in the direction of the vector** $\mathbf{v} = \langle 5, -12 \rangle$ **at the point** (0, –2).

First, we need to normalize the direction vector:

$$\mathbf{u} = \frac{\mathbf{v}}{\|\mathbf{v}\|} = \frac{\langle 5, -12 \rangle}{\sqrt{5^2 + 12^2}} = \frac{\langle 5, -12 \rangle}{\sqrt{169}} = \left\langle \frac{5}{13}, \frac{-12}{13} \right\rangle$$

The directional derivative is the dot product of the normalized direction vector with the gradient (at the point given):

$$D_{\mathbf{u}}f(5, -12) = \nabla f(5, -12) \cdot \mathbf{u} = \left\langle -4, -\frac{4}{5} \right\rangle \cdot \left\langle \frac{5}{13}, \frac{-12}{13} \right\rangle = (-4)\left(\frac{5}{13}\right) + \left(-\frac{4}{5}\right)\left(\frac{-12}{13}\right)$$

$$= -\frac{20}{13} + \frac{48}{65} = \frac{-4}{5}$$

(This represents the slope of the surface at the point given in the direction given.)

CALCULUS III   **Practice Midterm 3 KEY, CONTINUED**

(c)   **Find the maximum value of the directional derivative at the point (0, –2).**

The maximum value of a directional derivative at a given point is simply equal to the magnitude of the gradient at that point:

$$\textit{Maximum Value} = \|\nabla f(0, -2)\| = \sqrt{(-4)^2 + \left(\frac{-4}{5}\right)} = \sqrt{16 + \frac{16}{25}} = \sqrt{\frac{416}{25}} = \frac{4\sqrt{26}}{5} \approx 4.08$$

This represents the maximum slope at the point given.

15.   **Consider the surface given by** $x^2 + 4y^2 = 10z$.

(a)   **Find an equation of the tangent plane (in general form) to the surface at the point (2, –2, 2).**

We first need to rewrite the equation as $F(x, y, z) = 0$. We do this now:

$$F(x, y, z) = x^2 + 4y^2 - 10z = 0$$

Next, we get a vector normal to the tangent plane. This is exactly the gradient for the function of *three* variables we found.

Therefore, we have $\mathbf{n} = \nabla F(x, y, z) = \left\langle F_x, F_y, F_z \right\rangle = \left\langle 2x, 8y, -10 \right\rangle$.

Now, evaluate this gradient at the point given:

$$\mathbf{n} = \nabla F(2, -2, 2) = \left\langle 2(2), 8(-2), -10 \right\rangle = \left\langle 4, -16, -10 \right\rangle$$

Then, we use the equation for a tangent plane in standard form: $a(x - x_0) + b(y - y_0) + c(z - z_0) = 0$.

$4(x - 2) + -16(y - (-2)) + -10(z - 2) = 0 \ \Rightarrow \ 4x - 8 - 16y - 32 - 10z + 20 = 0 \ \Rightarrow 4x - 16y - 10z = 20$

OR : $2x - 8y - 5z = 10$

(b)   **Find the angle of inclination** θ **(in degrees) of the tangent plane at the point (2, –2, 2). Round answer to nearest tenths.**

We use the formula that gives the angle that a tangent plane makes with the horizontal that is:

$$\cos\theta = \frac{|\nabla F \cdot \mathbf{k}|}{\|\nabla F\|} = \frac{|\langle 4, -16, -10 \rangle \cdot \langle 0, 0, 1 \rangle|}{\|\langle 4, -16, -10 \rangle\|} = \frac{|(4)(0) + (-16)(0) + (-10)(1)|}{\sqrt{4^2 + 16^2 + 10^2}} = \frac{|-10|}{\sqrt{372}} = \frac{10}{\sqrt{372}}$$

$$\theta = \cos^{-1}\left(\frac{10}{\sqrt{372}}\right) \approx 58.8°$$

(c)   **Find a set of parametric equation for the normal line to the surface at the point (2, –2, 2).**

The direction numbers for the line are exactly the components for the normal vector we found above. In other words, $a = 4$, $b = -16$, and $c = -10$. Using the point given on the surface as a point on the line as well and the parametric equations for a line in space, we have:

$$\begin{cases} x = x_0 + at \\ y = y_0 + bt \\ z = z_0 + ct \end{cases} \Rightarrow \begin{cases} x = 2 + 4t \\ y = -2 - 16t \\ z = 2 - 10t \end{cases}$$

Calculus III   **Practice Midterm 3 KEY, continued**

---

16. **Test the function** $f(x,y) = x^3 - 3xy + y^2 + y - 5$ **for relative extrema and saddle points. Give answer(s) using ordered triples to get full credit.**

First, we find the first partials, set them equal to zero, and solve the resulting system of equations to get the critical point(s):

$$f_x(x,y) = 3x^2 - 3y = 0$$
$$f_y(x,y) = -3x + 2y + 1 = 0$$

This is a *non*-linear system of equations. To solve, we'll use the substitution method. We first solve for y using the first equation. Then we'll substitute this into the second equation to obtain a quadratic equation in the variable x.

Solving for y using equation #1, we have:

$$3y = 3x^2 \implies y = x^2$$

Substitute this expression for y into the second equation:

$$-3x + 2(x^2) + 1 = 0 \implies 2x^2 - 3x + 1 = 0.$$

To solve, you may use either the quadratic formula or factor and apply the ZPP. (I will use the latter method.):

$$2x^2 - 3x + 1 = 0 \implies 2x^2 - 2x - x + 1 = 0 \implies 2x(x-1) - 1(x-1) = 0 \implies (2x-1)(x-1) = 0$$

$$2x - 1 = 0 \quad or \quad x - 1 = 0 \implies x = \frac{1}{2}, \ x = 1$$

Now, we get the corresponding y-coordinates by using the fact that $y = x^2$.

So, our critical points are:

$$\left(\frac{1}{2}, \frac{1}{4}\right) \quad and \quad (1, 1)$$

We will now perform the *D*-Test on these critical points to determine the type of extrema. We form the expression for *D*:

$$D = f_{xx}f_{yy} - \left(f_{xy}\right)^2 = (6x)(2) - (-3)^2 = 12x - 9$$

Evaluate *D* at each of the critical points:

$$D\left(\frac{1}{2}, \frac{1}{4}\right) = 12\left(\frac{1}{2}\right) - 9 = 6 - 9 = -3 < 0$$

When *D* is negative, we have a saddle point:

$$D(1,1) = 12(1) - 9 = 12 - 9 = 3 > 0 \quad \Longrightarrow$$

Since f_{xx} is positive, we have a relative minimum at this point.

When D is positive, we need to do further testing to see if it is a maximum or a minimum. Plug the critical point into the second partial, f_{xx}, now:

$$f_{xx}(1,1) = 6(1) > 0$$

Final answer: Saddle point at $\left(\frac{1}{2}, \frac{1}{4}, -\frac{79}{16}\right)$ and a relative minimum at $(1, 1, -5)$

CALCULUS III   Practice Midterm 3 KEY, CONTINUED

17. **Find the point on the surface where the tangent plane is horizontal:** $z = 3x^2 + 2y^2 - 3x + 4y - 5$.

A tangent plane will be horizontal when the gradient, or its normal vector, $\mathbf{n} = \nabla F(x,y,z) = \langle F_x, F_y, F_z \rangle$, is parallel to the unit standard vector $\mathbf{k}$. In other words, it will be a multiple of the unit standard vector $\mathbf{k} = \langle 0, 0, 1 \rangle$.

We first need to rewrite the equation as $F(x,y,z) = 0$. We do this now:

$$F(x,y,z) = 3x^2 + 2y^2 - 3x + 4y - 5 - z = 0$$

Therefore, our gradient will be $\mathbf{n} = \nabla F(x,y,z) = \langle F_x, F_y, F_z \rangle = \langle 6x - 3, 4y + 4, -1 \rangle$.

This vector will be a multiple of the standard unit vector $\mathbf{k} = \langle 0, 0, 1 \rangle$ when its first two components are simultaneously equal to zero:

$$6x - 3 = 0 \quad \Rightarrow \quad x = \frac{1}{2}$$

$$4y + 4 = 0 \quad \Rightarrow \quad y = -1$$

We find the corresponding z-coordinate by plugging these values for x and y into the original equation given for the surface:

$$z = 3\left(\frac{1}{2}\right)^2 + 2(-1)^2 - 3\left(\frac{1}{2}\right) + 4(-1) - 5 = -\frac{31}{4}$$

The final answer is: The point $\left(\frac{1}{2}, -1, -\frac{31}{4}\right)$ is a point on the surface where the tangent plane is horizontal.

18. **Find the path of a heat-seeking particle placed at the point in space (2, 2, 5) with a temperature field** $T(x,y,z) = 100 - 3x - y - z^2$. **(*Hint:* Use Differential Equations.)**

The path will be a vector-valued function $\mathbf{r}(t) = \langle x(t), y(t), z(t) \rangle$. This is the form for our final answer.

Since the particle always travels in the direction of maximum temperature increase, this means it will always be traveling in the direction of the gradient, $\nabla T(x,y,z) = \langle T_x, T_y, T_z \rangle = \langle -3, -1, -2z \rangle$.

A vector that is tangent to the curve at all times is the velocity vector:

$$\mathbf{r}'(t) = \langle x'(t), y'(t), z'(t) \rangle = \left\langle \frac{dx}{dt}, \frac{dy}{dt}, \frac{dz}{dt} \right\rangle$$

That is, the velocity vector is always pointing in the direction of the motion. These two vectors mentioned will be parallel to each other (i.e., one is a multiple of the other). Let us assume that they are equivalent to make this problem a little easier. So, we have that:

$$\mathbf{r}'(t) = \nabla T(x,y,z)$$

$$\left\langle \frac{dx}{dt}, \frac{dy}{dt}, \frac{dz}{dt} \right\rangle = \langle -3, -1, -2z \rangle$$

ANSWER 18, CONTINUED...

...*ANSWER 18, CONTINUED*

CALCULUS III   Practice Midterm 3 KEY, CONTINUED

We equate the components and this gives us two separate differential equations to solve. I will solve them simultaneously using the separation of variables technique:

$$\frac{dz}{dt} = -2z$$

$$\frac{dz}{z} = -2dt$$

$$\frac{dx}{dt} = -3 \qquad\qquad \frac{dy}{dt} = -1 \qquad\qquad \int \frac{dz}{z} = \int -2dt$$

$$dx = -3dt \qquad\qquad dy = -1dt \qquad\qquad \ln|z| = -2t + C_3$$

$$\int dx = \int -3dt \qquad \int dy = \int -dt \quad\Longrightarrow\quad e^{-2t+C_3} = z$$

$$x = -3t + C_1 \qquad\qquad y = -t + C_2 \qquad\qquad e^{-2t}e^{C_3} = z$$

$$C_3 e^{-2t} = z$$

To find the constants of integration, we use the initial condition that tells us at time $t = 0$, the particle is at the point (2, 2, 5). In other words, $x = 2$, and $y = 2$, and $z = 5$.

$$x(0) = 2 = -3(0) + C_1 \qquad y(0) = 2 = -(0) + C_2 \qquad z(0) = 5 = C_3 e^{-2(0)}$$

$$2 = C_1 \qquad\qquad\qquad 2 = C_2 \qquad\qquad\qquad 5 = C_3$$

Thus, our final answer describing the path of the heat-seeking particle is the vector-valued function $\mathbf{r}(t) = \left\langle -3t + 2, -t + 2, 5e^{-2t} \right\rangle$.

19. **Determine the relative extrema and saddle points (if any) for the function** $f(x, y) = 3x^2 + 6xy + 7y^2 - 2x + 4y$. **Give answer(s) as ordered triples.**

First, we find the first partials, set them equal to zero, and solve the resulting system of equations to get the critical point(s):

$$f_x(x, y) = 6x + 6y - 2 = 0$$

$$f_y(x, y) = 6x + 14y + 4 = 0$$

This is a linear system of equations. To solve, we'll use the elimination method. Subtract equation #2 from equation #1 to eliminate the variable x.

$$6x + 6y = 2$$

$$-6x - 14y = 4$$

$$-8y = 6 \quad\Rightarrow\quad y = -\frac{6}{8} = -\frac{3}{4}$$

Now, we get the corresponding x-coordinate by substituting this into either of the original equations. (I will use equation #1.)

$$6x + 6\left(-\frac{3}{4}\right) - 2 = 0 \quad\Rightarrow\quad 6x = 2 + \frac{9}{2} = \frac{13}{2} \quad\Rightarrow\quad x = \frac{13}{2}\left(\frac{1}{6}\right) = \frac{13}{12}$$

So, our critical point is $\left(\frac{13}{12}, -\frac{3}{4}\right)$.

CALCULUS **III**   **Practice Midterm 3 KEY,** CONTINUED

We will now perform the *D*-Test on this critical point to determine the type of extrema. We form the expression for *D*:

$$D = f_{xx}f_{yy} - \left(f_{xy}\right)^2 \quad = \quad (6)(14) - (6)^2 \quad = 48 > 0$$

Because *D* is positive, we examine the sign for f_{xx}: $f_{xx} = 6 > 0$.

Since f_{xx} is positive, we have a relative minimum at this point.

Final answer:
$$\text{A relative minimum at } \left(\frac{13}{12}, -\frac{3}{4}, -\frac{31}{12}\right)$$

20. **Suppose you are** *not* **given the original function, but you are given the first partial derivatives** $f_x(x, y) = x^2 - 121$ **and** $f_y(x, y) = 8y^3 + 14y^2 - 15y$. **Determine the relative extrema and saddle points (if any) for the function. Give answer(s) as ordered** *pairs,* **not ordered triples, since you do not have access to the original function. Please do not attempt to find the original function. You do not require it to complete this problem!**

Find the critical point(s) by setting the first partials equal to zero and solving the system. It turns out that this is not a genuine system of equations since the first partial w.r.t. *x* contains only the variable *x*, and the first partial w.r.t. *y* contains only the variable *y*. So, we solve the equations separately, then "mix and match" the results as follows:

$$f_x(x, y) = x^2 - 121 = 0 \quad \Rightarrow \quad x = \pm\sqrt{121} = \pm 11$$

$$f_y(x, y) = 8y^3 + 14y^2 - 15y = 0 \quad \Rightarrow \quad y\left(8y^2 + 14y - 15\right) = 0$$

We could use the quadratic formula on the trinomial factor, but I claim it factors! The Key Number is $(8)(-15) = -120$. Factors of -120 that sum up to 14 are 20 and -6. Therefore, we have:

$$y\left(8y^2 + 20y - 6y - 15\right) = 0 \quad \Rightarrow \quad y\left[4y(2y + 5) - 3(2y + 5)\right] = 0$$

$$y(4y - 3)(2y + 5) = 0$$

Apply the ZPP now to solve $y = 0$, $4y - 3 = 0$, $2y + 5 = 0$.

The three solutions are $y = 0$, $y = \dfrac{3}{4}$, $y = -\dfrac{5}{2}$.

Now, we "mix and match" all of these different *x* and *y* coordinates together to get all of the critical points:

$$(11, 0), (-11, 0), \left(11, \frac{3}{4}\right), \left(-11, \frac{3}{4}\right), \left(11, -\frac{5}{2}\right), \text{ and } \left(-11, -\frac{5}{2}\right)$$

WOW! We need to perform the *D*-test on all of these:

$$D = f_{xx}f_{yy} - \left(f_{xy}\right)^2 \quad = \quad (2x)(24y^2 + 28y - 15) - (0)^2 \quad = \quad 48xy^2 + 56xy - 30x$$

We also know that $f_{xx} = 2x$.

ANSWER 20, CONTINUED...

CALCULUS III   **Practice Midterm 3 KEY, CONTINUED**

Now, test each point separately:

$$D(11, 0) \quad 48(11)(0)^2 + 56(11)(0) - 30(11) < 0$$

Since D is negative, we have a saddle point at $(11, 0)$:

$$D(-11, 0) \quad 48(-11)(0)^2 + 56(-11)(0) - 30(=11) > 0$$

Since D is positive, we examine f_{xx}:

$$f_{xx}(-11, 0) = 2(-11) < 0$$

We have a relative maximum at the point $(-11, 0)$:

$$D\left(11, \frac{3}{4}\right) \quad 48(11)\left(\frac{3}{4}\right)^2 + 56(11)\left(\frac{3}{4}\right) - 30(11) = 429 > 0$$

Since D is positive, we examine f_{xx}:

$$f_{xx}\left(11, \frac{3}{4}\right) = 2(11) > 0$$

We have a relative minimum at the point $\left(11, \frac{3}{4}\right)$:

$$D\left(-11, \frac{3}{4}\right) \quad 48(-11)\left(\frac{3}{4}\right)^2 + 56(-11)\left(\frac{3}{4}\right) - 30(-11) = -429 < 0$$

Since D is negative, we have another saddle point at $\left(-11, \frac{3}{4}\right)$:

$$D\left(11, -\frac{5}{2}\right) \quad 48(11)\left(-\frac{5}{2}\right)^2 + 56(11)\left(-\frac{5}{2}\right) - 30(11) = 1430 > 0$$

Since D is positive, we examine f_{xx}:

$$f_{xx}\left(11, -\frac{5}{2}\right) = 2(11) > 0$$

We have another relative minimum at the point $\left(11, -\frac{5}{2}\right)$:

$$D\left(-11, -\frac{5}{2}\right) \quad 48(-11)\left(-\frac{5}{2}\right)^2 + 56(-11)\left(-\frac{5}{2}\right) - 30(-11) = -1430 < 0$$

Since D is negative, we have another saddle point at $\left(-11, -\frac{5}{2}\right)$.

So, we have three saddle points, two relative minimums, and one relative maximum in all!

$$\sqrt{}$$ **PART III** **TEST ANSWER KEY**

CALCULUS III   Practice Midterm 3 KEY, CONTINUED

21. **Determine the relative extrema and saddle points (if any) for the function** $f(x,y) = \dfrac{1}{x^2 + y^2 - 1}$. **Give answer(s) as ordered triples.**

First, we rewrite the function in order to make it a little easier to differentiate:

$$f(x,y) = \left(x^2 + y^2 - 1\right)^{-1}$$

Now we take the first partials and set them equal to zero to find critical points:

$$f_x(x,y) = -1\left(x^2 + y^2 - 1\right)^{-2}(2x) = \frac{-2x}{\left(x^2 + y^2 - 1\right)^2} = 0$$

$$f_y(x,y) = -1\left(x^2 + y^2 - 1\right)^{-2}(2y) = \frac{-2y}{\left(x^2 + y^2 - 1\right)^2} = 0$$

The point $(0, 0)$ is a critical point. To form D, we need the second partials. It will take the Quotient Rule to find these:

$$f_{xx}(x,y) = \frac{-2\left(x^2 + y^2 - 1\right)^2 - 2\left(x^2 + y^2 - 1\right)(2x)(-2x)}{\left(x^2 + y^2 - 1\right)^4} = \frac{\left(x^2 + y^2 - 1\right)\left[-2\left(x^2 + y^2 - 1\right) + 8x^2\right]}{\left(x^2 + y^2 - 1\right)^4}$$

$$= \frac{6x^2 - 2y^2 + 2}{\left(x^2 + y^2 - 1\right)^3}$$

$$f_{yy}(x,y) = \frac{-2\left(x^2 + y^2 - 1\right)^2 - 2\left(x^2 + y^2 - 1\right)(2y)(-2y)}{\left(x^2 + y^2 - 1\right)^4} = \frac{\left(x^2 + y^2 - 1\right)\left[-2\left(x^2 + y^2 - 1\right) + 8y^2\right]}{\left(x^2 + y^2 - 1\right)^4}$$

$$= \frac{6y^2 - 2x^2 + 2}{\left(x^2 + y^2 - 1\right)^3}$$

$$f_{xy}(x,y) = \frac{\partial\left[(-2x)(x^2 + y^2 - 1)^{-2}\right]}{\partial y} = (-2x)(-2)(x^2 + y^2 - 1)^{-3}(2y) = \frac{8xy}{(x^2 + y^2 - 1)^3}$$

Evaluating D at the critical point, we have $D(0,0) = (-2)(-2) - 0^2 = 4 > 0$.

Next, since $f_{xx}(0,0) = -2 < 0$, we have a relative maximum at the point.

Final answer: Relative maximum at the point $(0, 0, -1)$.

CALCULUS III　　　　　　　　　　　　　　　　**Practice Midterm 3 KEY, CONTINUED**

22.　**Find the *absolute* extrema of** $f(x,y) = (4x - x^2)\cos y$ **on the rectangular plate** $1 \le x \le 3,\ -\pi/4 \le y \le \pi/4$.

The boundary of the region is simply a rectangle. We will first determine the relative extrema (if any) by finding critical points:

$$f_x(x,y) = (4 - 2x)\cos y = 0 \quad and \quad f_y(x,y) = (4x - x^2)(-\sin y)$$

The critical point is $(2, 0)$. At this point, the function has the value $f(2,0) = (4(2) - (2)^2)\cos(0) = 4$. We log this info filling into our table:

(x, y)	$f(x, y)$
$(2, 0)$	4
$(1, 0)$	3
$\left(1, -\dfrac{\pi}{4}\right)$	$\dfrac{3\sqrt{2}}{2}$
$\left(1, \dfrac{\pi}{4}\right)$	$\dfrac{3\sqrt{2}}{2}$

We now look at the boundaries of our rectangular region. The left-most side of the rectangle is the vertical line $x = 1$. At $x = 1$, the function will be:

$$f(1, y) = (4(1) - (1)^2)\cos y = 3\cos y$$

We do first-semester calculus on this single-variable function and determine its critical points:

$$f'(1, y) = -3\sin y = 0 \quad \Rightarrow \quad y = 0$$

Thus, the point $(1, 0)$ is a critical point and we find the corresponding value for the function at this point and put the info into our table: $f(1,0) = 3\cos(0) = 3$. We finish this edge of our rectangle by examining the vertices of the rectangle:

$$\left(1, -\frac{\pi}{4}\right) \text{ and } \left(1, \frac{\pi}{4}\right). \quad f\left(1, -\frac{\pi}{4}\right) = 3\cos\left(-\frac{\pi}{4}\right) = \frac{3\sqrt{2}}{2} \quad \text{and} \quad f\left(1, \frac{\pi}{4}\right) = 3\cos\left(\frac{\pi}{4}\right) = \frac{3\sqrt{2}}{2}$$

We now do all of the same things to the right-most edge of our rectangle, given by the vertical line $x = 3$. At $x = 3$, the function will be:

$$f(3, y) = (4(3) - (3)^2)\cos y = 3\cos y$$

We do first-semester calculus on this single-variable function and determine its critical points:

$$f'(3, y) = -3\sin y = 0 \quad \Rightarrow \quad y = 0$$

Thus, the point $(3, 0)$ is a critical point and we find the corresponding value for the function at this point and put the info into our table:

$$f(3,0) = 3\cos(0) = 3$$

CALCULUS **III**   **Practice Midterm 3 KEY,** CONTINUED

We finish this edge of our rectangle by examining the vertices of the rectangle. That is, the points:

$\left(3, -\dfrac{\pi}{4}\right)$ and $\left(3, \dfrac{\pi}{4}\right)$. $f\left(3, -\dfrac{\pi}{4}\right) = 3\cos\left(-\dfrac{\pi}{4}\right) = \dfrac{3\sqrt{2}}{2}$ and $f\left(3, \dfrac{\pi}{4}\right) = 3\cos\left(\dfrac{\pi}{4}\right) = \dfrac{3\sqrt{2}}{2}$

(x, y)	$f(x, y)$
$(2, 0)$	4
$(1, 0)$	3
$\left(1, -\dfrac{\pi}{4}\right)$	$\dfrac{3\sqrt{2}}{2}$
$\left(1, \dfrac{\pi}{4}\right)$	$\dfrac{3\sqrt{2}}{2}$
$\left(3, -\dfrac{\pi}{4}\right)$	$\dfrac{3\sqrt{2}}{2}$
$\left(3, \dfrac{\pi}{4}\right)$	$\dfrac{3\sqrt{2}}{2}$
$(3, 0)$	3
$\left(2, \dfrac{\pi}{4}\right)$	$2\sqrt{2}$
$\left(2, -\dfrac{\pi}{4}\right)$	$2\sqrt{2}$

We will examine the upper edge of the rectangle given by the horizontal line $y = \dfrac{\pi}{4}$. The function will be:

$$f\left(x, \dfrac{\pi}{4}\right) = (4x - x^2)\cos\left(\dfrac{\pi}{4}\right) = \dfrac{\sqrt{2}}{2}(4x - x^2)$$

We do first-semester calculus on this single-variable function and determine its critical points:

$$f'\left(x, \dfrac{\pi}{4}\right) = \dfrac{\sqrt{2}}{2}(4 - 2x) = 0 \quad \Rightarrow \quad x = 2$$

Thus, the point $\left(2, \dfrac{\pi}{4}\right)$ is a critical point and we find the corresponding value for the function at this point and put the info into our table:

$$f\left(2, \dfrac{\pi}{4}\right) = \dfrac{\sqrt{2}}{2}(4(2) - (2)^2) = 2\sqrt{2}$$

We already did the vertices for this edge, so we will now move on to the final side length of our rectangle, given by the horizontal line $y = -\dfrac{\pi}{4}$.
The function will be:

$$f\left(x, -\dfrac{\pi}{4}\right) = (4x - x^2)\cos\left(-\dfrac{\pi}{4}\right) = \dfrac{\sqrt{2}}{2}(4x - x^2)$$

We already did the calculus on this single-variable function above, and we have that:

$$f'\left(x, -\dfrac{\pi}{4}\right) = \dfrac{\sqrt{2}}{2}(4 - 2x) = 0 \quad \Rightarrow \quad x = 2$$

Thus, the point $\left(2, -\dfrac{\pi}{4}\right)$ is a critical point and we find the corresponding value for the function at this point and put the info into our table: $f\left(2, -\dfrac{\pi}{4}\right) = \dfrac{\sqrt{2}}{2}(4(2) - (2)^2) = 2\sqrt{2}$.

ANSWER 22, CONTINUED...

...ANSWER 22, CONTINUED

CALCULUS III   Practice Midterm 3 KEY, CONTINUED

We look at the second column of our table. The largest value is our absolute maximum and the smallest is our absolute minimum. We have a four-way tie for the absolute minimum that has the value $\dfrac{3\sqrt{2}}{2}$.

Final answer: Absolute maximum of 4 at the point $(2, 0)$, and absolute minimum is $\dfrac{3\sqrt{2}}{2}$ at the points:

$$\left(3, -\frac{\pi}{4}\right), \left(3, \frac{\pi}{4}\right), \left(1, -\frac{\pi}{4}\right), \text{ and } \left(1, \frac{\pi}{4}\right)$$

23. **Find the critical points and test for relative extrema. List the critical points for which the Second Partials Test fails** $f(x, y) = x^{2/3} + y^{2/3}$.

The first partials are:

$$f_x(x, y) = \frac{2}{3} x^{-1/3} \quad and \quad f_y(x, y) = \frac{2}{3} y^{-1/3}$$

The critical point is $(0,0)$ that makes the first partials undefined.

$$D = f_{xx} f_{yy} - \left(f_{xy}\right)^2 = \left(\frac{-2}{9} x^{-4/3}\right)\left(\frac{-2}{9} y^{-4/3}\right)$$

The value of D is undefined at the critical point. Therefore, the Second Partials test fails. It is actually a minimum since every value for the function is greater that 0 at any other point.

24. **Determine the relative extrema and saddle points (if any) for the function** $f(x, y) = 9x^3 + \dfrac{y^3}{3} - 4xy$. **Give answer(s) as ordered triples.**

The first partials are:

$$f_x(x, y) = 27x^2 - 4y \quad and \quad f_y(x, y) = y^2 - 4x$$

Setting both partials equal to 0 simultaneously, and solving the resulting non-linear system:

$$\begin{cases} 27x^2 - 4y = 0 \\ y^2 - 4x = 0 \end{cases} \Rightarrow y^2 = 4x \Rightarrow x = \frac{y^2}{4} \Rightarrow 27\left(\frac{y^2}{4}\right)^2 - 4y = 0 \Rightarrow \frac{27}{16} y^4 - 4y = 0$$

$$y\left(\frac{27}{16} y^3 - 4\right) = 0 \qquad \text{Applying the Zero Product Property, we have:}$$

$$y = 0 \quad or \quad \frac{27}{16} y^3 - 4 = 0 \Rightarrow y^3 = 4\left(\frac{16}{27}\right) \Rightarrow y = \sqrt[3]{\frac{64}{27}} = \frac{4}{3}$$

Substitute this into one of the original equations for our non-linear system to find the corresponding x-coordinate of our critical point:

For $y = \dfrac{4}{3}$, we have $x = \dfrac{(4/3)^2}{4} = \left(\dfrac{16}{9}\right)\left(\dfrac{1}{4}\right) = \dfrac{4}{9}$ For $y = 0$, $x = 0$

CALCULUS III   **Practice Midterm 3 KEY,** CONTINUED

The two critical points are $(0, 0)$ and $\left(\dfrac{4}{9}, \dfrac{4}{3}\right)$.

Next, we find the second partials and form the expression for D:

$$D = f_{xx}f_{yy} - \left[f_{xy}\right]^2 = (54x)(2y) - (-4)^2 = 108xy - 16$$

Performing the Second Partials Test for each of the critical points $D(0, 0) = -16 < 0$, so we have a saddle point at $(0, 0)$.

$$D\left(\frac{4}{9}, \frac{4}{3}\right) = 108\left(\frac{4}{9}\right)\left(\frac{4}{3}\right) - 16 = 48 > 0, \quad f_{xx}\left(\frac{4}{9}, \frac{4}{3}\right) = 54\left(\frac{4}{9}\right) > 0$$

Thus, we have a relative minimum. Final answer: Saddle point at $(0, 0, 0)$ and Relative minimum at:

$$\left(\frac{4}{9}, \frac{4}{3}, -\frac{64}{81}\right)$$

25. **Find the minimum distance from the point $(0, 0, 0)$ to the plane given by $x + 2y + z = 6$. You are forbidden from using the formula for calculating the distance from a point to a plane from Chapter 1.**

    (a) **Use the techniques from section 3.8 and the Second Partials Test to find this minimum distance.**

    A point on the plane is given by $(x, y, 6 - x - 2y)$. Use the distance formula to find the distance from this point to the origin:

    $$d = \sqrt{x^2 + y^2 + (6 - x - 2y)^2}$$

    Minimizing this function is the same as minimizing the radicand. That is, we will do calculus on the function of two variables:

    $$R(x, y) = x^2 + y^2 + (6 - x - 2y)^2$$

    The first partials are:

    $$R_x(x, y) = 2x + 2(6 - x - 2y)(-1) = 4x + 4y - 12$$
    $$R_y(x, y) = 2y + 2(6 - x - 2y)(-2) = 4x + 10y - 24$$

    Set these equal to zero and solve the (linear) system:

    $$\begin{cases} 4x + 4y - 12 = 0 \\ 4x + 10y - 24 = 0 \end{cases} \Rightarrow \begin{cases} 4x + 4y = 12 \\ 4x + 10y = 24 \end{cases}$$

    Subtract the second equation $\Rightarrow$ from the first to eliminate the x-variable.

    $$-6y = -12 \quad \Rightarrow \quad y = 2 \quad \Rightarrow \quad x = 1$$

    The point $(1, 2)$ is a critical point. Perform the D-Test:

    $$D = R_{xx}R_{yy} - \left[R_{xy}\right]^2 = (4)(10) - 4^2 > 0, \quad R_{xx} = 4 > 0$$

    We do indeed have a relative minimum. The z-coordinate on the plane is $z = 6 - 1 - 2(2) = 1$. Therefore, the minimum distance is:

    $$d = \sqrt{x^2 + y^2 + (z)^2} = \sqrt{1^2 + 2^2 + 1^2} = \sqrt{6}$$

ANSWER 25, CONTINUED...

CALCULUS III   Practice Midterm 3 KEY, CONTINUED

(b) **Do the problem all over again, only now use the Method of Lagrange Multipliers to find the minimum distance.**

Let the point on the plane be given by the ordered triple (x, y, z) be the desired point.

We use the distance formula to find the distance between this point (x, y, z) and the origin $(0, 0, 0)$:

$$\sqrt{(x-0)^2 + (y-0)^2 + (z-0)^2} = \sqrt{x^2 + y^2 + z^2}$$

We want to minimize this distance. But if we minimize the radicand we will automatically minimize the entire square root value. So, the function we are trying to minimize will simply be the radicand of this distance formula. That is, we want to minimize the function $f(x,y,z) = x^2 + y^2 + z^2$ subject to the constraint equation $g(x,y,z) = 0$, or $x + 2y + z - 6 = 0$. Next, we form the Lagrange function:

$$L(x,y,z,\lambda) = f(x,y,z) - \lambda g(x,y,z)$$
$$L(x,y,z,\lambda) = x^2 + y^2 + z^2 - \lambda(x + 2y + z - 6)$$

Take all four first partial derivatives for this Lagrange function, and set them all simultaneously equal to zero:

$$L_x = 2x - \lambda = 0$$
$$L_y = 2y - \lambda(2) = 0$$
$$L_z = 2z - \lambda = 0$$
$$L_\lambda = -(x + 2y + z - 6) = 0$$

This is a linear system of four equations and four unknowns. Solve for the Lagrange Multiplier λ using equations one and two and three. Then set those quantities equal to each other.

$$\lambda = \lambda$$
$$\lambda = 2x \qquad 2x = 2z$$
$$\lambda = y \implies x = z$$
$$\lambda = 2z \qquad y = 2x$$

Next, plug these into the constraint:

$$x + 2(2x) + x = 6$$
$$6x = 6$$
$$x = 1$$

This means that $y = 2x = 2$, $z = x = 1$. The actual minimum distance will be:

$$\text{Distance} = \sqrt{1^2 + 2^2 + 1^2} = \sqrt{6} \, .$$

26. **Use Lagrange multipliers to locate the maximum for the function** $z = 3xy + x$ **subject to the constraint** $11x + 15y = 215$.

We want to maximize the function $f(x, y) = 3xy + x$ subject to the constraint equation $g(x, y) = 0$, or $11x + 15y - 215 = 0$. Next, we form the Lagrange function:

$$L(x, y, \lambda) = f(x, y) - \lambda g(x, y)$$
$$L(x, y, \lambda) = 3xy + x - \lambda(11x + 15y - 215)$$

Take all three first partial derivatives for this Lagrange function, and set them all simultaneously equal to zero:

$$L_x = 3y + 1 - \lambda(11) = 0$$
$$L_y = 3x - \lambda(15) = 0$$
$$L_\lambda = -(11x + 15y - 215) = 0$$

This is a linear system of three equations and three unknowns. Solve for the Lagrange Multiplier λ using both equations one and two. Then set those quantities equal to each other.

$$3y + 1 = 11\lambda$$
$$\lambda = \frac{3y + 1}{11}$$
$$3x = 15\lambda$$
$$\lambda = \frac{3x}{15} = \frac{x}{5}$$

Since $\lambda = \lambda$, we now have:

$$\frac{3y + 1}{11} = \frac{x}{5}$$
$$3y + 1 = \frac{11x}{5}$$
$$3y = \frac{11x}{5} - 1$$
$$y = \left(\frac{1}{3}\right)\left(\frac{11x}{5} - 1\right)$$
$$y = \frac{11x}{15} - \frac{1}{3}$$

ANSWER 26, CONTINUED...

...*ANSWER 26, CONTINUED*

CALCULUS III   Practice Midterm 3 KEY, CONTINUED

Next, plug this into the constraint to solve for *x*:

$$11x + 15\left(\frac{11x}{15} - \frac{1}{3}\right) = 215$$

$$11x + 11x - 5 = 215$$

$$22x = 220$$

$$x = \frac{220}{22} = 10$$

Since $y = \dfrac{11x}{15} - \dfrac{1}{3}$, this means that $y = \dfrac{11(10)}{15} - \dfrac{1}{3} = \dfrac{11(2)}{3} - \dfrac{1}{3} = \dfrac{22}{3} - \dfrac{1}{3} = \dfrac{21}{3} = 7$.

So, the maximum of *f* occurs when *x* = 10, *y* = 7 and is equal to:

$$f(10, 7) = 3(10)(7) + 10 = 210 + 10 = 220$$

27. **Use the Method of Lagrange Multipliers to find the *minimum* temperature on the surface given by** $x^2 + y^2 + z^2 = 9$ **if the temperature is given by** $T(x, y, z) = x + y + z$**. (Give exact answer.)**

We want to minimize the temperature $T(x, y, z) = x + y + z$ subject to the constraint equation, which is $x^2 + y^2 + z^2 = 9$. Rewriting the constraint equation in standard form $g(x, y, z) = x^2 + y^2 + z^2 - 9 = 0$, we form the Lagrange function:

$$L(x, y, z, \lambda) = f(x, y, z) - \lambda g(x, y, z)$$
$$L(x, y, z, \lambda) = x + y + z - \lambda\left(x^2 + y^2 + z^2 - 9\right)$$

Take all *four* first partial derivatives for this Lagrange function, and set them all simultaneously equal to zero:

$$L_x = 1 - \lambda(2x) = 0$$
$$L_y = 1 - \lambda(2y) = 0$$
$$L_z = 1 - \lambda(2z) = 0$$
$$L_\lambda = -(x^2 + y^2 + z^2 - 9) = 0$$

This non-linear system of four equations and four unknowns won't be too hard to solve. Solve for the Lagrange Multiplier λ using equations one and two and three. Then set those quantities equal to each other.

$$\lambda = \frac{1}{2x}$$

$$\lambda = \frac{1}{2y}$$

$$\lambda = \frac{1}{2z}$$

PART III TEST ANSWER KEY

CALCULUS **III**   **Practice Midterm 3 KEY,** CONTINUED

Since $\lambda = \lambda = \lambda$, we have that $x = y = z$. Substitute $y = x$ and $z = x$ into the constraint equation and solve for x:

$$x^2 + x^2 + x^2 = 9$$
$$3x^2 = 9$$
$$x^2 = 3$$
$$x = \pm\sqrt{3}$$

Since we would like to minimize the temperature, we will choose the negative value, that is $x = -\sqrt{3} = y = z$, and we find the minimum temperature T is given by:

$$T(-\sqrt{3}, -\sqrt{3}, -\sqrt{3}) = \left(-\sqrt{3}\right) + \left(-\sqrt{3}\right) + \left(-\sqrt{3}\right) = -3\sqrt{3}$$

CALCULUS III — QUIZ 10 KEY

1. **Set up and evaluate the double integral that would calculate the volume of the solid bounded above by $f(x,y) = y\cos(xy)$ and the region in the xy-plane $0 \le x \le \pi$, $0 \le y \le 1$.**

$$V = \int_0^{\pi}\int_0^{1} y\cos(xy)\,dy\,dx$$

Attempting to use u-substitution, where $u = xy, du = xdy$, we cannot use the cosine rule, since we have a factor of y in the integrand. We *could* try integration by parts, but it turns out it is much easier to reverse the order of integration instead. Because the region of integration is simply a rectangle, we can simply interchange the integral symbols.

$$V = \int_0^{1}\int_0^{\pi} y\cos(xy)\,dx\,dy$$

We use u-substitution where $u = xy, du = ydx$, and $\int \cos u\,du = \sin u$:

$$= \int_0^1 \sin(xy)\Big]_0^{\pi} dy = \int_0^1 [\sin(\pi y) - \sin(0)]dy = \int_0^1 \sin(\pi y)dy$$

Using u-substitution again, $u = \pi y, du = \pi dy$ means that we need to put in a factor of π (provided we divide it out):

$$= \frac{1}{\pi}\int_0^1 \sin(\pi\, y)\pi dy = \frac{-\cos(\pi\, y)\Big]_0^1}{\pi} = \frac{-\cos(\pi) - (-\cos(0))}{\pi} = \frac{-(-1)+1}{\pi} = \frac{2}{\pi}$$

2. **Evaluate $\int_0^8\int_{\sqrt[3]{x}}^{2} \frac{dydx}{1+y^4}$ after reversing the order of integration. *(Hint:* It might help to first sketch the region.)**

We begin by sketching the graphs of $x = 0,\ x = 8,\ y = \sqrt[3]{x},\ and\ y = 2$:

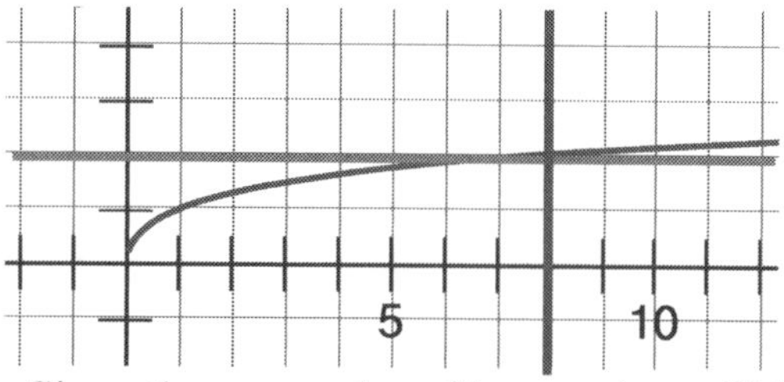

The region of integration lies above the graph of $y = \sqrt[3]{x},\ and\ below\ y = 2$.

Since the new order of integration will be *dxdy* which is horizontally simple, we need to determine the limits for x that will be functions of y. Examining our region and solving for x with the equation $y = \sqrt[3]{x} \Rightarrow x = y^3$, the double integral becomes:

$$\int_0^2\int_0^{y^3} \frac{dxdy}{1+y^4} = \int_0^2 \left(\frac{1}{1+y^4}\right)[x]_0^{y^3} dy = \int_0^2 \frac{y^3}{1+y^4}dy$$

We can use u-substitution and the Log Rule where u = $u = 1 + y^4,\ du = 4y^3dy$.

CALCULUS III   Quiz 10 KEY, CONTINUED

Putting in the factor of 4 we require (and dividing it out), we have:

$$= \frac{1}{4}\int_0^2 \frac{4y^3}{1+y^4}dy = \frac{1}{4}\int \frac{1}{u}du = \frac{1}{4}\ln|u| = \frac{1}{4}\ln\left(1+y^4\right)\bigg]_0^2 = \frac{1}{4}\left\{\ln(1+2^4) - \ln(1+0^4)\right\} = \frac{1}{4}\ln 17$$

3.   **Evaluate** $\displaystyle\int_0^{1/16}\int_{y^{1/4}}^{1/2}\cos\!\left(16\,\pi\,x^5\right)dxdy$ **after reversing the order of integration.** (*Hint:* **It might help to first sketch the region.**)

First note that the integrand is not even integrable if we had to integrate right now. When this

happens, always reverse the order of integration immediately. We begin by sketching the graphs of

$y = 0,\ y = \dfrac{1}{16},\ x = y^{1/4},\ and\ x = 1/2$.

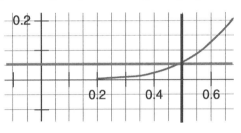

The graph of $x = y^{1/4}$ is the same as $x^4 = y$:

The region of integration is to the right of the curve $x^4 = y$
and to the left of the vertical line $x = 0.5$. The new order of
integration $dydx$ corresponds to a vertically simple region.

This is usually easier for most students to set up than the horizontally simple regions, since most of
us are more comfortable with functions of x rather than functions of y.

The double integral becomes $\displaystyle\int_0^{1/2}\int_0^{x^4}\cos\!\left(16\pi x^5\right)dydx$, which is easy to integrate now since the

expression $\cos(16\pi x^5)$ is a constant with respect to the variable y.

$$\int_0^{1/2}\int_0^{x^4}\cos\!\left(16\pi x^5\right)dydx = \int_0^{1/2}\cos\!\left(16\pi x^5\right)\big[y\big]_0^{x^4}dx = \int_0^{1/2}\cos\!\left(16\pi x^5\right)x^4dx$$

Using u-substitution and the cosine rule, where $u = 16\pi x^5$, $du = 80\pi x^4dx$, we put in a factor of 80π:

$$= \frac{1}{80\pi}\int_0^{1/2}\cos\!\left(16\pi x^5\right)80\pi x^4dx = \frac{1}{80\pi}\left[\sin\!\left(16\pi x^5\right)\right]_0^{1/2} = \frac{1}{80\pi}\left\{\sin\!\left(16\pi\!\left(\frac{1}{2}\right)^5\right) - \sin(0)\right\} = \frac{1}{80\pi}$$

4.   **Given the iterated integral** $\displaystyle\int_0^{\ln 2}\int_{e^y}^{2}dxdy$ ...

(a)   **Sketch the region.**

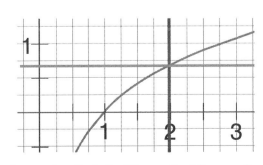

We first sketch the graphs of
$y = 0,\ y = \ln 2,\ x = e^y,\ x = 2$. We can
sketch $x = e^y$ easier if we solve for y. Start
by taking the natural logarithm of both sides:

$$\ln x = \ln(e^y) \ \Rightarrow\ \ln x = y\ln e \ \Rightarrow\ y = \ln x$$

The region of integration is to the right of the curve $y = \ln x$ and to the left of the vertical line $x = 2$.

ANSWER 4, CONTINUED...

THIRD SEMESTER CALCULUS STUDENT SUPPLEMENT, 4TH EDITION

...*ANSWER 4, CONTINUED*

CALCULUS III

(b) **Evaluate the integral:**

$$\int_0^{\ln 2}\int_{e^y}^2 dxdy = \int_0^{\ln 2}[x]_{e^y}^2\, dy = \int_0^{\ln 2}(2-e^y)dy = [2y-e^y]_0^{\ln 2} = (2\ln 2 - e^{\ln 2}) - (2(0)-e^0) = 2\ln 2 - 2 + 1 = 2\ln 2 - 1$$

(c) **Switch the order of integration and show that the value matches your answer to part (b).** **(*Hint:* Integration by parts will be needed.)**

Integration by parts is required now, where:

$$u = \ln x, \quad dv = 1dx$$

$$du = \frac{1}{x}dx, \quad v = x$$

$$\int_1^2\int_0^{\ln x} dydx = \int_1^2[y]_0^{\ln x}\, dx = \int_1^2 \ln xdx$$

$$\int_1^2 \ln xdx = uv - \int vdu = x\ln x\Big]_1^2 - \int_1^2 x\frac{1}{x}dx = 2\ln 2 - 1\ln 1 - \int_1^2 dx = 2\ln 2 - [x]_1^2 = 2\ln 2 - (2-1) = 2\ln 2 - 1$$

5. **Use a double integral to find the volume of the solid bounded above by the cylinder $z = x^2$, and below by the region enclosed by the parabola $y = 2 - x^2$, and the line $y = x$ in the *xy*-plane. (*Hint:* The region in the *xy*-plane includes *more* than just the first quadrant!)**

First we sketch the base of the solid in the *xy*-plane:

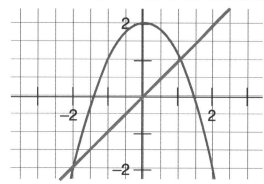

We find the points of intersection for the two curves by setting them equal:

$$y = y$$
$$2 - x^2 = x$$
$$0 = x^2 + x - 2$$
$$0 = (x+2)(x-1)$$
$$x + 2 = 0, x - 1 = 0$$
$$x = -2, \quad x = 1 \quad \Rightarrow \quad (-2,-2) \quad and \quad (1,1)$$

The height of the solid is given by the equation $z = x^2$. The volume of the solid is:

$$\int_{-2}^1\int_x^{2-x^2} x^2 dydx = \int_{-2}^1 x^2[y]_x^{2-x^2}\, dx = \int_{-2}^1 x^2(2-x^2-x)dx = \int_{-2}^1 (2x^2 - x^4 - x^3)dx = \left[\frac{2}{3}x^3 - \frac{x^5}{5} - \frac{x^4}{4}\right]_{-2}^1 = \frac{63}{20}$$

6. **Convert the double integral to polar, and then evaluate:**

(a) $\displaystyle \int_0^1\int_0^{\sqrt{1-x^2}} e^{-(x^2+y^2)}dydx$

First, we sketch the region of integration in order to help us convert to polar. We sketch the graphs of $y = 0$, $y = \sqrt{1-x^2}$, $x = 0, x = 1$.

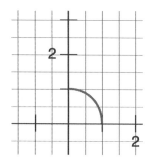

CALCULUS III   **Quiz 10 KEY, CONTINUED**

The region of integration is the quarter of a circle of radius 1 in the first quadrant. Using the fact that

$x^2 + y^2 = r^2$, the integral now becomes $\displaystyle\int_0^{\pi/2} \int_0^1 e^{-\left(r^2\right)} r\,dr\,d\theta$.

We will use u-substitution and the exponential rule where $u = -r^2$, $du = -2r\,dr$, so we need a factor of -2 in the integrand (provided we divide it out):

$$= \int_0^{\pi/2} \left(-\frac{1}{2}\right)\int_0^1 e^{-\left(r^2\right)}(-2)r\,dr\,d\theta = -\frac{1}{2}\int_0^{\pi/2}\left[\int e^u\,du\right]d\theta = -\frac{1}{2}\int_0^{\pi/2}\left[e^{-r^2}\right]_0^1 d\theta = -\frac{1}{2}\int_0^{\pi/2}\left[e^{-1} - e^0\right]d\theta = -\frac{1}{2}\left(e^{-1}-1\right)\left[\theta\right]_0^{\pi/2}$$

$$= \left(-\frac{1}{2}\right)\left(\frac{\pi}{2}\right)\left(e^{-1}-1\right) = \left(-\frac{\pi}{4}\right)\left(e^{-1}-1\right) = \frac{\pi}{4}\left(1-e^{-1}\right)$$

(b) $\displaystyle\int_{-1}^0 \int_{-\sqrt{1-x^2}}^0 \frac{2}{1+\sqrt{x^2+y^2}}\,dy\,dx$

     Double Hint: **1)** This region is tricky to convert to polar. First, sketch the region!

                              2) You will need either change-of-variables or long division to integrate.

First, we sketch the region of integration in order to help us convert to polar. We sketch the graphs of $y = 0$, $y = -\sqrt{1-x^2}$, $x = 0$, $x = -1$:

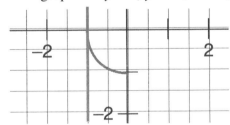

The limits for the angle θ are $\pi \le \theta \le \dfrac{3\pi}{2}$. *Be careful!* Many students make the mistake of writing $\pi \le \theta \le -\dfrac{\pi}{2}$ instead. This is incorrect because the smaller value in the interval must be the lower limit.

Another common mistake students make is writing that r ranges from $r = 0$ to $r = -1$—probably by confusing the fact that $x = -1$. Please keep your r positive at all times! The integral becomes:

$$\int_\pi^{3\pi/2} \int_0^1 \frac{2}{1+\sqrt{r^2}}\,r\,dr\,d\theta = 2\int_\pi^{3\pi/2} \int_0^1 \frac{r}{1+r}\,dr\,d\theta$$

...because we have an improper rational function in the integrand.

(Reminder: A rational function has the form $f(x) = \dfrac{polynomial}{polynomial}$, and an improper rational function is where the degree of the polynomial in the numerator is either equal to or greater than the degree of the polynomial in the denominator. When this happens in an integrand, always perform polynomial long division before integrating.):

$$\begin{array}{r} 1 \\ r+1\overline{)r} \\ \underline{r+1} \\ -1 \end{array}$$

Always write the result of polynomial long division as $Quotient + \dfrac{remainder}{divisor}$.

ANSWER 6, CONTINUED...

CALCULUS III   Quiz 10 KEY, CONTINUED

The double integral now becomes:

$$= 2\int_{\pi}^{3\pi/2}\int_{0}^{1}\left(1-\frac{1}{r+1}\right)drd\theta = 2\int_{\pi}^{3\pi/2}[r-\ln(r+1)]_{0}^{1}\,d\theta = 2\int_{\pi}^{3\pi/2}[1-\ln(1+1)-(0-\ln(0+1))]d\theta$$

$$= 2(1-\ln 2)\int_{\pi}^{3\pi/2}d\theta = 2(1-\ln 2)[\theta]_{\pi}^{3\pi/2} = 2(1-\ln 2)\left(\frac{3\pi}{2}-\pi\right) = \pi(1-\ln 2)$$

(c) $\displaystyle\int_{0}^{2}\int_{-\sqrt{1-(y-1)^2}}^{0}xy^2\,dxdy$

Be careful, since this region of integration is a circle *not* centered at the origin. We will also need to use our knowledge from Chapter 1 in this book on converting rectangular equations to polar equations. We need to sketch the graphs of $x=0$, $x=-\sqrt{1-(y-1)^2}$, $y=0, y=2$.

The equation $x=-\sqrt{1-(y-1)^2}$ needs to be rewritten. First, we square both sides of this equation:

$[x]^2 = \left[-\sqrt{1-(y-1)^2}\right]^2$

$x^2 = 1-(y-1)^2$

$x^2+(y-1)^2 = 1$

This is the equation for a circle centered at the point (0, 1) with a radius of 1. The sketch of the region looks like:

To convert this equation to polar, we use the conversion formulas $x=r\cos\theta$ *and* $y=r\sin\theta$ followed by substitution:

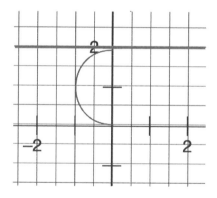

$$(r\cos\theta)^2+(r\sin\theta-1)^2 = 1$$

$$(r\cos\theta)^2+(r\sin\theta-1)(r\sin\theta-1) = 1$$

$$(r\cos\theta)^2+r^2\sin^2\theta-2r\sin\theta+1 = 1$$

$$r^2\cos^2\theta+r^2\sin^2\theta = 2r\sin\theta$$

$$r^2(\cos^2\theta+\sin^2\theta) = 2r\sin\theta$$

$$r^2(1) = 2r\sin\theta$$

We now solve for r by dividing both sides of the equation by r, assuming r is non-zero. No information will be lost as the resulting equation has $r=0$ as one of its solution. That is the final resulting polar equation is $r=2\sin\theta$. This means the range for the limits for r will be $0\le r\le 2\sin\theta$. The challenge now is to determine the limits for θ. It is tempting to look at the diagram for the region and make the wrong decision about the limits. An easier way is to use your graphing calculator (in POLAR mode) and graph the equation $r=2\sin\theta$. Then, "play"

√ *PART III* Test Answer Key

with the limits on θ by changing θmin and θmax in your calculator's range. Using this trial and error method, we can find that the limits on θ to re-create our sketch of the region need to be $\frac{\pi}{2} \leq \theta \leq \pi$. We can now write our double integral in polar coordinates as:

$$= \int_{\pi/2}^{\pi} \int_{0}^{2\sin\theta} (r\cos\theta)(r\sin\theta)^2 \, r \, dr \, d\theta$$

$$= \int_{\pi/2}^{\pi} \int_{0}^{2\sin\theta} r^4 \cos\theta \sin^2\theta \, dr \, d\theta$$

$$= \int_{\pi/2}^{\pi} \cos\theta \sin^2\theta \left[\frac{r^5}{5} \right]_0^{2\sin\theta} d\theta$$

$$= \int_{\pi/2}^{\pi} \cos\theta \sin^2\theta \left[\frac{(2\sin\theta)^5}{5} \right] d\theta$$

$$= \frac{32}{5} \int_{\pi/2}^{\pi} \cos\theta \sin^7\theta \, d\theta$$

We will use u-substitution and the General Power Rule to integrate, where $u = \sin\theta$, $du = \cos\theta$:

$$= \frac{32}{5} \int u^7 \, du = \frac{32}{5} \left(\frac{u^8}{8} \right) = \frac{32}{5} \left(\frac{1}{8} \right) [\sin^8\theta]_{\pi/2}^{\pi} = \frac{4}{5} \left(\sin^8(\pi) - \sin^8\left(\frac{\pi}{2}\right) \right) = \frac{4}{5}(0 - 1) = -\frac{4}{5}$$

7.   ***Converting to a polar integral*: Integrate** $f(x,y) = \dfrac{\ln(x^2 + y^2)}{x^2 + y^2}$ **over the region** $1 \leq x^2 + y^2 \leq e^2$.

The region of integration is the area between two concentric circles centered at the origin, one with radius of 1 and the other with radius e. The double integral will be:

$$\int_{0}^{2\pi} \int_{1}^{e} \frac{\ln(r^2)}{r^2} \, r \, dr \, d\theta = \int_{0}^{2\pi} \int_{1}^{e} \frac{2\ln(r)}{r} \, dr \, d\theta$$

We used log properties to simplify in that last step. To integrate, we will use u-substitution and the General Power Rule to integrate, where $u = \ln r$, $du = \frac{1}{r} dr$:

$$= 2 \int_{0}^{2\pi} \int_{1}^{e} (u \, du) \, d\theta$$

$$= 2 \int_{0}^{2\pi} \left[\frac{u^2}{2} \right] d\theta$$

$$= 2 \int_{0}^{2\pi} \left(\frac{1}{2} \right) [(\ln r)^2]_1^e \, d\theta$$

$$= \int_{0}^{2\pi} [(\ln e)^2 - (\ln 1)^2] \, d\theta$$

$$= \int_{0}^{2\pi} (1 - 0) \, d\theta$$

$$= [\theta]_0^{2\pi} = 2\pi$$

8.   Find the mass of a thin plate covering the region outside the circle $r = 3$ and inside the circle $r = 6\sin\theta$ if the plate's density function is: $\rho(x,y) = \dfrac{1}{r}$. *(**Hint:** To find the range of the angle, θ, you will need to find the point(s) of intersection of these two graphs by equating the equations. It also might help to graph the equations using the **Polar** mode on your calculator so you can identify the region.)*

Using the polar mode on your graphing calculator, we graph the polar equations $r = 3$ and $r = 6\sin\theta$.

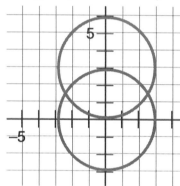

Next, we equate the two equations to find the points of intersection:

Thus, the limits for θ are $\dfrac{\pi}{6} \le \theta \le \dfrac{5\pi}{6}$ and the limits for r are $3 \le r \le 6\sin\theta$.

$r = r$

$6\sin\theta = 3$

$\sin\theta = \dfrac{1}{2}$

$\sin^{-1}(\sin\theta) = \sin^{-1}\left(\dfrac{1}{2}\right)$

$\theta = \dfrac{\pi}{6} \quad and \quad \theta = \dfrac{5\pi}{6}$

The double integral representing mass of the planar lamina is:

$$\textbf{\textit{mass}} = \iint\limits_{R} \rho(x,y)\,dA = \iint\limits_{R} \rho(r,\theta)\,dA = \int_{\pi/6}^{5\pi/6} \int_{3}^{6\sin\theta} \frac{1}{r}\,r\,dr\,d\theta = \int_{\pi/6}^{5\pi/6} \left[r\right]_{3}^{6\sin\theta} d\theta = \int_{\pi/6}^{5\pi/6} (6\sin\theta - 3)\,d\theta$$

$$= \left[-6\cos\theta - 3\theta\right]_{\pi/6}^{5\pi/6} = \left[-6\cos\left(\frac{5\pi}{6}\right) - 3\left(\frac{5\pi}{6}\right)\right] - \left[-6\cos\left(\frac{\pi}{6}\right) - 3\left(\frac{\pi}{6}\right)\right]$$

$$= (-6)\left(-\frac{\sqrt{3}}{2}\right) - \frac{5\pi}{2} + 6\left(\frac{\sqrt{3}}{2}\right) + \frac{\pi}{2}$$

$$= 3\sqrt{3} - 2\pi + 3\sqrt{3} = 6\sqrt{3} - 2\pi \approx 4.1$$

9.   Given the triple integral $\displaystyle\int_{0}^{6}\int_{\frac{3x-18}{2}}^{0}\int_{0}^{\frac{18+2y-3x}{6}} dz\,dy\,dx$, **rewrite the order of the integral in the order** $dx\,dy\,dz$ **.** *(No need to evaluate!)*

This triple integral represents the volume of the solid cut from the first octant by the plane:

$$z = \frac{18+2y-3x}{6} \quad\Rightarrow\quad 6z = 18+2y-3x \quad\Rightarrow\quad 3x-2y+6z = 18$$

Since z will be the outer variable of integration, the limits of z will be constants. If we set x and y equal to 0 for this plane equation, we have that $z = 3$. Thus the limits for z will be $0 \le z \le 3$. The middle of integration is y, so we set only $x = 0$ in the equation for the plane to get that:

$$3(0) - 2y + 6z = 18 \quad\Rightarrow\quad -2y = 18 - 6z \quad\Rightarrow\quad y = \frac{18-6z}{-2} \quad\Rightarrow\quad y = \frac{18}{-2} - \frac{6z}{-2} \quad\Rightarrow\quad y = -9 + 3z$$

CALCULUS **III**   **Quiz 10 KEY,** CONTINUED

For the inside integral, the variable of integration will be x. We simply solve the plane equation for x to get:

$$3x - 2y + 6z = 18 \quad \Rightarrow \quad 3x = 18 + 2y - 6z \quad \Rightarrow \quad x = \frac{18 + 2y - 6z}{3}$$

$$\textit{Final answer: } \int_0^3 \int_{3z-9}^0 \int_0^{\frac{18+2y-6z}{3}} dx\,dy\,dz$$

10.   **Consider the solid can be found by the bounds of the three coordinate planes and the plane** $x + 2y + z = 6$ ...

(a)   **Set up (but do not evaluate) the** *double* **integral that would calculate the volume of the solid.**

The "height" of the solid is given by the plane equation. Solve this for z to obtain the integrand for the double integral $z = f(x, y) = 6 - 2y - x$. To use a double integral for volume, we use the fact that volume $= \iint_R f(x, y)\,dy\,dx$. To find the limits for y, let $z = 0$ in the equation for the plane:

$$x + 2y + 0 = 6 \quad \Rightarrow \quad y = \frac{1}{2}(6 - x) \quad \Rightarrow y = 3 - 0.5x$$

To find the limits for x, set both x and y equal to 0 and get that $x = 6$.

$$\text{Now, we have the iterated integral } \int_0^6 \int_0^{3-0.5x} (6 - 2y - x)\,dy\,dx\,.$$

(b)   **Set up (but do not evaluate) the** *triple* **integral that would calculate the volume of the solid.**

$$\textit{Volume} = \int_0^6 \int_0^{3-0.5x} \int_0^{6-2y-x} dz\,dy\,dx$$

(c)   **Set up** *and* **evaluate the integral that would calculate the surface area of the top of the solid (i.e., the portion of the plane).**

We use the surface area formula that tells us:

$$\textit{Surface Area} = \iint_R \sqrt{1 + [f_x(x, y)]^2 + [f_y(x, y)]^2}\,dA \text{, where } z = f(x, y) = 6 - 2y - x$$

Since $f_x(x, y) = -1$ *and* $f_y(x, y) = -2$, we have:

$$\int_0^6 \int_0^{3-0.5x} \sqrt{1 + (-1)^2 + (-2)^2}\,dy\,dx = \int_0^6 \int_0^{3-0.5x} \sqrt{6}\,dy\,dx = \sqrt{6} \int_0^6 [y]_0^{3-0.5x}\,dx = \sqrt{6} \int_0^6 (3 - 0.5x)\,dx = \sqrt{6}\left[3x - \frac{x^2}{4}\right]_0^6$$

$$= \sqrt{6}\left(3(6) - \frac{6^2}{4}\right) = \sqrt{6}(18 - 9) = 9\sqrt{6} \qquad \text{SQUARE UNITS}$$

11. **Set up (but do not evaluate) the *triple* integral to calculate the volume of the solid bounded by** $z = 36 - x^2 - y^2$ **and** $z = 0$ ...

   (a) **using rectangular coordinates.**

   We need to first determine the region in the xy-plane that represents the base of the solid. This will be the projection of the graph of $z = 36 - x^2 - y^2$ onto the xy-plane. Set $z = 0$ for the equation $0 = 36 - x^2 - y^2 \Rightarrow x^2 + y^2 = 36$. This is the equation for a circle of radius 6 centered at the origin. If we choose our order of integration to be "$dzdydx$," then:

   $$y^2 = 36 - x^2 \Rightarrow y = \pm\sqrt{36 - x^2}$$

   $$\textbf{\textit{Volume}} = \int_{-6}^{6} \int_{-\sqrt{36-x^2}}^{\sqrt{36-x^2}} \int_{0}^{36-x^2-y^2} dzdydx$$

   (b) **using cylindrical coordinates.**

   Cylindrical coordinates mean that only r, θ, and z may appear in the integral. The base of the solid is in the $r\theta$-plane where the circle gives us the limits $0 \le \theta \le 2\pi$ *and* $0 \le r \le 6$. Then...

   $$z = 36 - x^2 - y^2 \Rightarrow z = 36 - (x^2 + y^2) \Rightarrow z = 36 - r^2$$

   $$\textbf{\textit{Volume}} = \int_{0}^{2\pi} \int_{0}^{6} \int_{0}^{36-r^2} dzrdrd\theta$$

   (Note: Remember to always replace the "$dydx$" with: "$rdrd\theta$.")

12. **Set up (no need to evaluate) the *triple* integral that would calculate the volume of the solid bounded by the three coordinate planes, the plane** $y + z = 2$ **, and the cylindrical surface** $x = 4 - y^2$ **.**

   If we choose our order of integration to be $dzdydx$, then we need to consider the projection of the solid onto the xy-plane so that we may iterate the integral. The plane $y + z = 2$ will intersect the xy-plane as a line with equation $y = 2$. The cylindrical surface given by $x = 4 - y^2$ will be a parabola opening to the left in the xy-plane. A sketch of the region in the xy-plane is shown:

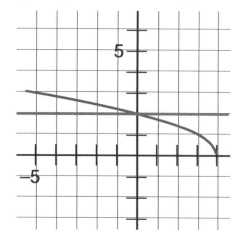

   The range for the outer variable of integration (which is x) will be $0 \le x \le 4$. Solving the equation $x = 4 - y^2$ for y gives $y = \pm\sqrt{4 - x}$.

   We take the "+" version for this equation since the solid is contained in the first octant, and therefore the first quadrant in the xy-plane.

   The triple integral representing the volume of the solid will be:

   $$\textbf{\textit{Volume}} = \iiint_Q dV = \int_{0}^{4} \int_{0}^{\sqrt{4-x}} \int_{0}^{2-y} dzdydx \quad \textit{(answer)}$$

CALCULUS III   **Quiz 10 KEY,** CONTINUED

13. **Find the surface area for that portion of the surface** $x^2 + y^2 - z = 0$ **that is inside the cylinder** $x^2 + y^2 = 2$.
     (Hint: **Convert to polar.)**

$$\textbf{\textit{Surface Area}} = \iint\limits_{R} \sqrt{1+[f_x(x,y)]^2 + [f_y(x,y)]^2}\, dA, \text{ where } z = f(x,y) = x^2 + y^2$$

And the region of integration R is the circle given by $x^2 + y^2 = 2$ that is a circle centered at the origin of radius $\sqrt{2}$. We have the first partials $f_x(x,y) = 2x$ *and* $f_y(x,y) = 2y$:

$$\iint\limits_{R} \sqrt{1+[2x]^2 + [2y]^2}\, dA = \iint\limits_{R} \sqrt{1 + 4x^2 + 4y^2}\, dA = \iint\limits_{R} \sqrt{1 + 4(x^2 + y^2)}\, dA$$

We convert to polar:

$$= \int\limits_{0}^{2\pi}\int\limits_{0}^{\sqrt{2}} \sqrt{1 + 4r^2}\, r\, dr\, d\theta$$

To integrate, we will use u-substitution and the General Power Rule to integrate, where $u = 1 + 4r^2$, $du = 8r\, dr$. We need a factor of 8 in the integrand, provided we divide it out:

$$= \frac{1}{8}\int\limits_{0}^{2\pi}\int\limits_{0}^{\sqrt{2}} \sqrt{1+4r^2}\,(8r)\,dr\,d\theta = \frac{1}{8}\int\limits_{0}^{2\pi}\Big[\int \sqrt{u}\,du\Big]d\theta = \frac{1}{8}\int\limits_{0}^{2\pi}\Big[\int u^{1/2}\,du\Big]d\theta = \frac{1}{8}\int\limits_{0}^{2\pi}\Big[\frac{2u^{3/2}}{3}\Big]d\theta = \frac{1}{8}\int\limits_{0}^{2\pi}\Big[\frac{2(1+4r^2)^{3/2}}{3}\Big]_{0}^{\sqrt{2}}d\theta$$

$$= \frac{1}{8}\int\limits_{0}^{2\pi}\left\{\Big[\frac{2(1+4(\sqrt{2})^2)^{3/2}}{3}\Big] - \Big[\frac{2(1+4(0)^2)^{3/2}}{3}\Big]\right\}d\theta$$

$$= \frac{1}{8}\int\limits_{0}^{2\pi}\frac{2}{3}\big(9^{3/2} - 1^{3/2}\big)d\theta$$

Simplify $9^{3/2} = \big(\sqrt{9}\big)^3 = 3^3 = 27$:

$$= \frac{1}{8}\int\limits_{0}^{2\pi}\frac{2}{3}(26)d\theta = \frac{26}{12}\big[\theta\big]_{0}^{2\pi} = \frac{13}{6}(2\pi) = \frac{13\pi}{3} \text{ SQUARE UNITS}$$

14. **Set up (but don't evaluate) the iterated triple integral (in cylindrical coordinates) that would find the volume of a solid** Q **if** Q **is the right circular cylinder whose base is the circle** $r = 2\sin\theta$ **in the** *xy***-plane and whose top lies in the plane** $z = 4 - y$ **as shown.** *(Hint:* **Graph the polar equation using your graphing calculator to determine the limits of** θ. **(It's not what you'd expect!))**

The base of the solid is the circle given by $r = 2\sin\theta$. It is easy to see this using your graphing calculator in POLAR mode. Although it is tempting to say that the limits for θ are between 0 and 2 π, this is not the case. Set your θ*min* and θ*max* on your calculator's range to find

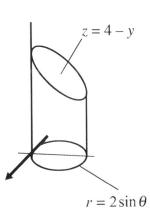

ANSWER 14, CONTINUED...

...Answer 14, continued

Calculus III   Quiz 10 KEY, continued

that the circle is sketched out only once if θ ranges from 0 to π, not 2π.
The limits for z are $0 \le z \le 4 - y$. Using the conversion formula for y,
we have that $4 - y = 4 - r\sin\theta$

So, the triple integral representing the volume of this solid is: $\textbf{\textit{Volume}} = \int_0^\pi \int_0^{2\sin\theta} \int_0^{4-r\sin\theta} dz\,r\,dr\,d\theta$

15. **Evaluate this triple integral:** $\int_{-1}^{1} \int_{-\sqrt{1-x^2}}^{\sqrt{1-x^2}} \int_{-\sqrt{x^2+y^2}}^{\sqrt{x^2+y^2}} \dfrac{1}{\left[1 + \left[\sqrt{\left(x^2+y^2\right)}\right]^3\right]^2} dz\,dy\,dx$

Convert to cylindrical coordinates.

If we graph the equations $x = -1$, $x = 1$, $y = -\sqrt{1-x^2}$ $\;$ and $\;$ $y = \sqrt{1-x^2}$, we have that the region
in the xy-plane is a circle centered at the origin of radius 1.

Using the conversion formula $r = \sqrt{x^2 + y^2}$, we have the limits for z are $-r \le z \le r$. After conversion,
the triple integral becomes:

$$\int_0^{2\pi}\int_0^1\int_{-r}^{r} \dfrac{1}{\left[1+r^3\right]^2}dz\,r\,dr\,d\theta = \int_0^{2\pi}\int_0^1 \dfrac{r}{\left[1+r^3\right]^2}\,[z]_{-r}^{r}\,dr\,d\theta = \int_0^{2\pi}\int_0^1 \dfrac{r}{\left[1+r^3\right]^2}(r-(-r))\,dr\,d\theta$$

$$= \int_0^{2\pi}\int_0^1 \dfrac{r}{\left[1+r^3\right]^2}(2r)\,dr\,d\theta = 2\int_0^{2\pi}\int_0^1 \dfrac{r^2}{\left[1+r^3\right]^2}\,dr\,d\theta$$

To integrate, we will use u-substitution and the General Power Rule to integrate, where
$u = 1 + r^3$, $du = 3r^2 dr$. We need a factor of 3 in the integrand, provided we divide it out:

$$= \dfrac{2}{3}\int_0^{2\pi}\int_0^1 \dfrac{3r}{\left[1+r^3\right]^2}\,dr\,d\theta = \dfrac{2}{3}\int_0^{2\pi}\left[\int\dfrac{du}{u^2}\right]d\theta = \dfrac{2}{3}\int_0^{2\pi}\left[\int u^{-2}\,du\right]d\theta = \dfrac{2}{3}\int_0^{2\pi}\left[\dfrac{u^{-1}}{-1}\right]d\theta = \dfrac{2}{3}\int_0^{2\pi}\left[-\dfrac{1}{u}\right]d\theta$$

$$= \dfrac{2}{3}\int_0^{2\pi}\left[-\dfrac{1}{1+r^3}\right]_0^1 d\theta = \dfrac{2}{3}\int_0^{2\pi}\left[\dfrac{-1}{1+1^3} - \dfrac{-1}{1+0^3}\right]d\theta = \dfrac{2}{3}\int_0^{2\pi}\left[\dfrac{-1}{2} - (-1)\right]d\theta = \dfrac{2}{3}\int_0^{2\pi}\dfrac{1}{2}\,d\theta = \dfrac{1}{3}[\theta]_0^{2\pi} = \dfrac{2\pi}{3}$$

16. **Evaluate the triple integral** $\displaystyle\int_0^{3\pi/2}\int_0^{\pi}\int_0^1 5\rho^3 \sin^3\phi\,d\rho\,d\phi\,d\theta$. **(Hint: You will need to review the integration technique back in 2ᴺᴰ-semester calculus covering powers of trig functions in the integrand.**

We integrate with respect to r first (which is easy): $= 5\int_0^{3\pi/2}\int_0^{\pi}\sin^3\phi\left[\dfrac{\rho^4}{4}\right]_0^1 d\phi\,d\theta$

When the power of sine is odd and positive, we save a sine factor and convert the remaining sines to
cosines. That is, $\sin^3\phi = \sin\phi\sin^2\phi = \sin\phi(1-\cos^2\phi) = \sin\phi - \sin\phi\cos^2\phi$:

$$= \dfrac{5}{4}\int_0^{3\pi/2}\int_0^{\pi}\left(\sin\phi - \sin\phi\cos^2\phi\right)d\phi\,d\theta$$

Calculus III   **Quiz 10 KEY,** continued

The first term is easy to integrate, and the second term requires u-substitution and the general power rule where $u = \cos\phi,\ du = -\sin\phi$:

$$= \frac{5}{4}\int_0^{3\pi/2}\left[-\cos\phi\Big|_0^\pi + \int u^2\,du\right]d\theta$$

$$= \frac{5}{4}\int_0^{3\pi/2}\left\{-\cos\pi - (-\cos 0) + \left[\frac{u^3}{3}\right]\right\}d\theta$$

$$= \frac{5}{4}\int_0^{3\pi/2}\left\{-(-1)+1+\left[\frac{\cos^3\phi}{3}\right]_0^\pi\right\}d\theta$$

$$= \frac{5}{4}\int_0^{3\pi/2}\left\{2+\left[\frac{\cos^3\phi}{3}\right]_0^\pi\right\}d\theta$$

$$= \frac{5}{4}\int_0^{3\pi/2}\left\{2+\left[\frac{\cos^3\pi}{3}-\frac{\cos^3 0}{3}\right]\right\}d\theta$$

$$= \frac{5}{4}\int_0^{3\pi/2}\left\{2+\left[\frac{(-1)^3}{3}-\frac{1^3}{3}\right]\right\}d\theta$$

$$= \frac{5}{4}\int_0^{3\pi/2}\left\{2-\frac{2}{3}\right\}d\theta$$

$$= \frac{5}{4}\left(\frac{4}{3}\right)[\theta]_0^{3\pi/2} = \frac{5}{3}\left(\frac{3\pi}{2}\right) = \frac{5\pi}{2}$$

17. Evaluate $\displaystyle\iiint_Q \frac{1}{\sqrt{x^2+y^2+z^2}}\,dV$ **using spherical coordinates if Q is...**

(a) **the upper hemisphere of** $x^2+y^2+z^2=25$.

This is the equation for a sphere centered at the origin with radius 5. The limits for ρ and θ are $0 \le \rho \le 5,\ and\ 0 \le \theta \le 2\pi$.

Since we are considering the upper hemisphere, the "drop-down" angle ϕ will go from 0 (that's in alignment with the positive z-axis down to the xy-plane, that gives $\phi = \frac{\pi}{2}$. Using the fact that $\sqrt{x^2+y^2+z^2} = \sqrt{\rho^2} = \rho$, the integral becomes:

$$\int_0^{2\pi}\int_0^{\pi/2}\int_0^5 \frac{1}{\rho}\rho^2\sin\phi\,d\rho\,d\phi\,d\theta = \int_0^{2\pi}\int_0^{\pi/2}\int_0^5 \rho\sin\phi\,d\rho\,d\phi\,d\theta = \int_0^{2\pi}\int_0^{\pi/2}\sin\phi\left[\frac{\rho^2}{2}\right]_0^5 d\phi\,d\theta = \int_0^{2\pi}\int_0^{\pi/2}\sin\phi\left(\frac{25}{2}\right)d\phi\,d\theta$$

$$= \frac{25}{2}\int_0^{2\pi}[-\cos\phi]_0^{\pi/2}\,d\theta = \frac{25}{2}\int_0^{2\pi}\left[-\cos\left(\frac{\pi}{2}\right)-(-\cos 0)\right]d\theta = \frac{25}{2}[\theta]_0^{2\pi} = \frac{25}{2}(2\pi) = 25\pi$$

Answer 17, continued...

...Answer 17, continued

Calculus III   Quiz 10 KEY, continued

(b)   **the lower hemisphere of** $x^2 + y^2 + z^2 = 25$.

The only difference is the drop-down angle of ϕ here. It will have a range going from $\phi = \dfrac{\pi}{2}$ (the xy-plane) and the negative z-axis, $\phi = \pi$. Everything else stays the same.

$$\int_0^{2\pi} \int_{\pi/2}^{\pi} \int_0^5 \frac{1}{\rho} \rho^2 \sin\phi \, d\rho d\phi d\theta = \int_0^{2\pi} \int_{\pi/2}^{\pi} \int_0^5 \rho \sin\phi \, d\rho d\phi d\theta = \int_0^{2\pi} \int_{\pi/2}^{\pi} \sin\phi \left[\frac{\rho^2}{2} \right]_0^5 d\phi d\theta = \pi \int_0^{2\pi} \int_{\pi/2}^{\pi} \sin\phi \left(\frac{25}{2} \right) d\phi d\theta$$

$$= \frac{25}{2} \int_0^{2\pi} \left[-\cos\phi \right]_{\pi/2}^{\pi} d\theta = \frac{25}{2} \int_0^{2\pi} \left[-\cos\pi - \left(-\cos\left(\frac{\pi}{2} \right) \right) \right] d\theta = \frac{25}{2} \left[\theta \right]_0^{2\pi} = \frac{25}{2} (2\pi) = 25\pi$$

18.   **Let Q be the sphere $x^2 + y^2 + z^2 = 9$.**

(a)   **Use cylindrical coordinates to set up the triple integral to calculate the volume of the upper hemisphere of Q. (*No need to evaluate.*)**

Solve for z: $x^2 + y^2 + z^2 = 9$ $\Rightarrow$ $z = \pm\sqrt{9 - x^2 - y^2}$ $\Rightarrow$ $z = \sqrt{9 - r^2}$

$$\textbf{\textit{Volume}} = \int_0^{2\pi} \int_0^3 \int_0^{\sqrt{9-x^2-y^2}} dz \, r \, dr \, d\theta$$

(b)   **Use spherical coordinates to set up the triple integral to calculate the volume of the upper hemisphere of Q. (*No need to evaluate.*)**

$$\textbf{\textit{Volume}} = \int_0^{2\pi} \int_0^{\pi/2} \int_0^3 \rho^2 \sin\phi \, d\rho d\phi d\theta$$

19.   **(Review) Test the function $f(x,y) = x^3 + 3x^2 + y^3 - 3y^2$ for relative extrema and saddle points. Give answer(s) as ordered triples.**

$$\begin{cases} f_x = 3x^2 + 6x = 0 & \Rightarrow \quad 3x(x+2) = 0 \quad \Rightarrow 3x = 0, \ or \ x + 2 = 0 \quad \Rightarrow \quad x = 0, \ x = -2 \\ f_y = 3y^2 - 6y = 0 & \Rightarrow \quad 3y(y-2) = 0 \quad \Rightarrow 3y = 0, \ or \ y - 2 = 0 \quad \Rightarrow \quad y = 0, \ y = 2 \end{cases}$$

The critical points are $(0, 0)$, $(0, 2)$, $(-2, 0)$, and $(-2, 2)$.

Form the expression for D and perform the D-Test (Second Partials Test) on all four critical points:

$$D = f_{xx} f_{yy} - \left[f_{xy} \right]^2 \quad \Rightarrow \quad D = (6x+6)(6y-6) - 0^2$$

$$D(0,0) = (6)(-6) = -36 < 0 \quad \Rightarrow \quad \text{saddle point}$$

$$D(0,2) = (6)(6) > 0, \ f_{xx}(0,2) > 0 \quad \Rightarrow \quad \text{rel. minimum}$$

$$D(-2,0) = (-6)(-6) > 0, \ f_{xx}(-2,0) < 0 \quad \Rightarrow \quad \text{rel. maximum}$$

$$D(-2,2) = (-6)(6) = -36 < 0 \quad \Rightarrow \quad \text{saddle point}$$

Final answer: Saddle points at $(0, 0, 0)$ and $(-2, 2, 0)$,
Relative Minimum at $(0, 2, 4)$,
and a Relative Maximum at $(2, 0, -8)$

CALCULUS **III**   **Quiz 10 KEY,** CONTINUED

EXTRA CREDIT (optional): Convert to the triple integral to spherical coordinates. *(No need to evaluate.)*

$$\int\limits_{-1}^{1}\int\limits_{-\sqrt{1-x^2}}^{\sqrt{1-x^2}}\int\limits_{\sqrt{x^2+y^2}}^{1}dzdydx$$

By examining the limits for x and y, we find that the projection of the solid we are integrating over onto the xy-plane is a circle centered at the origin with radius 1. Thus, the limits for θ are $0\le\theta\le 2\pi$.

Next, we consider the equations of $z=1$ *and* $z=\sqrt{x^2+y^2}$ to help us obtain the limits on both ρ and ϕ. The graphs of these equations would be a plane parallel to the xy-plane and a cone, respectively. Thus, the solid we are integrating over would be a cone with the vertex at the origin bounded above by a plane. So the lower limit for ρ would be 0. To get the upper limit for ρ, we consider the plane given by $z=1$. This is when ρ is the farthest from the origin. Convert this equation to spherical coordinates:

$z=1$ *and* $z=\rho\cos\phi$

$1=\rho\cos\phi$

$\dfrac{1}{\cos\phi}=\rho$

$\sec\phi=\rho$

So, the upper limit for ρ is $\sec\phi$.

The lower limit for ϕ is starting from the positive z-axis, or $\phi=0$.

The upper limit for ϕ is the most "drop-down" we get, which will be at the boundary of the cone.

To get this angle, we examine the equation $z=\sqrt{x^2+y^2}$. Use all of the spherical coordinates conversion formulas on this equation and solve for ϕ:

$z=\sqrt{x^2+y^2}$

$\rho\cos\phi=\sqrt{(\rho\sin\phi\cos\theta)^2+(\rho\sin\phi\sin\theta)^2}$

$\rho\cos\phi=\sqrt{\rho^2\sin^2\phi\cos^2\theta+\rho^2\sin^2\phi\sin^2\theta}$

$\rho\cos\phi=\sqrt{\rho^2\sin^2\phi(\cos^2\theta+\sin^2\theta)}$

$\rho\cos\phi=\sqrt{\rho^2\sin^2\phi(1)}$

$\rho\cos\phi=\rho\sin\phi$

$\cos\phi=\sin\phi$

$1=\dfrac{\sin\phi}{\cos\phi}=\tan\phi$

$\tan^{-1}(1)=\tan^{-1}(\tan\phi)$

$\dfrac{\pi}{4}=\phi$

Final answer: $\displaystyle\int\limits_{0}^{2\pi}\int\limits_{0}^{\pi/4}\int\limits_{0}^{\sec\phi}\rho^2\sin\phi\,d\rho d\phi d\theta$

Calculus III — Practice Midterm 4 KEY

1. **Evaluate** $\displaystyle\int_0^{\frac{\pi}{2}}\int_y^{2y}\sin(x+y)\,dxdy$.

   Use *u*-substitution and the Sine Rule to integrate where $u = x+y,\ du = dx$:

$$\int_0^{\pi/2}[-\cos(x+y)]_y^{2y}\,dy = \int_0^{\pi/2}[-\cos(2y+y)-(-\cos(y+y))]dy = \int_0^{\pi/2}[-\cos(3y)+\cos(2y)]dy$$

$$=\left[-\frac{1}{3}\sin(3y)+\frac{1}{2}\sin(2y)\right]_0^{\pi/2} = \left[-\frac{1}{3}\sin\left(3\cdot\frac{\pi}{2}\right)+\frac{1}{2}\sin\left(2\cdot\frac{\pi}{2}\right)\right]-\left[-\frac{1}{3}\sin(3\cdot 0)+\frac{1}{2}\sin(2\cdot 0)\right] = -\frac{1}{3}(-1) = \frac{1}{3}$$

2. **Evaluate** $\displaystyle\iint_R \frac{y}{1+x^2}\,dA$**, where *R* is the region bounded by the graphs of** $y = 0,\ y = \sqrt{x},$ and $x = 4$.

   To iterate the integral, it helps to sketch the region of integration first.

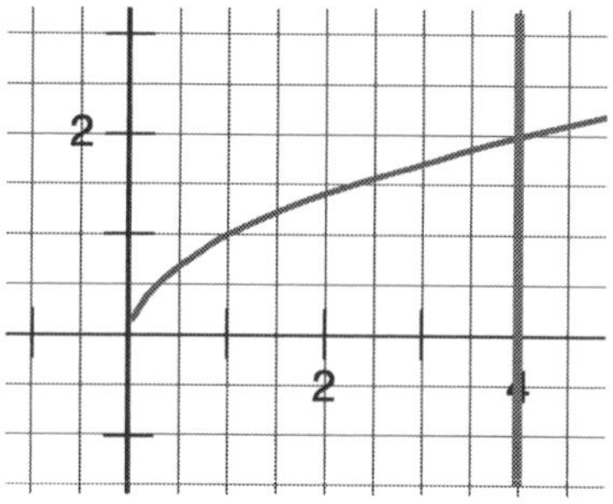

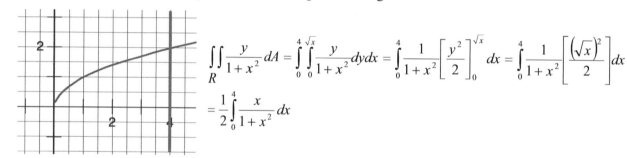

$$\iint_R \frac{y}{1+x^2}\,dA = \int_0^4\int_0^{\sqrt{x}}\frac{y}{1+x^2}\,dydx = \int_0^4\frac{1}{1+x^2}\left[\frac{y^2}{2}\right]_0^{\sqrt{x}}dx = \int_0^4\frac{1}{1+x^2}\left[\frac{\left(\sqrt{x}\right)^2}{2}\right]dx$$

$$=\frac{1}{2}\int_0^4\frac{x}{1+x^2}\,dx$$

   To continue, we use *u*-substitution and the Log Rule to integrate where $u = 1+x^2,\ du = 2xdx$. This means we need to put a factor of 2 into the integrand (provided we divide it out):

$$=\frac{1}{2}\left(\frac{1}{2}\right)\int\frac{2x}{1+x^2}\,dx = \frac{1}{4}\int\frac{1}{u}\,du = \frac{1}{4}\ln|u| = \frac{1}{4}\ln\left(1+x^2\right)\Big]_0^4 = \frac{1}{4}\left[\ln(1+4^2)-\ln(1+0^2)\right] = \frac{1}{4}\ln 17$$

3. **Evaluate the integral** $\displaystyle\int_0^3\int_0^{x^3} e^{y/x^3}\,dydx$

   We start by using *u*-substitution with the exponential rule. We identify *u* as the exponent for the integrand's exponential function $u = \dfrac{y}{x^3}$.

   Next we determine *du*, keeping in mind that the "*du*" will have us do partial differentiation with respect to the variable *y* (because we are currently integrating with respect to the *y*.)

   So, we have that $du = \dfrac{1}{x^3}dy$.

$\sqrt{}$ *PART III* Test Answer Key

CALCULUS **III**   **Practice Midterm 4 KEY,** continued

Because x is fixed when differentiating with respect to y, the expression $\dfrac{1}{x^3}$ is considered a constant.

We are permitted to insert factors into an integrand so long as they represent constants and provided we multiply the front of the integral by its reciprocal. This may seem very strange because you were most likely told in both first- and second-semester calculus that it was forbidden to put a variable factor into an integrand in order to integrate. We are allowed to do this here in third-semester calculus, provided the "variable" we are putting in as a factor is considered fixed, that is, constant, with respect to the variable of integration!

At this point the integral $\displaystyle\int_0^3\int_0^{x^3} e^{y/x^3}\,dy\,dx$ becomes:

$$\int_0^3 x^3 \int_0^{x^3} \frac{1}{x^3} e^{y/x^3}\,dy\,dx \;=\; \int_0^3 x^3 \left[\int e^u\,du\right]dx \;=\; \int_0^3 x^3\left[e^u\right]dx \;=\; \int_0^3 x^3\left[e^{y/x^3}\right]_0^{x^3}dx$$

Next, we evaluate the anti-derivative between the limits:

$$\left[e^{y/x^3}\right]_0^{x^3} \;=\; e^{x^3/x^3} - e^0 \;=\; e^1 - 1 \;=\; e-1$$

But, this is just a constant factor that we can pull in front of the last integral we still need to evaluate:

$$(e-1)\int_0^3 x^3\,dx \;=\; (e-1)\left[\frac{x^4}{4}\right]_0^3 \;=\; (e-1)\left(\frac{3^4}{4}-\frac{1}{4}\right) \;=\; (e-1)\frac{80}{4} = \; 20(e-1)\ \ \textit{(answer)}$$

4. **Evaluate the integral:** $\displaystyle\int_0^\infty\int_0^\infty xy\,e^{-(x^2+3y^2)}\,dx\,dy$

The integral $\displaystyle\int_0^\infty\int_0^\infty xy\,e^{-(x^2+3y^2)}\,dx\,dy$ is considered an "improper integral" because of the infinite limits.

This particular improper integral is taken from the study of probability theory, where x and y are considered two random variables, and where the integrand is referred to as the "probability density function" for the random variables.

You probably learned about improper integrals when you took first- or second-semester calculus. You may also recall that there are two possibilities when evaluating an improper integral: 1) the improper integral converges, that is, the result is a finite value, or 2) the improper integral diverges, that is, it "blows up."

We start by integrating with respect to the variable x, and again use u-substitution with the exponential rule. We identify u as the exponent for the integrand's exponential function:

$$u = -\left(x^2 + y^2\right)$$

Next, we determine du, keeping in mind that the "du" will have us do partial differentiation with respect to the variable x (because we are currently integrating with respect to the x.) So, we have that $du = -2x\,dx$. The good news is that the variable "x" already exists in the integrand. So, we only need to insert a factor of –2, provided we multiply the front of the integral by the reciprocal of –2, or $-\dfrac{1}{2}$.

Calculus III

Practice Midterm 4 KEY, continued

So the integral $\displaystyle\int_0^\infty\int_0^\infty xye^{-(x^2+3y^2)}dxdy$ now becomes:

$$\int_0^\infty\left(-\frac{1}{2}\right)\int_0^\infty(-2x)ye^{-(x^2+3y^2)}dxdy \;=\; \int_0^\infty\left(-\frac{1}{2}\right)y\left[\int e^u du\right]dy \;=\; \int_0^\infty\left(-\frac{1}{2}\right)y\left[e^u\right]dy \;=\int_0^\infty\left(-\frac{1}{2}\right)y\left[e^{-(x^2+3y^2)}\right]_0^\infty dy$$

Notice that we also pulled the variable "y" out of the inner integral in the above calculations, since it is considered a constant with respect to the variable of integration x.

Evaluating the exponential anti-derivative between its limits gives us:

$$\left[e^{-(x^2+3y^2)}\right]_0^\infty \;=\; e^{-(\infty+3y^2)}-e^{-(0+3y^2)}=e^{-\infty}-e^{-3y^2}$$

From your study of limits in first-semester calculus, we know that $\displaystyle\lim_{x\to\infty}e^{-x}=0$.

So, that result is simply $-e^{-3y^2}$. To complete this problem, we still need to integrate once more. The problem at this point looks like $\displaystyle\int_0^\infty\left(-\frac{1}{2}\right)y\left(-e^{-3y^2}\right)dy$. This is now a simple first-semester calculus problem.

Again we apply u-substitution with the exponential rule. We identify u as the exponent for the integrand's exponential function $u=-3y^2$. Next, we determine du, keeping in mind that the "du" will have us do (regular) differentiation with respect to the variable y (because we are currently integrating with respect to the y.) So, we have that $du=-6ydy$. Continuing on to solve this problem:

$$\int_0^\infty\left(-\frac{1}{2}\right)y\left(-e^{-3y^2}\right)dy \;=\; \left(-\frac{1}{2}\right)\left(\frac{1}{6}\right)\int_0^\infty-6ye^{-3y^2}dy \;=\; \left(-\frac{1}{12}\right)\int e^u du \;=\; \left(-\frac{1}{12}\right)e^u=\left(-\frac{1}{12}\right)\left[e^{-3y^2}\right]_0^\infty$$

We now finish the problem.

$$-\left(\frac{1}{12}\right)\left[e^{-3y^2}\right]_0^\infty \;=\; -\left(\frac{1}{12}\right)\left(e^{-\infty}-e^0\right) \;=\; -\left(\frac{1}{12}\right)(-1) \;=\; \frac{1}{12}$$

Thus, the improper double integral converges. This result makes sense in probability theory, where probabilities for a particular event must be a value in the range of 0 to 1 (inclusive).

5. **Evaluate** $\displaystyle\int_0^1\int_x^1 \cos(y^2)dydx$ **by reversing the order of integration.**

First, I want everyone to notice that this integral is not integrable. Even if we try to use u-substitution we will "run into a brick wall!" That is, if we let $u=y^2$, then $du=2ydy$. However, we are missing the variable "y" from the integrand, so we cannot continue to integrate. When this happens with a multiple integral, always reverse the order of integration. Then, we hope that it will be integrable after that. (Unfortunately, it won't always.)

Always sketch the region of integration before reversing the order of integration. We sketch the graphs of $x=0$, $x=1$, $y=x$, and $y=1$:

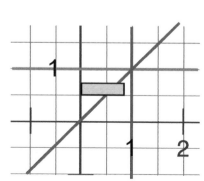

CALCULUS III   **Practice Midterm 4 KEY, CONTINUED**

The region of integration lies above the line $y = x$, but below the line given by $y = 1$. The new order of integration will be $dxdy$. This means that we treat the region as horizontally simple. A representative rectangle is drawn on the region. The left-most edge of this representative rectangle lies on the lower limit for x. This edge lies on the y-axis, so the lower limit for x is $x = 0$. The right-most edge of the representative rectangle lies on the diagonal line $y = x$, so this is the upper limit for x. The limits for y will be constants, since y will be the variable of integration for the outermost integral $0 \le y \le 1$.

The new iterated integral is:

$$\int_0^1 \int_0^y \cos(y^2)\,dxdy = \int_0^1 \cos(y^2)[x]_0^y\,dy = \int_0^1 \cos(y^2)\,ydy$$

Now, we can use u-substitution where $u = y^2$, then $du = 2ydy$. We put in the missing factor of 2 (provided we also divide it out):

$$\frac{1}{2}\int_0^1 \cos(y^2)2ydy = \frac{1}{2}\int \cos u\,du = \frac{1}{2}\sin u = \frac{1}{2}\sin(y^2)\Big|_0^1 = \frac{1}{2}\sin(1)$$

Note: $\sin(1)$ is not a "nice" value. If we had to write our final answer as a decimal value, we would need to use our calculator. (Be sure you are in RADIANS mode, otherwise you will get an incorrect value!) That is, $\sin(1) \approx 0.8415$

6.   **Use a double integral to find the volume of the solid in the first octant bounded above by the plane** $x + y + z = 4$, **and below by the rectangle on the** xy**-plane** $\{(x,y): 0 \le x \le 1,\ 0 \le y \le 2\}$.

$$\int_0^1 \int_0^2 (4 - x - y)\,dydx = \int_0^1 \left[4y - xy - \frac{y^2}{2}\right]_0^2 dx = \int_0^1 \left(4(2) - x(2) - \frac{(2)^2}{2}\right)dx = \int_0^1 (8 - 2x - 2)dx$$

Volume =
$$= \int_0^1 (6 - 2x)dx = \left[6x - x^2\right]_0^1 = 6(1) - (1)^2 = 5$$

7.   **Evaluate the integral…**

$$\int_0^\pi \int_0^{6\sin\theta} r\,drd\theta = \int_0^\pi \left[\frac{r^2}{2}\right]_0^{6\sin\theta} d\theta = \int_0^\pi \frac{1}{2}(6\sin\theta)^2\,d\theta = 18\int_0^\pi (\sin^2\theta)d\theta$$

We will now need to use the Power-Reduction formula to integrate. In other words, we substitute:

$$\sin^2\theta = \frac{1 - \cos(2\theta)}{2} = \frac{1}{2}(1 - \cos(2\theta))$$

$$\frac{1}{2}18\int_0^\pi (1 - \cos(2\theta))d\theta = 9\left[\theta - \frac{1}{2}\sin(2\theta)\right]_0^\pi = 9\left\{\left[\pi - \frac{1}{2}\sin(2\cdot\pi)\right] - \left[0 - \frac{1}{2}\sin(2\cdot0)\right]\right\} = 9\pi$$

CALCULUS III   **Practice Midterm 4 KEY, CONTINUED**

8. **Find the limits of integration for calculating the volume of the solid Q enclosed by the graphs of $y = x^2$, $z = 0$, and $y + z = 2$ if $V = \iiint\limits_Q dz\,dy\,dx$. Just set up the integral; *no need to evaluate!***

The height of the solid ranges from $z = 0$ to $z = 2 - y$. These will be the limits for the inner-most integral. We need to determine what the base of the solid looks like in the xy-plane in order to iterate the two outer integrals. We already have the equation $y = x^2$ in the xy-plane. Since we know that $z = 0$ in the xy-plane, we set $z = 0$ in the equation $y + z = 2$ to get the equation $y = 2$. The two equations together—$y = 2$ and $y = x^2$—together make the base of the solid:

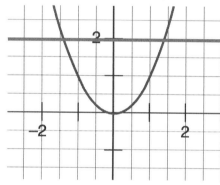

We need to find the x-coordinates of the intersection points for these two graphs in order to get the limits for the variable x (which will be constants).

The lower limit for y is the curve $y = x^2$, and the upper limit for y is $y = 2$.

$y = y$

$x^2 = 2$

$x = \pm\sqrt{2}$

Final answer:

$$\text{Volume} = \int_{-\sqrt{2}}^{\sqrt{2}} \int_{x^2}^{2} \int_{0}^{2-y} dz\,dy\,dx$$

9. **Rewrite the order of the integral $\displaystyle\int_0^4 \int_0^{\frac{4-x}{2}} \int_0^{\frac{(12-3x-6y)}{4}} dz\,dy\,dx$ in the order $dy\,dx\,dz$. *No need to evaluate the integral; just set it up!***

The triple integral represents the volume of the solid bounded by the three coordinate planes and the plane $z = \dfrac{12 - 3x - 6y}{4} \;\Rightarrow\; 3x + 6y + 4z = 12$.

If y needs to be the innermost variable of integration, we solve the plane equation for y:

$$3x + 6y + 4z = 12 \;\Rightarrow\; y = \frac{12 - 3x - 4z}{6}$$

Then, for the middle variable of integration (x), we set $y = 0$ and get:

$$3x + 6(0) + 4z = 12 \;\Rightarrow\; x = \frac{12 - 4z}{3}$$

Finally, we set both $x = 0$ and $y = 0$ to get the upper limit for the outermost variable of integration (z):

$$3(0) + 6(0) + 4z = 12 \;\Rightarrow\; z = 3$$

Final answer: $\displaystyle\int_0^3 \int_0^{\frac{12-4z}{3}} \int_0^{\frac{12-3x-4z}{6}} dy\,dx\,dz$

CALCULUS III   **Practice Midterm 4 KEY, CONTINUED**

10.   Evaluate the integral $\displaystyle\int_1^3\int_x^{x^2}\int_0^{\ln x} xe^y\,dy\,dz\,dx$.

$$\int_1^3\int_x^{x^2}\int_0^{\ln x} xe^y\,dy\,dz\,dx = \int_1^3\int_x^{x^2} x\left[e^y\right]_0^{\ln x}dz\,dx = \int_1^3\int_x^{x^2} x\left[e^{\ln x}-e^0\right]dz\,dx = \int_1^3\int_x^{x^2} x\left[x-1\right]dz\,dx$$

$$= \int_1^3\int_x^{x^2}\left(x^2-x\right)dz\,dx = \int_1^3\left(x^2-x\right)\left[z\right]_x^{x^2}dx = \int_1^3\left(x^2-x\right)\left(x^2-x\right)dx = \int_1^3\left(x^4-2x^3+x^2\right)dx$$

$$= \left[\frac{x^5}{5}-\frac{x^4}{2}+\frac{x^3}{3}\right]_1^3 = \left[\frac{3^5}{5}-\frac{3^4}{2}+\frac{3^3}{3}\right]-\left[\frac{1^5}{5}-\frac{1^4}{2}+\frac{1^3}{3}\right] = \frac{256}{15}$$

11.   **Set up the triple integral to calculate the volume of the solid bounded by** $z=9-x^2-y^2$ **and** $z=0$ …

   (a)   **using rectangular coordinates.**

The base of the solid will be in the xy-plane. The base of the solid will be the projection of the graph of $z=9-x^2-y^2$ onto the xy-plane. Set $z=0$, to get $0=9-x^2-y^2 \Rightarrow x^2+y^2=9$, which is a circle centered at the origin of radius 3:

$$\textbf{\textit{Volume}} = \int_{-3}^{3}\int_{-\sqrt{9-x^2}}^{\sqrt{9-x^2}}\int_0^{9-x^2-y^2} dz\,dy\,dx$$

   (b)   **using cylindrical coordinates.**

Convert to cylindrical: $z=9-x^2-y^2=9-(x^2+y^2)=9-r^2$

Also, remember that $dy\,dx$ is replaced with $r\,dr\,d\theta$.

$$\textbf{\textit{Volume}} = \int_0^{2\pi}\int_0^3\int_0^{9-r^2} dz\,r\,dr\,d\theta$$

12.   **Consider the triple integral** $\displaystyle\int_{-5}^{5}\int_0^{\sqrt{25-x^2}}\int_0^{\frac{1}{x^2+y^2}} \sqrt{x^2+y^2}\,dz\,dy\,dx$.

   (a)   **Set up the integral using cylindrical coordinates.**

First, we sketch the graphs of the equations $x=-5$, $x=5$, $y=0$, and $y=\sqrt{25-x^2}$ to get a picture of the region of integration in the $r\theta$-plane:

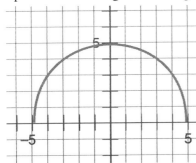

The region of integration the semi-circle above the x-axis where $0\le\theta\le\pi$ and $0\le r\le 5$.

Thus, the integral becomes:

$$\int_0^{\pi}\int_0^5\int_0^{\frac{1}{r^2}} \sqrt{r^2}\,dz(r)\,dr\,d\theta = \int_0^{\pi}\int_0^5\int_0^{\frac{1}{r^2}} r^2\,dz\,dr\,d\theta$$

ANSWER 10, CONTINUED…

...ANSWER 10, CONTINUED

CALCULUS III

Practice Midterm 4 KEY, CONTINUED

(b) Evaluate the integral you set up in part (a):

$$\int_0^\pi \int_0^5 \int_0^{\frac{1}{r^2}} r^2 \, dz \, dr \, d\theta = \int_0^\pi \int_0^5 r^2 [z]_0^{1/r^2} \, dr \, d\theta = \int_0^\pi \int_0^5 r^2 \left(\frac{1}{r^2}\right) dr \, d\theta = \int_0^\pi \int_0^5 dr \, d\theta = \int_0^\pi [r]_0^5 \, d\theta = 5[\theta]_0^\pi = 5\pi$$

13. Consider the solid bounded above by the plane $x + y + z = 3$ in the first octant.

(a) Find the volume of the solid. (Use any method you prefer.)

If we use a double integral to represent the volume of the solid, then we use the formula:

$$\textit{Volume} = \iint_R f(x,y)dA, \text{ where } f(x,y) = z = 3 - x - y$$

To see the base of the solid (that is in the xy-plane), we set $z = 0$, to get $x + y + 0 = 3 \implies y = 3 - x$:

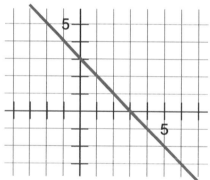

Thus, the limits for y are $0 \le y \le 3 - x$ and the limits for x are $0 \le x \le 3$. The iterated integral is:

$$\int_0^3 \int_0^{3-x} (3 - x - y)dydx = \int_0^3 \left[3y - xy - \frac{y^2}{2} \right]_0^{3-x} dx$$

$$= \int_0^3 \left[3(3-x) - x(3-x) - \frac{(3-x)^2}{2} \right] dx$$

$$= \int_0^3 \left[(3-x)(3-x) - \frac{(3-x)^2}{2} \right] dx = \int_0^3 \left[(3-x)^2 - \frac{(3-x)^2}{2} \right] dx = \int_0^3 \left[\frac{(3-x)^2}{2} \right] dx$$

Use u-substitution and the General Power Rule to integrate where $u = 3 - x$, $du = (-1)dx$. So, we need a factor of -1 in the integrand (provided we divide it out):

$$-\frac{1}{2} \int_0^3 (3-x)^2(-1)dx = -\frac{1}{2} \int u^2 \, du = -\frac{1}{2} \left[\frac{u^3}{3} \right] = -\frac{1}{6} \left[(3-x)^3 \right]_0^3 = -\frac{1}{6} \left\{ (3-3)^3 - (3-0)^3 \right\} = -\frac{1}{6}(0 - 3^3) = \frac{9}{2}$$

CALCULUS III   Practice Midterm 4 KEY, CONTINUED

(b)   **Find the surface area of the top of the solid (i.e., the portion of the plane).**

We use the formula:

Surface Area $= \iint\limits_{R} \sqrt{1 + [f_x(x,y)]^2 + [f_y(x,y)]^2}\, dA$, where $f(x,y) = z = 3 - x - y$,

so the first partials are $f_x(x,y) = -1$ *and* $f_x(x,y) = -1$.

So, we have **Surface Area** =

$$\int\limits_{0}^{3}\int\limits_{0}^{3-x} \sqrt{1 + [-1]^2 + [-1]^2}\, dydx = \sqrt{3}\int\limits_{0}^{3} [y]_0^{3-x}\, dx = \sqrt{3}\int\limits_{0}^{3} (3-x)dx = \sqrt{3}\left[3x - \frac{x^2}{2}\right]_0^3 = \sqrt{3}\left[3(3) - \frac{(3)^2}{2}\right] = \frac{9\sqrt{3}}{2}$$

14.   **Find the surface area for that portion of the surface** $z = xy$ **that is inside the cylinder** $x^2 + y^2 = 1$.

We use the formula:

Surface Area $= \iint\limits_{R} \sqrt{1 + [f_x(x,y)]^2 + [f_y(x,y)]^2}\, dA$, where $f(x,y) = z = xy$,

so the first partials are $f_x(x,y) = y$ *and* $f_x(x,y) = x$.

The region of integration is the graph of the equation $x^2 + y^2 = 1$ in the *xy*-plane.

$$\textbf{Surface Area} = \int\limits_{-1}^{1}\int\limits_{-\sqrt{1-x^2}}^{\sqrt{1-x^2}} \sqrt{1 + [x]^2 + [y]^2}\, dydx$$

We convert to the polar coordinates system to make it easier for us to integrate. This is a perfect candidate for polar coordinates because the region of integration is a circle. Using the formulas to convert from rectangular to polar, we now have the surface area as:

$$\int\limits_{0}^{2\pi}\int\limits_{0}^{1} \sqrt{1 + r^2}\, r\, dr\, d\theta$$

We will use *u*-substitution and the general Power Rule to integrate where $u = 1 + r^2$, $du = 2r\, dr$. Of course, we need a factor of 2 in the integrand (provided we divide it out):

$$\frac{1}{2}\int\limits_{0}^{2\pi}\int\limits_{0}^{1} \sqrt{1+r^2}\,(2r)dr d\theta = \frac{1}{2}\int\limits_{0}^{2\pi}\left[\int \sqrt{u}\,du\right]d\theta = \frac{1}{2}\int\limits_{0}^{2\pi}\left[\int u^{1/2}\,du\right]d\theta = = \frac{1}{2}\int\limits_{0}^{2\pi}\frac{2}{3}\left[u^{3/2}\right]d\theta = \frac{1}{3}\int\limits_{0}^{2\pi}\left[(1+r^2)^{3/2}\right]_0 d\theta$$

$$= \frac{1}{3}\int\limits_{0}^{2\pi}\left[(1+1^2)^{3/2} - (1+0^2)^{3/2}\right]d\theta = \frac{1}{3}\int\limits_{0}^{2\pi}\left[(2)^{3/2} - (1)^{3/2}\right]d\theta = \frac{1}{3}\left(2^{3/2} - 1\right)\int\limits_{0}^{2\pi} d\theta = \frac{1}{3}\left(2^{3/2} - 1\right)[\theta]_0^{2\pi} = \frac{2\pi}{3}\left(2^{3/2} - 1\right)$$

This value is approximately 3.8294.

15. Evaluate $\displaystyle\iiint_Q \frac{1}{\sqrt{x^2+y^2+z^2}}\,dV$ using spherical coordinates if Q is the sphere $x^2+y^2+z^2=25$.

$$\int_0^{2\pi}\int_0^{\pi}\int_0^5 \frac{1}{\rho}\,\rho^2\sin\phi\,d\rho\,d\phi\,d\theta = \int_0^{2\pi}\int_0^{\pi}\int_0^5 \rho\sin\phi\,d\rho\,d\phi\,d\theta = \int_0^{2\pi}\int_0^{\pi}\sin\phi\left[\frac{\rho^2}{2}\right]_0^5 d\phi\,d\theta = \int_0^{2\pi}\int_0^{\pi}\sin\phi\left[\frac{25}{2}\right]d\phi\,d\theta$$

$$= \frac{25}{2}\int_0^{2\pi}\left[-\cos\phi\right]_0^{\pi} d\theta = \frac{25}{2}\int_0^{2\pi}\left[-\cos\pi-(-\cos 0)\right]d\theta = \frac{25}{2}(2)\int_0^{2\pi} d\theta = 50\pi$$

16. Let Q be the sphere $x^2+y^2+z^2=9$.

   (a) **Use cylindrical coordinates to set up the triple integral to calculate the volume of the upper hemisphere of Q. (No need to evaluate.)**

   We solve for z in the equation $x^2+y^2+z^2=9$:

   $$x^2+y^2+z^2=9 \quad\Rightarrow\quad z^2=9-x^2-y^2 \quad\Rightarrow\quad z=\pm\sqrt{9-x^2-y^2}$$

   We take the positive version, since we are considering the *upper* hemisphere. Next, we convert to cylindrical coordinates:

   $$z=\sqrt{9-x^2-y^2} \quad\Rightarrow\quad z=\sqrt{9-\left(x^2+y^2\right)} \quad\Rightarrow\quad z=\sqrt{9-r^2} \qquad \textbf{\textit{Volume}} = \int_0^{2\pi}\int_0^3\int_0^{\sqrt{9-r^2}} r\,dz\,dr\,d\theta$$

   (b) **Use spherical coordinates to set up the triple integral to calculate the volume of the upper hemisphere of Q. (No need to evaluate.)** $\textbf{\textit{Volume}} = \displaystyle\int_0^{2\pi}\int_0^{\pi/2}\int_0^3 \rho^2\sin\phi\,d\rho\,d\phi\,d\theta$

17. **Evaluate the integral** $\displaystyle\int_0^1\int_0^{e^{-9x}} xy\,dy\,dx$. **(Hint: Integration by parts will be involved somehow, somewhere in this problem.)**

   We integrate with respect to y first (where we do *not* require Integration By Parts):

   $$\int_0^1 x\left[\frac{y^2}{2}\right]_0^{e^{-9x}} dx = \frac{1}{2}\int_0^1 x\left(e^{-9x}\right)^2 dx = \frac{1}{2}\int_0^1 xe^{-18x}dx$$

   Now, we need integration by parts where $\begin{cases} u=x, & dv=e^{-18x}dx \\ du=dx, & v=-\dfrac{1}{18}e^{-18x} \end{cases}$. Recall the integration by parts formula that tells us $\int u\,dv = uv - \int v\,du$. Apply the formula:

   $$\frac{1}{2}\int_0^1 xe^{-18x}dx = \frac{1}{2}\left\{\left[-\frac{1}{18}xe^{-18x}\right]_0^1 - \int_0^1\left(-\frac{1}{18}e^{-18x}\right)dx\right\} = \frac{1}{2}\left[-\frac{1}{18}xe^{-18x} - \left(\frac{1}{18}\right)\left(\frac{1}{18}\right)e^{-18x}\right]_0^1$$

   $$= \frac{1}{2}\left[-\frac{1}{18}xe^{-18x} - \left(\frac{1}{324}\right)e^{-18x}\right]_0^1 = \frac{1}{2}\left\{\left[-\frac{1}{18}(1)e^{-18(1)} - \left(\frac{1}{324}\right)e^{-18(1)}\right] - \left[-\frac{1}{18}(0)e^{-18(0)} - \left(\frac{1}{324}\right)e^{-18(0)}\right]\right\}$$

   $$= \frac{1}{2}\left\{\left(-\frac{1}{18}-\frac{1}{324}\right)e^{-18} + \frac{1}{324}\right\} = -\frac{19}{648}e^{-18} + \frac{1}{648} \approx 0.0155$$

CALCULUS III   **Practice Midterm 4 KEY, CONTINUED**

18.   Given a solid Q enclosed by the graphs of $x = y^4$, $z = 0$, *and* $x + z = 16$.

   (a)   Set up but do not evaluate the double integral that would calculate the *area* of the *base* of the solid region Q. (*Hint:* Identify what is going on in the *xy*-plane.)

   We graph the equation $x = y^4$. We set $z = 0$ in the equation $x + z = 16$ and graph $x = 16$:

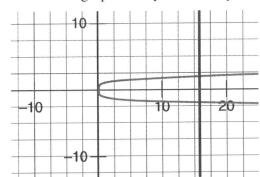

Then, we have the area of the base of the solid:

$$Area = \int_{0}^{16}\int_{-\sqrt[4]{x}}^{\sqrt[4]{x}} dy\,dx$$

Or, if we want to consider the region as horizontally simple, we could write:

$$Area = \int_{-2}^{2}\int_{y^4}^{16} dx\,dy$$

   (b)   Set up but do not evaluate the double integral that would calculate the *volume* of the solid region Q.

$$\boldsymbol{Volume} = \int_{0}^{16}\int_{-\sqrt[4]{x}}^{\sqrt[4]{x}}(16-x)\,dy\,dx \quad \mathbf{OR} \quad \boldsymbol{Volume} = \int_{-2}^{2}\int_{y^4}^{16}(16-x)\,dx\,dy$$

   (c)   Set up but do not evaluate the *triple* integral that would calculate the *volume* of the solid region Q.

$$\boldsymbol{Volume} = \int_{0}^{16}\int_{-\sqrt[4]{x}}^{\sqrt[4]{x}}\int_{0}^{16-x} dz\,dy\,dx$$

19.   Find the surface area for that portion of the surface $z = \dfrac{x^2}{2} - y$ that lies above the triangle in the first quadrant of the *xy*-plane bounded by the lines $y = 3x$, $y = 0$, and $x = 2$.

We use the formula:

$$\boldsymbol{Surface\,Area} = \iint_{R} \sqrt{1 + \left[f_x(x,y)\right]^2 + \left[f_y(x,y)\right]^2}\,dA, \text{ where } f(x,y) = z = \dfrac{x^2}{2} - y, \text{ so the first partials}$$

are $f_x(x,y) = x$ *and* $f_x(x,y) = -1$.

The region of integration R is the triangle in the first quadrant.

$$\boldsymbol{Surface\,Area} = \int_{0}^{2}\int_{0}^{3x}\sqrt{1 + [x]^2 + [-1]^2}\,dy\,dx = \int_{0}^{2}\int_{0}^{3x}\sqrt{2 + x^2}\,dy\,dx = \int_{0}^{2}\sqrt{2 + x^2}\,[y]_0^{3x}\,dx = \int_{0}^{2}\sqrt{2 + x^2}\,(3x)\,dx$$

We will use *u*-substitution and the general Power Rule to integrate where $u = 2 + x^2$, $du = 2x\,dx$.

Of course, we need a factor of 2 in the integrand (provided we divide it out):

$$= 3\left(\dfrac{1}{2}\right)\int_{0}^{2}\sqrt{2 + x^2}\,(2x)\,dx = \dfrac{3}{2}\int\sqrt{u}\,du = \dfrac{3}{2}\int u^{1/2}\,du = \dfrac{3}{2}\left[\dfrac{2}{3}u^{3/2}\right] = \left[(2 + x^2)^{3/2}\right]_0^2 = \left(2 + 2^2\right)^{3/2} - \left(2 + 0^2\right)^{3/2}$$

$$= 6^{3/2} - 2^{3/2} \approx 11.9$$

CALCULUS III   **Practice Midterm 4 KEY,** CONTINUED

---

20.   **Evaluate** $\iint\limits_{R} \dfrac{x^2}{\sqrt{1+y^2}}\, dA$ **, where** R **is the region bounded by the graphs of** $y = x^3$, $y = 8$, **and** $x = 0$.

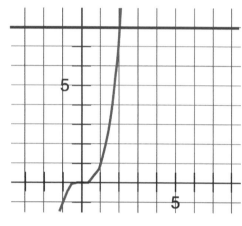

If we let the order of integration be "*dydx*," then the iterated integral will be:

$$\int_0^2 \int_{x^3}^8 \frac{x^2}{\sqrt{1+y^2}}\, dy\, dx$$

However, to integrate with respect to y will require trigonometric substitution, which requires effort. Instead, let's switch the order of integration to make it easier:

$$\int_0^8 \int_0^{\sqrt[3]{y}} \frac{x^2}{\sqrt{1+y^2}}\, dx\, dy = \int_0^8 \frac{1}{\sqrt{1+y^2}}\left[\frac{x^3}{3}\right]_0^{\sqrt[3]{y}} dy = \int_0^8 \frac{1}{\sqrt{1+y^2}}\left[\frac{\left(\sqrt[3]{y}\right)^3}{3}\right] dy$$

$$= \frac{1}{3}\int_0^8 \frac{y}{\sqrt{1+y^2}}\, dy$$

We will use u-substitution and the general Power Rule to integrate where $u = 1 + y^2$, $du = 2y\, dy$. Of course, we need a factor of 2 in the integrand (provided we divide it out):

$$= \frac{1}{3}\left(\frac{1}{2}\right)\int_0^8 \frac{2y}{\sqrt{1+y^2}}\, dy = \frac{1}{6}\int \frac{du}{\sqrt{u}} = \frac{1}{6}\int u^{-1/2}\, du = \frac{1}{6}\left[2u^{1/2}\right] = \frac{1}{3}\left[\left(1+y^2\right)^{1/2}\right]_0^8 = \frac{1}{3}\left\{\left(1+8^2\right)^{1/2} - \left(1+0^2\right)^{1/2}\right\}$$

$$= \frac{1}{3}\left(65^{1/2} - 1\right) \approx 2.35$$

21.   **Convert to polar coordinates and then evaluate the double integral:** $\displaystyle\int_{-6}^{6} \int_{-\sqrt{36-y^2}}^{0} \frac{\sqrt{x^2+y^2}}{1+\sqrt{x^2+y^2}}\, dx\, dy$

First, determine the region of integration by sketching the graphs

of $y = -6$, $y = 6$, $x = -\sqrt{36-y^2}$ and $x = 0$. To sketch the

graph of $x = -\sqrt{36-y^2}$, it might help to solve for y first:

$$(x)^2 = \left(-\sqrt{36-y^2}\right)^2 \;\Rightarrow\; x^2 = 36 - y^2 \;\Rightarrow\; x^2 + y^2 = 36$$

The region of integration is the left half of a circle centered at the origin with radius 6. We convert the double integral to polar coordinates:

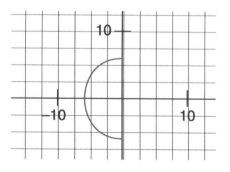

$$\int_{\pi/2}^{3\pi/2} \int_0^6 \frac{r}{1+r}\, r\, dr\, d\theta = \int_{\pi/2}^{3\pi/2} \int_0^6 \frac{r^2}{1+r}\, dr\, d\theta$$

CALCULUS **III**   **Practice Midterm 4 KEY,** CONTINUED

The integrand is an improper rational function (a polynomial divided by a polynomial, where the degree of the numerator's polynomial is greater than or equal to the degree of the denominator polynomial). This means we must perform polynomial long division before integrating:

$$r+1\overline{)r^2}\;\;\begin{array}{l}r-1\\\hline\end{array}\qquad\Rightarrow\qquad r-1+\dfrac{1}{r+1}$$

$$\begin{array}{r}-\underline{r^2+r}\\-r\\-\underline{-r-1}\\1\end{array}$$

$$= \int_{\pi/2}^{3\pi/2}\int_0^6\left[r-1+\frac{1}{1+r}\right]drd\theta = \int_{\pi/2}^{3\pi/2}\left[\frac{r^2}{2}-r+\ln|1+r|\right]_0^6 d\theta = \int_{\pi/2}^{3\pi/2}\left[\left(\frac{6^2}{2}-6+\ln|1+6|\right)-\left(\frac{0^2}{2}-0+\ln|1+0|\right)\right]d\theta$$

$$= \int_{\pi/2}^{3\pi/2}\left(12+\ln|7|\right)d\theta = (12+\ln 7)[\theta]_{\pi/2}^{3\pi/2} = (12+\ln 7)\pi \approx 43.8$$

22.   **Set up the triple integral using cylindrical coordinates, and then evaluate:** $\displaystyle\int_0^1\int_{-\sqrt{1-x^2}}^0\int_{-(x^2+y^2)}^{(x^2+y^2)}21xy^2\,dzdydx$

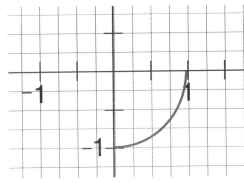

First, sketch the region of integration in the xy-plane by sketching the graphs of: $x=0$, $x=1$, $y=-\sqrt{1-x^2}$ and $y=0$.

The limits for r are $0 \le r \le 1$. The limits for θ can either be

$$-\frac{\pi}{2}\le\theta\le 0 \quad OR \quad \frac{3\pi}{2}\le\theta\le 2\pi.$$

Please do *not* write $\dfrac{3\pi}{2}\le\theta\le 0$. This is wrong because the number $\dfrac{3\pi}{2}$ is *greater* than the number 0.

Use the conversion formulas $x=r\cos\theta$, $y=r\sin\theta$ and $x^2+y^2=r^2$ to rewrite the integral:

$$\int_{-\pi/2}^0\int_0^1\int_{-r^2}^{r^2}21(r\cos\theta)(r\sin\theta)^2\,dzdrd\theta$$

We integrate with respect to the variable z first:

$$21\int_{-\pi/2}^0\int_0^1 r^4\sin^2\theta\cos\theta[z]_{-r^2}^{r^2}drd\theta = 21\int_{-\pi/2}^0\int_0^1 r^4\sin^2\theta\cos\theta\left(r^2-\left(-r^2\right)\right)drd\theta = 21\int_{-\pi/2}^0\int_0^1 r^4\sin^2\theta\cos\theta\left(2r^2\right)drd\theta$$

$$= 42\int_{-\pi/2}^0\int_0^1 r^6\sin^2\theta\cos\theta drd\theta = 42\int_{-\pi/2}^0\sin^2\theta\cos\theta\left[\frac{r^7}{7}\right]_0^1 d\theta = \frac{42}{7}\int_{-\pi/2}^0\sin^2\theta\cos\theta d\theta$$

Next, we use u-substitution and the General Power Rule where $u=\sin\theta$, $du=\cos\theta$:

$$= 6\int u^2\,du = 6\left[\frac{u^3}{3}\right] = 2\left[\sin^3\theta\right]_{-\pi/2}^0 = 2\left(\sin^3 0-\sin^3\left(-\frac{\pi}{2}\right)\right) = 2\left(0-(-1)^3\right) = 2$$

CALCULUS III — Practice Midterm 4 KEY, CONTINUED

23. Set up and evaluate the triple integral using spherical coordinates the would calculate the mass of a sphere with radius 7 if the density function is given by $f(x,y,z)=k\sqrt{x^2+y^2+z^2}$, where *k* is a constant.

Use the formula: Mass = $\iiint\limits_Q f(x,y,z)dV$ and the conversion formula $\sqrt{x^2+y^2+z^2}=\rho$.

$$\int_0^{2\pi}\int_0^{\pi}\int_0^{7} k\rho\cdot\rho^2\sin\phi\, d\rho\, d\phi\, d\theta = k\int_0^{2\pi}\int_0^{\pi}\int_0^{7}\rho^3\sin\phi\, d\rho\, d\phi\, d\theta = k\int_0^{2\pi}\int_0^{\pi}\sin\phi\left[\frac{\rho^4}{4}\right]_0^7 d\phi\, d\theta$$

$$\textit{Mass}=\frac{2401k}{4}\int_0^{2\pi}\left[-\cos\phi\right]_0^{\pi}d\theta=\frac{2401k}{4}(2)\left[\theta\right]_0^{2\pi}=2401k\pi$$

CALCULUS III — QUIZ 11 KEY

1. Sketch several representative vectors in the vector field.

(a) $\mathbf{F}(x,y)=\langle 0,-2\rangle$

No matter what *x* or *y* are, the vectors in the field will all be constant.

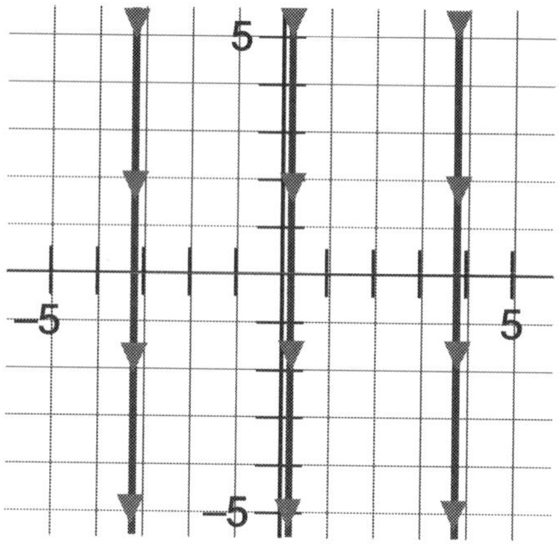

(b) $\mathbf{F}(x,y)=\langle x,0\rangle$

Determine the level curves for the vector field by setting the magnitude of vector field equal to a constant:

$$\|F(x,y)\|=c$$
$$\sqrt{x^2+0^2}=c$$
$$x=c$$

So, the level curves fall on vertical lines. Make a table:

(x,y)	<x, 0>
(1, 2)	<1, 0>
(1, 1)	<1, 0>
(1, –1)	<1, 0>
(2, 2)	<2, 0>
(2, 1)	<2, 0>
(2, 0)	<2, 0>

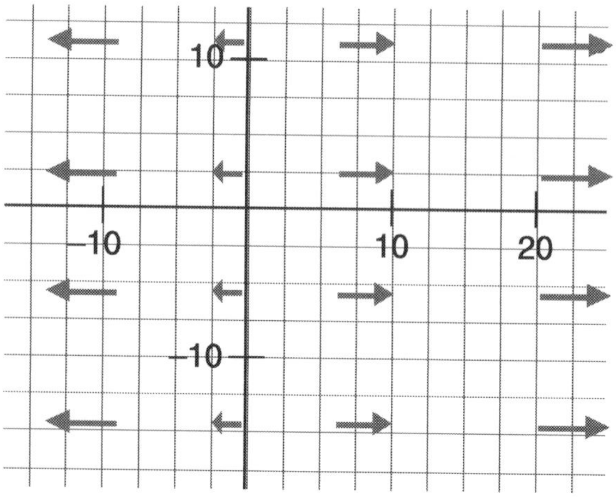

CALCULUS III   **Quiz 11 KEY, CONTINUED**

2.   **Find a three-dimensional vector field that has the potential function $f(x,y,z) = x^2 e^{yz}$. (Hint: Differentiate.)**

$$\mathbf{F}(x,y,z) = \nabla f(x,y,z) = \langle f_x, f_y, f_z \rangle = \langle 2xe^{yz}, \; x^2 z e^{yz}, \; x^2 y e^{yz} \rangle$$

3.   **Given the three-dimensional vector field $\mathbf{F}(x,y,z) = \langle e^y, xe^y + y, 0 \rangle$, find the potential function for F.**
     (Hint: Anti-differentiate.)

     We integrate each component separately, then add the results, disregarding "duplicates:"

$$f(x,y,z) = \begin{cases} \int M dx = \int e^y dx = xe^y + g(y,z) \\[2mm] \int N dy = \int (xe^y + y) dy = xe^y + \dfrac{y^2}{2} + h(x,z) \\[2mm] \int P dz = \int 0 dz = k \end{cases} \Rightarrow \quad f(x,y,z) = xe^y + \dfrac{y^2}{2} + k$$

4.   ***Multiple choice:* Determine which of the following vector fields is *not* conservative. *(Please circle your choice.)***

     Two-dimensional vector fields are conservative if $\dfrac{\partial M}{\partial y} = \dfrac{\partial N}{\partial x}$.

(a) $\mathbf{F}(x,y) = \left\langle -\dfrac{y}{x^2}, \dfrac{1}{x} \right\rangle$ (b) $\mathbf{F}(x,y) = \langle e^y, xe^y + y \rangle$

$\dfrac{\partial M}{\partial y} = \dfrac{\partial N}{\partial x} = -\dfrac{1}{x^2}$ $\dfrac{\partial M}{\partial y} = \dfrac{\partial N}{\partial x} = e^y$

(c) $\mathbf{F}(x,y) = \langle -2y^3 \sin 2x, 3y^2(1 + \cos 2x) \rangle$ (d) $\mathbf{F}(x,y) = \langle 4x^2 - 4y^2, 8xy - \ln y \rangle$

$\dfrac{\partial M}{\partial y} = \dfrac{\partial N}{\partial x} = -6y^2 \sin(2x)$ $\dfrac{\partial M}{\partial y} = -8y \neq \dfrac{\partial N}{\partial x} = 8y$

                                                                     Not conservative!

(e) $\mathbf{F}(x,y) = \left\langle \dfrac{x}{\sqrt{x^2 + y^2}}, \dfrac{y}{\sqrt{x^2 + y^2}} \right\rangle$

$\dfrac{\partial M}{\partial y} = \dfrac{\partial [x(x^2+y^2)^{-1/2}]}{\partial y} = x\left(-\dfrac{1}{2}\right)(x^2 + y^2)^{-3/2}(2y) = \dfrac{-xy}{(x^2 + y^2)^{3/2}} = \dfrac{\partial N}{\partial x}$

5.   **Let C be the line segment from the point (0, 0, 0) to the point (1, 3, –2). Find $\int_C (x + y^2 - 2z) ds$.**
     We first need to parameterize the curve C:

$$\mathbf{r}(t) = \langle t, 3t, -2t \rangle \qquad 0 \le t \le 1 \; \Rightarrow \; \mathbf{r}'(t) = \langle 1, 3, -2 \rangle \; \Rightarrow \; \|\mathbf{r}'(t)\| = \sqrt{1^2 + 3^2 + 2^2} = \sqrt{14}$$

We rewrite the line integral using the substitutions $ds = \|\mathbf{r}'(t)\| dt$, and $x = t, y = 3t, z = -2t$:

$$\int_C (x + y^2 - 2z) ds = \int_0^1 (t + (3t)^2 - 2(-2t)) \sqrt{14} dt = \sqrt{14} \int_0^1 (t + 9t^2 + 4t) dt = \sqrt{14} \int_0^1 (9t^2 + 5t) dt$$

$$= \sqrt{14} \left[3t^3 + \frac{5t^2}{2} \right]_0^1 = \sqrt{14} \left(3 + \frac{5}{2} \right) = \frac{11\sqrt{14}}{2}$$

Calculus III   **Quiz 11 KEY, continued**

6. **Let $\mathbf{F}(x,y,z) = \langle 2x - y, 2z, y - z \rangle$. Find the work done by the force F on an object moving along the straight line from the point (0, 0, 0) to the point (1, 1, 1).**

We first need to parameterize the curve C:

$$\mathbf{r}(t) = \langle t, t, t \rangle \qquad 0 \le t \le 1 \quad \Rightarrow \quad \mathbf{r}'(t) = \langle 1, 1, 1 \rangle \quad \Rightarrow \quad dr = \mathbf{r}'(t)dt$$

We use the formula for Work:

$$\mathbf{Work} = \int_C \mathbf{F} \cdot d\mathbf{r} \text{ and that } \mathbf{F}(x(t), y(t), z(t)) = \langle 2t - t, 2t, t - t \rangle = \langle t, 2t, 0 \rangle$$

$$\mathbf{Work} = \int_C \mathbf{F} \cdot d\mathbf{r} = \int_0^1 \langle t, 2t, 0 \rangle \cdot \langle 1,1,1 \rangle dt = \int_0^1 (t + 2t + 0)dt = \int_0^1 3t\,dt = \left[\frac{3t^2}{2}\right]_0^1 = \frac{3}{2}$$

7. **A particle moves along a path parameterized by $\mathbf{r}(t) = \langle t, t^2, t^3 \rangle$ from the point (0, 0, 0) to the point (1, 1, 1) under a force given by $\mathbf{F}(x,y,z) = \langle 2xz, -yz, yz^2 \rangle$. Calculate the work done on the particle by the force.**

We use the formula for Work:

$$\mathbf{Work} = \int_C \mathbf{F} \cdot d\mathbf{r} \text{ and that } \mathbf{F}(x(t), y(t), z(t)) = \langle 2t(t^3), -t^2(t^3), (t^2)(t^3)^2 \rangle = \langle 2t^4, -t^5, t^8 \rangle$$

And also, that $\mathbf{r}'(t) = \langle 1, 2t, 3t^2 \rangle$ where t ranges from $t = 0$ to $t = 1$.

$$\int_C \mathbf{F} \cdot d\mathbf{r} = \int_0^1 \langle 2t^4, -t^5, t^8 \rangle \cdot \langle 1, 2t, 3t^2 \rangle dt = \int_0^1 \left(2t^4 + -t^5(2t) + t^8(3t^2)\right)dt = \int_0^1 \left(2t^4 - 2t^6 + 3t^{10}\right)dt$$

$$= \left[\frac{2t^5}{5} - \frac{2t^7}{7} + \frac{3t^{11}}{11}\right]_0^1 = \frac{149}{385}$$

8. **Let $\mathbf{F}(x,y,z) = \langle y, x, z^2 \rangle$ and evaluate $\int_C \mathbf{F} \cdot d\mathbf{r}$ for the curve $r(t) = \langle t, \cos t, \sin t \rangle$ for $0 \le t \le 2\pi$.** *(Hint: You need Integration By Parts to do this problem.)*

$$\int_C \mathbf{F} \cdot d\mathbf{r} = \int_0^{2\pi} \langle \cos t, t, \sin^2 t \rangle \cdot \langle 1, -\sin t, \cos t \rangle dt = \int_0^{2\pi} \left(\cos t - t\sin t + \sin^2 t \cos t\right)dt$$

The second term is the one that requires integration by parts where $u = t, dv = \sin t\,dt, du = dt,$ and $v = -\cos t$. The last term needs u-substitution and the General Power Rule, where $u = \sin t, du = \cos t\,dt$. We do these separately right now:

$$-\int t\sin t\,dt = -\left\{uv - \int v\,du\right\} = -\left\{-t\cos t - \int -\cos t\,dt\right\} = t\cos t - \sin t + C$$

$$\int \sin^2 t \cos t\,dt = \int u^2 du = \frac{u^3}{3} + C = \frac{\sin^3 t}{3} + C$$

Putting it all together we have that:

$$\int_0^{2\pi} \left(\cos t - t\sin t + \sin^2 t \cos t\right)dt = \left[\sin t + t\cos t - \sin t + \frac{\sin^3 t}{3}\right]_0^{2\pi} = \left[t\cos t + \frac{\sin^3 t}{3}\right]_0^{2\pi}$$

$$= \left(2\pi \cos 2\pi + \frac{\sin^3 2\pi}{3}\right) - \left(0\cos 0 + \frac{\sin^3 0}{3}\right) = 2\pi$$

CALCULUS III   **Quiz 11 KEY,** CONTINUED

9.   Let $\int_C y\,ds$, where $C = C_1 \cup C_2$ is the path given in the figure shown.

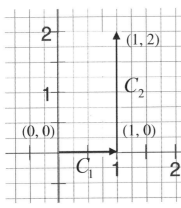

We first need to individually parameterize the two curves:

$C_1: \quad x = t, \; y = 0, \quad 0 \le t \le 1 \quad \Rightarrow \quad \mathbf{r}(t) = \langle t, 0 \rangle, 0 \le t \le 1 \quad \Rightarrow \quad \mathbf{r}'(t) = \langle 1, 0 \rangle$

$\|r'(t)\| = \sqrt{1^2 + 0^2} = 1 \quad \Rightarrow \quad ds = \|r'(t)\|dt = dt$

Therefore, the line integral on the first curve looks like: $\quad \int_{C_1} y\,ds = \int_0^1 (0)(1)dt = 0$

The second curve has:

$C_2: \quad y = t, \; x = 1, \quad 0 \le t \le 2 \quad \Rightarrow \quad \mathbf{r}(t) = \langle 1, t \rangle, 0 \le t \le 2 \quad \Rightarrow \quad \mathbf{r}'(t) = \langle 0, 1 \rangle$

$\|r'(t)\| = \sqrt{0^2 + 1^2} = 1 \quad \Rightarrow \quad ds = \|r'(t)\|dt = dt$

Therefore, the line integral on the first curve looks like:

$$\int_{C_2} y\,ds = \int_0^2 (t)(1)dt = \left[\frac{t^2}{2}\right]_0^2 = 2$$

Final answer: $\int_C y\,ds = \int_{C_1} y\,ds + \int_{C_2} y\,ds = 0 + 2 = 2$

10.   **Evaluate** $\int_C x\,dx - xy\,dy$ **over the path** C **given by** $x = t^2, y = 3t, 0 \le t \le 1$.

Because we have $x = t^2$, *and* $y = 3t$, then we can write that $dx = 2t\,dt$, *and* $y = 3\,dt$, and substitute all of these into the integrand:

$$\int_C x\,dx - xy\,dy = \int_0^1 \left[t^2(2t\,dt) - t^2(3t)(3\,dt)\right] = \int_0^1 \left[2t^3 - 9t^3\right]dt = \int_0^1 \left(-7t^3\right)dt = \left[\frac{-7t^4}{4}\right]_0^1 = -\frac{7}{4}$$

11.   **Let** $\mathbf{F}(x,y) = \left\langle \dfrac{y}{x^2 + y^2}, \dfrac{-x}{x^2 + y^2} \right\rangle$. **Calculate** $\int_C \mathbf{F} \cdot d\mathbf{r}$ **where** C **is the semi-circle** $\mathbf{r}(t) = \langle \cos t, \sin t \rangle$ **for** $0 \le t \le \pi$.

We find $d\mathbf{r} = \mathbf{r}'(t)dt = \langle -\sin t, \; \cos t \rangle dt$ and

$$\mathbf{F}(x(t), y(t)) = \left\langle \frac{\sin t}{\cos^2 t + \sin^2 t}, \frac{-\cos t}{\cos^2 t + \sin^2 t} \right\rangle = \langle \sin t, \; -\cos t \rangle$$

Then, the integral becomes:

$$\int_C \mathbf{F} \cdot d\mathbf{r} = \int_0^\pi \langle \sin t, \; -\cos t \rangle \cdot \langle -\sin t, \; \cos t \rangle dt = \int_0^\pi \left(-\sin^2 t - \cos^2 t\right)dt = \int_0^\pi -(\sin^2 t + \cos^2 t)dt = \int_0^\pi (-1)dt$$

$$= \left[-t\right]_0^\pi = -\pi$$

Calculus III   Quiz 11 KEY, continued

12. Use the **Fundamental Theorem of Line Integrals** to evaluate $\int_C \left(y^2 - 3x^2 \right) dx + \left(2xy + 2 \right) dy$ where C is a smooth curve from (1, 1) to (−1, 0).

First, we need to test whether the vector field is conservative. If it is, then we are permitted to use the Fundamental Theorem of Line Integrals. If it is not, then we would need more information about the path in order to parameterize it and then evaluate the line integral the "hard way."

The field is indeed conservative because $\dfrac{\partial M}{\partial y} = \dfrac{\partial N}{\partial x} = 2y$. We then find the potential function for the vector field by integrating twice and summing the results (ignoring duplicates, if any).

$$f(x,y) = \begin{cases} \int M dx = \int \left(y^2 - 3x^2 \right) dx = xy^2 - x^3 + g(y) \\ \int N dy = \int \left(2xy + 2 \right) dy = xy^2 + 2y + h(x) \end{cases} \Rightarrow \quad f(x,y) = xy^2 - x^3 + 2y + k$$

$$\int_C \left(y^2 - 3x^2 \right) dx + \left(2xy + 2 \right) dy$$

$$= \left[xy^2 - x^3 + 2y \right]_{(1,1)}^{(-1,0)} = \left((-1)(0)^2 - (-1)^3 + 2(0) \right) - \left((1)(1)^2 - (1)^3 + 2(1) \right) = 1 - 2 = -1$$

13. Evaluate $\int_C e^x \sin y \, dx + e^x \cos y \, dy$ where C is the square with vertices (0, 0), (1, 1), (0, 2), (−1, 1).

The field is conservative because $\dfrac{\partial M}{\partial y} = \dfrac{\partial N}{\partial x} = e^x \cos y$. Since the curve is closed, we can conclude by the Theorem that the line integral is equal to 0 since $\int_C \mathbf{F} \cdot d\mathbf{r} = 0$ for every closed curve C if $\mathbf{F}$ is a conservative vector field.

14. Find the work done by the force field $\mathbf{F}(x, y) = \langle y, x \rangle$ in moving a particle from the point (0, 4) to the point (3, 1) along the following paths:

(a) $C_1: y = -x + 4$

The field is conservative because $\dfrac{\partial M}{\partial y} = \dfrac{\partial N}{\partial x} = 1$.

This means we can use the Fundamental theorem of Line Integrals for both of these paths. Further, since both of these paths have the same beginning and ending points, the values will be the same.

We then find the potential function for the vector field by integrating twice and summing the results (ignoring duplicates, if any).
$$f(x,y) = \begin{cases} \int M dx = \int y \, dx = xy + g(y) \\ \int N dy = \int x \, dy = xy + h(x) \end{cases} \Rightarrow \quad f(x,y) = xy + k$$

$$\textbf{Work} = \int_C \mathbf{F} \cdot d\mathbf{r} = \left[xy \right]_{(0,4)}^{(3,1)} = (3)(1) - (0)(4) = 3$$

(b) $C_2: y = (x - 2)^2$

Same answer as for part **(a)** above.

CALCULUS III   **Quiz 11 KEY, CONTINUED**

15.   Let C be the closed curve shown to the right, oriented counter-clockwise. Evaluate $\int_C 2y^3 dx + \left(x^4 + 6y^2 x\right)dy$ using Green's Theorem.

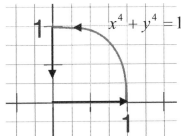

$M = 2y^3$ *and* $N = x^4 + 6y^2 x$,

so we have that $\dfrac{\partial M}{\partial y} = 6y^2$ *and* $\dfrac{\partial N}{\partial x} = 4x^3 + 6y^2$.

Green's Theorem states that:

$$\oint_C Mdx + Ndy = \iint_R \left(\frac{\partial N}{\partial x} - \frac{\partial M}{\partial y}\right)dA = \int_0^1 \int_0^{\sqrt[4]{1-x^4}}\left(4x^3 + 6y^2 - 6y^2\right)dydx$$

We first integrate w.r.t. y:

$$\int_0^1 \int_0^{\sqrt[4]{1-x^4}}\left(4x^3\right)dydx = \int_0^1 \left[4x^3 y\right]_0^{\sqrt[4]{1-x^4}} dx = \int_0^1 4x^3 \sqrt[4]{1-x^4}dx$$

We will use u-substitution and the General Power Rule where $u = 1-x^4, du = -4x^3$. This means we need to put a factor of -1 into the integrand (provided we multiply the outside of the integral as well):

$$= -\int_0^1 \left(-4x^3\right)\sqrt[4]{1-x^4}d = -\int \sqrt[4]{u}du = -\int u^{1/4} du = -\frac{4}{5}u^{5/4} = -\frac{4}{5}\left[\left(1-x^4\right)^{5/4}\right]_0 = -\frac{4}{5}\left\{\left(1-1^4\right)^{5/4} - \left(1-0^4\right)^{5/4}\right\} = \frac{4}{5}$$

16.   Use Green's Theorem to evaluate the line integral $\int_C \left(x^2 + 2y\right)dx + \left(\frac{1}{2}x^2 - y^3\right)dy$ where C is the path from $(0, 0)$ to $(1, 0)$ along the path $y = 0$, and then from $(1, 0)$ to $(1, 1)$ along $x = 1$, and finally from $(1, 1)$ to $(0, 0)$ along the path $y = \sqrt{x}$.

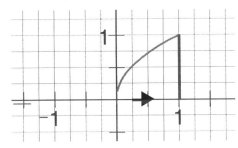

We start by sketching the region of integration that is enclosed by the graphs of $y = \sqrt{x}$, $x = 1$, $y = 0$.

Since $M = x^2 + 2y$ *and* $N = \frac{1}{2}x^2 - y^3$, we have that $\dfrac{\partial M}{\partial y} = 2$ *and* $\dfrac{\partial N}{\partial x} = x$.

Green's Theorem states that:

$$\oint_C Mdx + Ndy = \iint_R \left(\frac{\partial N}{\partial x} - \frac{\partial M}{\partial y}\right)dA = \int_0^1 \int_0^{\sqrt{x}}(x-2)dydx = \int_0^1 (x-2)[y]_0^{\sqrt{x}} dx = \int_0^1 (x-2)\sqrt{x}dx$$

$$= \int_0^1 \left(x\sqrt{x} - 2\sqrt{x}\right)dx = \int_0^1 \left(x^{3/2} - 2x^{1/2}\right)dx = \left[\frac{2}{5}x^{5/2} - 2\cdot\frac{2}{3}x^{3/2}\right]_0^1 = \frac{2}{5} - \frac{4}{3} = -\frac{14}{15}$$

CALCULUS III   Quiz 11 KEY, CONTINUED

17.   Use Green's Theorem to evaluate the line integral $\int_C \left(\sin x^2 + y\right)dx + \left(3x - \arctan(e^y)\right)dy$ where *C* is the square with vertices **(0, 0), (2, 0), (2, 2), and (0, 2).**

Since $M = \sin x^2 + y$ *and* $N = 3x - \arctan(e^y)$, we have that: $\dfrac{\partial M}{\partial y} = 1$ *and* $\dfrac{\partial N}{\partial x} = 3$.

Green's Theorem states that:

$$\oint_C Mdx + Ndy = \iint_R \left(\frac{\partial N}{\partial x} - \frac{\partial M}{\partial y}\right)dA = \int_0^2\int_0^2 (3-1)dydx = \int_0^2 (2)[y]_0^2\, dx = \int_0^2 (2)(2)dx$$

$$= 4[x]_0^2 = 8$$

18.   **Show that the work done by a force** $\mathbf{F}(x, y) = \left\langle e^x \sin y, \, e^x \cos y \right\rangle$ **around a simple closed curve is zero.**

Since $M = e^x \sin y$ *and* $N = e^x \cos y$, we have that $\dfrac{\partial M}{\partial y} = e^x \cos y = \dfrac{\partial N}{\partial x}$.

Green's Theorem states that:

$$\oint_C Mdx + Ndy = \iint_R \left(\frac{\partial N}{\partial x} - \frac{\partial M}{\partial y}\right)dA = \iint_R \left(e^x \cos y - e^x \cos y\right)dA = \iint_R (0)dA = 0$$

19.   **Use Green's Theorem to evaluate the line integral** $\int_C y^3 dx + \left(x^3 + 3xy^2\right)dy$ **, where *C* is a circle of radius 3 centered around the origin, oriented counter-clockwise.** *(Hint:* **Polar coordinates may be helpful.)**

Since $M = y^3$ *and* $N = x^3 + 3xy^2$, we have that $\dfrac{\partial M}{\partial y} = 3y^2$ *and* $\dfrac{\partial N}{\partial x} = 3x^2 + 3y^2$.

Green's Theorem states that:

$$\oint_C Mdx + Ndy = \iint_R \left(\frac{\partial N}{\partial x} - \frac{\partial M}{\partial y}\right)dA = \iint_R \left(3x^2 + 3y^2 - 3y^2\right)dA = \iint_R 3x^2\, dA = \int_{-3}^{3}\int_{-\sqrt{9-x^2}}^{\sqrt{9-x^2}} 3x^2\, dydx$$

$$= \int_0^{2\pi}\int_0^3 3(r\cos\theta)^2\, rdrd\theta = 3\int_0^{2\pi}\int_0^3 (r^2\cos^2\theta)rdrd\theta = 3\int_0^{2\pi}\int_0^3 (r^3\cos^2\theta)drd\theta = 3\int_0^{2\pi}\cos^2\theta\left[\frac{r^4}{4}\right]_0^3 d\theta$$

$$= \frac{243}{4}\int_0^{2\pi}\left(\frac{1+\cos(2\theta)}{2}\right)d\theta = \frac{243}{4}\int_0^{2\pi}\left(\frac{1}{2} + \frac{\cos(2\theta)}{2}\right)d\theta = \frac{243}{4}\left[\frac{1}{2}\theta + \frac{1}{4}\sin(2\theta)\right]_0^{2\pi} = \frac{243}{4}\pi$$

EXTRA CREDIT (optional): **Use back of page if you need more room to answer these.**

I.   **Could Green's Theorem be used in Problem #11? (Yes or no.) If yes, use Green's Theorem and show answers are the same. If no, explain why not.**

No, because Green's Theorem requires a closed curve.

CALCULUS III   **Quiz 11 KEY,** CONTINUED

II.   **For problem #19: This problem involves a closed curve path, yet the answer is not zero. Why not?**

The answer is not 0 because the field is not conservative.

III.   **Suppose in problem #11, the path was the entire circle and not just the semi-circle. Then the path is closed *and* the vector field is conservative, yet the answer is not zero. (It will be –2π.) Why isn't the answer zero? (This one is hard.)**

There is a singularity inside the path. In other words, the field is not defined at the origin. Green's Theorem requires that the field be defined and its partial derivatives be continuous on the region.

20.   **Given the vector field** $\mathbf{F}(x,y,z) = \langle 2xy + z^2, \ x^2, \ 2xz + \pi\cos(\pi z)\rangle$ ...
       Is this a conservative vector field? (yes or no) If yes, find the potential function of F.

To determine if a 3-dimensional vector field is conservative, we require that:

$$\frac{\partial M}{\partial y} = \frac{\partial N}{\partial x}, \quad \frac{\partial P}{\partial y} = \frac{\partial N}{\partial z}, \quad and \quad \frac{\partial M}{\partial z} = \frac{\partial P}{\partial x}$$

This field is indeed conservative because:

$$\frac{\partial M}{\partial y} = \frac{\partial N}{\partial x} = 2x, \quad \frac{\partial P}{\partial y} = \frac{\partial N}{\partial z} = 0, \quad and \quad \frac{\partial M}{\partial z} = \frac{\partial P}{\partial x} = 2z$$

We then find the potential function for the vector field by integrating three times and summing the results (ignoring duplicates, if any).

$$f(x,y,z) = \begin{cases} \int M dx = \int (2xy + z^2)dx = x^2 y + xz^2 + g(y,z) \\ \int N dy = \int (x^2)dy = x^2 y + h(x,z) \\ \int P dz = \int (2xz + \pi\cos(\pi z))dz = xz^2 + \sin(\pi z) + t(x,y) \end{cases} \Rightarrow f(x,y,z) = x^2 y + xz^2 + \sin(\pi z) + k$$

21.   **Let** $\mathbf{F}(x,y,z) = \langle x^3 \ln z, \ xe^{-y}, \ -(y^2 + 2z)\rangle$.

  (a)   Calculate the divergence of F.

$$\text{div}\mathbf{F} = \nabla \cdot \mathbf{F} = \left\langle \frac{\partial}{\partial x}, \frac{\partial}{\partial y}, \frac{\partial}{\partial z} \right\rangle \cdot \langle M, N, P\rangle = \frac{\partial M}{\partial x} + \frac{\partial N}{\partial y} + \frac{\partial P}{\partial z} = \frac{\partial[x^3 \ln z]}{\partial x} + \frac{\partial[xe^{-y}]}{\partial y} + \frac{\partial[-(y^2 + 2z)]}{\partial z}$$

$$= 3x^2 \ln z - xe^{-y} - 2$$

  (b)   Evaluate the divergence at the point (2, ln2, 1).

$$\nabla \cdot \mathbf{F}(2, \ln 2, 1) = 3(2)^2 \ln(1) - (2)e^{-\ln 2} - 2 = 0 - 2(e^{\ln 2})^{-1} - 2 = -2(2)^{-1} - 2 = -\frac{2}{2} - 2 = -1 - 2 = -3$$

ANSWER 20, CONTINUED...

...Answer 20, continued

Calculus III   Quiz 11 KEY, continued

(c)   Calculate the curl of F.

$$\text{CURL } \mathbf{F} =$$

$$\nabla \times \mathbf{F} = \begin{vmatrix} \mathbf{i} & \mathbf{j} & \mathbf{k} \\ \dfrac{\partial}{\partial x} & \dfrac{\partial}{\partial y} & \dfrac{\partial}{\partial z} \\ x^3 \ln z & xe^{-y} & -y^2 - 2z \end{vmatrix} = \begin{vmatrix} \dfrac{\partial}{\partial y} & \dfrac{\partial}{\partial z} \\ xe^{-y} & -y^2 - 2z \end{vmatrix} \mathbf{i} - \begin{vmatrix} \dfrac{\partial}{\partial x} & \dfrac{\partial}{\partial z} \\ x^3 \ln z & -y^2 - 2z \end{vmatrix} \mathbf{j} + \begin{vmatrix} \dfrac{\partial}{\partial x} & \dfrac{\partial}{\partial y} \\ x^3 \ln z & xe^{-y} \end{vmatrix} \mathbf{k}$$

$$= \left(\dfrac{\partial\left[-y^2 - 2z \right]}{\partial y} - \dfrac{\partial\left[xe^{-y} \right]}{\partial z} \right)\mathbf{i} - \left(\dfrac{\partial\left[-y^2 - 2z \right]}{\partial x} - \dfrac{\partial\left[x^3 \ln z \right]}{\partial z} \right)\mathbf{j} + \left(\dfrac{\partial\left[xe^{-y} \right]}{\partial x} - \dfrac{\partial\left[x^3 \ln z \right]}{\partial y} \right)\mathbf{k}$$

$$= (-2y - 0)\mathbf{i} - \left(0 - \dfrac{x^3}{z} \right)\mathbf{j} + \left(e^{-y} - 0 \right)\mathbf{k}$$

$$= \left\langle -2y, \dfrac{x^3}{z}, e^{-y} \right\rangle$$

(d)   Evaluate the curl of F at the point (1, 1, 1).

$$\nabla \times \mathbf{F}(1, 1, 1) = \left\langle -2(1), \dfrac{(1)^3}{1}, e^{-1} \right\rangle = \left\langle -2, 1, \dfrac{1}{e} \right\rangle$$

22.   Let $\mathbf{F}(x,y,z) = \left\langle \cos x, \; \sin y, \; e^{xy} \right\rangle$. Show that div (curl F) = 0.

First, we find curl **F**:

$$\nabla \times \mathbf{F} = \begin{vmatrix} \mathbf{i} & \mathbf{j} & \mathbf{k} \\ \dfrac{\partial}{\partial x} & \dfrac{\partial}{\partial y} & \dfrac{\partial}{\partial z} \\ \cos x & \sin y & e^{xy} \end{vmatrix} = \begin{vmatrix} \dfrac{\partial}{\partial y} & \dfrac{\partial}{\partial z} \\ \sin y & e^{xy} \end{vmatrix} \mathbf{i} - \begin{vmatrix} \dfrac{\partial}{\partial x} & \dfrac{\partial}{\partial z} \\ \cos x & e^{xy} \end{vmatrix} \mathbf{j} + \begin{vmatrix} \dfrac{\partial}{\partial x} & \dfrac{\partial}{\partial y} \\ \cos x & \sin y \end{vmatrix} \mathbf{k}$$

$$= \left(\dfrac{\partial\left[e^{xy} \right]}{\partial y} - \dfrac{\partial\left[\sin y \right]}{\partial z} \right)\mathbf{i} - \left(\dfrac{\partial\left[e^{xy} \right]}{\partial x} - \dfrac{\partial\left[\cos x \right]}{\partial z} \right)\mathbf{j} + \left(\dfrac{\partial\left[\sin y \right]}{\partial x} - \dfrac{\partial\left[\cos x \right]}{\partial y} \right)\mathbf{k}$$

$$= \left(xe^{xy} - 0 \right)\mathbf{i} - \left(ye^{xy} - 0 \right)\mathbf{j} + (0 - 0)\mathbf{k}$$

$$= \left\langle xe^{xy}, -ye^{xy}, 0 \right\rangle$$

Next, we take the divergence of this vector:

$$\nabla \cdot \mathbf{F} = \left\langle \dfrac{\partial}{\partial x}, \dfrac{\partial}{\partial y}, \dfrac{\partial}{\partial z} \right\rangle \cdot \langle M, N, P \rangle = \dfrac{\partial M}{\partial x} + \dfrac{\partial N}{\partial y} + \dfrac{\partial P}{\partial z} = \dfrac{\partial\left[xe^{xy} \right]}{\partial x} + \dfrac{\partial\left[-ye^{xy} \right]}{\partial y} + \dfrac{\partial[0]}{\partial z}$$

We need to use the Product Rule in order to differentiate the first and second terms of this sum:

$$= \left((1)e^{xy} + xye^{xy} \right) + \left((-1)e^{xy} - xye^{xy} \right) + 0 = 0$$

Note: When **div (curl F) = 0,** the vector is conservative.

CALCULUS III   �֎ PRACTICE **FINAL** EXAM **KEY** ✖

1. **(a)** **The vector v has magnitude 8 and direction $\theta = 120°$. Find its component form.**

Formula says: $\mathbf{v} = \langle \|\mathbf{v}\|\cos\theta, \|\mathbf{v}\|\sin\theta \rangle = \langle 8\cos 120°, 8\sin 120° \rangle = \left\langle 8\left(-\dfrac{1}{2}\right), 8\left(\dfrac{\sqrt{3}}{2}\right) \right\rangle = \langle -4, 4\sqrt{3} \rangle$ *(answer)*

    (b) **Suppose this same vector has as its initial point $\left(-2, 7\sqrt{3}\right)$. Use your answer from part (a) to find its terminal point.**

Always use "terminal minus initial point" as your mantra! So, if the terminal point has coordinates (x, y), then we need $x - (-2) = -4$, and we need $y - 7\sqrt{3} = 4\sqrt{3}$.

Solve for both x and y to get the point:
$$\left(-6, 11\sqrt{3}\right) \textit{ (answer)}$$

2. **Determine if the following pairs of vector are orthogonal, parallel, or neither. Show your work.**

    (a)   **v = <3, –2> and w = <–1, 2>**   **(b)**   **v = < –2, 0> and w = <0, 5>**

    (c)   **v = < –1, 2> and w $= \left\langle 0, -\dfrac{1}{2} \right\rangle$**   **(d)**   **v = < 2, –3> and w = <–2, 3>**

If they are orthogonal, then the dot product will be equal to zero. If they are parallel, then one must be a multiple of the other. That is, if **u** is parallel to **v**, then $\mathbf{u} = c\mathbf{v}$. The pair in **(b)** is orthogonal, since the dot product is zero. (Try it!) The pair in **(d)** is parallel since $\mathbf{v} = -\mathbf{w}$. The other pairs are "neither."

3. **Given the vectors u = < 2, –1, 1> and w = <–3, 2, 2>...**

    (a) **Calculate the angle (in degrees) between the vectors. Round your answer to the nearest hundredth.**

Use the formula $\cos\theta = \dfrac{\mathbf{u}\cdot\mathbf{w}}{\|\mathbf{u}\|\|\mathbf{w}\|} = \dfrac{\langle 2,-1,1\rangle\cdot\langle -3,2,2\rangle}{\sqrt{2^2+1^2+1^2}\,\sqrt{3^2+2^2+2^2}} = \dfrac{-6-2+2}{\sqrt{6}\sqrt{17}} = -\dfrac{6}{\sqrt{102}}$

So, $\theta = \cos^{-1}\left(\dfrac{-6}{\sqrt{102}}\right) \approx 126.45°$ *(answer)*

    (b) **Find** $\operatorname{proj}_{\mathbf{u}}\mathbf{w} = \left(\dfrac{\mathbf{u}\cdot\mathbf{w}}{\|\mathbf{u}\|^2}\right)\mathbf{u} = \left(\dfrac{-6}{\left(\sqrt{6}\right)^2}\right)\mathbf{u} = (-1)\mathbf{u} = -\langle 2,-1,1\rangle = \langle -2,1,-1\rangle$ *(answer)*

4. *Orthogonal Vectors:*

    (a) **Find a vector orthogonal to the *yz*-plane.**

Answer: The standard unit vector $\mathbf{i} = \langle 1,0,0\rangle$

    (b) **Find a vector orthogonal to the two given lines:**

$$\text{line \#1:} \begin{cases} x = -1+3t \\ y = 3-2t \\ z = 1+t \end{cases} \qquad \text{line \#2:} \begin{cases} x = 4+5t \\ y = 2-t \\ z = -1-2t \end{cases}$$

The vectors $\mathbf{v}_1 = \langle 3,-2,1\rangle$ and $\mathbf{v}_2 = \langle 5,-1,-2\rangle$ are the direction vectors for the lines, respectively.

ANSWER 4, CONTINUED...

CALCULUS III   Practice FINAL EXAM KEY, CONTINUED

The cross-product of these two vectors will be a vector orthogonal to both vectors, and hence, both:

$$\mathbf{v}_1 \times \mathbf{v}_2 = \begin{vmatrix} \mathbf{i} & \mathbf{j} & \mathbf{k} \\ 3 & -2 & 1 \\ 5 & -1 & -2 \end{vmatrix} = \begin{vmatrix} -2 & 1 \\ -1 & -2 \end{vmatrix}\mathbf{i} - \begin{vmatrix} 3 & 1 \\ 5 & -2 \end{vmatrix}\mathbf{j} + \begin{vmatrix} 3 & -2 \\ 5 & -1 \end{vmatrix}\mathbf{k} = (4+1)\mathbf{i} - (-6-5)\mathbf{j} + (-3+10)\mathbf{k} = \langle 5,11,7 \rangle \ \ (answer)$$

(c)   **Find a vector orthogonal to the plane given by** $2x - 3y + z = 11$.

Very simply, the coefficients of the variables in the equation for a plane give the components for the vector normal to the plane $\mathbf{v} = \langle 2,-3,1 \rangle$ *(answer)*

5.   **Determine the parametric equations for the line passing through the points (–3, 2, 0) and (4, 2, 3).**

$\overrightarrow{PQ} = $ terminal $-$ initial $= \langle 4-(-3), 2-2, 3-0 \rangle = \langle 7, 0, 3 \rangle$ This is the direction vector for the line. The parametric equations for a line are $x = x_1 + at$, $y = y_1 + bt$, $z = z_1 + ct$, where the direction vector is $\mathbf{v} = \langle a,b,c \rangle$. Use either point to substitute into these equations for $P(x_1, y_1, z_1)$. I'll use the first point:

$$x = -3 + 7t, \ y = 2, \ z = 3t \ \ (answer)$$

6.   **Three forces with magnitudes 3, 4, and 5 pounds act on a machine part at angles of –15°, 150°, and 220° respectively, with the positive x-axis. Find both the magnitude and direction of the resultant force. Round all answers to the nearest tenths of a unit. Be sure and include units in your final answer to get full credit.**

First, we need to write all three forces in component form:

$$\mathbf{F}_1 = \langle 3\cos(-15°), 3\sin(-15°) \rangle \approx \langle 2.90, -0.78 \rangle$$
$$\mathbf{F}_2 = \langle 4\cos(150°), 4\sin(150°) \rangle \approx \langle -3.46, 2 \rangle$$
$$\mathbf{F}_3 = \langle 5\cos(220°), 5\sin(220°) \rangle \approx \langle -3.83, -3.21 \rangle$$

Next, we sum all of the forces to get the resultant vector:

$$\mathbf{F}_1 + \mathbf{F}_2 + \mathbf{F}_3 = \langle 2.90 + (-3.46) + (-3.83), -0.78 + 2 + (-3.21) \rangle \approx \langle -4.39, -1.99 \rangle$$

Note that this resultant vector is in Quadrant III. This is important, especially when we try and find the direction (angle) that it makes with the positive x-axis. The magnitude of this resultant vector is given by:

$$\|\mathbf{F}_1 + \mathbf{F}_2 + \mathbf{F}_3\| = \sqrt{(-4.39)^2 + (-1.99)^2} \approx 4.8 \ lbs.$$

To find the direction, we first take the inverse tangent of the quotient of the y-component divided by the x-component:

$$\theta = \tan^{-1}\left(\frac{y}{x}\right) = \tan^{-1}\left(\frac{-4.39}{-1.00}\right) \approx 24.4°$$

This is not correct, since the angle of 24.4° is in Quadrant I. The reason why our calculator gives us this answer is because the range of the arctangent function lies in the interval $\theta \in \left[\dfrac{-\pi}{2}, \dfrac{\pi}{2}\right]$. To get the correct angle (lying in Quadrant III), we add 180° to our result to get:

$$\text{Direction} = \text{angle} = \theta = 24.4° + 180° = 204.4°$$

Final answer: The magnitude is 4.8 pounds and the direction is 204.4°

$$\sqrt{}$$ **PART III** TEST ANSWER KEY

CALCULUS III   **Practice FINAL EXAM KEY,** CONTINUED

---

7. **Consider the following plane curves. Eliminate the parameter and represent each curve by a vector-valued function $\mathbf{r}(t) = \langle x(t), y(t) \rangle$.**

    (a) $x = y^2 - 1$

    Let $y = t$, then $x = t^2 - 1$.

    So, $\mathbf{r}(t) = \langle t^2 - 1, t \rangle$ *(answer)*

    Caution! Do *not* use $x = t$, because then there will be two answers for y:
    $$y = \pm\sqrt{x+1}$$
    (and that's not very "nice.")

    (b) $y = x^2 - 1$

    Let $x = t$, then $y = t^2 - 1$, so we have the vector-valued function $\mathbf{r}(t) = \langle t, t^2 - 1 \rangle$.

8. **Find the domain of the vector-valued function $\mathbf{r}(t) = \left\langle \ln(3-t),\ \dfrac{\sqrt[4]{2+t}}{\ln|t|},\ \dfrac{t+1}{6t^2 - 7t - 3} \right\rangle$.**

    Write your answer answer using interval notation to get full credit.

    The first component contains a logarithm, so we "need to worry." The argument of the logarithm must be positive, so we require that:
    $$3 - t > 0$$
    $$3 > t$$
    $$or: \ t < 3$$

    The second component has two "issues." First, we need to guarantee that the radicand of the even root is never negative. In other words $2 + t \geq 0$, or $t \geq -2$.

    Also, we never want the denominator to be equal to zero. The denominator $\ln|t|$ will not equal zero so long as the argument t is never equal to 1. That is, since $\ln|1| = 0$, we want $t \neq 1$. Further, we need the argument to be non-zero, so $t \neq 0$ also. We do *not* need to worry about this particular argument being negative as we did for the logarithm in the first component because of the absolute value symbol. In other words, this logarithm is OK with negative numbers since they will "automatically" become positive due to the absolute value.

    Finally, the last component has a denominator we need to worry about becoming zero. First we set the denominator equal to zero, solve the quadratic equation, and then eliminate the solutions for this equation from the domain. We will solve by factoring:

    $$6t^2 - 7t - 3 = 0$$
    $$6t^2 + 2t - 9t - 3 = 0$$
    $$2t(3t + 1) - 3(3t + 1) = 0$$
    $$(2t - 3)(3t + 1) = 0$$
    $$2t - 3 = 0 \quad or \quad 3t + 1 = 0$$
    $$t = \frac{3}{2}, \qquad t = \frac{-1}{3}$$

ANSWER 8, CONTINUED...

...Answer 8, continued

Calculus III Practice Final Exam KEY, continued

This means we exclude the solutions from our domain. So we have $t \neq \dfrac{3}{2}$, $t \neq \dfrac{-1}{3}$.

Finally, we intersect all four of our sets of domains for individual functions, and we have the final result:

$$t \in \left[-2, -\frac{1}{3}\right) \cup \left(-\frac{1}{3}, 0\right] \cup (0, 1) \cup \left(1, \frac{3}{2}\right) \cup \left(\frac{3}{2}, 3\right] \quad (answer)$$

9. **Evaluate** $\displaystyle \int \left\langle t \ln t, \ \sqrt{1+5t}, \ \frac{t}{t+1} \right\rangle dt$.

This is an indefinite integral, so the final answer will have a constant *vector* of **C**. If it were a definite integral, we would evaluate each anti-derivative at the limits.

The integral $\int t \ln t\, dt$ requires integration by parts, where $u = \ln t$, $dv = t\, dt$, $du = \dfrac{1}{t} dt$, $v = \dfrac{t^2}{2}$.

Then, we have $\int t \ln t\, dt = uv - \int v\, du = (\ln t)\left(\dfrac{t^2}{2}\right) - \int \left(\dfrac{t^2}{2}\right)\left(\dfrac{1}{t}\right) dt = \dfrac{t^2 \ln t}{2} - \dfrac{1}{2}\int t\, dt = \dfrac{t^2 \ln t}{2} - \dfrac{t^2}{4} + C_1$.

The integral $\int \sqrt{1+5t}\, dt$ can be evaluated using the general power rule, where $u = 1+5t$, $du = 5\, dt$.

So, we have $\int \sqrt{1+5t}\, dt = \dfrac{1}{5}\int \sqrt{1+5t}\,(5) dt = \dfrac{1}{5}\int \sqrt{u}\, du = \dfrac{1}{5}\int u^{1/2}\, du = \dfrac{1}{5}\left(\dfrac{2}{3}\right)u^{3/2} = \dfrac{2}{15}(1+5t)^{3/2} + C_2$.

The last integral has an improper rational function in its integrand, and so requires polynomial long division first:

$$\begin{array}{r} 1 \\ t+1\overline{)t} \\ \underline{t+1} \\ -1 \end{array}$$

Write in the form of:

$$Quotient + \frac{rem}{divisor} = 1 + \frac{-1}{t+1}$$

The integral now looks like $\displaystyle \int \left(1 - \frac{1}{t+1}\right) dt = t - \ln|t+1| + C_3$.

The final answer must be written as a vector:

$$\left\langle \frac{t^2 \ln t}{2} - \frac{t^2}{4}, \ \frac{2}{15}(1+5t)^{3/2}, \ t - \ln|t+1| \right\rangle + \langle C_1, C_2, C_3 \rangle$$

Or, you can write the constant vector differently:

$$\left\langle \frac{t^2 \ln t}{2} - \frac{t^2}{4}, \ \frac{2}{15}(1+5t)^{3/2}, \ t - \ln|t+1| \right\rangle + \mathbf{C} \quad (answer)$$

10. **An object starts from rest at the point (0, 1, 1) and moves with an acceleration $\mathbf{a}(t) = \langle 1, \cos t, 0 \rangle$. Find the position, $\mathbf{r}(t)$, at time $t = 4$.**

Starting from rest means the initial velocity is $\mathbf{v}(0) = \langle 0, 0, 0 \rangle$. We find velocity first by integrating acceleration:

$$\mathbf{v}(t) = \int \mathbf{a}(t) dt = \int \langle 1, \cos t, 0 \rangle dt = \langle t, \sin t, 0 \rangle + \langle C_1, C_2, C_3 \rangle$$

To find the constants, we use the initial condition for velocity:

$$\mathbf{v}(0) = \langle 0, 0, 0 \rangle = \langle 0 + C_1, \sin(0) + C_2, 0 + C_3 \rangle$$

We equate the components and find that $C_1 = 0, C_2 = 0, C_3 = 0$.

Now, we have the complete velocity function $\mathbf{v}(t) = \langle t, \sin t, 0 \rangle$.

CALCULUS III   Practice FINAL EXAM KEY, CONTINUED

We integrate velocity now to find the position function:

$$\mathbf{r}(t) = \int \mathbf{v}(t)dt = \int \langle t, \sin t, 0 \rangle dt = \left\langle \frac{t^2}{2} + C_{11}, -\cos t + C_{22}, 0 + C_{33} \right\rangle$$

We use the initial condition for position in order to find the new sets of constants of integration:

$$\mathbf{r}(0) = \langle 0, 1, 1 \rangle = \left\langle \frac{0^2}{2} + C_{11}, -\cos(0) + C_{22}, C_{33} \right\rangle$$

Equating the components for these vectors gives us that $C_{11} = 0, C_{22} = 2, C_{33} = 1$.

Our final answer is now $\mathbf{r}(t) = \left\langle \dfrac{t^2}{2}, -\cos(t) + 2, 1 \right\rangle$

11.   **Given the vector-valued function $\mathbf{r}(t) = \left\langle t + 4, 1 - t^2 \right\rangle$ ...**

(a)   **Sketch the graph, be sure and identify all intercepts.**

The easiest way to graph this vector-valued function is to use your graphing calculator in parametric mode. Another way is to make a table with two columns: one for time t, and the other for the corresponding ordered pair, (x, y).

The x-intercept(s) can be found by setting the y-component equal to 0, and solving for t:

$$1 - t^2 = 0$$
$$(1 + t)(1 - t) = 0$$
$$t = \pm 1$$

The position for when $t = \pm 1$ occurs when the vector-valued function is evaluated at these two points: $\mathbf{r}(1) = \langle 1 + 4, 0 \rangle$ and $\mathbf{r}(-1) = \langle -1 + 4, 0 \rangle$. That is, at the points $(5, 0)$ and $(3, 0)$.

The y-intercept(s) can be found by setting the x-component equal to 0: $t + 4 = 0$. So we have $t = -4$. This occurs at the position $\mathbf{r}(-4) = \langle 0, 1 - (-4)^2 \rangle = \langle 0, -15 \rangle$. So, the y-intercept occurs at the point $(0, -15)$.

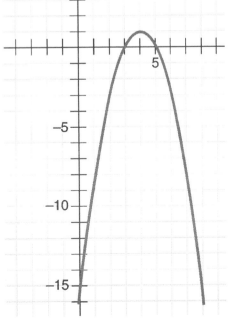

ANSWER 11, CONTINUED...

CALCULUS III   **Practice FINAL EXAM KEY, CONTINUED**

(b)   **Evaluate the velocity vector when $t = 2$, and sketch it on the same graph at the appropriate position.**

First we find the velocity vector, which is the first derivative of the vector-valued function $\mathbf{r}'(t) = \langle 1, -2t \rangle$.

Next, we evaluate this at the specified time, $t = -2$: $\mathbf{r}'(-2) = \langle 1, -2(-2) \rangle = \langle 1, 4 \rangle$. We sketch this vector at the appropriate position. This position is found by evaluating the position vector given, $\mathbf{r}(t)$ at the time $t = -2$. That is, the position is $\mathbf{r}(-2) = \langle -2 + 4, 1 - (-2)^2 \rangle = \langle 2, -3 \rangle$.

So, sketch the vector <1, 4> where the initial point of the vector is located at the point $(2, -3)$ on the graph. Remember: velocity vectors are always *tangent* to the line of motion. So, if your velocity vector does not appear to be tangent to the curve, you probably did something wrong!

12. *Projectile Motion.* **A projectile is fired at a height of 2 meters above the ground with an initial velocity of 100 meters per second at an angle of 35° with the horizontal. Round each result to the nearest tenths of a unit.**

(a)   **Find the vector-valued function describing the motion.** *Hint:* **Use $g = 9.8$ meters per second per second.**

Use the Projectile Motion Theorem $\mathbf{r}(t) = \left\langle v_0 \cos\theta\, t, \ h + v_0 \sin\theta\, t - \dfrac{1}{2} g t^2 \right\rangle$.

So using $v_0 = 100 m/\sec, h = 2m, \ g = 9.8 m/\sec^2, \theta = 35°$, we have:

$$\mathbf{r}(t) = \left\langle 100\cos(35°)t, \ 2 + 100\sin(35°)t - \frac{1}{2}(9.8)t^2 \right\rangle \approx \langle 81.915t, \ 2 + 57.358t - 4.9t^2 \rangle$$

(b)   **Find the maximum height.**

Maximum height occurs when vertical component of velocity is equal to zero. The velocity vector is given by $\mathbf{v}(t) = \mathbf{r}'(t) = \langle 81.915, \ 57.358 - 9.8t \rangle$. Set the vertical component equal to zero. Be careful! The solution for this equation will give the *time* it takes for the projectile to achieve its maximum height, and *not* the actual maximum height. You then plug this value for time into the vertical component of the position function to obtain the maximum height:

$$57.358 - 9.8t = 0$$

$$t = \frac{57.358}{9.8} \approx 5.85 \text{ seconds}$$

Please note that you *do not* want to round to tenths just yet. Wait until the final result to round your answer to the nearest tenths, else your answer will not be as accurate.

Then, plug $t = 5.85$ seconds into the y-component of $\mathbf{r}(t)$ to get:

$$2 + 57.35(5.85) - 4.9(5.85)^2 \approx 169.8 \text{ meters}$$

(c)   **How long was the projectile in the air?**

The projectile will hit the ground and end its journey when height = 0. So, we will set the vertical component of the position function equal to zero and solve for time t. This will give us how long the projectile was in the air.

Set the y-component of $\mathbf{r}(t)$ equal to zero to get $2 + 57.35t - 4.9t^2 = 0$.

CALCULUS III   **Practice FINAL EXAM KEY,** CONTINUED

This is a quadratic equation. We will solve it using the quadratic formula, where

$t = \dfrac{-b \pm \sqrt{b^2 - 4ac}}{2a}$, where $a = -4.9$, $b = 57.35$, and $c = 2$.

We get two solutions: $t = 11.739$ seconds and $t = -0.034$ seconds (which is impossible here).

Answer: The projectile was in the air for a total of time $t = 11.739$ seconds.

(d)   **Find the range.**

To get the range, we plug our result from part (c) into the horizontal component of the position function to get $(81.915)(11.739)$ which is approximately equal to 961.6 meters. *(answer)*

13.   **Find the unit tangent vector, T(t), for the curve given by** $\mathbf{r}(t) = \langle 4\cos t, -3\sin t, 1 \rangle$ **when** $t = \dfrac{3\pi}{2}$.

We use the definition:

$$\mathbf{T}(t) = \frac{\mathbf{r}'(t)}{\|\mathbf{r}'(t)\|} = \frac{\langle -4\sin t, -3\cos t, 0 \rangle}{\sqrt{(-4\sin t)^2 + (-3\cos t)^2 + (0)^2}} = \frac{\langle -4\sin t, -3\cos t, 0 \rangle}{\sqrt{16\sin^2 t + 9\cos^2 t}}$$

Evaluating at time $t = \dfrac{3\pi}{2}$, we have that:

$$\mathbf{T}\left(\frac{3\pi}{2}\right) = \frac{\langle -4\sin(3\pi/2), -3\cos(3\pi/2), 0 \rangle}{\sqrt{16\sin^2(3\pi/2) + 9\cos^2(3\pi/2)}} = \frac{\langle (-4)(-1), (-3)(0), 0 \rangle}{\sqrt{16(1) + 9(0)}} = \frac{\langle 4, 0, 0 \rangle}{\sqrt{16}} = \langle 1,0,0 \rangle \; \textit{(answer)}$$

14.   **Find the length of the curve** $\mathbf{r}(t) = \langle e^t, e^{-t}, \sqrt{2}t \rangle$, **when [0, 2].**

Use the formula for arclength:

$$s = \int_a^b \|\mathbf{r}'(t)\| \, dt = \int_0^2 \left\| \langle e^t, -e^{-t}, \sqrt{2} \rangle \right\| dt = \int_0^2 \sqrt{(e^t)^2 + (-e^{-t})^2 + (\sqrt{2})^2} \, dt = \int_0^2 \sqrt{e^{2t} + e^{-2t} + 2} \, dt = \int_0^2 \sqrt{e^{2t} + \frac{1}{e^{2t}} + 2} \, dt = \int_0^2 \sqrt{\frac{e^{4t} + 1 + 2e^{2t}}{e^{2t}}} \, dt$$

The numerator in the radicand factors: We integrate now:

$$\int_0^2 \sqrt{\frac{(e^{2t} + 1)^2}{e^{2t}}} \, dt = \int_0^2 \frac{e^{2t} + 1}{e^t} \, dt = \int_0^2 \left(\frac{e^{2t}}{e^t} + \frac{1}{e^t} \right) dt = \int_0^2 (e^t + e^{-t}) \, dt$$

$$\left[e^t - e^{-t} \right]_0^2 = (e^2 - e^{-2}) - (e^0 - e^0) = e^2 - e^{-2}$$
(answer)

15.   *Domain for a function of 2 variables.* **Find the domain for the given function and write the answer using set notation:**

$$f(x, y) = \frac{\ln(x - y)}{e^{1/x}} + \sin(x - y^2) + \sqrt[3]{x + 3y} + \sqrt{x^2 + 2y^3} + \frac{x}{6x - 7y}$$

Answer: $\left\{ (x, y) \mid x - y > 0, \, x \neq 0, \, x^2 + 2y^3 \geq 0, \, 6x - 7y \neq 0 \right\}$

CALCULUS III   **Practice FINAL EXAM KEY**, CONTINUED

---

16. **Find the f_{xy} for $f(x,y) = x^2y + 2y^2x^2 + 4x$.**

We first find the first partial: $f_x = 2xy + 4xy^2 + 4$.

To find the second partial, f_{xy}, we integrate the first partial w.r.t. y and get:

$$f_{xy} = 2x + 8xy \quad (answer)$$

17. **Find the first partial derivative with respect to x: $F(x,y,z) = xe^{xyz}$.**

We need to use the Product Rule, and take the derivative with respect to x. This means that we treat both the variables y and z as constants:

$$F_x(x,y,z) = (1)e^{xyz} + x(yz)e^{xyz} = e^{xyz}(1 + xyz)$$

18. **Use the total differential dz to approximate the change in $z = \dfrac{y}{x}$ as (x, y) moves from the point $(2, 1)$ to the point $(2.1, 0.8)$. Then, calculate the actual change Δz.**

First, we find the total differential that is given by $dz = \dfrac{\partial z}{\partial x}dx + \dfrac{\partial z}{\partial y}dy$.

So, we find the partial derivatives first: $\dfrac{\partial z}{\partial x} = (-1)yx^{-2} = -\dfrac{y}{x^2}$ and $\dfrac{\partial z}{\partial y} = \dfrac{1}{x}$.

Next, we note that $dx = \Delta x = 2.1 - 2 = 0.1$ and also that $dy = \Delta y = 0.8 - 1 = -0.2$.

Then we have the total differential evaluated at the starting point of $(2,1)$:

$$dz = -\dfrac{1}{(2)^2}(0.1) + \dfrac{1}{2}(-0.2) = -0.125 = -\dfrac{1}{8}$$

The actual change for z if given by:

$$\Delta z = f(2.1, 0.8) - f(2, 1) = \dfrac{0.8}{2.1} - \dfrac{1}{2} = -0.1190476 = -\dfrac{5}{42} \quad \text{(A pretty close approximation!)}$$

19. **The radius of a right circular cylinder is decreasing at the rate of 4 inches per minute and the height is increasing at the rate of 8 inches per minute. What is the rate of change of the volume when r = 4 inches and h = 8 inches? (*Hint:* Use Chain Rule for function of several variables.)**

The volume of a right circular cylinder is a function of two variables: $V(r,h) = \pi r^2 h$.

Using the Chain Rule for Functions of Several Variables, we differentiate both sides of this "equation" with respect to time t:

$$\dfrac{d[V(r,h)]}{dt} = \dfrac{\partial V}{\partial r}\cdot\dfrac{dr}{dt} + \dfrac{\partial V}{\partial h}\cdot\dfrac{dh}{dt}$$

Taking all of the partial derivatives as well as the "regular" ones, we find that:

$$\dfrac{dV}{dt} = (2\pi hr)\cdot\dfrac{dr}{dt} + (\pi r^2)\cdot\dfrac{dh}{dt}$$

Now, we use all of the info given, that is:

$$\dfrac{dr}{dt} = -4in/\min, \quad \dfrac{dh}{dt} = 8in/\min, \quad r = 4in, \quad h = 8in$$

$$\dfrac{dV}{dt} = (2\pi)(8)(4)\cdot(-4) + \pi(4)^2\cdot(8) = -128\pi \text{ cubic inches per minute } (answer)$$

CALCULUS III   **Practice FINAL EXAM KEY,** CONTINUED

---

20. **Find the directional derivative of** $f(x, y) = x^2 y$ **at the point (1, –3) in the direction < –2, 1>.**

We first need to normalize the direction vector. That is, form a unit vector out of the direction vector:

$$\mathbf{u} = \frac{\langle -2,1 \rangle}{\sqrt{2^2 + 1^2}} = \left\langle \frac{-2}{\sqrt{5}}, \frac{1}{\sqrt{5}} \right\rangle$$

Next, we find the first partials for the function: $f_x = 2xy,\ f_y = x^2$.

Then, we evaluate these partials at the point given, that is (1, –3):

$$f_x\big|_{(1,-3)} = 2(1)(-3) = -6, \quad f_y\big|_{(1,-3)} = (1)^2 = 1$$

We form the gradient now:

$$\nabla f(x, y) = \langle f_x, f_y \rangle = \langle 2xy, x^2 \rangle$$

At the point (1, –3), the gradient is the vector $\nabla f(1,-3) = \langle -6,1 \rangle$.

Finally, we use the definition for the directional derivative that is:

$$D_{\mathbf{u}} f(x, y) = \nabla f(x, y) \cdot \langle \cos \theta, \sin \theta \rangle = \langle -6,1 \rangle \cdot \left\langle \frac{-2}{\sqrt{5}}, \frac{1}{\sqrt{5}} \right\rangle = (-6)\left(\frac{-2}{\sqrt{5}}\right) + (1)\left(\frac{1}{\sqrt{5}}\right) = \frac{13}{\sqrt{5}} \ \textit{(answer)}$$

Note: this result gives the *slope* of the surface in the direction of the direction vector!

21. **Given the surface** $2x^2 + 3y^2 + 4z^2 = 18$ ...

(a) **Use implicit differentiation and find the slope in the x-direction,** $\dfrac{\partial z}{\partial x}$**, at the point (–1, 2, 1).**

Use the definition $\dfrac{\partial z}{\partial x} = -\dfrac{F_x}{F_z}$, where $F(x, y, z) = 2x^2 + 3y^2 + 4z^2 - 18 = 0$.

Therefore, $\dfrac{\partial z}{\partial x} = -\dfrac{F_x}{F_z} = -\dfrac{4x}{8z}$.

Then at the point (–1, 2, 1), we have the slope as:

$$\frac{\partial z}{\partial x}\bigg|_{(-1,2,1)} = -\frac{4(-1)}{8(1)} = \frac{1}{2} \ \textit{(answer)}$$

(b) **Find an equation of the tangent plane (in general form) to the surface at the point (–1, 2, 1).**

The gradient of the function $F(x, y, z)$ found in part **(a)** above will be a vector orthogonal to the surface that we will use to form the equation of the plane:

$$\nabla F(x, y, z) = \langle F_x, F_y, F_z \rangle = \langle 4x, 6y, 8z \rangle$$

Evaluate at the point given: $\nabla F(-1,2,1) = \langle 4(-1), 6(2), 8(1) \rangle = \langle -4, 12, 8 \rangle$.

Now we use the format for the equation of a plane, which is: $a(x - x_0) + b(y - y_0) + c(z - z_0) = 0$, where the vector orthogonal to the plane has components a, b, and c.

So, $-4(x - (-1)) + 12(y - (2)) + 8(z - (1)) = 0$.

Distribute and collect like terms to get the equation into general form:

$$-x + 3y + 2z = 9 \ \textit{(answer)}$$

Calculus III — Practice FINAL EXAM KEY, continued

22. **Find a set of parametric equations for the normal line to the surface given by** $z = f(x, y) = x^2 y$ **at the point (2, 1, 4).**

First rewrite the equation in the form $F(x, y, z) = 0$. We will do this by subtracting $x^2 y$ from both sides: $z - x^2 y = 0$.

We then find the gradient $\nabla F(x, y, z) = \langle F_x(x, y, z), F_y(x, y, z), F_z(x, y, z) \rangle$.

This gradient vector will be a vector that is normal to the surface. It will also be a vector that will serve as the direction vector for our normal line to the surface.

So, finding the first partials we obtain $\nabla F(x, y, z) = \langle -2xy, -x^2, 1 \rangle$.

We evaluate the gradient at the given point, (2, 1, 4):
$$\nabla F(2, 1, 4) = \langle -2(2)(1), -(2)^2, 1 \rangle = \langle -4, -4, 1 \rangle$$

Now we have all of the information we need to write the set of parametric equations for our line that is normal to the surface. We use where the direction vector is given by $\mathbf{v} = \langle a, b, c \rangle$ and a point on the line given by (x_0, y_0, z_0).

$$\Longrightarrow \begin{cases} x = x_0 + at \\ y = y_0 + bt \\ z = z_0 + ct \end{cases}$$

The answer is: $\begin{cases} x = 2 - 4t \\ y = 1 - 4t \\ z = 4 + t \end{cases}$

23. **Find extrema and saddle point(s), if any, for the function** $f(x, y) = x^2 + x - 3xy + y^3 - 5$. **Write your answer(s) in ordered triple(s) to get full credit.**

We first find the first partials. Then we form a system of equations by setting both partial simultaneously equal to zero, and solve the system.

The solution(s) of this system give the critical point(s) $f_x = 2x + 1 - 3y = 0$, $f_y = -3x + 3y^2 = 0$.

This is a non-linear system, which is most easily solved using the method of substitution. Using the second equation, solve for x, then substitute into the first equation:
$$3y^2 = 3x \Rightarrow x = y^2 \Rightarrow 2(y^2) - 3y + 1 = 0$$

This is a quadratic equation. We can solve by either using the quadratic formula, or by factoring:
$$(2y - 1)(y - 1) = 0 \Rightarrow 2y - 1 = 0, \quad or \quad y - 1 = 0 \Rightarrow y = \frac{1}{2}, 1$$

Since $x = y^2$, we get the corresponding x-coordinates, and find that the solutions to the system, which are exactly the critical points $(1, 1)$ *and* $\left(\frac{1}{4}, \frac{1}{2} \right)$.

Now we form D for the D-Test: $D = (f_{xx})(f_{yy}) - (f_{xy})^2 = (2)(6y) - (-3)^2 = 12y - 9$

We perform the D-test by evaluating D at each of the critical points: $D(1,1) = 12(1) - 9 = 3 > 0$.

Since $D > 0$, we have an extrema. To find out what type, we evaluate the second partial f_{xx} at the point $f_{xx}(1,1) = 2 > 0$.

Since it's also positive, we have a relative minimum at the point $(1, 1, -5)$.

Answer 23, continued...

CALCULUS III   **Practice FINAL EXAM KEY, CONTINUED**

Now, test the other point by evaluating D:

$$D\left(\frac{1}{4},\frac{1}{2}\right)=12\left(\frac{1}{2}\right)-9=-3<0$$

We have a saddle point at $\left(\frac{1}{4},\frac{1}{2},-\frac{79}{16}\right)$. *(answer)*

24. **Use Lagrange Multipliers to find the dimensions of a rectangular box of maximum volume with one vertex at the origin and the opposite vertex lying in the plane given by $6x+4y+3z=24$. Then, give the actual maximum volume.**

The function we are trying to maximize is the volume function for a rectangular box. For this problem where one vertex is at the origin, we will have width $= x$, length $= y$, and height $= z$. So, the volume of the box is given by $f(x,y,z)=xyz$.

The constraint equation needs to be rewritten as $g(x,y,z)=0$. The plane equation is our constraint equation. We rewrite it $6x+4y+3z-24=0$.

Next, we form the Lagrange function:

$$L(x,y,z,\lambda)=f(x,y,z)-\lambda g(x,y,z)$$

Using our function and constraint, our Lagrange function looks like:

$$L(x,y,z,\lambda)=xyz-\lambda(6x+4y+3z-24)$$

Next, we find all four first partial derivatives for the Lagrange function. We set all four of these partials equal to zero, and then solve the resulting (non-linear in this case) system of equations:

$$L_x=yz-6\lambda=0$$
$$L_y=xz-4\lambda=0$$
$$L_z=xy-3\lambda=0$$
$$L_\lambda=-(6x+4y+3z-24)=0$$

Notice that the last partial is simply our constraint equation $g(x,y,z)=0$, if you go ahead and rewrite it. By the way, this will always be the case when using the method of Lagrange Multipliers for the last partial (with respect to lambda—λ).

There are several approaches in attempting to solve this system. The strategy I will use is the following: Solve for λ using the first three equations. Use these to get both y and z in terms of x. Substitute these into the constraint equation to find x:

$$\lambda=\frac{yz}{6}$$

$$\lambda=\frac{xz}{4}$$

$$\lambda=\frac{xy}{4}$$

ANSWER 24, CONTINUED...

...ANSWER 24, CONTINUED

CALCULUS III   Practice FINAL EXAM KEY, CONTINUED

Setting the first form for l equal to the second one: $1 = \dfrac{yz}{6} = \dfrac{xz}{4}$. Next divide both sides by z to get that $y = \dfrac{3}{2}x$.

Next, set the second and third equations equal to each other: $\dfrac{xz}{4} = \dfrac{xy}{3}$.

Then divide both sides by x to get that $z = \dfrac{4}{3}y = \dfrac{4}{3}\left(\dfrac{3}{2}x\right) = 2x$.

Substitute both $y = \dfrac{3}{2}x$ and $z = 2x$ into the constraint equation and solve for x:

$$6x + 4\left(\frac{3}{2}x\right) + 3(2x) = 24$$

$$6x + 6x + 6x = 24$$

$$18x = 24$$

$$x = \frac{24}{18} = \frac{4}{3}$$

Now, we get the other dimensions for the box:

$$y = \frac{3}{2}x = \left(\frac{3}{2}\right)\left(\frac{4}{3}\right) = 2$$

$$z = 2x = 2\left(\frac{4}{3}\right) = \frac{8}{3}$$

Therefore the dimensions are $x = \dfrac{4}{3}$, $y = 2$, $z = \dfrac{8}{3}$.

The maximum volume for the box is $f\left(\dfrac{4}{3}, 2, \dfrac{8}{3}\right) = \left(\dfrac{4}{3}\right)(2)\left(\dfrac{8}{3}\right) = \dfrac{64}{9}$ cubic units. *(answer)*

25. (a) **Evaluate the integral:**

$$\int_0^{\ln 2}\int_{e^y}^2 x\,dx\,dy = \int_0^{\ln 2}\left[\frac{x^2}{2}\right]_{e^y}^2 dy = \int_0^{\ln 2}\left(\frac{2^2}{2} - \frac{(e^y)^2}{2}\right)dy = \int_0^{\ln 2}\left(2 - \frac{e^{2y}}{2}\right)dy = \left[2y - \frac{e^{2y}}{4}\right]_0^{\ln 2} = \left[2(\ln 2) - \frac{e^{2(\ln 2)}}{4}\right] - \left(2(0) - \frac{e^{(2)(0)}}{4}\right)$$

$$= 2\ln 2 - \frac{2^2}{4} + \frac{1}{4} = 2\ln 2 - \frac{3}{4} \; \textit{(answer)}$$

Note: We used the fact that $e^{\ln x} = x$ in the last step.

ANSWER 25, CONTINUED...

...ANSWER 25, CONTINUED

CALCULUS III   Practice FINAL EXAM KEY, CONTINUED

(b) Evaluate the integral $\int\limits_{0}^{1}\int\limits_{y}^{1} x^2 e^{xy}\, dxdy$. (*Hint:* Reverse order of integration first.)

We sketch the region of integration by graphing the equations $y = 0$, $y = 1$, $x = y$, *and* $x = 1$. It is a triangle with vertices: $(0, 0)$, $(1, 0)$, and $(1, 1)$.

To switch the order of integration, we'll require the variable x to have constants as limits, so, $0 \le x \le 1$, and $0 \le y \le x$.

We have $\int\limits_{0}^{1}\int\limits_{0}^{x} x^2 e^{xy}\, dydx$.

To integrate, we use the exponential rule, where $u = xy$, and $du = xdx$. Save one factor of x^2 for the u-substitution, and the other we'll treat as a constant multiple:

$$\int\limits_{0}^{1} x\int\limits_{0}^{x} xe^{xy}\, dydx = \int\limits_{0}^{1} x\left[\int e^u du\right]dx = \int\limits_{0}^{1} x\left[e^{xy}\right]_0^x dx = \int\limits_{0}^{1} x\left[e^{x^2} - e^0\right]dx = \int\limits_{0}^{1} \left(xe^{x^2} - x\right)dx .$$

We'll use u-substitution again with the exponential rule, where $u = x^2$, $du = 2xdx$. Now, we have:

$$\frac{1}{2}\int\limits_{0}^{1} 2xe^{x^2}\, dx - \int\limits_{0}^{1} xdx = \left[\frac{1}{2}e^{x^2} - \frac{x^2}{2}\right]_0^1 = \frac{1}{2}e - \frac{1}{2} - \left(\frac{1}{2}e^0 - 0\right) = \frac{1}{2}e - 1 \ (answer)$$

26. Evaluate the double integral $\int\limits_{-1}^{1}\int\limits_{0}^{\sqrt{1-x^2}} e^{x^2+y^2}\, dydx$ by changing to polar coordinates.

First, we'll sketch the region of integration in order to be able to see what the limits for both r and θ will be. We see that $0 \le \theta \le \pi$, and that $0 \le r \le 1$.

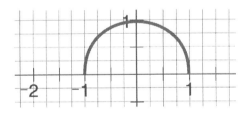

$$\int\limits_{0}^{\pi}\int\limits_{0}^{1} e^{r^2}\, rdrd\theta = \frac{1}{2}\int\limits_{0}^{\pi}\int\limits_{0}^{1} e^{r^2}\, 2rdrd\theta = \frac{1}{2}\int\limits_{0}^{\pi}\left[e^{r^2}\right]_0^1 d\theta = \frac{1}{2}\int\limits_{0}^{\pi}(e-1)d\theta$$

$$= \frac{1}{2}(e-1)[\theta]_0^\pi = \frac{1}{2}(e-1)\pi \ (answer)$$

27. Set up the triple integrals (do not evaluate) that would calculate the volume of the solid bounded by the graphs of $z = 0$, $x^2 + y^2 = 16$, and $z = 5 - y$ using...

(a) rectangular coordinates

$$V = \int\limits_{-4}^{4}\int\limits_{-\sqrt{16-x^2}}^{\sqrt{16-x^2}}\int\limits_{0}^{5-y} dzdydx \ (answer)$$

(b) cylindrical coordinates

$$V = \int\limits_{0}^{2\pi}\int\limits_{0}^{4}\int\limits_{0}^{5-r\sin\theta} dzrdrd\theta \ (answer)$$

CALCULUS III   Practice FINAL EXAM KEY, CONTINUED

28. **Find work done by the force** $\mathbf{F}(x, y, z) = \langle y - x^2, z - y^2, x - z^2 \rangle$ **over the curve** $\mathbf{r}(t) = \langle t, t^2, t^3 \rangle$ **from the point (0, 0, 0) to (1, 1, 1).**

We cannot use the Fundamental Theorem of Line Integrals because the vector field is not conservative. Instead, we use $Work = \int_C \mathbf{F} \cdot d\mathbf{r}$, where we have that $d\mathbf{r} = \mathbf{r}'(t) = \langle 1, 2t, 3t^2 \rangle$.

We then replace x, y, and z in the vector field with the components from $\mathbf{r}(t) = \langle t, t^2, t^3 \rangle$, and have:

$$\int_C \mathbf{F} \cdot d\mathbf{r} = \int_0^1 \langle t^2 - t^2, t^3 - (t^2)^2, t - (t^3)^2 \rangle \cdot \langle 1, 2t, 3t^2 \rangle dt = \int_0^1 \langle 0, t^3 - t^4, t - t^6 \rangle \cdot \langle 1, 2t, 3t^2 \rangle dt = \int_0^1 (2t^4 - 2t^5 + 3t^3 - 3t^8) dt$$

Integrate and evaluate, and get: $= \left[\frac{2}{5}t^5 - \frac{1}{3}t^6 + \frac{3}{4}t^4 - \frac{1}{3}t^9 \right]_0^1 = \frac{29}{60}$ *(answer)*

29. **Evaluate** $\int_C \frac{x + y^2}{\sqrt{1 + x^2}} ds$, **where** $C = C_1 + C_2$.

The curve C_1 is the straight line segment from the point $(1, 0)$ to the point $\left(1, \frac{1}{2}\right)$.

C_2 is the curve along the graph of $y = \frac{x^2}{2}$ from the point $\left(1, \frac{1}{2}\right)$ to $(0, 0)$.

We first parameterize each of the curves: $C_1 : \mathbf{r}_1(t) = \langle 1, t \rangle, 0 \le t \le \frac{1}{2}$.

So, the arclength element for this parameterization is $ds = \left\| \mathbf{r}_1'(t) \right\| dt = \sqrt{0^2 + 1^2} = 1 dt$.

Parameterizing the second curve for the path, we have that:

$$C_2 : \mathbf{r}_2(t) = \left\langle 1 - t, \frac{(1-t)^2}{2} \right\rangle, \quad 0 \le t \le 1$$

Taking the derivative of $\mathbf{r}_2$, we get that:

$$\mathbf{r}_2'(t) = \left\langle -1, \frac{2(1-t)}{2}(-1) \right\rangle = \langle -1, t - 1 \rangle$$

Next, we calculate the arclength element for this parameterization:

$$ds = \left\| \mathbf{r}_2'(t) \right\| dt = \sqrt{(-1)^2 + (t-1)^2} dt = \sqrt{1 + (t-1)^2} dt$$

Then, we evaluate the line integral where we rewrite it as a sum of two separate integrals (with all of the appropriate replacements and substitutions):

$$\int_C \frac{x + y^2}{\sqrt{1 + x^2}} ds = \int_{C_1} \frac{x + y^2}{\sqrt{1 + x^2}} ds + \int_{C_2} \frac{x + y^2}{\sqrt{1 + x^2}} ds = \int_0^{1/2} \frac{1 + t^2}{\sqrt{1 + (1)^2}} \sqrt{0^2 + 1^2} dt \quad + \quad \int_0^1 \frac{(1-t) + \left[(1-t)^2/2 \right]^2}{\sqrt{1 + (1-t)^2}} \sqrt{1 + (t-1)^2} dt$$

ANSWER 29, CONTINUED...

PART III   TEST ANSWER KEY

CALCULUS III   Practice FINAL EXAM KEY, CONTINUED

Please note that $(1-t)^2 = (t-1)^2$ ("FOIL" each and see!) So, the second integral can be reduced to lower terms, i.e., the denominator "cancels" with its equivalent numerator factor.

$$= \frac{1}{\sqrt{2}} \int_0^{1/2} (1+t^2) dt + \int_0^1 \left((1-t) + \frac{(1-t)^4}{4} \right) dt = \frac{1}{\sqrt{2}} \left[t + \frac{t^3}{3} \right]_0^{1/2} + \left[t - \frac{t^2}{2} - \frac{(1-t)^5}{20} \right]_0^1$$

$$= \frac{1}{\sqrt{2}} \left(\frac{1}{2} + \frac{(1/2)^2}{3} \right) + \left[1 - \frac{1}{2} - 0 \right] - \left[0 - \frac{1}{20} \right]$$

$$= \frac{13}{24\sqrt{2}} + \frac{11}{20}$$

30.   **Given the field** $\mathbf{F} = \left\langle 2x, -y^2, -\frac{4}{1+z^2} \right\rangle$ ...

(a)   **Show that the field is conservative.**

Because we have $\dfrac{\partial P}{\partial y} = \dfrac{\partial N}{\partial z} = 0$, $\dfrac{\partial P}{\partial x} = \dfrac{\partial M}{\partial z} = 0$, $\dfrac{\partial N}{\partial x} = \dfrac{\partial M}{\partial y} = 0$, the field is conservative.

This means that there exists a potential function, $f(x,y,z)$ for which $\mathbf{F}$ is its gradient. In other words, $\nabla f(x,y,z) = \left\langle f_x, f_y, f_z \right\rangle = \mathbf{F}(x,y,z)$.

We find this potential function by integrating three times, and summing the results (but disregarding the "duplicates"):

$$f(x,y,z) = \int M dx = \int 2x \, dx = x^2 + g(y,z)$$

So, we have that:

$$= \int N dy = \int (-y^2) dy = -\frac{y^3}{3} + h(x,z)$$

$$f(x,y,z) = x^2 - \frac{y^3}{3} - 4\arctan z + C$$

$$= \int P dz = \int -\frac{4}{1+z^2} dz = -4\arctan z + f(x,y)$$

(answer)

(b)   **Evaluate** $\displaystyle\int_{(0,0,0)}^{(3,3,1)} 2x\,dx - y^2\,dy - \frac{4}{1+z^2}dz$.

We use the Fundamental Theorem of Line Integrals. We already proved the field was conservative and found its potential function. We use this now:

$$\int_{(0,0,0)}^{(3,3,1)} 2x\,dx - y^2\,dy - \frac{4}{1+z^2}dz = \left[f(x,y,z) \right]_{(0,0,0)}^{(3,3,1)} = \left[x^2 - \frac{y^3}{3} - 4\arctan z \right]_{(0,0,0)}^{(3,3,1)} = \left(3^2 - \frac{3^3}{3} - 4\arctan 1 \right) - (0)$$

$$= 9 - 9 - 4\left(\frac{\pi}{4} \right) = -\pi \quad \text{(answer)}$$

CALCULUS III   **Practice FINAL EXAM KEY, CONTINUED**

---

31.   **Find the work done by the field** $\mathbf{F} = \left\langle 2\cos y, \dfrac{1}{y} - 2x\sin y \right\rangle$ **on the object that follows a path from the point**

   (2, 0), to the point (2, 1), and then to the point $\left(1, \dfrac{\pi}{2} \right)$ **.**

The force field is conservative since $\dfrac{\partial N}{\partial x} = \dfrac{\partial M}{\partial y} = -2\sin y$.

This means that the work done is independent of path and we can use the Fundamental Theorem of Line Integrals. In other words, there is no need to parameterize the curves and we only care about the starting and ending points! We find this potential function by integrating two times, and summing the results (but disregarding the "duplicates"):

$$\int 2\cos y\, dx = 2x\cos y + g(y); \quad and \quad \int\left(\tfrac{1}{y} - 2x\sin y\right) dy = \ln|y| + 2x\cos y + h(x) \Rightarrow f(x,y) = \ln|y| + 2x\cos y + C$$

Next, we evaluate the line integral using the Fundamental Theorem of Line Integrals:

$$\boldsymbol{Work} = \int_C \mathbf{F}\cdot d\mathbf{r} = \ln|y| + 2x\cos y\Big]_{(2,0)}^{(1,\pi/2)} = \left[\ln(\pi/2) + 2(1)\cos(\pi/2)\right] - \left[\ln(0) + 2(2)\cos(0)\right] = \ln(\pi/2) - 4 \quad (answer)$$

32.   **Use Green's Theorem to evaluate the line integral** $\displaystyle\int_C \left(x - y^3\right) dx + x^3\, dy$ **, where C is the right half of a circle of radius 2, $x^2 + y^2 = 4$.**

Since the path is closed, we may use Green's Theorem. First, sketch the path in order to determine the region of integration:

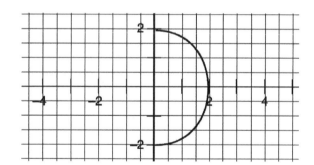

First, find $\dfrac{\partial N}{\partial x}$ and $\dfrac{\partial M}{\partial y}$,

which are

$$\dfrac{\partial N}{\partial x} = 3x^2 \; and \; \dfrac{\partial M}{\partial y} = -3y^2.$$

Next, apply Green's Theorem:

$$\int_C (x - y^3)dx - x^3 dy = \iint_R \left(\dfrac{\partial N}{\partial x} - \dfrac{\partial M}{\partial y}\right) dA = \int_0^2 \int_{-\sqrt{4-x^2}}^{\sqrt{4-x^2}} \left(3x^2 - \left(-3y^2\right)\right) dy\, dx$$

This double integral will be easier to evaluate if we convert to polar coordinates.

Please note that the range for the angle θ is $-\dfrac{\pi}{2} \le \theta \le \dfrac{\pi}{2}$ and *not* $\dfrac{3\pi}{2} \le \theta \le \dfrac{\pi}{2}$. If you do this, you will lose *lots* of points due to the fact that the lower limit is actually a number larger than the upper limit—so the double inequality is in fact *meaningless!*

CALCULUS III   Practice FINAL EXAM KEY, CONTINUED

So now, we have that:

$$\int_{-\pi/2}^{\pi/2}\int_0^2 3(x^2+y^2)\,r\,dr\,d\theta = \int_{-\pi/2}^{\pi/2}\int_0^2 3(r^2)\,r\,dr\,d\theta = \int_{-\pi/2}^{\pi/2}\int_0^2 3(r^3)\,dr\,d\theta = \int_{-\pi/2}^{\pi/2}\left[\frac{3}{4}r^4\right]_0^2 d\theta = \int_{-\pi/2}^{\pi/2}\left[\frac{3}{4}2^4\right]d\theta$$

$$= 12\int_{-\pi/2}^{\pi/2}d\theta = 12[\theta]_{-\pi/2}^{\pi/2} = 12\left(\frac{\pi}{2}-\left(-\frac{\pi}{2}\right)\right) = 12\pi$$

33. **Evaluate the surface integral $\iint_S y\,dS$ if S is the part of the plane $z = 6 - 3x - 2y$ in the first octant.**

We have that $g(x, y) = z = 6 - 3x - 2y$.

We find $dS = \sqrt{1+[g_x]^2+[g_y]^2} = \sqrt{1+(-3)^2+(-2)^2} = \sqrt{14}$.

The region of integration in the xy-plane is the projection of the surface
(the plane: $(z = 6 - 3x - 2y)$) onto the xy-plane. Let $z = 0$ and sketch the region if you need to:

$$\iint_S y\,dS = \iint_R y\sqrt{14}\,dA = \int_0^2\int_0^{(6-3x)/2} y\sqrt{14}\,dy\,dx = \sqrt{14}\int_0^2\left[\frac{y^2}{2}\right]_0^{(6-3x)/2}dx = \sqrt{14}\int_0^2\frac{(6-3x)^2}{8}dx$$

Use u-substitution, where $u = 6 - 3x$, $du = -3dx$:

$$\frac{\sqrt{14}}{8}\left(-\frac{1}{3}\right)\int_0^2 (6-3x)^2(-3)dx = -\frac{\sqrt{14}}{24}\left[\frac{u^3}{3}\right] = -\frac{\sqrt{14}}{72}\left[(6-3x)^3\right]_0^2 = -\frac{\sqrt{14}}{72}\left[(6-3(2))^3 - (6-3(0))^3\right] = -\frac{\sqrt{14}}{72}(-6^3) = 3\sqrt{14}$$

34. **Find the flux integral if $\iint_S \mathbf{F}\cdot\mathbf{N}\,dS$ if $\mathbf{F} = \langle x, y, z\rangle$ where S is the surface $z = 1 - x^2 - y^2$ above the xy-plane.**

Use the fact that $\iint_S \mathbf{F}\cdot\mathbf{N}\,dS = \iint_R \mathbf{F}\cdot\langle -g_x, -g_y, 1\rangle\,dA$, where $g(x, y) = z = 1 - x^2 - y^2$.

So, we get the vector $\langle -g_x, -g_y, 1\rangle = \langle -(-2x), -(-2y), 1\rangle = \langle 2x, 2y, 1\rangle$. The region of integration will be in the xy-plane. We let $z = 0$ in the equation $z = 1 - x^2 - y^2$, and find that we have a circle of radius 1. Now, our double integral is of the form:

$$\iint_R \mathbf{F}\cdot\langle -g_x, -g_y, 1\rangle\,dA = \int_{-1}^1\int_{-\sqrt{1-x^2}}^{\sqrt{1-x^2}}\langle x, y, z\rangle\cdot\langle 2x, 2y, 1\rangle\,dy\,dx = \int_{-1}^1\int_{-\sqrt{1-x^2}}^{\sqrt{1-x^2}}(2x^2 + 2y^2 + z)\,dy\,dx$$

$$= \int_{-1}^1\int_{-\sqrt{1-x^2}}^{\sqrt{1-x^2}}(2x^2 + 2y^2 + (1 - x^2 - y^2))\,dy\,dx$$

We replaced "z" with the equation for the surface $z = 1 - x^2 - y^2$ in that last step. This is because you are only allowed to have two different variables for double integrals, not three variables! Also, we will convert to polar coordinates to make the integral easier to evaluate:

$$\int_0^{2\pi}\int_0^1 (2r^2 + 1 - r^2)\,r\,dr\,d\theta = \int_0^{2\pi}\int_0^1 (r^3 + r)\,dr\,d\theta = \int_0^{2\pi}\left[\frac{r^4}{4}+\frac{r^2}{2}\right]_0^1 d\theta = \frac{3}{4}\int_0^{2\pi}d\theta = \frac{3}{4}(2\pi) = \frac{3\pi}{2} \quad (answer)$$

CALCULUS III

Practice FINAL EXAM KEY, CONTINUED

35. **Let Q be the cube bounded by the planes $x = \pm 1$, $y = \pm 1$, and $z = \pm 1$, and let $F(x, y, z) = \langle x^2, y^2, z^2 \rangle$. Use the Divergence Theorem to evaluate $\iint_S F \cdot N dS$. Source, sink, or neither?**

First, we'll find the divergence of **F**:

$$div\mathbf{F} = \nabla \cdot \mathbf{F} = \left\langle \frac{\partial}{\partial x}, \frac{\partial}{\partial y}, \frac{\partial}{\partial z} \right\rangle \cdot \left\langle x^2, y^2, z^2 \right\rangle = \frac{\partial[x^2]}{\partial x} + \frac{\partial[y^2]}{\partial y} + \frac{\partial[z^2]}{\partial z} = 2x + 2y + 2z$$

Next, we apply the Theorem:

$$\iint_S \mathbf{F} \cdot \mathbf{N} dS = \iiint_Q div\mathbf{F} dV = \int_{-1}^{1}\int_{-1}^{1}\int_{-1}^{1}(2x + 2y + 2z)dzdydx = \int_{-1}^{1}\int_{-1}^{1}\left[2xz + 2yz + z^2\right]_{-1}^{1} dydx$$

$$= \int_{-1}^{1}\int_{-1}^{1}\left[2x(1) + 2y(1) + (1)^2\right] - \left[2x(-1) + 2y(-1) + (-1)^2\right] dydx = \int_{-1}^{1}\int_{-1}^{1}(4x + 4y)dydx = \int_{-1}^{1}\left[4xy + 2y^2\right]_{-1}^{1} dx$$

$$= \int_{-1}^{1}\left[4x(1) + 2(1)^2\right] - \left[4x(-1) + 2(-1)^2\right]dx = \int_{-1}^{1}8xdx = \left[4x^2\right]_{-1}^{1} = 4(1)^2 - 4(-1)^2 = 0 \quad (answer)$$

Since the answer is zero, the outward flux of the field **F** through the surface is *neither* a source or a sink (i.e., incompressible).

36. **Let Q be the solid bounded by the cylinder $x^2 + y^2 = 1$ and the planes $z = 0$ and $z = 1$. Use the Divergence Theorem to evaluate $\iint_S F \cdot N dS$ and calculate the outward flux of F through S where S is the surface of Q and $F(x, y, z) = \langle x, y, z \rangle$. Source, sink, or neither?**

The divergence of **F** is: $div\mathbf{F} = \dfrac{\partial M}{\partial x} + \dfrac{\partial N}{\partial y} + \dfrac{\partial P}{\partial z} = \dfrac{\partial[x]}{\partial x} + \dfrac{\partial[y]}{\partial y} + \dfrac{\partial[z]}{\partial z} = 1 + 1 + 1 = 3$

We now apply the Divergence Theorem: $\iint_S \mathbf{F} \cdot \mathbf{N} dS = \iiint_Q div\mathbf{F} dV = \int_{-1}^{1}\int_{-\sqrt{1-x^2}}^{\sqrt{1-x^2}}\int_{0}^{1} 3dzdydx$

We will convert to cylindrical coordinates to make it easier to integrate:

$$= \int_0^{2\pi}\int_0^1\int_0^1 3dzrdrd\theta = \int_0^{2\pi}\int_0^1 3[z]_0^1 rdrd\theta = 3\int_0^{2\pi}\int_0^1 rdrd\theta = 3\int_0^{2\pi}\left[\frac{r^2}{2}\right]_0^1 d\theta = \frac{3}{2}\int_0^{2\pi} d\theta = \frac{3}{2}[\theta]_0^{2\pi} = \frac{3}{2}(2\pi) = 3\pi \ (answer)$$

Since the answer is positive, we have a *source*.

$\sqrt{}$ *PART III* TEST ANSWER KEY

CALCULUS III   Practice FINAL EXAM KEY, CONTINUED

37. Let Q be the region bounded above by the sphere $x^2 + y^2 + z^2 = 9$ and below by the plane $z = 0$ in the first quadrant. Use the Divergence Theorem to evaluate $\iint_S \mathbf{F} \cdot \mathbf{N} dS$ and find the outward flux of F through S, where S is the surface of the solid and $\mathbf{F}(x, y, z) = \langle xy, 4x, 2y \rangle$. Source, sink, or neither?

First we get the divergence of **F**: $div\mathbf{F} = \dfrac{\partial M}{\partial x} + \dfrac{\partial N}{\partial y} + \dfrac{\partial P}{\partial z} = \dfrac{\partial [xy]}{\partial x} + \dfrac{\partial [4x]}{\partial y} + \dfrac{\partial [2y]}{\partial z} = y$

Next, we apply the theorem: $\iint_S \mathbf{F} \cdot \mathbf{N} dS = \iiint_Q div\mathbf{F} dV = \displaystyle\int_0^3 \int_0^{\sqrt{9-x^2}} \int_0^{\sqrt{9-x^2-y^2}} y\, dz\, dy\, dx$

We will convert to spherical coordinates to make it easier to integrate. Keep in mind that the "drop-down" angle, ϕ, ranges from $0 \le \phi \le \dfrac{\pi}{2}$, since we are bounded below by the xy-plane:

$= \displaystyle\int_0^{\pi/2} \int_0^{\pi/2} \int_0^3 (\rho \sin\phi \sin\theta)\rho^2 \sin\phi\, d\rho\, d\phi\, d\theta = \int_0^{\pi/2} \int_0^{\pi/2} \int_0^3 (\rho^3 \sin^2\phi \sin\theta)\, d\rho\, d\phi\, d\theta = \int_0^{\pi/2} \int_0^{\pi/2} \sin^2\phi \sin\theta \left[\dfrac{\rho^4}{4}\right]_0^3 d\phi\, d\theta$

$= \dfrac{81}{4}\displaystyle\int_0^{\pi/2} \int_0^{\pi/2} \sin^2\phi \sin\theta\, d\phi\, d\theta = \dfrac{81}{4}\int_0^{\pi/2} \int_0^{\pi/2} \sin\theta \left[\dfrac{1 - \cos(2\phi)}{2}\right] d\phi\, d\theta = \dfrac{81}{4}\left(\dfrac{1}{2}\right)\int_0^{\pi/2} \sin\theta \left[\phi - \dfrac{1}{2}\sin(2\phi)\right]_0^{\pi/2} d\theta$

$= \dfrac{81}{8}\displaystyle\int_0^{\pi/2} \sin\theta\left(\dfrac{\pi}{2}\right) d\theta = \dfrac{81\pi}{16}\left[-\cos\theta\right]_0^{\pi/2} = \dfrac{81\pi}{16}\left[-\cos\dfrac{\pi}{2} - (-\cos 0)\right] = \dfrac{81\pi}{16}(0 + 1) = \dfrac{81\pi}{16}$

SOURCE *(answer)*

38. Find the curl of the vector field $\mathbf{F}(x, y, z) = \langle z^2, -x^2, y \rangle$. Is the field conservative?

$curl\mathbf{F} = \nabla \times \mathbf{F} = \begin{vmatrix} \mathbf{i} & \mathbf{j} & \mathbf{k} \\ \dfrac{\partial}{\partial x} & \dfrac{\partial}{\partial y} & \dfrac{\partial}{\partial z} \\ z^2 & -x^2 & y \end{vmatrix} = \begin{vmatrix} \dfrac{\partial}{\partial y} & \dfrac{\partial}{\partial z} \\ -x^2 & y \end{vmatrix}\mathbf{i} - \begin{vmatrix} \dfrac{\partial}{\partial x} & \dfrac{\partial}{\partial z} \\ z^2 & y \end{vmatrix}\mathbf{j} + \begin{vmatrix} \dfrac{\partial}{\partial x} & \dfrac{\partial}{\partial y} \\ z^2 & -x^2 \end{vmatrix}\mathbf{k}$

$= \left(\dfrac{\partial[y]}{\partial y} - \dfrac{\partial[-x^2]}{\partial z}\right)\mathbf{i} - \left(\dfrac{\partial[y]}{\partial x} - \dfrac{\partial[z^2]}{\partial z}\right)\mathbf{j} + \left(\dfrac{\partial[-x^2]}{\partial x} - \dfrac{\partial[z^2]}{\partial y}\right)\mathbf{k} = (1 - 0)\mathbf{i} - (0 - 2z)\mathbf{j} + (-2x - 0)\mathbf{k} = \langle 1, 2z, -2x \rangle$

Since the curl is not equal to zero, the field is *not* conservative. The curl of a conservative vector field is always equal to zero.

CALCULUS III   **Practice FINAL EXAM KEY, CONTINUED**

39. Use Stokes's Theorem to evaluate $\int_C \mathbf{F} \cdot d\mathbf{r}$, where $\mathbf{F}(x, y, z) = \langle z, 2x, 2y \rangle$ and S is the surface of the paraboloid (oriented upward) of $z = 4 - x^2 - y^2$, $z \geq 0$, and C is its boundary.

First, we have that $g(x, y) = z = 4 - x^2 - y^2$.

So, $\mathbf{N}dS = \langle -g_x, -g_y, 1 \rangle dA = \langle -(-2x), -(-2y), 1 \rangle dA = \langle 2x, 2y, 1 \rangle dA$.

Next, we find the curl of the field:

$$curl\mathbf{F} = \nabla \times \mathbf{F} = \begin{vmatrix} \mathbf{i} & \mathbf{j} & \mathbf{k} \\ \dfrac{\partial}{\partial x} & \dfrac{\partial}{\partial y} & \dfrac{\partial}{\partial z} \\ z & 2x & 2y \end{vmatrix} = \begin{vmatrix} \dfrac{\partial}{\partial y} & \dfrac{\partial}{\partial z} \\ 2x & 2y \end{vmatrix}\mathbf{i} - \begin{vmatrix} \dfrac{\partial}{\partial x} & \dfrac{\partial}{\partial z} \\ z & 2y \end{vmatrix}\mathbf{j} + \begin{vmatrix} \dfrac{\partial}{\partial x} & \dfrac{\partial}{\partial y} \\ z & 2x \end{vmatrix}\mathbf{k}$$

$$= \left(\dfrac{\partial[2y]}{\partial y} - \dfrac{\partial[2x]}{\partial z} \right)\mathbf{i} - \left(\dfrac{\partial[2y]}{\partial x} - \dfrac{\partial[z]}{\partial z} \right)\mathbf{j} + \left(\dfrac{\partial[2x]}{\partial x} - \dfrac{\partial[z]}{\partial y} \right)\mathbf{k} = (2 - 0)\mathbf{i} - (0 - 1)\mathbf{j} + (2 - 0)\mathbf{k} = \langle 2, 1, 2 \rangle$$

Next, we apply Stokes' Theorem:

$$\int_C \mathbf{F} \cdot d\mathbf{r} = \iint_S curl\mathbf{F} \cdot \mathbf{N}dS = \iint_R \langle 2, 1, 2 \rangle \cdot \langle 2x, 2y, 1 \rangle dA = \int_{-2}^{2} \int_{-\sqrt{4-x^2}}^{\sqrt{4-x^2}} (4x + 2y + 2)\,dy\,dx$$

Convert to polar:

$$= \int_0^{2\pi} \int_0^2 (4r\cos\theta + 2r\sin\theta + 2)r\,dr\,d\theta = \int_0^{2\pi} \int_0^2 (4r^2\cos\theta + 2r^2\sin\theta + 2r)\,dr\,d\theta = \int_0^{2\pi} \left[\dfrac{4r^3}{3}\cos\theta + \dfrac{2r^3}{3}\sin\theta + r^2 \right]_0^2 d\theta$$

$$= \int_0^{2\pi} \left[\dfrac{4(2)^3}{3}\cos\theta + \dfrac{2(2)^3}{3}\sin\theta + (2)^2 \right]_0^2 d\theta = \int_0^{2\pi} \left[\dfrac{32}{3}\cos\theta + \dfrac{16}{3}\sin\theta + 4 \right]_0^2 d\theta = \left[\dfrac{32}{3}\sin\theta - \dfrac{16}{3}\cos\theta + 4\theta \right]_0^{2\pi} = 8\pi$$

NOTES